20th International Summer School on Vacuum, Electron and Ion Technologies (VEIT 2017)

Journal of Physics: Conference Series Volume 992

Sozopol, Bulgaria
25 - 29 September 2017

Editors:

M.C.M. van de Sanden
Miglena Dimitrova
Chavdar Ghelev

ISBN: 978-1-5108-6124-4
ISSN: 1742-6588

Printed from e-media with permission by:

Curran Associates, Inc.
57 Morehouse Lane
Red Hook, NY 12571

CURRAN ASSOCIATES INC. proceedings .com

Some format issues inherent in the e-media version may also appear in this print version.

Printed by Curran Associates, Inc. (2018)

For permission requests, please contact the Institute of Physics
at the address below.

Institute of Physics
Dirac House, Temple Back
Bristol BS1 6BE UK

Phone: 44 1 17 929 7481
Fax: 44 1 17 920 0979

techtracking@iop.org

Additional copies of this publication are available from:

Curran Associates, Inc.
57 Morehouse Lane
Red Hook, NY 12571 USA
Phone: 845-758-0400
Fax: 845-758-2633
Email: curran@proceedings.com
Web: www.proceedings.com

Table of contents

Volume 992

20th International Summer School on Vacuum, Electron and Ion Technologies

25–29 September 2017, Sozopol, Bulgaria

Accepted papers received: 19 March 2018
Published online: 5 April 2018

Preface

Papers

IOP Conf. Series: Journal of Physics: Conf. Series **992** (2018) 011001 doi:10.1088/1742-6596/992/1/011001

VEIT 2017

The International Summer School on Vacuum, Electron and Ion Technologies (VEIT) has been organized biennially since 1978, when the series of VEIT Schools was launched by the Institute of Electronics, Bulgarian Academy of Sciences, with the aim to act as a forum for exchange and dissemination of knowledge and ideas on the latest developments in electron-, ion-, and plasma-assisted technologies. The organizers of the 2017 edition of the event were the Institute of Electronics, Bulgarian Academy of Sciences, Sofia, Bulgaria and the Dutch Institute for Fundamental Energy Research, Eindhoven, The Netherlands. While the school has initially been providing a meeting place for researchers mainly from Eastern and Central European countries, its importance has grown issue by issue. The school is now a major scientific event and a meeting place for young scientists from Eastern and Western Europe involved in research and development associated with high-tech industries. Many former school participants have gone on to become leading scientists in research establishments and companies throughout the world. Leading international companies, such as High Voltage Engineering, Balzers, Varian, and Hauzer have used the VEIT forum to present their products through oral presentations, poster contributions or exhibits. The School Proceedings have been published in special issues of the international journals Vacuum, Plasma Processes and Polymers, Journal of Physics: Conference Series.

The twentieth jubilee edition of VEIT was held in the Black Sea resort Sozopol, Bulgaria, from 25 to 29 September 2017. It was attended by 149 participants (36 students and young researchers) from 20 countries on three continents: Austria, Bulgaria, France, Germany, Iran, Japan, Mexico, Portugal, Romania, Russia, Spain, Serbia, Slovakia, Slovenia, Sweden, The Czech Republic, The Netherlands, Ukraine, UK, USA.

The School Program covered a large spectrum of fundamental problems of the nanotechnologies, the interaction of charged particles with solid surfaces, the deposition and characterization of thin films, which were balanced out by reports on a wide range of significant high-technology applications – from equipment for deposition of hard coatings or thin optical/protective films to nanostructures prepared via evaporation, sputtering or irradiation. In accordance with their subjects, the lectures and reports were grouped for presentation and discussion in three sessions, each of them including oral presentations at plenary sessions and posters. Thus, the program comprised three tutorial topic reviews of 90 minutes, four topic lectures of 45 minutes, 24 progress reports of 30 minutes, and eleven 20-minute reports by young attendees. Also, 90 original contributions were presented at the three poster sessions. Marking VEIT's 20th anniversary, the Organizing Committee awarded three best-poster prizes (one per poster session) to young scientists (up to five years after PhD graduation).

Following the tradition of publishing the VEIT Proceedings, a selection of 64 manuscripts presented at the event will be published in a special issue of Journal of Physics: Conference Series, under the originality and quality criteria of acceptance by the journal, including peer reviewing.

Despite the busy scientific program, the atmosphere was relaxed and informal. The early afternoons of most conference days were free to stimulate both scientific and social interaction between participants, which often took place on the beach. The social program included a welcome party, an official dinner, and an outing to a winery in the town of Medovo, near the city of Burgas.

VEIT 2015 owes its success to many people. The International Advisory Committee shaped the scientific program and ensured high-quality plenary presentations by careful selection of invited speakers. The Local Committee bore the brunt of the organization both at the conference site and in dealing with correspondence, abstracts, and manuscripts for these proceedings. We are grateful to our sponsors, the Dutch Institute for Fundamental Energy Research, Eindhoven, The Netherlands, IUVSTA, the company HAUZER and the Ministry of Education and Science of Republic of Bulgaria for there generosity that enabled us to support the attendance of students and provided support to deal with mailing, printing, renting the conference site, etc. We are also grateful to the members of the International Advisory Committee – N. Guerassimov (Bulgarian Academy of

Sciences, Sofia, Bulgaria), M. Dimitrova (Czech Academy of Sciences, Prague, The Czech Republic), R. Panek (Czech Academy of Sciences, Prague, The Czech Republic), Th. Czerwiec (Institut Jean Lamour, Ecole des Mines de Nancy, France), W. Möller (Forschungszentrum Dresden-Rossendorf, Germany), B. Rauschenbach (IOM and Leipzig University, Germany), H. Kersten (IEAP University of Kiel, Germany), D. Mataras (University of Patras, Greece), V. Guerra (Instituto Superior Tecnico, Lisbon, Portugal), G. Dinescu (National Institute for Laser, Plasma and Radiation Physics, Magurele-Bucharest, Romania), Z. Petrovic (Institute of Physics, Belgrade, Serbia), M. Mozetic (Jozef Stefan Institute, Ljubljana, Slovenia), K. Larsson (Uppsala University, Sweden), I. Katardjiev (Uppsala University, Sweden), M.C.M. (Richard) van de Sanden (Dutch Institute for Fundamental Energy Research, Eindhoven, The Netherlands), G.J. van Rooij (Dutch Institute for Fundamental Energy Research, Eindhoven, The Netherlands), M. Ürgen (Istanbul Technical University, Turkey), A. Ehiasarian, (Sheffield Hallam University, UK), I. Petrov (University of Illinois, Urbana, IL USA) for their continuous concern about the high scientific level and the future of the event, as well as for their contribution to the formation of a new generation of scientists and experts in the field. And, last but not least, we would also like to thank all authors for their valuable contributions to these proceedings and to the school, and all reviewers for their hard and tedious, but very important, work.

The next conference in the series will be held in September 2019.

M.C.M. van de Sanden, Miglena Dimitrova, Chavdar Ghelev,
Guest Editors

IOP Conf. Series: Journal of Physics: Conf. Series **992** (2018) 011002 doi:10.1088/1742-6596/992/1/011002

Peer review statement

All papers published in this volume of *Journal of Physics: Conference Series* have been peer reviewed through processes administered by the proceedings Editors. Reviews were conducted by expert referees to the professional and scientific standards expected of a proceedings journal published by IOP Publishing.

IOP Conf. Series: Journal of Physics: Conf. Series **992** (2018) 012001 doi:10.1088/1742-6596/992/1/012001

Modelling, simulation and computer-aided design (CAD) of gyrotrons for novel applications in the high-power terahertz science and technologies

S Sabchevski[1,4], T Idehara[2], M Damyanova[1], I Zhelyazkov[3], E Balabanova[1] and E Vasileva[1]

[1]E. Djakov Institute of Electronics, Bulgarian Academy of Sciences,
72 Tzarigradsko Chaussee, 1784 Sofia, Bulgaria
[2]Research Center for Development of Far-Infrared Region, University of Fukui,
910-8507 Fukui, Japan
[3]Faculty of Physics, St. Kl. Ohridski University of Sofia,
5 James Bourchier Blvd., 1164 Sofia, Bulgaria

E-mail: sabch@ie.bas.bg

Abstract. Gyrotrons are the most powerful sources of CW coherent radiation in the sub-THz and THz frequency bands. In recent years, they have demonstrated a remarkable potential for bridging the so-called THz-gap in the electromagnetic spectrum and opened the road to many novel applications of the terahertz waves. Among them are various advanced spectroscopic techniques (e.g., ESR and DNP-NMR), plasma physics and fusion research, materials processing and characterization, imaging and inspection, new medical technologies and biological studies. In this paper, we review briefly the current status of the research in this broad field and present our problem-oriented software packages developed recently for numerical analysis, computer-aided design (CAD) and optimization of gyrotrons.

1. Introduction

The family of gyro-devices includes both amplifiers (e.g., gyro-klystrons, gyro-TWT) and oscillators (gyrotron, *aka*, gyro-monotron, gyro-BWO, etc.). In fact, they are fast-wave free-electron vacuum tubes in which an annular beam of electrons gyrates (thus the prefix "gyro" in their names) following helical orbits and interacts with the high-frequency field having a phase velocity exceeding the speed of light in the resonant structure (cavity). Unlike the slow-wave Cherenkov devices (for example BWO, TWT) the dimensions of the resonant structure (e.g., cavity radius R_{cav}) can be much larger than the wavelength of the generated radiation. This allows powerful electron beams (formed by an electron-optical system (EOS) with a magnetron-injection gun (MIG) and an adiabatically increasing magnetic field provided by a superconducting magnet) to be injected into the cavity, where the electrons exercise rotational motion with a cyclotron frequency $\Omega_{\mathrm{c}} = \frac{eB}{m} = \frac{eB}{m_0\gamma}$ on a circle with a Larmor radius $r_{\mathrm{L}} = \frac{v_\perp}{\Omega_c}$. Here $\frac{e}{m}$ is the electron charge-to-mass ratio, B is the magnetic field strength

[4] To whom any correspondence should be addressed.

and $v_\perp$ is the transverse (rotational) velocity. In the resonator, where B attains its maximum value, most of the energy of the electrons is associated with their rotational motion and the velocity pitch factor $\alpha = \frac{v_\perp}{v_z} > 1$, v_z being the axial velocity. During the interaction of the electrons with the transverse electric field, some electrons lose energy to the wave, while others gain energy. Due to the relativistic dependence of the electron mass $m = m_0\gamma$ on the energy E (m_0 and $\gamma = E/m_0c^2$ being the rest mass and the relativistic Lorenz factor, respectively, and c is the speed of light), the cyclotron frequency also changes accordingly (i.e., increases for deaccelerated electrons and decreases for accelerated ones) leading to a non-isochronous circular motion, which results in an azimuthal bunching of the electrons. This mechanism is responsible for an efficient beam-wave interaction with a positive gain at the cyclotron resonance provided that proper synchronization conditions are satisfied. Such resonance occurs at a frequency $\omega = s\Omega_c + v_z k_z$, where s and k_z are the harmonic number and the axial wave number, respectively. The gyrotrons are distinguished from other gyro-devices by the fact that they operate at frequencies close to the cut-off frequency of the cavity ($\omega_{\text{cut}} = c\frac{\chi_{mn}}{R_{\text{cav}}}$, where χ_{mn} is the n-th zero of the Bessel function J_m for TM modes and of its derivative J'_m for the TE modes), i.e. the Doppler shift term $v_z k_z$ is negligibly small and thus $\omega \approx s\Omega_c$ [1]. A helpful engineering estimate gives for the fundamental resonances (s=1) approximately 28 GHz per 1 T. In recent years, the progress in the development of superconducting magnets that deliver field intensity up to 20 T and the operation at higher harmonics of the cyclotron frequency have facilitated the advancement of the gyrotrons from the sub-THz to the THz range. At present, they are the most powerful sources of CW coherent radiation in this region of the electromagnetic spectrum and contribute to bridging the so-called THz gap [2, 3]. The remarkable breakthroughs demonstrated by them, alongside with the significant improvement in their operational performance utilizing more compact cryo-free superconducting magnets, not only increased their number in many traditional fields, but have also opened an avenue to many novel and prospective applications in the fundamental physical research and most notably in the emerging high-power terahertz science and technology [4–7].

Although all gyrotrons operate on the same principle (known as electron cyclotron resonance maser (ECRM) instability) and can be described in the framework of a unified approach using identical physical models, each particular application imposes specific requirements to the operational output parameters and sets some constructive restraints. In order to elucidate them, first, we present briefly (Sect. 2) the most prominent applications of the gyrotrons. Then, in Sect. 3 we depict our tools (computational infrastructure, problem-oriented software packages and stand-alone codes) developed by us for numerical analysis, computer-aided design (CAD), and optimization of powerful gyrotrons for fusion research and sub-THz gyrotrons for various advanced spectroscopic techniques, materials processing, and novel medical technologies. We conclude the paper with an outlook on the envisaged future work in this field.

2. Traditional, novel, and prospective applications of the gyrotrons

The advantages mentioned above (most notably the unprecedented output power in the sub-THz to the THz frequency range in CW mode) makes the gyrotrons versatile and appropriate radiation sources for many physical studies and technological processes. Some of the most prominent traditional, novel and prospective applications are represented as branches of an ever-growing tree in figure 1.

Among the traditional and most matured applications is their usage for electron cyclotron resonance heating (ECRH) and current drive (ECCD) of magnetically confined plasma in various reactors (e.g. tokamaks and stellarators) for controlled thermonuclear fusion. The gyrotrons are used also for plasma initiation (start-up), plasma control (suppression of the neoclassical tearing modes) and for plasma diagnostics based on a CTS (collective Thomson scattering). The 170-GHz gyrotrons developed in Japan, the EU and the Russian Federation have demonstrated megawatt levels of the output power with an efficiency exceeding 50% at long pulse operation (3600 s) and have reached the requirements of ITER [1]. The next big step in the fusion research – DEMO, however, requires even higher frequency, power and efficiency (230 GHz, 1.5 MW, and 60%, respectively) and, additionally, multi-frequency operation. Compared with other tubes, the gyrotrons for fusion use higher beam currents, highly oversized cavities, and larger emitters. This makes the influence of such factors as space-charge effects, beam instabilities, nonuniformity of the emission, misalignment and thermal shift of the electrodes, etc., more severe. Since most of these effects are both non-stationary and non-axially symmetric they could be taken into account using time-dependent self-consistent physical models formulated in three dimensions (3D) and implemented in 3D numerical codes.

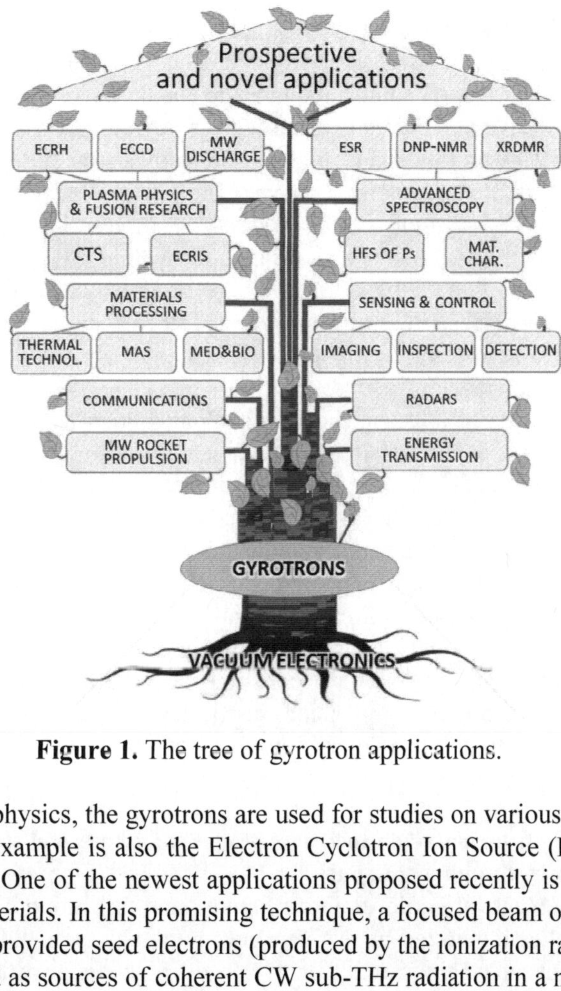

Figure 1. The tree of gyrotron applications.

In the field of plasma physics, the gyrotrons are used for studies on various gas discharges initiated by microwaves. A good example is also the Electron Cyclotron Ion Source (ECIS) for production of multi-charged ion beams. One of the newest applications proposed recently is the remote detection of concealed radioactive materials. In this promising technique, a focused beam of THz radiation initiates an avalanche breakdown provided seed electrons (produced by the ionization radiation) are present.

The gyrotrons are used as sources of coherent CW sub-THz radiation in a number of advanced and precise spectroscopic methods. Among them are ESR (Electron Spin Resonance) spectroscopy, and DNP-NMR (Nuclear Magnetic Resonance with a signal enhancement through Dynamic Nuclear Polarization) [5–8]. The latter technique is revolutionizing the studies of proteins and other bio-molecules offering significant reduction of the spectra acquisition time and an increased sensitivity. Its variety, the solid-state DNP-NMR, is a powerful method for studies on the molecular structure and surface characterization of many different materials. Another novel spectroscopy is based on the X-ray detected magnetic resonance (XDMR). In this element and edge-selective pump and probe technique, the X-ray magnetic circular dichroism is used to probe the resonant precision of the magnetization components in an intense field of a sub-THz wave produced by a gyrotron. XDMR spectroscopy is a unique tool for studies on the magnetization dynamics induced by the precision of both the spin and orbital moments. The newest high-precision spectroscopy that uses a gyrotron as a radiation source has been developed quite recently for a direct measurement of the ground state hyperfine splitting

(HFS) of the positronium (Ps). Recently, the whole Breit–Wigner resonance of the stimulated transition from o-Ps to p-Ps has been measured for the first time using a frequency tunable gyrotron in the range 201 GHz to 205 GHz. In the Research Center for Development of Far-Infrared Region at the University of Fukui, Japan many gyrotrons have been developed for the above mentioned spectroscopic studies [5–7].

When irradiated by sub-THz and THz waves, many substances acquire unique characteristic spectra ("fingerprints") that are used for materials identification and characterization, as well as for a security control.

Thermal treatment of advanced materials (e.g., ceramic sintering) using mm-wave radiation of gyrotrons is a mature industrial grade technology. It benefits from the well-known advantageous features of the microwave heating: (i) uniform volumetric and selective heating; (ii) rapid heat transfer and high heating rate from several hundred to thousands of degrees per second; (iii) precise control of the absorbed energy; (iii) lower losses and higher efficiency compared to other heating methods. Usually, the process is carried out inside an applicator in which steerable mirrors and reflectors are used for an additional homogenization of the microwave irradiation. The gyrotrons are used in industrial facilities for microwave melting, joining and sealing of glass; thermal processing of semiconductors; annealing of thin Si films for solar cells; polymer coating, molding, and curing; drying of porous materials, etc. (visit: **http://www.gyrotrontech.com**). The radiation of gyrotrons is promising also for several emerging technologies (most notably for MAS, microwave assisted synthesis) where both thermal and non-thermal effects are used.

Recently, several concepts of novel medical technologies utilizing gyrotron radiation have been developed. Among them is the local hyperthermia of cancerous tumors. As more promising, however, we consider its combinations with other methods that, hopefully, would instigate a synergy effect of the combined treatment. One of them is a hybrid dual-beam therapy, which combines the BNCT (boron neutron capture therapy) with hyperthermia by sub-THz wave irradiation [9]. The other hybrid technique combines such hyperthermia with photodynamic laser therapy (PDT) [10].

The gyrotrons have demonstrated a great potential for the development of many high-power terahertz technologies in other fields as well. For example, in the next generation mm-wave communications, advanced radars, nonlethal weapons (e.g. active denial systems), energy transmission (beaming) and even microwave rocket propulsion, just to name a few due to the limited space here. We anticipate that in the near future many other novel applications of sub-THz and THz gyrotrons will emerge.

3. Development of problem-oriented software packages for numerical studies, computer-aided design (CAD) and optimization of gyrotrons

Motivated by our involvement in the development, study, and utilization of gyrotrons for several of the above-mentioned application, in recent years we have built a hierarchy of modeling and simulation tools (see figure 2) that are being used for CAD of high-performance tubes. Our approach encompasses all levels of that hierarchy starting from the formulation of adequate physical models and selection of efficient numerical methods, programming libraries, and appropriate integrated development environments (IDE). Then, the mathematical formulations of the physical models are being implemented (codded) in computational modules, subroutines and other programing building blocks of our problem-oriented (i.e., specialized to different subsystems of the device) software packages and several stand-alone computer programs. Both the development (programming, debugging, benchmarking) and the numerical experiments are being carried out using the available computational infrastructure that consists of several workstations.

The structure of the developed software tools used for numerical studies and CAD of gyrotrons is presented in figure 3. Also shown here is the functional assignment of their components (computational modules) to the iterative design loop (bottom panel of the figure).

IOP Conf. Series: Journal of Physics: Conf. Series **992** (2018) 012001 doi:10.1088/1742-6596/992/1/012001

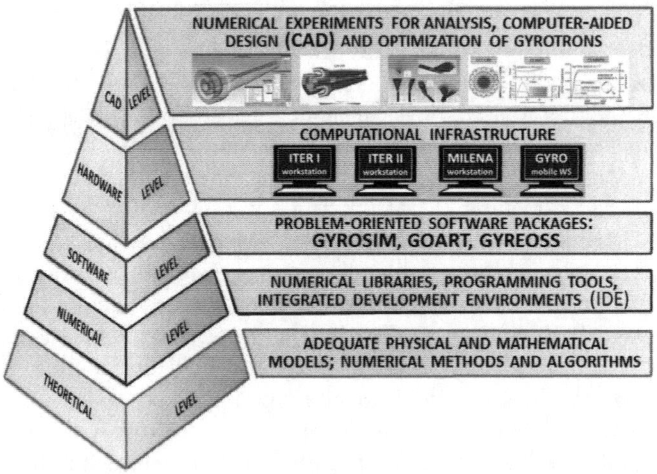

Figure 2. Hierarchy of the modelling and simulation tools.

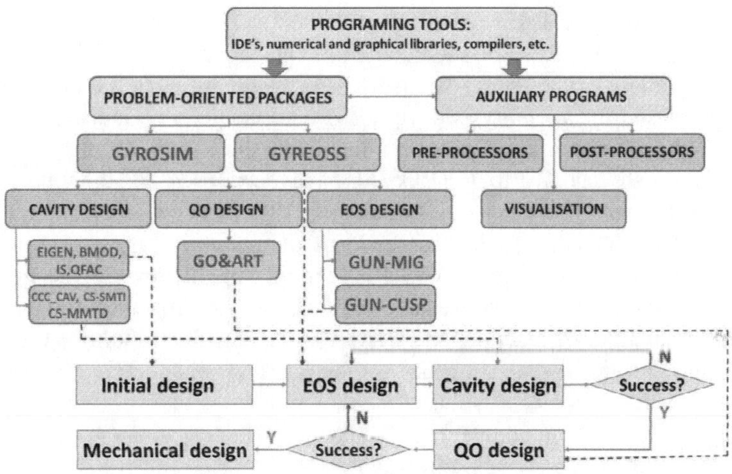

Figure 3. Structure of the software tools and their functional assignment to the iterative CAD.

The problem-oriented software package GYROSIM (an abbreviation from GYROtron SIMulation) contains computer codes for numerical analysis and CAD of three of the most important subsystems of the gyrotron tube, namely the EOS (including a MIG and a set of magnetic coils), the resonance cavity (electro-dynamical system), and the quasi-optical system. An important task during the initial design is the selection of an appropriate operating cavity mode and preliminary analysis of the operational parameters (beam voltage and current, velocity ratio (pitch factor), magnetic field intensity) while iterating different configurations and sizes of the resonant structure. On this stage, several programs (EIGEN, BMOD) for analysis of the cavity spectrum and scanning of the corresponding resonances are used. Furthermore, the codes IS and QFAC are used for an initial evaluation of the starting beam currents (based on the linear gyrotron theory) and the Q-factors (diffractive, Ohmic and loaded), respectively. This is followed by simulations carried out using cold-cavity and single-mode static (time-independent) codes (CCC_CAV, CS-SMTI) for simulation of the beam-wave interaction and for a preliminary estimation of the possible operational parameters (output power, efficiency, etc.). After the initial design, an iterative design loop starts, during which more sophisticated (based on self-consistent time-dependent multimode physical models) codes CS-MMTD are used.

GYROSIM contains two codes for CAD of the EOS, namely GUN-MIG and GUN-CUSP. In both of them self-consistent fully relativistic physical models that take into account the space-charge and initial velocity spread are used. They are formulated in 2.5D, i.e. in a 5D phase space of two spatial and three momentum coordinates in an axially-symmetric coordinate system. Their field solvers are based on the finite-difference method (FDM) with a successive over relaxation and an automatic selection of an optimal parameter of the relaxation. GUN-MIG is used for an analysis of conventional MIGs that form helical electron beams, in which the guiding centers of the electron orbits are off-axis. MIG-CUSP is specialized to EOS with a field reversal that form axis-encircling (aka uniaxial) beams that are used in the Large Orbit Gyrotrons (LOG). These codes for trajectory analysis (ray tracing) include auxiliary programs for calculation of the magnetic field distribution generated by a set of solenoids (together with the main superconducting coils) as well as for post-processing and visualization of the results of the numerical experiments.

Many applications of the gyrotrons require a well-collimated wave beam. In order to transform the wave of the high-order operating mode into a Gaussian-like beam an internal mode converter is used. It includes a launcher of either Vlasov or Denisov type and a system of mirrors that shape the output radiation before coupling it to the vacuum window. The module GO&ART (which stands for Geometric Optics and Analytical Ray Tracing) of the GYROSIM package has several codes (RAYS, COMODES, and TRACE) for CAD of such quasi-optical system. The code RAYS uses the analytical theory for designing Vlasov type antennas, while COMODES utilizes a model based on the coupled modes theory and is used for an analysis of dimpled-wall launchers (Denisov launchers). The system of reflectors and phase correcting mirrors is designed using the code TRACE.

Novel modules added recently to GYROSIM have been used for a numerical study on finite-bandwidth resonances of high-order axial modes (HOAM) in a gyrotron cavity [11]. The proposed approach allows one to study the frequency response of the cavity using different resonance curves obtained from the solution of the inhomogeneous Helmholtz equation for the field amplitude by the finite-difference method (FDM). The results give deeper insight into the continuous frequency tunability through excitation of a sequence of HOAM. As an illustration, in figure 4 we present the calculated axial field profiles of such sequence for the operating mode $TE_{4,1}$.

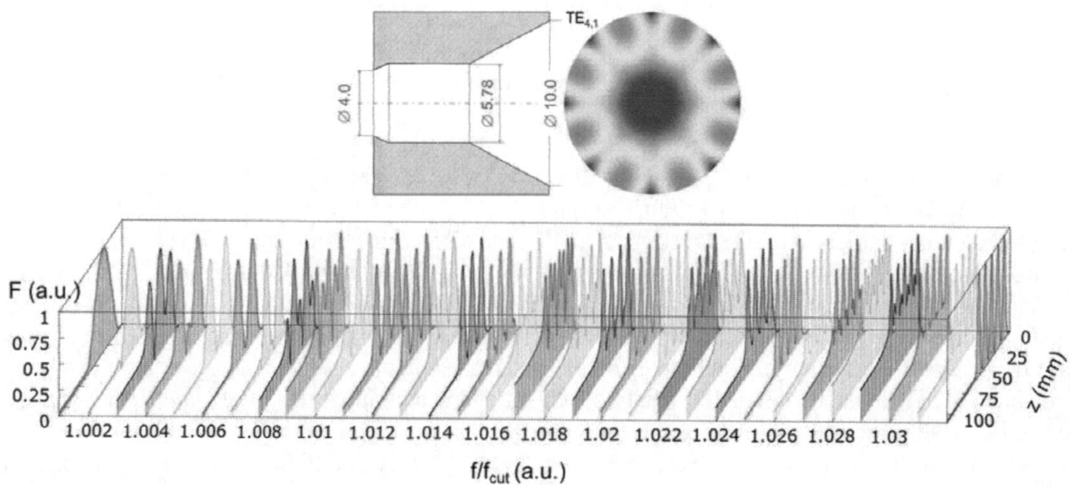

Figure 4. Configuration of the gyrotron cavity studied and field pattern of the $TE_{4,1}$ mode (upper panel); and axial profiles of the field intensity for a sequence of HOM (lower panel). Here $f/f_{\rm cut}$ is the frequency normalized to the cut-off frequency of the cavity.

IOP Conf. Series: Journal of Physics: Conf. Series **992** (2018) 012001 doi:10.1088/1742-6596/992/1/012001

As already mentioned, the departure from axial symmetry due to various inevitable misalignments and nonuniformities (for instance nonuniformity of the emission) requires the utilization of physical models formulated in 3D. In order to meet this challenge, we have started the development of the problem–oriented package GYREOSS (GYREOSS stands for Gyrotron Electron-Optical System Simulation) [12]. This PIC (particle-in-cell) code is implemented as a library of scripts for the partial differential equations solver FreeFEM+++ [13] and uses its IDE FreeFEM++-cs for the code development and execution. The computational domain is discretized by unstructured tetrahedral grids produced by gmsh [14] or by the mesh generator embedded in FreeFEM++. The Poisson equation for the potential of the electric field is solved by the finite element method (FEM). The field produced by the coils of the magnetic system is calculated using a paraxial expansion of the magnetic field along the axis of the tube. For scattering (allocation) of the space charge to the mesh an efficient and fast charge-conserving algorithm is used. In the relativistic particle pusher of GYREOSS, a second-order leap-frog Boris–Buneman scheme for tracing the motion of the particles is implemented. The code uses the advanced visualization capabilities of FreeFRM++-cs for post processing and analysis of the results of the numerical experiments. As an illustration, figure 5 shows a screenshot from the simulation of a coaxial gyrotron for plasma heating (2 MW at 170 GHz). A work on the parallelization of the current serial version of GYREOSS package will be the next step in its development.

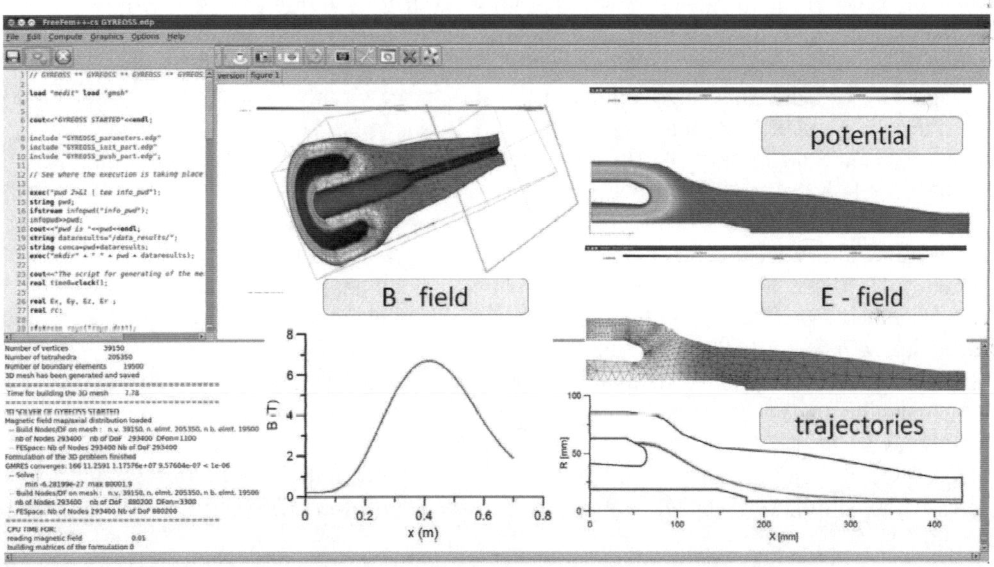

Figure 5. Screenshot taken during the simulation of a coaxial gyrotron for ECRH and ECCD.

4. Conclusions

The developed tools for modelling and simulation of gyrotrons have been used successfully for CAD, numerical studies, and optimization of the operational performance of a series of radiation sources for various applications presented briefly in this paper. Most of them have been constructed and manufactured at the Research Center for Development of Far-Infrared Region (FIR-UF) and are being currently used for advanced spectroscopic techniques (ESR, DNP-NMR, XDMR, measurement of the HFS of positronium), materials processing, novel medical technologies, etc. The reach set of computer codes integrated into the problem-oriented packages GYROSIM and GYREOSS is a solid base for further numerical studies and CAD of powerful gyrotrons many anticipated novel applications.

Besides the ongoing continuous work on the improvement of the program implementation, testing and benchmarking of the computational modules, upgrade of the underlying programming libraries, and so on, we plan a further extension of these packages adding new components that will take into account more physical factors. Among them are, for instance: the influence of the trapped (reflected) electrons; the impact ionization of the ambient high vacuum atmosphere in the tube by the beam electrons (and, thus, resulting space charge compensation); various instabilities (e.g. diocotron instability, electrostatic Bernstein instability, Langmuir instability, electrostatic cyclotron instability (aka negative-mass instability), trapped electron instabilities and others) affecting the beam formation and the beam-wave interaction.

References

[1] Kartikeyan M V, Borie E and Thumm M K A 2003 *GYROTRONS High Power Microwave and Millimeter Wave Technology* (Springer)

[2] Thumm M 2017 State-of-the-art of high power gyro-devices and free electron masers (Update 2016) *KIT Sci. Rep.* **7735** (KIT Sci. Publ.) pp 1–196

[3] Booske J H, Dobbs R J, Joye C D, Kory C L, Neil G R, Park G S, Park J and Temkin R J 2011 Vacuum electronic high power terahertz sources *IEEE Trans. Terahertz Sci. Technol.* **1** 54–75

[4] Idehara T, Saito T, Ogawa I, Mitsudo S, Tatematsu Y and Sabchevski S 2008 The potential of the gyrotrons for development of the sub-terahertz and the terahertz frequency range – A review of novel and prospective applications *Thin Solid Films* **517** 1503–06

[5] Idehara T and Sabchevski S 2012 Development and applications of high–frequency gyrotrons in FIR FU covering the sub–THz to THz range *J. Infrared, Millimeter and Terahertz Waves* **33** 667–94

[6] Glyavin M Y, Idehara T and Sabchevski S P 2015 Development of THz gyrotrons at IAP RAS and FIR UF and their applications in physical research and high-power THz technologies, *IEEE Trans. Terahertz Sci. Technol.* **5** 788–97

[7] Idehara T and Sabchevski S P 2017 Gyrotrons for high-power terahertz science and technology at FIR UF *J. Infrared, Millimeter, and Terahertz Waves* **38** 62–86

[8] Temkin R J 2014 Development of terahertz gyrotrons for spectroscopy at MIT *Terahertz Sci. Technol.* **7** 1–9

[9] Miyoshi N, Idehara T, Khutoryan E, Fukunaga Y, Bibin A B, Ito S and Sabchevski S P 2017 Combined hyperthermia and photodynamic therapy using a sub-THz gyrotron as a radiation source *J. Infrared, Millimeter, and Terahertz Waves* **37** 805–14

[10] Sabchevski S P, Idehara T, Ishiyama S, Miyoshi N and Tatsukawa T 2013 A dual–beam irradiation facility for a novel hybrid cancer therapy *J. Infrared, Millimeter and Terahertz Waves* **34** 71–87

[11] Sabchevski S P and Idehara T 2015 A numerical study on finite-bandwidth resonances of high-order axial modes (HOAM) in a gyrotron cavity *J. Infrared, Millimeter and Terahertz Waves* **36** 628–53

[12] Damyanova M, Sabchevski S, Zhelyazkov I, Vasileva E, Balabanova E, Dankov P and Malinov P 2014 Current status of the development of the problem-oriented software package GYREOSS, *J. Phys.: Conf. Series* **514** 012056

[13] Hecht F 2017 *FreeFem++, 3rd edition, Version 3.55-1* (Laboratorie Jacques-Louis Lions, Université Pierre et Marie Curie Paris) 1–392

[14] Geuzaine Ch and Remacle J-F 2017 *Gmsh: a three-dimensional finite element mesh generator with built-in pre- and post-processing facilities (Version 3.0.3, June 23 2017)*

IOP Conf. Series: Journal of Physics: Conf. Series **992** (2018) 012002 doi:10.1088/1742-6596/992/1/012002

Two-dimensional nanowires on homoepitaxial interfaces: Atomic-scale mechanism of breakdown and disintegration

Michail Michailov[1] and Bogdan Ranguelov

Acad. R. Kaishev Institute of Physical Chemistry, Bulgarian Academy of Sciences, Acad. G. Bonchev Str., Bl. 11, 1113 Sofia, Bulgaria

E-mail: mike@ipc.bas.bg

Abstract. We present a model for hole-mediated spontaneous breakdown of ahomoepitaxial two-dimensional (2D) flat nanowire based exclusively on random, thermally-activated motion of atoms. The model suggests a consecutive three-step mechanism driving the rupture and complete disintegration of the nanowire on a crystalline surface. The breakdown scenario includes: (i) local narrowing of a part of the stripe to a monatomic chain, (ii) formation of a recoverable single vacancy or a 2D vacancy cluster that causes temporary nanowire rupture, (iii) formation of a non-recoverable 2D hole leading to permanent nanowire breakdown. These successive events in the temporal evolution of the nanowire morphology bring the nanowire stripe into an irreversible unstable state, leading to a dramatic change in its peculiar physical properties and conductivity. The atomistic simulations also reveal a strong increase of the nanowire lifetime with an enlargement of its width and open up a way for a fine atomic-scale control of the nanowire lifetime and structural, morphological and thermodynamic stability.

1. Introduction

Metal nanowires excite remarkable academic curiosity as physical systems with confined geometry, reduced dimensionality and peculiar quantum properties, [1-10]. They also hold special attention in contemporary nanoscale material science because of the large number of exotic technological applications. Being physical objects with reduced space dimensions, the atomic nanowires demonstrate a variety of extraordinary features related mostly to their electronic conductivity. In these systems, which are considered as the ultimate quantum conductors, the electrons are able to propagate freely, without scattering, which leads to a dramatic variation of the nanowire conductivity, respectively, the electrical resistance. This is why in the last decades the basic mechanisms of nanowire growth have been studied in great detail by considering the formation of one or more linear atomic chains via different methods, including scanning tunneling microscopy (STM) tip-based methods, break-junction techniques, and epitaxial growth on vicinal and atomically smooth crystal surfaces. These theoretical and experimental studies have been generally focused on the problem of how to create nanowires [1-4]. Considerably less attention has been paid to the thermodynamic stability, rupture and complete disintegration of these exotic structures [11-13]. For this reason, the present work deals with the opposite issue of how nanowires spontaneously breakdown as a result exclusively of the random motion of their building atoms.

[1] To whom any correspondence should be addressed.

IOP Conf. Series: Journal of Physics: Conf. Series **992** (2018) 012002 doi:10.1088/1742-6596/992/1/012002

Recently we suggested a general atomic-scale scenario of a breakdown mechanism and complete disintegration of metal monatomic nanowires, freely standing in space or located on a crystalline substrate [11-13]. This research dealt with single atomic chains being exposed exclusively to the random thermal motions of their atoms without any impact of external fields (elastic strain, magnetic or electric forces, etc.). The specific linear atomic configuration of this ensemble of particles strongly deviates from that required for thermodynamic equilibrium and tends spontaneously to minimize its Gibbs free energy. Hence, at any finite non-zero temperature, the nanowire atomic configuration passes through a sequence of local energy states minima. This structural time evolution and atomic reorganization leads to a loss of continuous linear ordering of the nanowire via a multi-step mechanism of rupture at random sites and formation of small, differently sized clusters on the epitaxial interface. As a result, the ensemble of atoms finally loses its peculiar physical and electronic properties. This particular scenario of breakdown and complete disintegration quantifies the finite lifetime of the nanowire, which is intimately related to the nanowire conductivity as one of the most essential feature of this physical system.

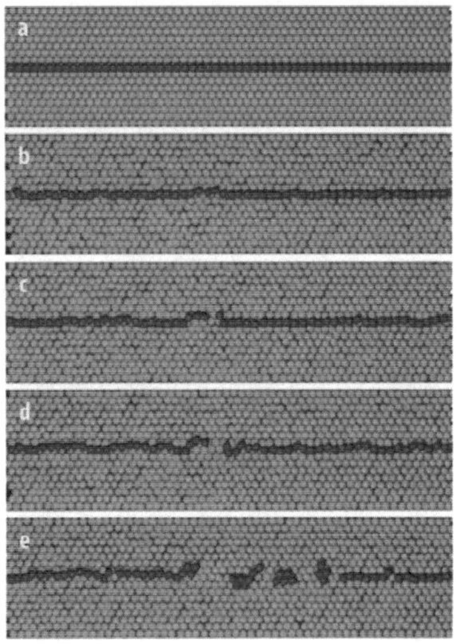

Figure 1. Hole-mediated breakdown of a nanowire on a crystalline surface. The time evolution reveals a multi-step rupture mechanism leading to complete nanowire dis-integration:
(a) - initial configuration;
(b) - formation of active sites (precursors);
(c) - generation of single vacancy;
(d) - generation of hole (double vacancy) and
(e) - formation of 2D atomic clusters ([12], [13]).

In the present study, our previously suggested physical and computational model for single monatomic chains is extended to flat, two-dimensional (2D) monolayer thick nanowire with variable width, figure 2. These flat nanowires are constituted of finite atomic chains located on crystalline substrates of the same material. Being ordered parallel to each other and touching themselves, they form 2D atomic stripes. Here, we discuss the impact of the nanowire's width and temperature on its structural stability affected by formation of atomic vacancies, vacancy clusters, step anisotropy, nanowire narrowing and generation of 2D wave-shaped atomic morphology. On that background we evaluate important physical quantities, such as the probability of breakdown, life time and the mechanism of complete disintegration. Since we focus our study on nanowires that are not exposed to any external fields, we exclude the impact of lattice mismatch between the nanowire and substrate. This is why in our model we use nanowires of the same material as that of the substrate. Hence, we consider the nanowire rupture and its disintegration as processes

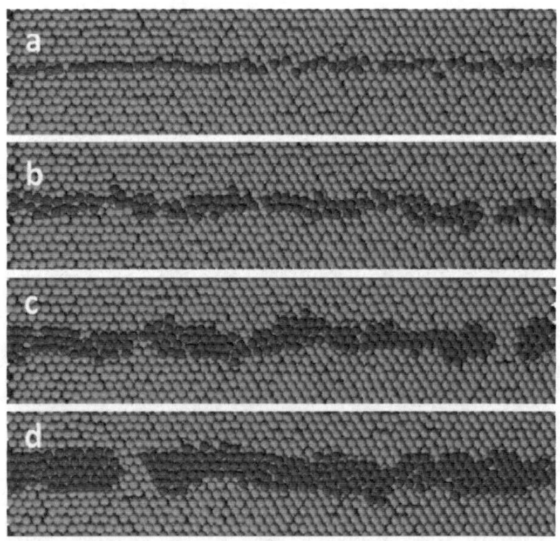

Figure 2. Non-recoverable vacancy cluster (hole) and permanent breakdown of nanowires with variable width. Atomically smooth fcc(111) homoepitaxial interface at 500 K:
(a) - monatomic nanowire;
(b), (c), (d) - nanowire 2D stripe consisting of 2, 3 and 4 monatomic chains, respectively. Generation of 2D wave-shaped atomic morphology on the stripe is clearly seen.

activated only by thermal motion of atoms. Let us point out here that the misfit between nanowire and substrate could stabilize the linear atomic chain or, in contrast, could facilitate its rupture, depending on the value of the lattice mismatch and the anisotropy of the epitaxial interface. This analysis is a subject of a separate study.

2. Physical and computational models

The analysis of the nanowire stability, breakdown and complete disintegration is based on an atomistic simulation experiment within the framework of the physical and computational models outlined below. The physical model in the present study is designed to reveal the temporal, structural and morphological evolution of 2D flat atomic nanowires at a constant temperature T. These flat nanowires are constituted of several atomic chains located on a homoepitaxial interface. Being ordered in a parallel way in contact with each other, they form atomic configuration similar to stripes, figure 2. The interface on which the nanowires are placed is atomically smooth, step free Cu (111) crystal surface. Being bound via many-body interacting potential, all atoms of the system (substrate and nanowire) are exposed to random thermal fluctuations only. The computational model is grounded on classical atomistic Monte Carlo (MC) sampling, with many-body tight-binding second-moment approximation potential between the interacting atoms [14-16, 18]. The statistical distribution in our model is canonical, i.e. an NVT ensemble with a constant number of particles, N, at a fixed temperature, T, and volume, V, [17, 18]. Following this approach, the total interaction energy E_i of atom i in the system is expressed by

$$E_i = E_i^r + E_i^b , \qquad (1)$$

where E_i^r and E_i^b are the repulsive and attractive energy terms, respectively. Explicitly, E_i is given by [13]

$$E_i = \sum_{j,r_{ij}<r_c} A\exp\left[-p\left(\frac{r_{ij}}{r_0}-1\right)\right] - \sqrt{\sum_{j,r_{ij}<r_c} \xi^2 \exp\left[-2q\left(\frac{r_{ij}}{r_0}-1\right)\right]} . \qquad (2)$$

Here, r_{ij} is the distance between atoms i and j, r_0 is the nearest-neighbor atomic distance of the corresponding pure metal, A and p are energy- and compressibility-related free parameters, and r_c is the cut-off distance for the interaction (3 atomic distances in our model). The first term in equation (2) is a Born-Meyer ion-ion type repulsion and the second term is the bond energy term, obtained in the form of the second-moment approximation of the electron density of states [14]. In equation (2), ξ is an effective hopping integral and q describes its dependence on the relative interatomic distance. In general, the expression for E_i^b, can be considered as a sum over the local electronic charge density induced at site i from the atoms at site j. Hence, ξ is expected to be sensitive to the number of surrounding atoms in all space directions. The energy calculation is performed over all atoms within a sphere having radius 3 times the Cu-Cu nearest-neighbors distance in the bulk crystal. The values of the hoping integral ξ and all related parameters A , p , and q in equation (2) are taken from [14,16]. In equilibrium, the system energy reaches its minimum and fluctuates around a constant value. Full lattice dynamics of all nanowire atoms ensures the complete relaxation of the system. All simulations are grounded on a three-dimensional continuous space model, i.e. substrate and nanowire atoms are able to change their positions in all space directions with a step of 0.05 units of the lattice constant. The initial nanowire configuration is a perfectly ordered set of linear monatomic chains with periodic boundary conditions along the chains axis. All simulation data are averaged over 10 runs with the nanowire stripe at specific width and temperature. The nanowire length is 76 atoms. The variation of the nanowire morphology and its atomic structure are monitored by a series of successive snapshots taken at every 100 Monte Carlo Steps (MCS). The generation of structural defects in the atomic stripe: single atomic vacancy, double vacancy, vacancy clusters, 2D atomic islands, etc. are detected during the simulations by measuring the distance between the atoms building the nanowire and via visual inspection of the snapshots taken on every run. In all simulations, the nanowire follows the classical thermodynamic pathway towards equilibrium of an ensemble of atoms by minimization of the system energy via clustering. Complete simulation details are described elsewhere [12, 13, 16].

3. Results and discussion

Before discussing the breakdown scenario of the flat 2D nanowire, let us point to an important detail related to the nanowire atomic morphology and anisotropy on an fcc(111) interface. The interface anisotropy leads to a different step free energy, relaxation ability, and diffusion barrier for displacement of the nanowire atoms in directions A and B, figure 3, [11, 12, 15]. This effect acts on the atomic step roughening transition, which is considered in our model as a precursor to the nanowire narrowing, breakdown and disintegration. Therefore, the breaking process is not symmetric in directions A and B and depends on the substrate surface orientation through the anisotropy of the surface atomic dynamics in different directions. Let us note that the temperature-dependent atomic roughening of B-steps costs less energy compared to that of A-steps. This is not the case at high temperatures, where this difference disappears since the thermal diffusion energy of atoms belonging to the steps is higher than the energy barrier for perpendicular translation in both directions, figure 3.

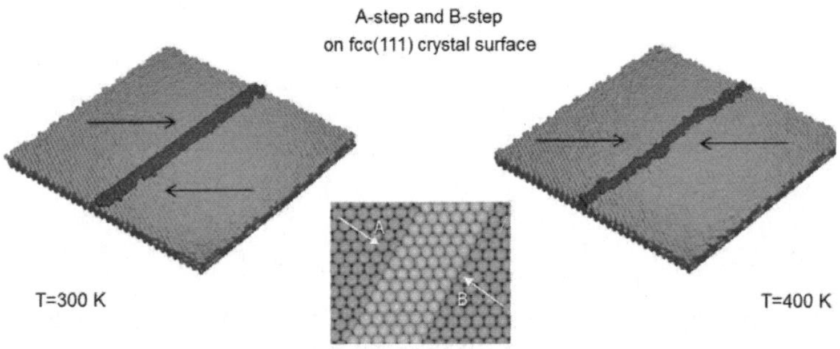

Figure 3. Anisotropy of thermal fluctuations of nanowire atoms and resulting equilibrium atomic structure of A- and B-steps at 300 K and 400 K on an fcc(111) substrate. The initial atomic configurations of A- and B-steps are presented in the middle panel of the figure ([11]).

The time evolution of the nanowire 2D stripe reveals a multi-step breakdown mechanism that could be considered as a sequence of well-defined structural changes and different morphological states. Let us note that our atomistic simulations are conducted at a constant temperature and the nanowire stripe build-up of variable number of monatomic chains follows a path toward its equilibrium atomic configuration. Our simulation results demonstrate that the nanowire narrowing at random places precedes the nanowire rupture. Hence, we observed that a nanowire breakdown starts preferentially from the step with a lower atomic roughening transition temperature. This is a result of the lower desorption energy for the atoms that belong to the atomically rough step and which facilitates the formation of locally concave areas on an initially linear ordering of the stripe. Since, however, the nanowire is wider than the monatomic chain, the formation of an entropy-driven wave-shaped structure is less favorable than that of the latter. In that case we consider the local nanowire narrowing as an active site for breakdown due to atomic roughening of the nanowire steps. The nanowire rupture occurs when the narrowing that has appeared randomly at one of the nanowire steps becomes greater than the nanowire width. This process is manifested in figures 4a-4d, which illustrate the time evolution of a nanowire having four monatomic chains.

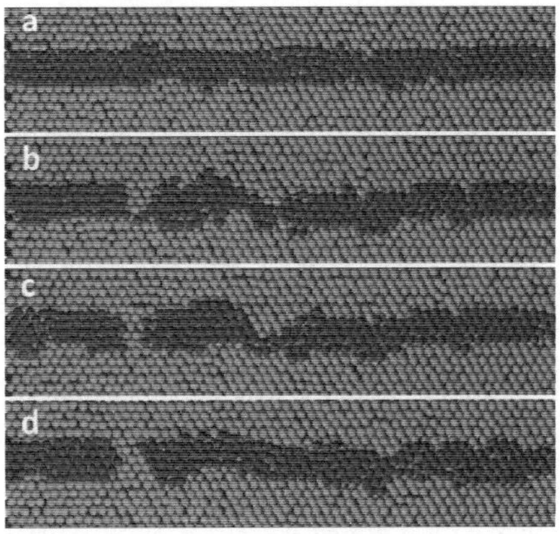

Figure 4. 500 K. Temporal evolution of a nanowire constituted of 4 monatomic chains - multi-step breakdown mechanism:
(a) local concave thinning - 60 0000 MCS and (b) 120 000 MCS;
(c) - formation of a vacancy - 180 000 MCS;
(d) - formation of hole (vacancy cluster) permanent nanowire breakdown - 300 000 MCS.

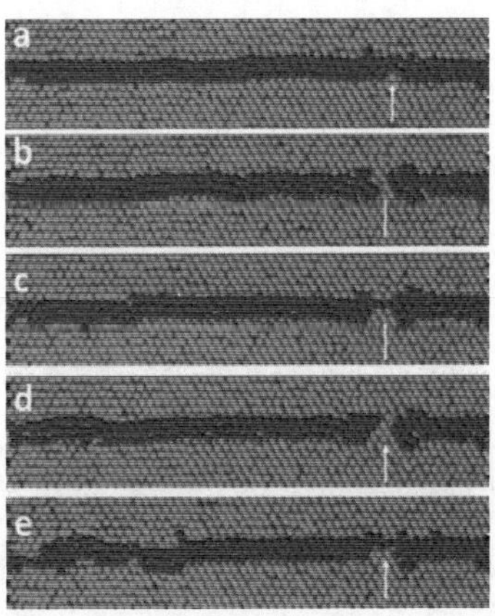

Figure 5. Recoverable vacancy formation and temporary breakdown of a nanowire constituted of 3 monatomic chains at 500 K:
(a) concave thinning – 26000 MCS;
(b) vacancy formation – 49000 MCS;
(c) vacancy recovering by randomly diffusing atom – 53000 MCS;
(d) vacancy formation on the same place – 57000 MCS;
(e) new recovering event – 73000 MCS.

The probability P_v of appearance of at least one vacancy in the nanowire until time t, defined as $P_v(t) = N_v(t)/N_{tot}$ is one of the parameters that characterize the temporary nanowire rupture at a certain temperature. Here $N_v(t)$ is the number of nanowires possessing at least one vacancy until time t, and N_{tot} is the total number of nanowires studied. Our results reveal that this probability dramatically decreases with enlargement of the nanowire stripe width, which is clearly seen in figure 6.

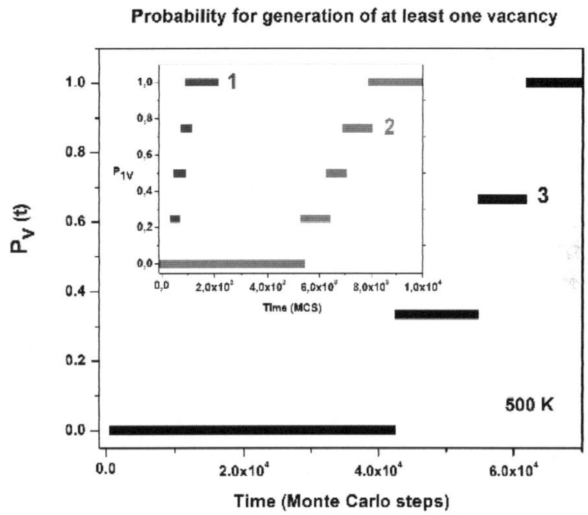

Figure 6. Time dependence of the probability $P_v(t)$ to form at least one vacancy at 500 K for nanowire consisting of 1, 2 and 3 monatomic chains.

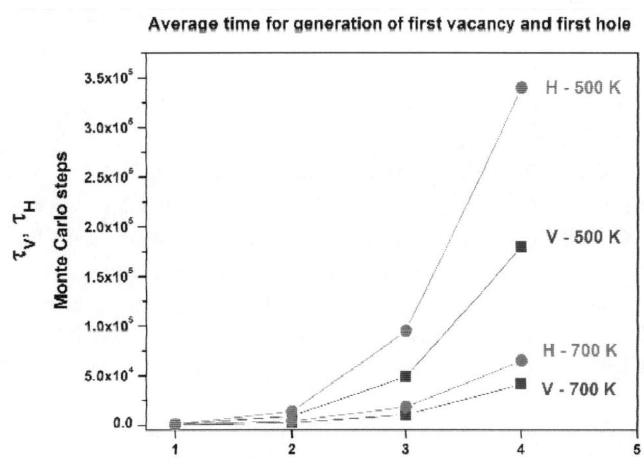

Figure 7. The average time (in MCS) for generation of single atomic vacancy, τ_V, and non-recoverable vacancy cluster (hole), τ_H, dependent on the nanowire width, at 500 K and 700 K.

The atomistic simulations reveal also that the enlargement of the nanowire stabilizes the linear chain structure and increases the lifetime. This is clearly seen in figure 7 where the average time for formation of vacancy and non-recoverable hole grow exponentially with increasing the nanowire width. As we have already pointed out, in the case of nanowire constituted of more than one monatomic chain, the breakdown probability will be complicated function of the probability for rupture of every individual monatomic chain that belongs to the nanowire. The evaluation of this probability is quite complex since two vacancies that are created in each monatomic chain have to be randomly generated very close to each other. This additional requirement is indispensable for generation of a non-recoverable atomic cluster that causes nanowire breakdown.

Similar to the case of monatomic nanowires is the formation observed by us of two types of atomic vacancies. The first type are recoverable vacancies, since they can be filled up back by the randomly walking atoms already detached from the nanowire in the vicinity of the single vacancy or vacancy cluster. This kind of vacancies cause only temporary breakdown of the integrity, and, hence of the conductivity, too, figure 5. Considerably more important and physically significant is the second type of vacancies, representing a vacancy cluster leading to permanent nanowire rupture. At a given size of this cluster, which we consider as a non-recoverable hole, the probability of its being filled up by randomly diffusing atoms is practically zero, figure 4. Therefore, the n-chain atomic nanowire follows three-step mechanism of breakdown and complete disintegration, as follows: (i) concave narrowing at a random location on the nanowire, figure 4b; (ii) generation of a single atomic recoverable vacancy, figure 4c; and (iii) formation of a non-recoverable vacancy cluster, figure 4d. Then this process can be schematized as:

$$\mathrm{S} \underset{\omega_{cs}}{\overset{\omega_{sc}}{\rightleftarrows}} \mathrm{C} \underset{\omega_{vc}}{\overset{\omega_{cv}}{\rightleftarrows}} \mathrm{V} \underset{\omega_{hv}}{\overset{\omega_{vh}}{\rightleftarrows}} \mathrm{H}$$

Here S, C, V and H stand for a straight morphology state (S), a concave morphology state (C), a vacancy state (V), and a hole state (H), and ω_{sc}, ω_{cv}, ω_{vh}, ω_{hv}, ω_{vc} and ω_{cs} are the corresponding frequencies of S-C, C-V, V-H, H-V, V-C and C-S transitions. This description accounts for the real dynamics of generation and annihilation of different structural and morphological defects and is essential in the kinetic treatment of nanowires breakdown.

It is important to compare the single-vacancy formation in a monatomic nanowire, figure 1, and a nanowire stripe comprising two, three, four, etc. single atomic chains, figure 2, [12,13]. The first difference is presented in figure 4. The probability of generation of two, three or more vacancies (depending on the nanowire stripe width) at the same time and place and, thus, of creation of a single atomic vacancy in the whole nanowire is very low. This probability is a function of the individual probabilities of single-vacancy formation in each monatomic chain that builds up the stripe. Moreover, these vacancies have to be generated very close to each other or at the same place in the nanowire. This is why, a flat 2D nanowire breaks down via the mechanism of narrowing at certain places, instead of waiting for the simultaneous random generation of single vacancies at this place at the same time. Apparently, this process of stripe narrowing is energetically more favorable and hence more probable than the simultaneous generation of individual vacancies at the same location. Let us point out that once a single atomic vacancy is generated at one of the atomic step (A or B) of the stripe, the enlargement and development of this vacancy into a vacancy cluster is favored compared to the flat atomic step. This is because the generation of kinks on the step and detachment of the atoms from these kink positions costs less energy than a detachment from the flat, atomically smooth step. Moreover, detachment of atoms from the kinks does not change the line step energy. Therefore, the generation even of a single atomic vacancy at a given step (A or B) favors the process of narrowing the nanowire by formation of a locally concave shape morphology in the stripe.

The second difference relates to the critical size of the vacancy cluster. Our results clearly demonstrate that the cluster size for non-recoverable breakdown is larger than that of the single monatomic nanowire (two neighboring vacancies). This critical size depends on the nanowire width.

Being with a larger width because of the several monatomic chains in it, the 2D nanowire has more atoms that can participate in the random fill up of the single vacancy or the vacancy cluster. The atomistic simulations reveal that at a fixed vacancy cluster size, the number of recoverable events in the temporal evolution of the flat 2D nanowire increases with the nanowire width, figure 6. Therefore, widening the stripe width increases the critical size of non-recoverable cluster. It should be pointed out also that the recovering mechanism kinetics is tightly dependent on the mean free diffusion path of atoms, respectively, on the interface temperature.

We can also define the quantity mean lifetime of recoverable vacancy and vacancy cluster. These quantities reveal for how long a nanowire will be in a state of temporary loss of conductivity before being recovered. Our qualitative evaluation shows that this lifetime decreases with the enlargement of the nanowire width. This indicates that at a fixed temperature the wider nanowires will recover more easily compared to the monatomic nanowires. This effect is due to the larger number of atoms taking part in random diffusion in the vicinity of the recoverable vacancy and vacancy cluster. If the distance between the newly created ends of the nanowire as a result of vacancy formation is equal to one atomic distance, figure 5b and figure 5c, then only one randomly walking atom could fill up the vacancy between these ends.

4. Conclusions

The present study provides insight into the mechanism of random thermal rupture of 2D flat homoepitaxial nanowire constituted of two, three or more atomic rows and located on an atomically smooth fcc(111) interface. The structural and morphological temporal evolutions of the nanowire reveal three consecutive steps of rupture, as follows: (i) local narrowing of a part of the stripe up to a monatomic chain, (ii) formation of a recoverable single vacancy or a vacancy cluster that cause temporary nanowire rupture, and (iii) formation of a non-recoverable hole leading to permanent nanowire breakdown. Analyzing these successive events in the temporal evolution of the nanowire morphology, we observe that the nanowire narrowing causes formation of randomly distributed locally concave areas in the nanowire, which could generate inhomogeneity in the nanowire conductivity, figures 4b, 5c and 5e. The increase of the stripe width leads to a significant enhancement of the nanowire lifetime. In contrast to a single monatomic nanowire, temporary rupture takes place not only upon single atomic vacancy generation, but also upon the generation of a vacancy cluster that can be also recoverable. This is an essential difference with the case of a monatomic nanowire, where only a single atomic vacancy can be recovered. The lifetime of the recoverable cluster is considerably smaller compared to that of the monatomic nanowire. The size of a non-recoverable hole depends on the nanowire width and increases with its enlargement. Experimental research in this direction could significantly contribute to a better understanding of the methods for atomic-scale control of the nanowire stability, lifetime and rupture kinetics.

References

[1] Agrait N, Yeyati A L and van Ruitenbeek J M 2003 *Phys. Rept.* **377** 81
[2] Yanson A I, Rubio-Bollinger G, van den Brom H E, Agrait N and van Ruitenbeek J M 1998 *Nature (London)* **395** 783
[3] Kiguchi M, Konishi T and Murakoshi K 2006 *Phys. Rev.* B **73** 125406
[4] Onishi H, Kondo Y and Takayanagi K 1998 *Nature (London)* **395** 780
[5] Landman U, Barnett R N, Scherbakov A G and Avouris P 2000 *Phys. Rev. Lett.* **85** 1958
[6] Tosatti E, Prestipino S, Kostlmeier S, Dal Corso A and Di Tolla F D 2001 *Science* **291** 288
[7] Robinson I K, Bennett P A and Himpsel F J 2002 *Phys. Rev. Lett.* **88** 096104
[8] Agrait N, Untiedt C, Rubio-Bollinger G and Vieira S 2002 *Phys. Rev. Lett.* **88** 216803
[9] Pelaez S, Guerrero C, Paredes R, Serena P A and Garcia-Mochales P 2010 *Electrodeposited nanowires and their applications* ed N Lupu (InTech, Rijeka) p. 35
[10] Picaud F, Dal Corso A and Tosatti E 2003 *Surface Sci.* **532-535** 544
[11] Michailov M and Kashchiev D 2012 *J. Phys.: Conf. Series* **398** 012010

[12] Michailov M and Kashchiev D 2015 *Physica* E **70** 21-7
[13] Michailov M, Ranguelov B, Giazitzidis P and Argyrakis P 2017 *J. Crystal Growth* **457** 92–7
[14] Cleri F and Rosato V 1993 *Phys. Rev.* B **48** 22
[15] Michailov M 2009 *Phys. Rev.* B **80** 035425
[16] Michailov M 2011 Classification order of surface-confined intermixing at epitaxial interface *Nanophenomena at Surfaces: Fundamentals of Exotic Condensed Matter Properties Springer series in Surface Science 47* ed Michailov M (Springer Berlin Heidelberg) chapter 6 pp 145-68
[17] Landau D P and Binder K 2000 *A Guide to Monte Carlo Simulations in Statistical Physics* (Cambridge University Press, Cambridge)
[18] Michailov M 2004 Monte Carlo simulations of metal monoatomic layers *Computer Simulations Studies in Condensed Matter Physics XVI Springer Series Proc. in Physics vol 95* ed Landau D P, Lewis S P and Schettler H B (Springer Berlin Heidelberg) pp 226-49

IOP Conf. Series: Journal of Physics: Conf. Series **992** (2018) 012003 doi:10.1088/1742-6596/992/1/012003

Investigation of archaeological metal artefacts by laser-induced breakdown spectroscopy (LIBS)

V Tankova[1,3], **G Malcheva**[1], **K Blagoev**[1] **and L Leshtakov**[2]

[1]G. Nadjakov Institute of Solid State Physics, Bulgarian Academy of Sciences, 72 Tzarigradsko Chaussee, 1784 Sofia, Bulgaria
[2]National Institute of Archaeology with Museum, Bulgarian Academy of Sciences, 2 Saborna Str., 1000 Sofia, Bulgaria

E-mail: vanitankova@gmail.com

Abstract. In this work, laser-induced breakdown spectroscopy was applied to determining the elemental composition of a set of ancient bronze artefacts dated from the Late Bronze Age and Early Iron Age (14th – 10th century BC). We used a Nd:YAG laser at 1064 nm with pulse duration of 10 ns and energy of 10 mJ and determined the elemental composition of the bronze alloy that was used in manufacturing the samples under study. The concentrations of tin and lead in the bulk of the examined materials was estimated after generating calibration curves for a set of four standard samples. The preliminary results of the analysis will provide information on the artefacts provenance and on the production process.

1. Introduction

The analysis of ancient artefacts provides archaeologists and restorers with valuable information about the use and origin of archaeological objects as well as on the materials that were used. Metal objects have been made and used by people since very ancient times. Their making, style and context of use constitute indicators of technology, culture and trade activities. That is why the identification of compositional materials in metal findings discovered on a particular site is very important.

The laser-induced breakdown spectroscopy (LIBS) is a spectroscopic analysis of the plasma formed as a result of the interaction of a focused light pulse from a high-power laser with a material. Atomic (or ionic) spectral emissions of the plasma thus generated can be used for real-time multi-element analysis of different type of materials – gaseous, liquid or solid [1, 2]. A large number of papers have been published concerning both the LIBS fundamentals and the application of LIBS to material analysis [3–6]. In general, a large number of elemental analysis techniques are well established in the field of material science and quite a few of them have enjoyed recognition in archaeometry and artwork analysis (SEM, XRF, PIXE, ICP-MC etc.). The number of analytical applications of LIBS has been continuously growing thanks to its intrinsic conceptual simplicity and versatility compared to the conventional techniques for analysis.

LIBS has features which make it a useful tool for real-time measurements. These include multi-element analysis, micro-destructiveness, instrumentation portability, non-contact optical nature of the measurements regardless of the material to be analyzed, high measuring speed, and no sample

[3] To whom any correspondence should be addressed.

IOP Conf. Series: Journal of Physics: Conf. Series **992** (2018) 012003 doi:10.1088/1742-6596/992/1/012003

preparation. LIBS measurements are generally carried out in ambient air at atmospheric pressure. The possibilities of making qualitative and quantitative analysis by LIBS and depth analysis provide useful information on the composition of the investigated materials. These advantages are of decisive importance when the objects examined are archaeological artefacts.

In this work, we applied LIBS to studying a set of ancient bronze objects belonging to the museum collection of the National Institute of Archaeology with Museum at the Bulgarian Academy of Sciences. The analytical information obtained from the LIBS measurements was used in order to determine the elemental composition of the alloy and to investigate the provenance of the objects under study.

2. Experimental

2.1. Samples

We present results from a study of five bronze finds dated from Late Bronze Age and Early Iron Age (14th – 10th century BC). The artefacts are spearheads and socketed axes. Figure 1 presents sketches of two of the objects under study.

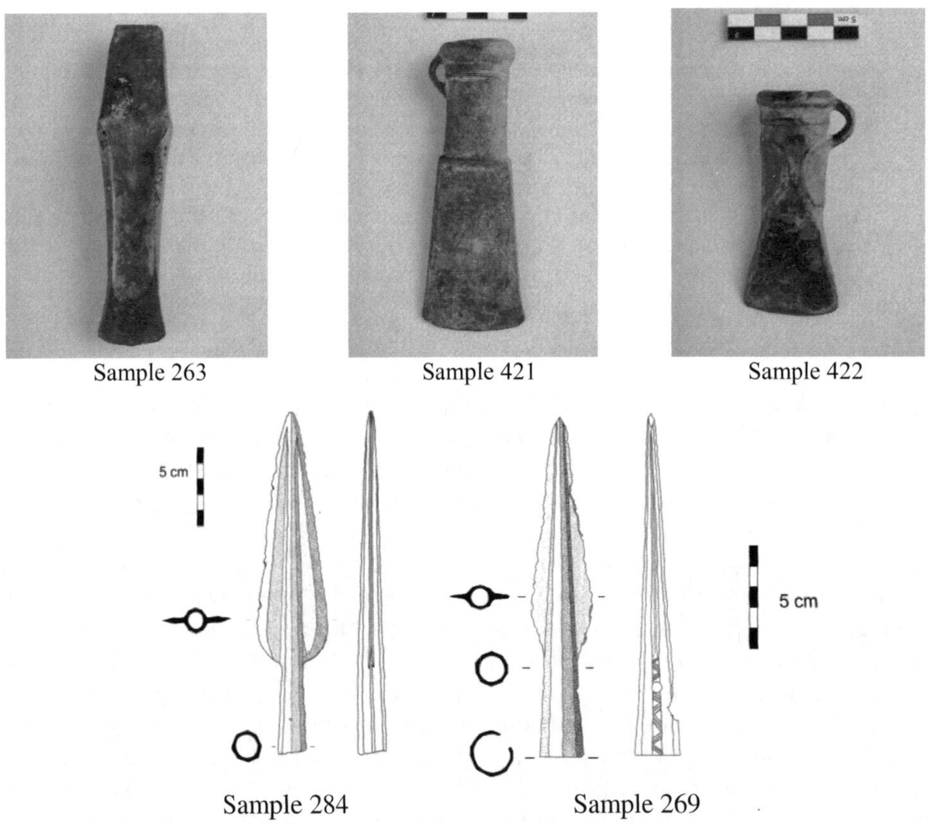

Figure 1. Analyzed archaeological bronze objects.

2.2. Experimental set-up

The laser beam of a nanosecond Q-switched Nd:YAG laser (Quanta Ray GC3) operating at the fundamental wavelength (1064 nm, pulse duration 10 ns, pulse energy 8 mJ) was focused onto thetarget by a lens with a focal length of 170 mm). The ablation took place in air at atmospheric

IOP Conf. Series: Journal of Physics: Conf. Series **992** (2018) 012003 doi:10.1088/1742-6596/992/1/012003

pressure. The plasma emission was collected using an optical fiber (with diameter 50 μm) and analyzed in an Echelle spectrometer (type Mechelle 5000, with a grating with 52 grooves/mm and a reverse linear dispersion of 1-4 nm/mm). The LIBS spectrum was recorded by an intensified charge coupled device (ICCD) detector (DH734-18F-03, Andor Technology) having 1024×1024 pixels. The detector was gated by a delay/pulse generator type Г5-56. In order to separate the atomic and ionic emissions from the continuum background of the plasma emission, delay time of 1 μs and gate of 1 μs were used. This equipment offers a large spectral range – from 220 nm to 850 nm; the spectrometer was calibrated by W/De and Hg/Ar standard lamps for intensity and wave-length, respectively.

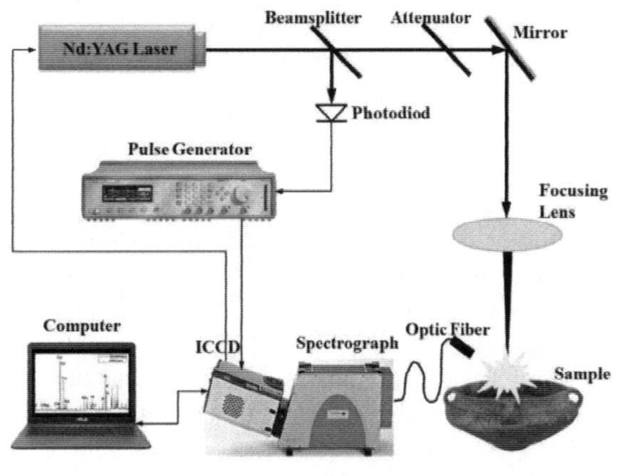

Figure 2. Experimental set-up.

The experimental parameters were optimized in order to obtain a good - signal – to - noise ratio, to ensure the best measurements reproducibility and, at the same time, to avoid damaging the samples. In depth profiling measurements, the emission spectra were collected and averaged for several successive laser shots. A schematic drawing of the experimental set-up is presented in figure 2.

3. Results and discussion

3.1. Qualitative analysis

Bronze is a metal alloy consisting primarily of copper (85-95%); commonly the remainder is tin, often with the addition of other metals (such as lead, manganese, nickel or zinc) and sometimes non-metals (such as arsenic, phosphorus, silicon) present in small amounts. Bronze is harder than copper as a result of alloying this metal with tin or other metals. During the Bronze Age, two distinct forms of bronze were commonly used: "classic bronze" (which contained about 10% tin and was used in casting) and "mild bronze" (about 6% tin, hammered into sheets from ingots).

During the experiments, the signals from 30 successive laser shots at a single position of the sample were accumulated and averaged. Before the measurements, 100 laser shots were directed at the same point in order to clean the surface and reach the pure metal.

A selected LIBS spectrum is shown in figure 3. As seen by the characteristic emission lines marked in the spectrum, the bronze alloy contains copper (Cu), tin (Sn) and lead (Pb) as main elements. Other elements, namely, iron (Fe), nickel (Ni), titanium (Ti), zinc (Zn), aluminium (Al) and antimony (Sb) were also detected. One of the aims of this work was to investigate the provenance of the finds. In view of the shape and decoration of the artefacts, it is assumed that they are more typical for Central Europe in this time period, being different from those Thracians were using during the Late Bronze Age. The concentration of antimony in the alloy could provide useful information about the origin of the artefacts. The copper-antimony alloys (usually containing between 2-12% Sb) were first discovered in the archaeological record during 1890's, primarily

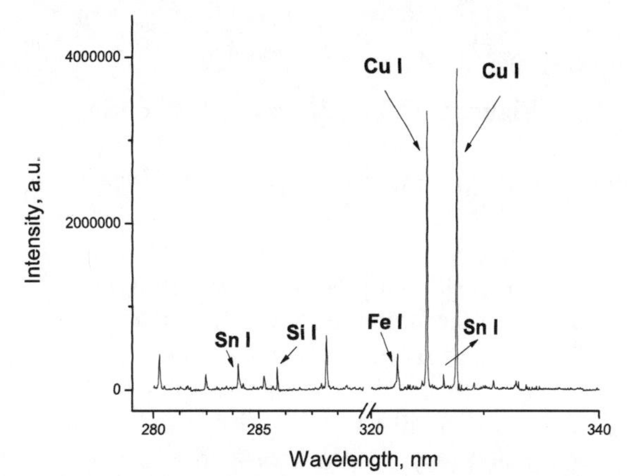

Figure 3. Typical spectrally resolved LIBS spectrum.

from Late Bronze Age/ Early Iron Age Hungarian contexts (Urnfield period) [7]. In the present experiment, we did find antimony in the objects (figure 4), but further experiments are needed to estimate the antimony concentration.

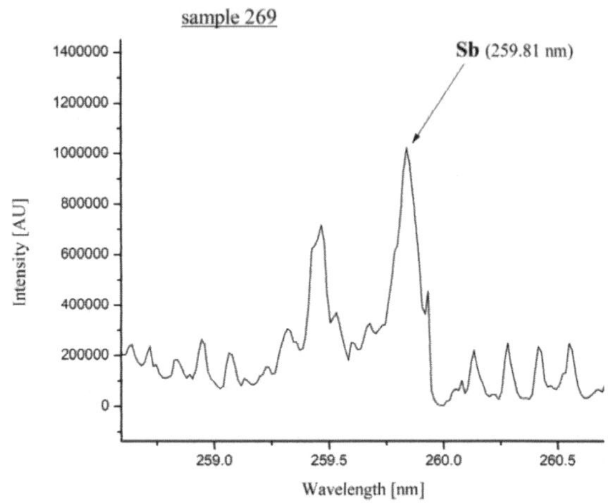

Figure 4. Detected Sb spectral line.

3.2. Quantitative analysis

We attempted to determine the tin and lead concentrations in the material used for manufacturing of the artefacts, because these elements define the mechanical properties of the alloy.

The elemental composition of the metal object can be affected by corrosion, which leads to material deterioration and a possible change of the weight ratio of certain elements, thus affecting the reliability of the results of the quantitative elemental analyses. This is why, in our experiments we applied 100 successive laser shots to reach the pure metal surface before recording the spectra of the alloy itself.

Four standard samples with a similar elemental composition and a suitable dominant element were used for quantitative determination of tin and lead in the artefacts. The sample properties were thus defined, bearing in mind that the matrix effects and the plasma emission are almost the same for all samples. The quantity of tin and lead in the real objects was established after plotting calibration curves (figures 5 and 6). The selected spectral lines used for the calibration of the elements were chosen so as to minimize the spectral interferences and the self-absorption. For good statistics, kinetic series of 10 spectra accumulated for 30 laser pulses were registered for the bronze artefacts and the standard samples. The results of the quantitative LIBS analysis are shown in table 1. According to the preliminary results for the tin and lead concentrations, in all samples (except for sample 263) the tin and lead concentrations are very high, the reason being the influence of the corrosion layer. Therefore, the surface of the artefacts should be cleaned mechanically

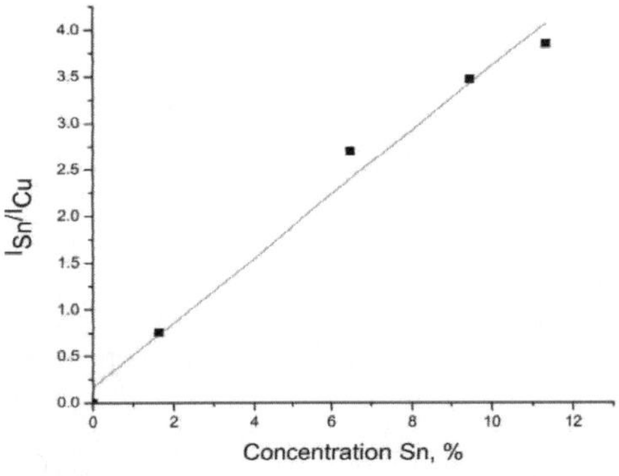

Figure 5. Calibration curve for tin (Sn).

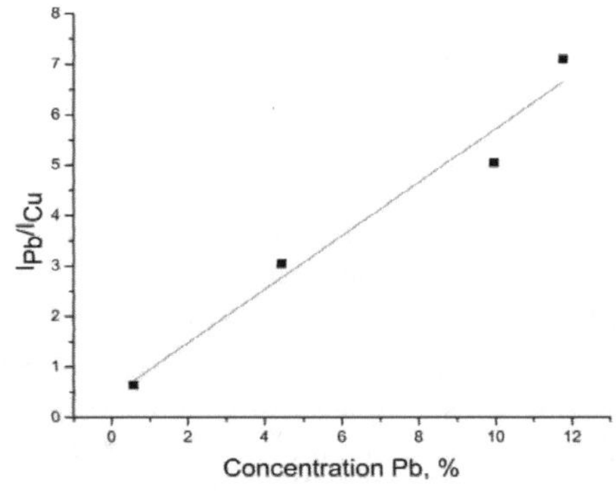

Figure 6. Calibration curve for lead (Pb).

Table 1. Concentrations of tin and lead in the bulk of the samples obtained by LIBS.

Sample	Concentration, %	
	Sn	Pb
263	18.3±4.8	0.03±0.006
269	47.2±10.9	7.26±0.7
284	37.5±3.7	0.9±0.09
421	55±11	1.7±0.3
422	31.7±6.6	4.3±0.6

IOP Conf. Series: Journal of Physics: Conf. Series **992** (2018) 012003 doi:10.1088/1742-6596/992/1/012003

to remove the corrosion layer before starting the experiments. For Sample 263, the concentrations of tin and lead are typical for the Late Bronze Age.

4. Conclusions

We present preliminary results of a study of five metal artefacts from the Late Bronze Age and the Iron Age. The quantitative analysis shows that the results obtained for tin and lead concentrations are affected by the thick corrosion layer, so that further experiments are planned.

Acknowledgements

This work was carried out in the framework of Project "THRACIANS – genesis and development of the ethnicity, cultural identities, civilization interactions and heritage from the antiquity", which is supported by Mr. P. Mandjukov. The authors wish to thank Krasimir Georgiev for providing the photographs of the artefacts.

References

[1] Cremers D A and Radziemski L J 2006 *Handbook of Laser-Induced Breakdown Spectroscopy* (West Sussex, England: John Wiley & Sons Ltd)
[2] Miziolek W, Palleschi V and Schechter I 2006 *Laser Induced Breakdown Spectroscopy (LIBS): Fundamental and Applications* (Cambridge: Cambridge University Press)
[3] Rusak D A, Castle B C, Smith B W and Winefordner J D 1997 *Crit. Rev. Anal. Chem.* **27** 257
[4] Ahmed I and Goddard B J 1993 *J. Fiz. Malays.* **14** 43
[5] Radziemski L J 1994 *Microchem. J.* **50** 218
[6] Radziemsky L J 2002 *Spectrochim. Acta* B **57** 1109
[7] Visy Z and Bartosiewicz L eds 2003*Hungarian Archaeology at the turn of the Millenium*

IOP Conf. Series: Journal of Physics: Conf. Series **992** (2018) 012004 doi:10.1088/1742-6596/992/1/012004

High-resolution laser spectroscopy of hot Cs and Rb vapor confined in a thin optical cell

P Todorov[1,4], A Krasteva[1], T Vartanyan[2], G Todorov[1], D Sarkisyan[3] and S Cartaleva[1]

[1]E. Djakov Institute of Electronics, Bulgarian Academy of Sciences, 72 Tzarigradsko Chaussee, 1784 Sofia, Bulgaria
[2]ITMO University, 49 Kronverkskiy Pr., 197101 St. Petersburg, Russian Federation
[3]Institute for Physical Research, National Academy of Sciences of Armenia, 0203 Ashtarak, Armenia

E-mail: petkoatodorov@yahoo.com

Abstract. We propose a novel use of an optical cell of micrometer thickness filled with Cs vapor in view of studying the collisions between two different alkali atoms of strongly different densities. We demonstrate narrow and good-contrast sub-Doppler resonances at the Rb D_2 line for a mean-free-path of the Cs atoms comparable to the optical cell longitudinal dimension; the resonances are completely destroyed when the mean-free-path of the Cs atoms is more than two orders of magnitude shorter than the longitudinal dimension of the thin cell.

1. Introduction

The field of photonics has undergone immense technological and scientific advancement in the past decade related to the miniaturization of various optical sensors. An important task in this respect is to induce atom-photon interactions in hot and dense atomic vapor confined in optical cells of sub-millimeter thickness. Such confinement makes it possible to build a high-density but optically thin atomic vapor layer that will be capable of producing diatomic Cs molecules alongside the atoms, while allowing one to perform linear transmission spectroscopy studies of the molecular ensemble.

In our experimental study, we used two cells of micrometer thickness, similar to the one described in [1] but with nearly parallel internal sides of the cell windows. Both cells were filled with Cs vapor, which naturally contains Rb atoms in an extremely small concentration. If heated to temperatures above 200 °C, such a cell will contain a mixture of alkali atoms: predominantly Cs atoms, a more than an order of magnitude lower concentration of Cs dimers and single Rb atoms. Under such conditions, the Cs atoms and dimers can be considered as a kind of buffer gas for the Rb atoms. Thus, such a mixture will make it possible to measure precisely the pressure effects on the Rb spectral profiles as the Cs density is raised.

2. Ehergy level diagram and experimental setup

The natural isotopic abundance of Rb is about 72.2% ^{85}Rb and about 27.8% ^{87}Rb [2]. Figure 1 presents their energy level diagrams, which should be involved in the analysis of the experimental results.

[4] To whom any correspondence should be addressed.

In our experiment, we explored the D_2 line of Rb in a 700-micrometer-thick cell and the D_1 line of Rb in a 60-micrometer-thick cell. The Rb D_1 line of consists of two ground state levels and two excited state levels [3]. The corresponding atomic numbers for the ground state levels of ^{87}Rb are Fg=1 and Fg=2; for excited state levels these are Fe=1 and Fe=2 (see figure 1). The D_2 line of Rb is slightly more complicated and consists of the same ground state levels as for D_1 line and four excited state levels. The corresponding atomic numbers for ^{87}Rb are Fg=1 and Fg=2 and for the excited state levels Fe=0, Fe=1, Fe=2 and Fe=3.

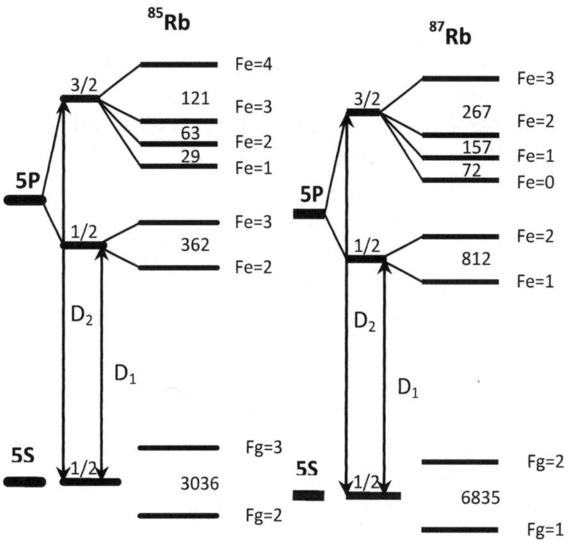

Figure 1. Rubidium isotopes energy levels diagram. The frequency separation between hyperfine levels is indicated in MHz.

The experimental setup is shown in figure 2. The light source was a distributed-feedback diode laser (DFBDL) emitting linearly polarized light with a short-term linewidth of 3 MHz and a maximal power of 150 mW. The laser frequency was scanned by varying the laser current. The laser beam was split in two by a beam splitter, its main part directed to the micrometric cell, and the second part used for reference. In the reference part of the setup we empoyed saturated absorption spectroscopy to observe the Rb resonances in a 5-cm-thick optical cell.

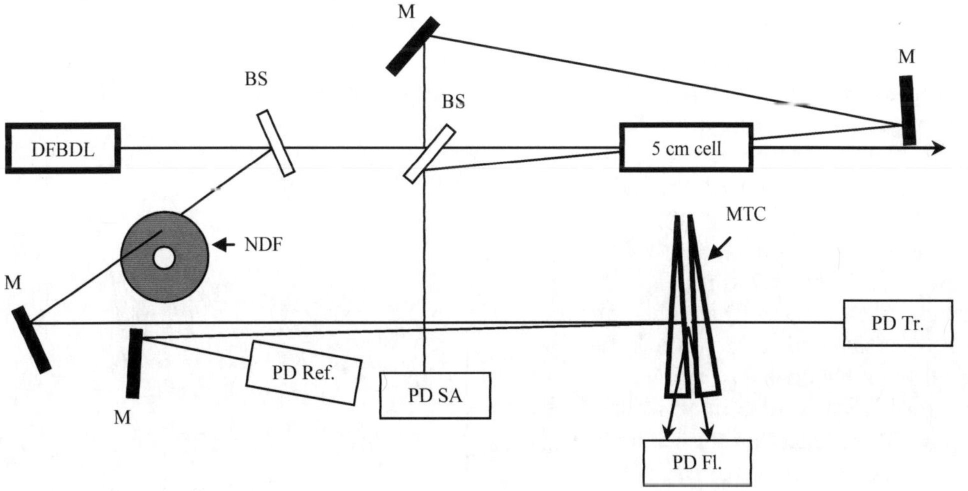

Figure 2. Experimental set up: DFBDL – distributed feedback diode laser; NDF – neutral density filter; MTS – micrometric thick cell; BS – beam splitter; M – mirror; PD Ref. – photodiode reflection signal; PD Tr. – photodiode transmission signal; PD Fl. – photodiode fluorescence.

Before entering the micrometric cell, the main part of the laser beam passes through a variable neutral density filter controlling the laser beam power. It is then directed by a mirror nearly perpendicularly to the cell internal window surface. The laser light interaction with the cell and the vapor gives rise to the three signals collected, namely, transmission, reflection and fluorescence. The reflection signal consists of three components – the back-surface reflection of the first window, the signal from the Rb atoms and the reflection from the front surface of the second window. The reflections from the external surfaces of the cell windows propagate in directions different from that of the reflection signal because the windows are formed as wedges. The signal reflected from the cell propagates close to the incident beam and is directed to the receiving photodiode by a small mirror. The fluorescence signal is collected perpendicularly to the laser beam. The transmission, reflection and fluorescence, as well as the saturated absorption signal from the 5-cm-thick cell, are registered by photodiodes and stored by a digital oscilloscope.

The micrometric cell is placed in an oven where it is heated to the temperature required of the atomic source. This temperature, which is indicative of the Cs atoms density, is measured by a thermocouple attached to the atomic source reservoir.

3. Experimental results and discussion

We first present the main experimental results obtained for the 700-μm-long cell filled with Cs. As seen in figure 3, the transmission spectrum is of lower amplitude and does not show narrow, sub-Doppler features in the Doppler profiles. It consists of four Doppler profiles that are due to the absorption at the four sets of hyperfine transitions starting from the following ground state levels: F_g=2 hyperfine level of the ^{87}Rb isotope, F_g=3 hyperfine level of the ^{85}Rb isotope, F_g=2 hyperfine level of the ^{85}Rb isotope and F_g=1 hyperfine level of the ^{87}Rb isotope (figure 1). Unlike the transmission profiles, in the fluorescence (not shown in figure 3) and reflection spectra, well-expressed narrow dips in the Doppler profiles are observed. The formation of the sub-Doppler structures is attributed to the beam reflected from the second window of the 700-μm cell, contributing to the formation of saturated absorption resonances centered at the hyperfine transitions.

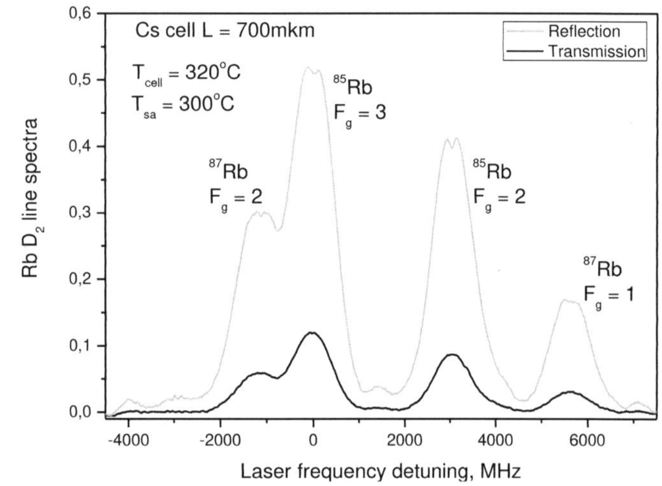

Figure 3. Transmission and reflection spectra of Rb measured in a 700-μm thick cell filled with Cs vapor.

We further studied the fluorescence spectrum dependence on the atomic source temperature (figure 4). In the figure, only two sets of hyperfine transitions are shown in order to present the fluorescence profiles with a better spectral resolution. Due to the formation of narrow sub-Doppler reduced-absorption resonances, all three hyperfine transitions, starting from the F_g=2 level of the ^{87}Rb isotope

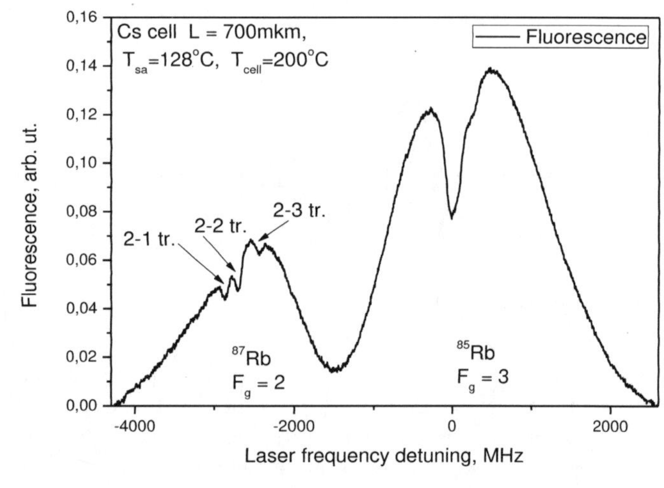

Figure 4. Well resolved sub-Doppler structure observation, for the Fg=2 set of transition of ^{87}Rb isotope, at low atomic source temperature T_{sa} = 128°C.

are very well resolved. The optimal spectral resolution is observed for relatively low cell side arm temperature $T_{sa} = 128\ ^{o}C$, i.e. a low Cs atoms density (7.31×10^{13} atoms/cm^3). For this atomic density, the Cs atoms mean-free-path (mfp) is estimated at 474 μm, which is close to the optical cell thickness of 700 μm. We studied in detail the sub-Doppler resonance behavior as a function of the atomic source temperature and found that the rise in the atomic density results in broadening the resonance and reducing its contrast, especially for T_{sa}=235 °C. The sub-Doppler structure disappears for T_{sa} = (276-280) °C (figure 5). This is attributed to the collisions of the high-density Cs atoms with the Rb atoms of much lower density. Indeed, at T = 276 °C, the Cs atoms density (2.01×10^{16} atoms/cm^3) is higher by more than two orders of magnitude than that for T_{sa} = 128 °C. The higher density determines a Cs atoms' mfp of 1.91 μm for T_{sa} = 276 °C. Hence, the sub-Doppler resonances at the ^{87}Rb F_g=2 set of transitions are destroyed at Cs atoms' mfp that is lower by more than two orders of magnitude than the longitudinal dimension of the thin cell. Under such conditions, the collisions of Rb atoms with the thin cell walls are strongly suppressed due to collisions with Cs atoms. Hence, our result supports the theoretical model in [4], where it was shown that a large number of atomic collisions with the optical cell surfaces result in amplifying the sub-Doppler features.

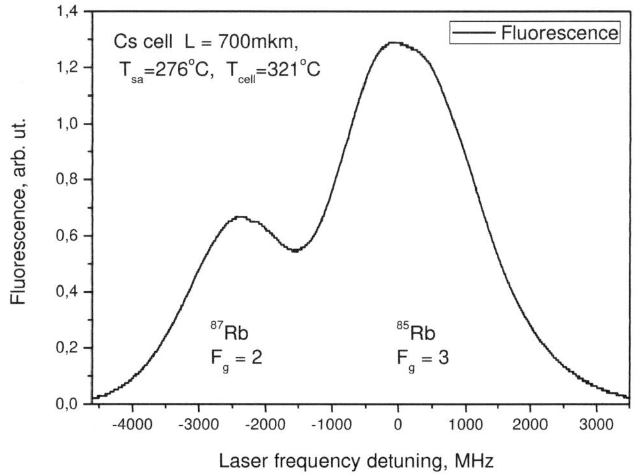

Figure 5. After spectral broadening and amplitude reduction, the sub-Doppler resonances disappear.

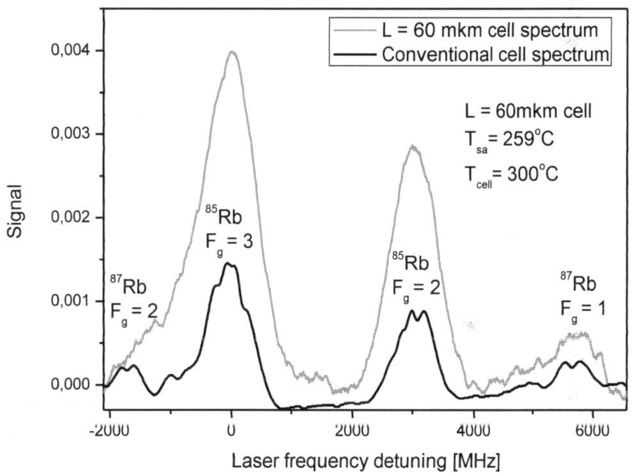

Figure 6. Fluorescence spectrum on the D_1 line observed for the significantly thinner optical cell.

Further on, we measured the fluorescence spectra of Rb vapor confined in a thin cell with L = 60 μm filled with Cs to explore the D_1 line of Rb (figure 6). Comparing the thin cell fluorescence spectra with the D_1 line spectrum in a conventional cell (i.e., one with a large longitudinal dimension), showed that even the 60-μm cell produces a Rb spectrum with a good signal/noise ratio. This result is promising for the development of a precise frequency reference to be further applied to Cs dimer study in thin optical cells.

4. Conclusions

In conclusion, we demonstrated a novel use of a micrometric thin cell filled with Cs vapor, which naturally contains Rb atoms in an extremely small concentration. The experimental results showed that it is possible to measure precisely the pressure broadening and destruction of sub-Doppler features in Rb spectral profiles as the Cs density is raised, in view of analyzing the processes involved in the formation of sub-Doppler resonances and studying the collisions between two different alkali atoms in thin optical cells.

Acknowledgments
The financial support is acknowledged of the Bulgarian National Science Fund under the following projects: DNTS/Russia 01/5 23.06.2017+RFBR grant No17-52-18037, “Nonlinear spectroscopy of spatially restricted alkali vapor: methodology and applications” and DN 08-19/ 14.12.2016, “New coherent and cooperative effects in hot alkali vapor”. T Vartanyan work was performed in the framework of state order #3.4903.2017/6.7.

References
[1] Andreeva C, Cartaleva S, Petrov L, Saltiel S, Sarkisyan D, Varzhapetyan T, Bloch D, and Ducloy M 2007 *Phys. Rev.* A **76** 013837
[2] Georges A, Bersillon O, Blachot J and Wapstra A H 2003 *Nuclear Phys.* A. *Atomic Mass Data Center* **729**/1 3–128
[3] Dancheva Y G Alzetta, S Cartaleva, M Taslakov and Ch Andreeva 2000 *Optics Comm.* **178**/1 103-10
[4] Todorov G, Polishchuk V, Krasteva A, Sargsyan A D, Cartaleva S and Vartanyan T A 2016 *J. Opt. Technol.* **83** 659-66

Schlieren flow visualization of helium atmospheric plasma jet and influence of the gas flow rate and applied voltage frequency

S M Borghei[1], N Vaziri and S Alibabaei

Physics Department, Karaj Branch, Islamic Azad University, Karaj, Iran

E-mail: majid.borghei@kiau.ac.ir

Abstract. We used schlieren photography to visualize the influence of gas flow rates of 1, 2.5, 5, 10 L/min and of the applied voltage frequency on a helium atmospheric plasma jet induced at the nozzle of a capillary tube. The expansion of the gas in the surrounding medium (air) was analyzed in the two different modes – plasma on/plasma off. Changes in the above parameters affect the gas flow regime and the hydrodynamics of the jet.

1. Introduction

Atmospheric pressure plasma jets are rapidly gaining importance as potential tools for plasma processing in a number of applications, such as film deposition [1], surface modification [2], sterilization [3], and biomedical processes [4]. The plasma jets are technologically simple, environmental friendly and very economical. Therefore, exploring the behavior of the exhaust gas and the effects of varying the operating parameters is very important.

The schlieren method introduced by Foucault [5] and Toepler [6] has long been an important optical tool for the study of different types of flows, especially compressible flows. Schlieren images are also used to investigate fluid dynamic characteristics of plasma jets. For example, Bradley *et al.* [7] studied the outflow of a helium plasma jet. Boselli *et al.* [8] investigated the fluid dynamic characteristics of a nanosecond plasma jet. Zheng *et al.* [9] also used schlieren photography to study the hydrodynamic of a helium plasma jet. In addition, many other related studies have been published. For example, the group at INP Greifswald applied laser schlieren deflectometry to the temperature analysis of filamentary plasmas [10, 11].

In the present study, schlieren photography was utilized to investigate the gas behavior of the plasma jet under different conditions. The plasma jet was generated in open air in a simple micro-discharge configuration by flowing helium through a coaxial dielectric barrier discharge chamber (DBD) excited by an AC voltage at frequencies of some tens of kHz.

2. Schlieren optical system

A schlieren system visualizes variations of the refraction index of light, thhus effectively depicting the density gradient variation in a fluid. Figure 1 shows the setup of the experiment. A small-size source illuminates a concave mirror of long focal length. The distance between the source and the mirror is

[1] To whom any correspondence should be addressed.

IOP Conf. Series: Journal of Physics: Conf. Series **992** (2018) 012005 doi:10.1088/1742-6596/992/1/012005

equal to two focal lengths of the mirror. About half of the reflected beam is blocked by the edge of a blade. A long-focal-length camera is focused on the mirror image to capture the shadow image of the plasma jet flow. In our experiment, the imaging system set-up consisted of a jet tube ($Ø_{int}$ = 2 mm), a concave mirror with a diameter of 130 mm (a focal length of 1470 mm), a six-megapixel compact digital camera (Nikon Coolpix L820 with a maximal focal length of 120 mm) and a 6 V continuous LED as a light source.

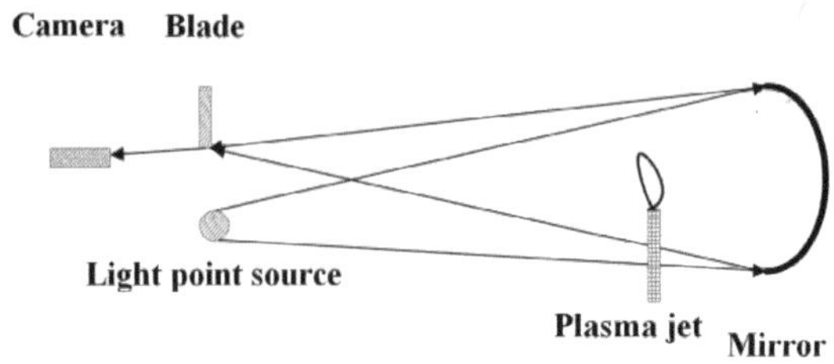

Figure1. Schematic diagram of the set-up used in the experiment.

3. Influence of the flow rate
Figure 2 shows schlieren images of the helium jet without (upper panel) and with (lower panel) plasma ignition, operating at 35 kHz and 5 kV, for different flow rates. The micro-discharge is visible on the bottom of each figure. It is clear that the laminar-turbulent transition occurs at flow rates exceeding 2.5 L/min.

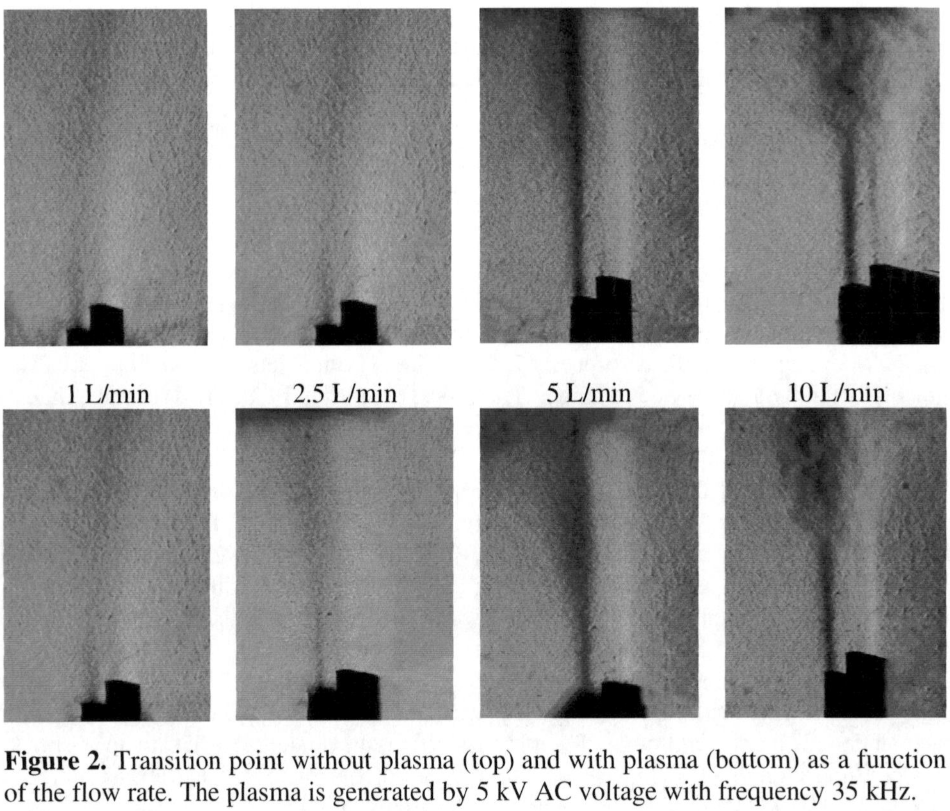

Figure 2. Transition point without plasma (top) and with plasma (bottom) as a function of the flow rate. The plasma is generated by 5 kV AC voltage with frequency 35 kHz.

Figure 3 shows schlieren imaging of a helium flow into ambient air at two different flow rates before plasma ignition. The buoyancy is vertically upward. The output diameter of the jet is 4 mm. Thus, in the case of a flow rate of 5 L/min, the Reynolds number at the exhaust section was about 225, and in the case of 2.5 L/min, about 112.5. It is clear that at the higher velocity cases, eddy flows (vortices) are created in the upper part of the imaged flow (the turbulence section) due to the buoyancy. For all flow rates below 5 L/ min, the displayed flow is laminar. Similar results were reported in earlier studies [7], [9].

IOP Conf. Series: Journal of Physics: Conf. Series **992** (2018) 012005 doi:10.1088/1742-6596/992/1/012005

The transition from laminar to turbulence flow depends on the plasma ignition. Figure 4 shows a comparison of these cases. The flow rate in both cases is 5 L/min. One can conclude that in case (a), vortices are created earlier than in case (b), in analogy with [12].

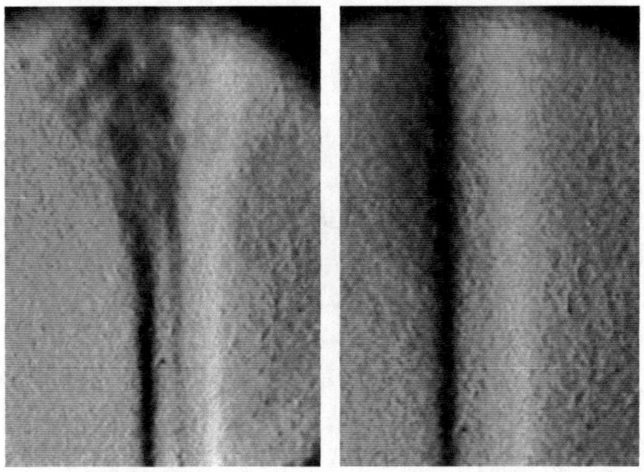

(a) (b)

Figure 3. Schlieren imaging of a helium flow into ambient air at different flow rates before plasma ignition: (a) 5 L/min, (b) 2.5 L/min.

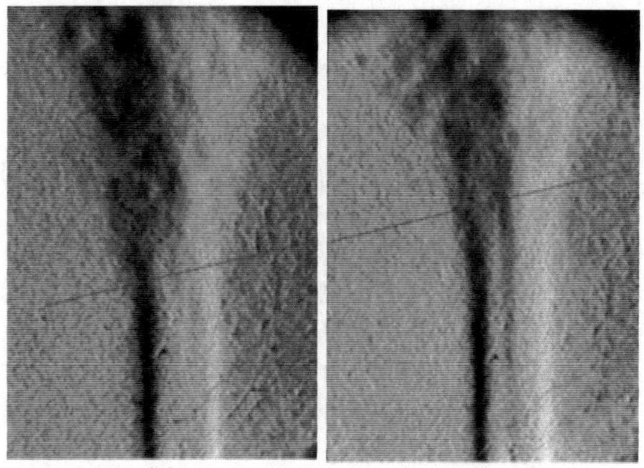

(a) (b)

Figure 4. Schlieren imaging of laminar to turbulent mode transition: (a) plasma on, (b) plasma off.

4. Influence of the operating voltage frequency

In addition, we examined the effect of the applied voltage frequency. The flow rate was set at 2.5 L/min and four sinusoidal AC voltages of 34.84 kHz, 35.15 kHz, 36.14 kHz and 36.84 kHz were considered. A comparison between these cases is presented in figure 5. It is clear that the laminar length decreases as the frequency of the AC voltage is raised.

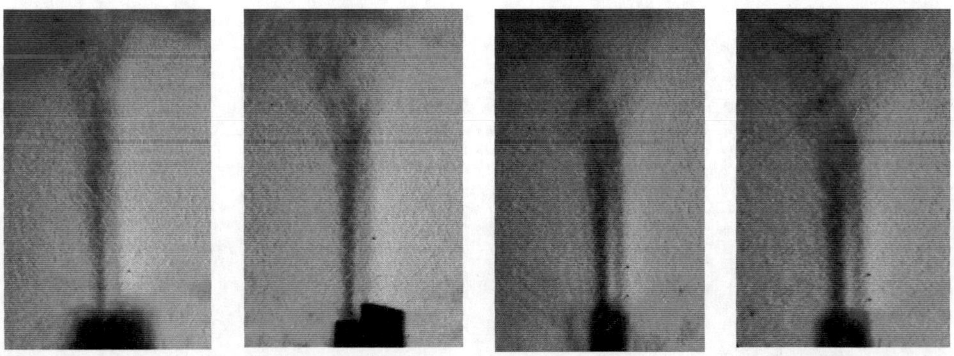

34.84 kHz 35.15 kHz 36.15 kHz 36.84 kHz

Figure 5. Transition point as function of the applied frequency for a flow rate of 2.5 L/min.

5. Conclusions

The gas behavior in a high-voltage driven single-electrode plasma jet (at atmospheric pressure) was studied by Schlieren imaging. First, the fluid dynamic behavior of the exhaust gas was investigated at different flow rates. The result showed that the transition from a laminar to turbulent mode takes place prior to the flow rate reaching 5 L/min. In addition, the plasma ignition increases the turbulent length. The influence of the plasma is more important at higher flow rates. Next, the effect of the operating frequency was studied. As one should expect, as the frequency is increased, the length of the turbulent flow also increases. The results presented could serve as a basis for the design, optimization and analysis of plasma processes in various applications.

IOP Conf. Series: Journal of Physics: Conf. Series **992** (2018) 012005 doi:10.1088/1742-6596/992/1/012005

References

[1] Ito Y, Urabe K, Takano N and Tachibana K 2008 *Appl. Phys. Express* **1** 067009

[2] Olabanji O T and Bradley J W 2012 *Plasma Process. Polym.* **9** 929

[3] Fridman G, Friedman G, Gutsol A, Shekhter A B, Vasilets V N and Fridman A 2008 *Plasma Process. Polym.* **5** 503

[4] Nastuta A V, Topala I, Grigoras C, Pohoata V and Popa G 2011 *J. Phys.* D: *Appl. Phys.* **44** 105204

[5] Foucault L 1859 Memoire sur la construction des telescopes en verre argente *Ann. de l'observatoire imp. de Paris, Tome V* **197**

[6] Toepler A 1906 Beobachtungen nach einer neuen optischen methode *Bonn: Max Cohen; Ostwalds Klassiker der ex. Wissenschaften* **157**

[7] Oh J S, Olabanji O T, Hale C, Mariani R, Kontis K and Bradley J W 2011 *J. Phys.* D: *Appl. Phys.* **44** 155206

[8] Boselli M, Colombo V, Ghedini E, Gherardi M, Laurita R, Liguori A, Sanibondi P and Stancampiano A 2014 *Plasma Chem. Plasma Process.* **34** 853

[9] Zheng Y, Wang L, Ning W and Jia S 2016 Schlieren imaging investigation of the hydrodynamics of atmospheric helium plasma jets *J. Appl. Phys.* **119** 123301

[10] Schafer J, Foest R, Kewitz T, Sperka J and Waltmann K D 2012 *Rev. Sci. Instrum.* **83** 10

[11] Schafer J, Bonaventura Z and Foest R 2015 *Eur. Phys. J. Appl. Phys.* **71** 20804

[12] Karakas E, Koklu M and Laroussi M 2010 *J. Phys.* D: *Appl. Phys.* **43** 155202

Self-induced optogalvanic effect in a segmented hollow-cathode discharge

V Steflekova and D Zhechev[1]

Acad. G. Nadjakov Institute of Solid State Physics, Bulgarian Academy of Sciences, 72 Tzarigradsko Chaussee, 1784 Sofia, Bulgaria

E-mail: spectron@issp.bas.bg

Abstract. Optogalvanic (OG) interaction is simulated and studied in a segmented hollow-cathode discharge (SHCD). HCD-lamps are used to induce an OG signal by their own emission or by that of another lamp. The efficiency of the OG of a Ne/Cu HCD lamp in the range 320-380 nm is estimated theoretically. An irregular galvanic peak arising near the inflection point in the *i-V* curve ($\partial V/\partial i<0$) is detected. Its origin is related to Penning ionization of the sputtered cathode material.

1. Introduction

A glow discharge modification, known as a hollow-cathode discharge (HCD), combines two important processes, namely *sputtering* and *excitation* of atoms. The commensurability of the large contact surfaces *negative glow (NG) –cathode dark space (CDS) – cathode surface* yields a sufficient density of sputtered atoms mostly residing in ground state, together with a characteristic electron energy distribution function (EEDF), which contains fast beam-like electrons that excite high-lying atomic states, including those of the ion buffer [1]. This spectroscopic advantage was developed in both emission and absorption spectroscopies.

Segmentation of the cathode extends the applications of HCDs to include laser media. The electrons motion, the magnetic field effects, the number of electrode segments, the cathode-anode surface ratio, the cathode length influence [2-4] are just a few of the specific aspects to be studied of a segmented HCD. On the other hand, HCD segmentation is a prerequisite for additional optical interactions that may arise between the different segments.

Penning was the first to observe *light-induced conductance* in a gas discharge [5]. It is the essence of the so-called opto-galvanic effect (OGE). Resonant light absorbed by atoms disturbs their steady-state population and, ultimately, changes the voltage across the discharge, the latter thus acting as an OG detector. This galvanic analogue of absorbed light can be measured more precisely than the absorbed light itself, making the OG spectroscopy a modern and promising alternative of absorption spectroscopy [6].

We report here manifestations of OG phenomena in a SHCD observed and analyzed, to the best of our knowledge, for the first time.

[1] To whom any correspondence should be addressed.

2. Experimental set-up

The experimental set-up used for measuring OG signals is presented schematically in figure 1. The optical interaction between two neighboring HCD-segments was simulated by using either: *i*) the same HCD lamp (HCDL) as a light source and an OG detector simultaneously (section *HCDL 2 – Ch – f - l – Mr* and back to *HCDL 2*) or *ii*) two HCDLs as a light source (*HCDL 1*) and an OG detector (*HCDL 2*) (section *HCDL 1 – l – f – Ch – HCDL 2*). In these simulations, different chemical *cathode-buffer* pairs were studied by using commercial Ne/Cu (Narva) and Ne/As (Pye Unicam) HCDLs, and Ne/Cd and Ar/Cd (Pye Unicam) HCDLs.

Simultaneously, controlling *i-V* curves were measured by using a Fluke 8062A multimeter. The discharge current values *i* were chosen taking into account the discharge stability. Also, He/Ne and Ar^+ lasers were used in comparative control measurements of the light-induced conductance.

An OG signal arises as an absorbed photon, $h\nu_{ik}$, causes a corresponding small perturbation of the atomic states populations $-\Delta n_i$ and $+\Delta n_k$; their relaxation brings about the OG effect. The latter manifests itself as a light-induced change in the gas discharge plasma resistance ΔR, discharge current Δi and voltage across the discharge ΔU. The quantity measured is the OG signal amplitude (AOGS), ΔU^{OG}, i.e. the OG-effect induced by the chopped light emission beam and averaged by the τ-constant of the amplifier. A 232B lock-in nanovoltmeter (Unipan) provided the $\Delta U(i)$ signal whose shape is close to that of the first derivative $\partial U(i)/\partial i$; A type 237 selective nanovoltmeter was also used.

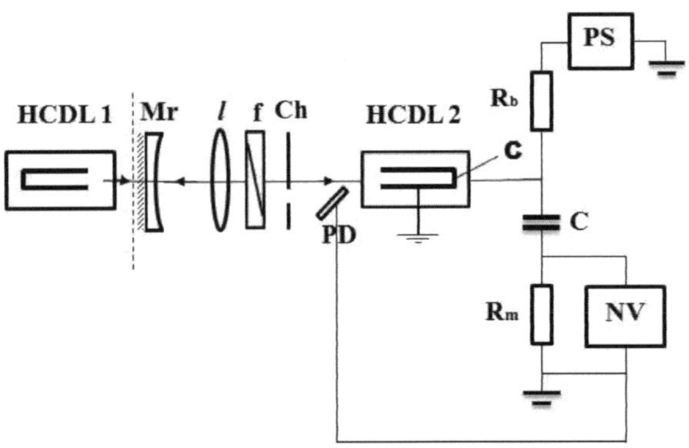

Figure 1. Experimental set-up: *HCDL* – hollow-cathode discharge lamp, *C* – hollow cathode, *l* – lens, *Ch* – chopper, *f* – interference filter, *Mr* – selective reflective mirror, *PS*–power supply, R_b – 100 kΩ ballast resistor, *C* – decoupling capacitor, R_m – 200 kΩ measuring resistor, *PD* – photodiode, *NV* – type 232B lock-in nanovoltmeter or type 237 selective nanovoltmeter.

3. Experimental results and discussions

3.1. On the optogalvanic efficiency of the own spectrum of Ne/Cu HCD in the 320-380 nm region, light-induced resistance and discharge current

The OG efficiency of a discharge-detector is characterized by the ratio *light-induced discharge current* Δi(A)/*absorbed light power* P'(W), where $P' = n_q$ is the number of quanta absorbed per second. We evaluated The OG efficiency vs. own spectrum for the Ne/Cu (Narva) (HCDL 2) emission within the 320-380 nm range. The OG response was simulated by the circuit *HCDL2 - Ch - f - l – Mr-* and back to *HCDL2* (figure 1). In the selected spectral band, the resonant lines CuI 324.8 nm and CuI 327.4 nm are one order of magnitude more intensive than the rest of the lines, including the NeI 352 nm line. Accordingly, the cumulative impact of the resonant CuI lines can be taken as a low limit only of both the emitted and absorbed light.

The value of Δi was estimated by the light-induced change ΔR^* of the discharge resistance R^*. A signal $\Delta U = 0.12$ mV was measured by the type 237 selective nanovoltmeter at $i = 10$ mA. This signal is integral in nature. If V_0 is the voltage across the discharge and R^*, the discharge resistance, the logical relation $\frac{\Delta U}{R_m} \propto \frac{V_0 \Delta R^*}{(R_b + R^*)^2}$ yields $\Delta R \infty \frac{\Delta U (R_b + R^*)^2}{R_m V_0}$. At R_b = 100 kΩ, V_0 = 400 V, R = 200 kΩ and $R^* << R_b$, the light-induced resistance ΔR^* = 0.15 Ω. Then $\Delta i = V_0\, \Delta R^*(R_b+R^*)^{-2} = 6\times10^{-9}\, A$.

The radiated light power P was calculated from the relation $P = N_k V \tau_{eff}$, where the effective lifetimes τ_{eff} (2.0×10^{-8} s, CuI $4^2P_{3/2}$ state, and 2.2×10^{-8} s, CuI $4^2P_{1/2}$ state) are taken from Refs. [7, 8], the population N_k of the upper resonant CuI levels ($N_k = n_e N_0 <v\sigma> = 2.4\times10^7$ cm^{-3} and 1.3×10^7 cm^{-3}) is estimated according to Refs. [9-12], and V is the discharge volume. The calculated values $n_q = 2.4\times10^{14}$ s^{-1} (CuI $4^2P_{1/2}$) and 1.2×10^{14} s^{-1} are underestimated, since a Maxwellian is used in the rate constant $<v\sigma>$ in [9, 10], σ being the cross-section of electron-atom interaction. The combined power of the two lines absorbed by the discharge proves to be 1.6×10^{-7} W, and the OG efficiency, 73.8×10^{-2} A W^{-1}. This value is based on the two dominating in terms of power CuI spectral lines only, so that this estimate provides an approximation of the order of magnitude of the OG interaction in a SHCD.

3.2. OGSs induced by parts of the CuI and NeI spectra. A non-resonant galvanic signal and HCD i-V inflection point

OGSs induced by own or an external spectrum were detected in the Ne/Cu HCD and the Ne/As HCD lamps. Figure 2(a) illustrates the signals in the Ne/Cu HCD lamp induced by two bands of its own spectrum when the discharge current was varied. The dielectric coating of mirror *Mr* (figure 1) was such as it returned to the Ne/Cu lamp 92 % of its spontaneous emission in the 320-380 nm region (curve b) and 96 % from the 594.5±20 nm region (curve c). For comparison, HCDL 2 was irradiated by the NeI 339.13 nm line of a He-Ne laser (curve a); curve d shows the corresponding cathode-anode (C-A) voltage across the HCD.

The monotonic curves $\Delta U(i)$ in figure 2(a) contain local peak-like OG (POG) extremums $\Delta U'(i)$ of type α and β, which do not follow from the analysis in section 3.1. To understand the nature of the POG appearance, we performed further measurements. The Ne/As HCD lamp was illuminated by the Ne/Cu lamp (figure 2(b)) directly (without filter *f*) and through interference filters of maximum transmittance at 590 nm and 654 nm. Simultaneously, the corresponding voltage on the HCD was measured and an inflection *i-V* point similar to those on curve d (figure 2(a)) was observed near the discharge current value of 4.2 mA.

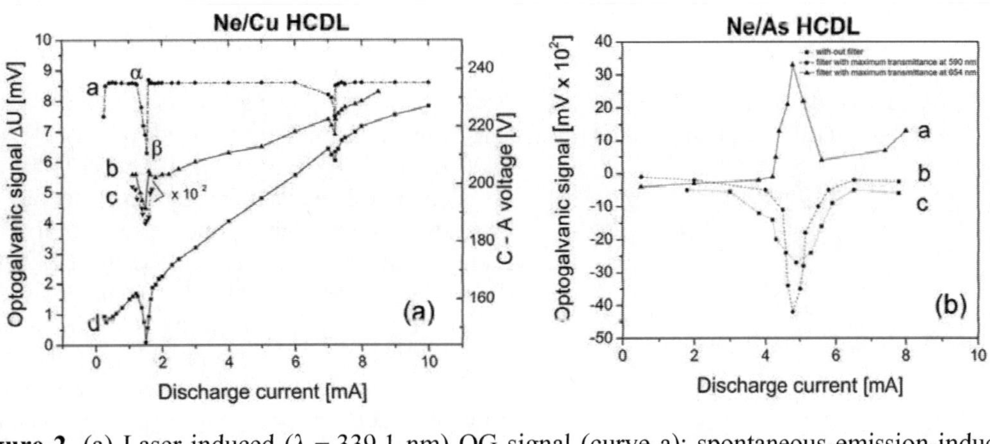

Figure 2. (a) Laser-induced (λ = 339.1 nm) OG signal (curve a); spontaneous-emission-induced (320-380 nm) (curve b) and (594.5 ± 20 nm) (curve c) integral OG signals in the Ne/Cu HCDL; *i-V* dependence (curve d); α, β – peak-like OGSs;
(b) Peak-like OG signal in the Ne/As HCDL illuminated by the Ne/Cu HCDL in the vicinity of the inflection point (4.2 mA) of the *i-V* curve. Curve a – filter of maximum transmittance at 654 nm; curve b – filter of maximum transmittance at 590 nm; curve c – without filter.

Let us now point to the characteristic properties of the POG extremums:

i/ they arise near an inflection *i-V* point, where $\partial U/\partial i < 0$ (pp α, β and the others) holds true in a narrow interval of about $\Delta i \approx (0.15 - 0.3)$ mA;

ii/ in the vicinity of the above Δi interval, the *i-V* curve forms bumps of increased HCD conductance (for example curve d (figure 2a); simultaneously, the phase has identical peak-like behavior and changes by up to 180^0 relative to the reference signal;

iii/ a rise in the noise level is detected in the vicinity of the region $\partial U/\partial i < 0$ (by more than a factor of three for the Ne/As HCD in the frequency range up to 20 kHz).

A symptomatic result concerning the POG effect nature was observed in the Ne/As HCDL illuminated by the 488.0 nm line of an Ar^+ laser. This line is in resonance with none of the Ne or As atomic transitions, so that no light-induced perturbations of the populations $\pm \Delta n_{ik}$ should occur. Nevertheless, a POG signal was observed. In this case, photoemission and space ionization only form the galvanic perturbation. Thus, POG signals in figure 2 (a, b) are non-resonant in nature.

The frequency non-selectivity of the peak-like galvanic responses and their various current values for each chemical pair of *buffer gas-cathode* represent a strong indication that the above facts are of more general nature. Based on these facts, a qualitative explanation can be given considering the Ohm's law for the discharge circuit. By analogy with the anomalous glow discharge [13], one may assume that $U(i) = V(i) + V'$, where $V(i)$ is an explicit function of the discharge current and V' depends on the other discharge parameters. Ohm's law can be expressed as

$$i(R_b + r_0) = E - V(i) - V', \tag{1}$$

where R_b is the ballast resistance, r_0 is the internal resistance of the current source, and E is the supply voltage. Then the total discharge current change Δi is

$$\Delta i = -V'(R_b + r_0 + \partial V/\partial i)^{-1}. \tag{2}$$

It is evident that the change Δi will increase at $\partial V/\partial i < 0$ (even more if $|\partial V/\partial i| \sim R_b + r_0$).

We consider the correlation *i-V* bump – POG signal as a cause-effect relationship. The at least two-component (buffer and sputtered atoms) HCD plasma is a prerequisite for *Penning ionization.* In some cases, this ionization dominates the stepwise one and an *i-V* bump arises. Having analyzed the energy levels of As, As^+ and Ne [14], the process of energy exchange between a metastable Ne^m and an As atom in ground state:

$$Ne^m\,(^3P_2) + As\,(^4S_{3/2}) \rightarrow Ne\,(^1S_0) + As^{+*}(^5S_2) + \Delta E + \vec{e} \tag{3}$$

could be considered as the most probable process, the reason being the negligible deficiency $\Delta E = 3.1\times10^{-3}$ eV. Therefore, complicated *i-V* and $\Delta U'(i)$ curve should be expected.

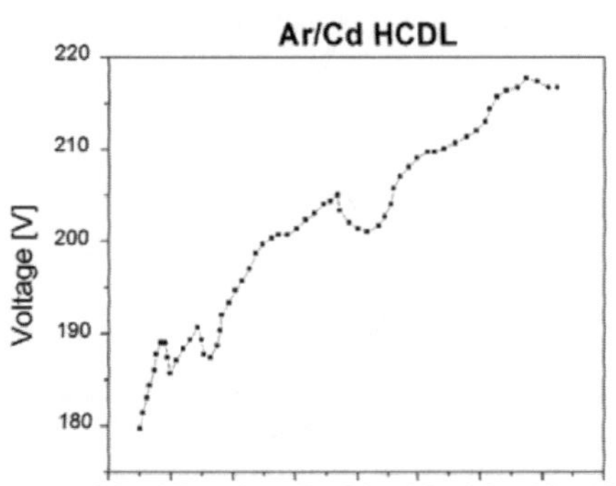

Figure 3. Part of the *i-V* curve of an Ar/Cd HCD spectral lamp.

The above analysis of the *cathode-buffer* chemical pair is relevant for some commercial lamps because of their highly nonlinear *i-V* curves. For example, at least six regions where $\partial V/\partial i < 0$ are observable in the spectrum of a commercial Ar/Cd HCDL (figure 3). Thus, the parts of the *i-V* curve where a stable mode for operation is possible are too limited. Moreover, self-sustained oscillations of different frequency were observed near each part where $\partial V/\partial i < 0$. At the same time, the *i-V* curve characterizing another chemical pair, namely of a Ne/Cd HCDL, is much smoother and no instabilities take place. However, the noise background grows when the Ne/Cd HCD lamp is irradiated by an Ar/Cd one. Here, the emission of Cd atoms may be the only optical channel of resonant interaction. This circumstances should also be taken into account with regard to the SHCD, namely the optical interaction between the HCD segments multiplies the instability near the inflection point in the *i-V* curve in a separate segment.

4. Conclusions

The mutual optical irradiation of two HCD segments induces an integral OG effect. This effect is an invariable property of a segmented cathode and depends on its effective length. The original emission of a Ne/Cu HCD lamp within the 320-380 nm range was calculated by us as inducing an OG signal of efficiency $\Delta i/W > 3.8\times10^{-2}$ AW^{-1}.

The correlation *i-V-bump – POG* signal is in a cause-effect relationship. The pair of *buffer atom–sputtered atom* in a HCD is a prerequisite for Penning ionization to occur. The latter changes the steepness of the *i-V* curve and forms a local *i-V* bump. A local peak-like OG (POG) signal arises near this inflection *i-V* point; the phase is of the same peak-like behavior and changes by up to 180^0 relative to the reference signal. The POG signal is a non-resonant galvanic response in nature and manifests itself as an instrumental effect due to the inflection *i-V* point, where $\partial U/\partial i < 0$.

One consequence of the negative resistance is the increased noise background of the OG signal in the inflection point vicinity. This frequency instability is multiplied optically in the SHCD.

References

[1] Jiménez-Redondo M, Carrasco E, Herrero Víctor J and Tanarro I 2013 *Plasma Sources Sci. Technol.* **22** 025022
[2] Donko Z, Rozsa K and Tobin R C 1996 *J. Phys.* D*: Appl. Phys.* **29** 105-14
[3] Mihailova D, van Dijk J, Grozeva M, Hagelaar G J M and van der Mullen J J A M 2010 *J. Phys.* D*: Appl. Phys.* **43** 465204
[4] Arslanbekov R R, Tobin R C and Kudryavtsev A A 1997 *J. Appl. Phys.* **81** 554
[5] Penning F M 1928 *Physica* **8** 137
[6] Barbieri B, Beverini N and Sasso A 1990 *Rev. Modern. Phys.* **62** 603-44
[7] Collegrove F D, Franken P A, Levis R R and Sands R H 1953 *Phys. Rev. Lett.* **3** 420
[8] Griem H R 1962 *Phys. Rev.* **128** 515
[9] Franken P A 1961 *Phys. Rev.* **121** 508
[10] Aleksandrov E B 1963 *Opt. Spectr.* **14** 436
[11] Bundaleska N, Tomova Z, Steflekova V and Zhechev D 2008 *Vacuum* **82** 455-8
[12] Zhechev D and Hvorostovsky S 1990 *Spectr. Lett.* **23** 911
[13] Ochkin V N, Preobragenskz N G and Shaparev N J 1991 *Optogalvanic effect in ionized gas* (Moscow: Nauka), in Russian
[14] Radzigand A and Smirnov B 1986 *Parameters of Atoms and Atoms Ions* (Moscow: Energoatomizdat), in Russian

IOP Conf. Series: Journal of Physics: Conf. Series **992** (2018) 012007 doi:10.1088/1742-6596/992/1/012007

Experimental demonstration of efficient and robust second harmonic generation using the adiabatic temperature gradient method

E Dimova[1], V Steflekova, S Karatodorov and E Kyoseva

Acad. G. Nadjakov Institute of Solid State Physics, Bulgarian Academy of Sciences, 72 Tzarigradsko Chaussee, 1784 Sofia, Bulgaria

E-mail: e.dimova@issp.bas.bg

Abstract. We propose a way of achieving efficient and robust second-harmonic generation. The technique proposed is similar to the adiabatic population transfer in a two-state quantum system with crossing energies. If the phase mismatching changes slowly, e.g., due to a temperature gradient along the crystal, and makes the phase match for second-harmonic generation to occur, then the energy would be converted adiabatically to the second harmonic. As an adiabatic technique, the second-harmonic generation scheme presented is stable to variations in the crystal parameters, as well as in the input light, crystal length, input intensity, wavelength and angle of incidence.

1. Introduction

Second-harmonic generation (SHG) was first reported for the most common nonlinear optical process for crystal materials with a $\chi^{(2)}$ nonlinearity [1]. It is a special case of sum-frequency generation [2-4]. SHG usually requires the phase matching to be efficient; the SHG efficiencies are typically extremely small when a phase matching does not occur. Traditionally, there are three ways to compensate the phase mismatching: by using different orientation of birefringent crystals, by temperature compensation, or by quasi-phase matching (QPM) [5, 6]. These ways of compensations are tunable for different wavelengths, either due to different orientation of the crystal or due to the temperature-controllable phase-matching of the crystal; unfortunately, these methods do not lead to a broadband SHG. A theoretical prediction of an efficient and stable SHG in the adiabatic limit was suggested first by Baranova *et al.* [7]. Later, chirped QPM gratings were used to achieve broadband SHG [8-10], but due to the lack of adiabaticity, the efficiency remained low.

In this paper, we propose a way of achieving an efficient, robust and stable SHG by adiabatically chirped phase-mismatching using a temperature gradient method. The technique proposed uses the analogue of adiabatic evolution in quantum physics [11-14]. A similar technique, one using chirped quasi-phase match gradients in the case of sum-frequency generation was recently demonstrated experimentally [15-18]. In contrast to the previous work, we show that an efficient and robust SHG is possible with a proper temperature gradient.

[1] To whom any correspondence should be addressed.

IOP Conf. Series: Journal of Physics: Conf. Series **992** (2018) 012007 doi:10.1088/1742-6596/992/1/012007

2. Experiment

The light source used was a Quanta-Ray GCR3 high-power Q-switched Nd:YAG laser system oscillating at the fundamental wavelength of 1064 nm, with pulse duration from 7 to 9 ns and maximal energy per pulse of 850 mJ. This laser system can also emit the second harmonic wavelength of 532 nm, pulse duration 5-7 ns, maximal energy per pulse 400 mJ; and the third harmonic at 355 nm, pulse duration 4-6 ns, maximal energy 200 mJ. The repetition rate is in the range 1-20 Hz. The crystal ($MgO:LiNbO_3$) chosen for SHG had a square cross-section (5×5 mm) and a length of 40 mm. The laser beam had a diameter of 6 mm; it was restricted by a diaphragm to fit the cross-section of the crystal (beam diameter of about 4 mm). The laser beam was well collimated to a distance of 4 m. The alignment into the active media was performed by means of two mirrors to ensure that the light beam passes through the middle of the sample. To separate the light of different wavelengths, a prism was placed in the output light beam path. We found that, together with the main frequency of 1064 nm polarized along the ordinary polarization axis, there was another light beam with a wavelength of 532 nm and polarized along the extraordinary polarization axis, as it is predicted theoretically [1]. The wavelengths were measured by an AvaSpec-2048 monochromator with AvaSoft 7.5 controlling software. The pulse energy of each beam was measured by laser power meter (Edmund),

$$1064\ \text{nm(e)} + 1064\ \text{nm(e)} \rightarrow 532\ \text{nm(o)}. \quad (1)$$

The temperature control was achieved by heating the crystal in an oven. Two ovens were built for this purpose, each made from a quartz cylinder with a diameter of 20 mm and a length of 60 mm. A tantalum wire was used to form a suitable heating coil on the outer surface of the cylinder as necessary to conduct the experiment. The coil turns on the first oven were placed equidistantly, while on the second oven the distance between the wire turns was decreased gradually from one end to the other. We could thus follow the effect of the temperature on the nonlinear response of the crystal medium at both a constant temperature and a temperature varying along the crystal sample. The temperature was measured by a thermocouple on the sample surface. To keep the temperature independent of the surrounding conditions, both sides of the oven were closed except for a hole in the middle for the input laser beam and the output light.

The experiment was carried out follows: the laser was oscillating at the IR wavelength of 1064 nm. The laser pulse energy was varied via the pumping lamp power. The losses due to the transfer of the laser beam through the whole optical set-up were accounted for by measuring the output power at the optical path end (figure 1) by using a prism separating the light with different wavelengths, and a power meter placed on the path of each of them.

As a first step of the experiment, we studied the effect of the temperature on the nonlinear response at a constant heating along the crystal medium. To ascertain that the effect measured was indeed due to the temperature difference, we examined the output light from the nonlinear crystal at room temperature (22 °C); no output light at 532 nm was detected. We followed the second harmonic generation at 532 nm in the temperature range 100 – 130 °C) for different pulse laser energies and found that the maximal nonlinear effect occurred at 110 °C (figure 2). Note that there is a non-critical phase matching along the propagation direction at this temperature.

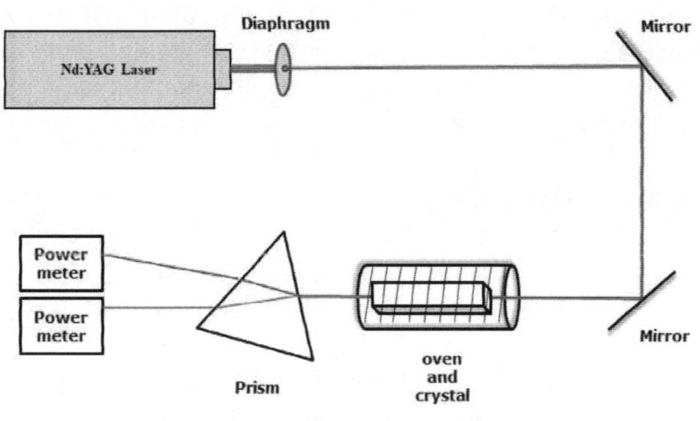

Figure 1. Experimental setup.

Bearing the theoretical model in mind, we applied a temperature gradient along the crystal by the second oven. In figure 3 one can see the effect of the temperature on the chirp along the nonlinear crystal at different temperature gradients. The optimal effect was measured was at the temperature gradient 121 °C – 110 °C. We were not able to achieve the expected full transfer of the laser light at 1064 nm to the SHG light at 532 nm. The reasons could be the high dispersion of the non-linear medium, the less-than-perfect temperature control along the crystal sample, or the limited laser pulse energy at 1064 nm due to technical problems.

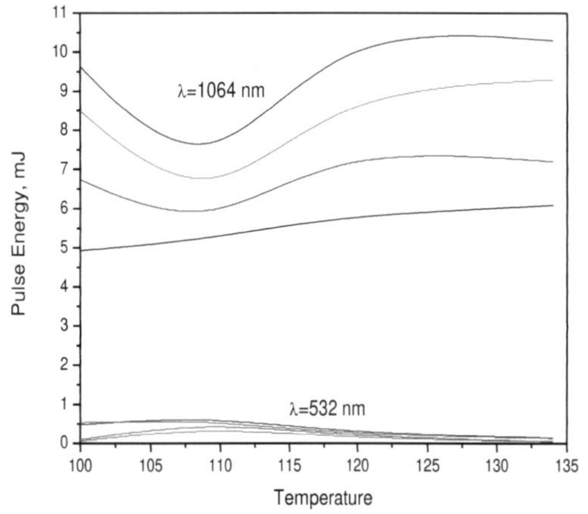

Figure 2. Second harmonic generation at 532 nm light in the temperature range 100 – 130 °C for different laser pulse energies. Non-critical phase matching along the propagation direction at 110 °C.

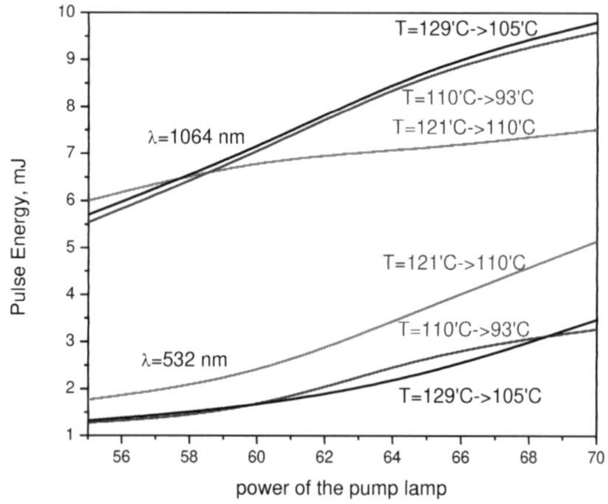

Figure 3. Second harmonic generation at 532 nm with a temperature gradient along the crystal at different laser pulse energies (achieved by varying the laser pump lamp power).

3. Conclusions

We used the analogy between the time-dependent Schrödinger equation and the SHG equations beyond the laser wave's regime. A temperature gradient at the two ends of the crystal produces a phase mismatch chirp along the light propagation path and creates a crossing in the phase matching, which in combination with adiabatic evolution conditions, allow for an efficient and robust SHG. The adiabatic temperature chirp offers stbility against variations of the parameters of both the crystal and the electromagnetic field, which include variation in the temperature, the wavelengths of the input electromagnetic field, the crystal length and the angle of incidence. The adiabatic SHG suggested is stable due to the absence of an inverse process (the fundamental signal becomes low when the phase mismatch is large and thus the inverse process that may start is very weak). The price that has to be paid for this stability, efficiency and robustness is a large crystal or/and high input-intensity at the fundamental wavelength.

Acknowledgments

The authors would like to thank Dr. A. Rangelov for the fruitful discussions and theoretical interpretations of the experimental results. Dr. E. Kyoseva acknowledges the funding from the European Union's Horizon 2020 Research and Innovation Program under the Marie Skłodowska-Curie grant agreement No 705256-COPQE. Dr. E. Dimova acknowledges the Polish-Bulgarian Joint Research Project (2015-2017) under the Scientific Agreement between the Bulgarian Academy of Sciences and the Polish Academy of Sciences.

References

[1] Franken P A, Hill A E, Peters C W and Weinreich G 1961 *Phys. Rev. Lett.* **7** 118
[2] Boyd R W 2007 *Nonlinear Optics* 3rd ed. (Academic, New York)
[3] Saleh B E A and Teich M C 2007 *Fundamentals of Photonics* (John Wiley & Sons, New Jersey)

[4] Yariv A and Yeh P 2007 *Photonics: Optical Electronics in Modern Communications*, 6th ed. (Oxford University Press, New York)
[5] Armstrong J, Bloembergen N, Ducuing J and Pershan P 1962 *Phys. Rev.* **127** 1918
[6] Arie A and Voloch N 2010 *Laser and Photon. Rev.* **4** 355-73
[7] Baranova N B, Bolshtyanskii M A and Zel'dovich B Ya 1995 *Quantum Electr.* **25** 638
[8] Arbore M A, Marco O and Fejer M M 1997 *Opt. Lett.* **22** 865-7
[9] Imeshev G, Arbore M, Fejer M, Galvanauskas A, Fermann M and Harter D 2000 *J. Opt. Soc. Am.* B **17** 304
[10] Hum D S and Fejer M M 2007 *C. R. Phys.* **8** 180
[11] Allen L and Eberly J H 1987 *Optical Resonance and Two-Level Atoms* (Dover, New York)
[12] Shore B W 1990 *The Theory of Coherent Atomic Excitation* (Wiley, New York)
[13] Vitanov N V, Halfmann T, Shore B W and Bergmann K 2001 *Annu. Rev. Phys. Chem.* **52** 763-809
[14] Vitanov N V, Fleischhauer M, Shore B W and Bergmann K 2001 *Adv. At. Mol. Opt. Phys.* **46** 55-190
[15] Suchowski H, Oron D, Arie A and Silberberg Y 2008 *Phys. Rev.* A **78** 063821
[16] Suchowski H, Prabhudesai V, Oron D, Arie A and Silberberg Y 2009 *Opt. Express* **17** 12731
[17] Suchowski H, Bruner B D, Ganany-Padowicz A, Juwiler I, Arie A and Silberberg Y 2011 *Appl. Phys.* B **105** 697
[18] Dahan A, Levanon A, Katz M and Suchowski H 2017 *J. Phys.: Condens. Matter.* **29** 084004

IOP Conf. Series: Journal of Physics: Conf. Series **992** (2018) 012008 doi:10.1088/1742-6596/992/1/012008

Theoretical study on some plasma parameters and thermophysical properties of various gas mixtures in gas-discharge lasers

K A Temelkov[1,3], S I Slaveeva[1], Yu I Fedchenko[1] and T P Chernogorova[2]

[1]Metal Vapor Lasers Laboratory, Georgi Nadjakov Institute of Solid State Physics, Bulgarian Academy of Sciences, 72 Tzarigradsko Chaussee, 1784 Sofia, Bulgaria
[2]Faculty of Mathematics and Informatics, St. Kl. Ohridski University of Sofia, 5 James Bourchier Blvd., 1164 Sofia, Bulgaria

E-mail: temelkov@issp.bas.bg

Abstract. Using the well-known Wassiljewa equation and a new simple method, the thermal conductivities of various 2- and 3-component gas mixtures were calculated and compared under gas-discharge conditions optimal for two prospective lasers excited in a nanosecond pulsed longitudinal discharge. By solving the non-stationary heat-conduction equation for electrons, a 2D numerical model was also developed for determination of the radial and temporal dependences of the electron temperature $T_e(r, t)$.

1. Introduction

Determination of basic plasma parameters, such as gas temperature, electron temperature, electron concentration, etc. is one of the main and relevant problems in the physics of gas discharges, gas-discharge lasers, laser-induced breakdown spectroscopy, gas-discharge mass-spectroscopy, plasma technologies and plasma in general. The characteristic constants of the heavy-particle interaction depend on the gas temperature, while the electron temperature and electron density determine the characteristic constants of elastic and inelastic electron-heavy particle collisions. For gas-discharge lasers in particular, the abovementioned plasma parameters influence the development of population inversion, and hence the output laser parameters. The gas temperature distribution, i.e., the thermal mode, is also important for the stability of the laser operation and the active particles density.

The experimental techniques for determining the gas temperature that make use of the Doppler broadening of spectral lines or the focal distance of thermal lens are definitely imprecise. Recently, the theoretical determination of the gas temperature spatial distribution by solving analytically or numerically the steady-state heat conduction equation was proved to be a quite simple and efficient technique [1]. The thermophysical properties of gases and gas mixtures are important parameters in solving the corresponding partial differential equation. For example, the heat conductivity influences the gas temperature distribution, as found by solving the steady-state heat conduction equation. Both research and industrial plasma devices, gas-discharge lasers inclusive, often comprise multicomponent gas mixtures rather than pure gases. The theory for calculating the heat conductivity of

[3] To whom any correspondence should be addressed.

IOP Conf. Series: Journal of Physics: Conf. Series **992** (2018) 012008 doi:10.1088/1742-6596/992/1/012008

multicomponent gas mixtures was developed around the beginning of the last century. Since then, various methods have been proposed. Most of them are empirical and reduce to some form of the original Wassiljewa equation [2, 3]. Due to the far-too-complex character of the Wassiljewa equation and its approximations, a new simple method based on iterative application of Brokaw's empirical equation for binary gas systems was proposed and applied to determining the thermal conductivity of multicomponent gas mixtures of our interest [4].

In this paper, using Wassiljewa equation and the new simple method quoted, the heat conductivities of various 2- and 3-component gas mixtures are calculated and compared under gas-discharge conditions optimal for two prospective lasers, namely deep-ultraviolet Cu^+ Ne-H_2-CuBr and middle-infrared Sr atom He-(Ne-)$SrBr_2$ lasers, which are excited in a nanosecond pulsed longitudinal discharge (NPLD).

The widely applied techniques for experimental determination of the electron temperature, such as Langmuir probes, laser Thomson scattering and the spectral-line-ratio method, are not suitable and, hence, are inapplicable to high-voltage, high-current and high-power NPLDs due to the unavoidable excessive noise. Unlike the gas temperature calculation, theoretical determination of the electron temperature requires years-consuming development of complex kinetic models, which include tens of energy levels per particle, and solving the corresponding ordinary or partial differential kinetic equations of dubious validity.

In this paper, the radial and temporal dependences of the electron temperature $T_e(r, t)$ are determined through numerically solving the nonstationary heat conduction equation for electrons.

2. Theory

Following [3], the heat conductivity of a gas mixture k_m is expressed by Wassiljewa empirical relation as follows:

$$k_m = \sum_{i=1}^{N} \frac{y_i k_i}{\sum_{j=1}^{N} y_j A_{ij}} \quad , \tag{1}$$

where k_m is the heat conductivity of the gas mixture in W m^{-1} K^{-1}, k_i is the heat conductivity of the pure i component in W m^{-1} K^{-1}, y_i and y_j are the mole fractions of components i and j, and A_{ij} is a function, which for $i = j$ is unit ($A_{ii} = 1$), while for $i \neq j$ remains to be obtained. Several methods have been proposed with varying complexity and accuracy for determination of A_{ij} [3, 5-8]. The simplest approach is to use the following expression [3, 5]:

$$A_{ij} = \frac{\left[1 + \left(\frac{k_i}{k_j}\right)^{\frac{1}{2}} \left(\frac{\mu_i}{\mu_j}\right)^{\frac{1}{4}}\right]^2}{\left[8\left(1 + \frac{\mu_i}{\mu_j}\right)\right]^{\frac{1}{2}}} \quad , \tag{2}$$

where μ_i and μ_j are the masses of particles i and j in amu, k_i and k_j are the heat conductivities of pure components i and j in W m^{-1} K^{-1}.

For binary gas mixtures, Brokaw's empirical equation [3, 9] is widely used. The thermal conductivity k_m of the gas mixture is calculated as follows:

$$\begin{aligned} k_m &= q k_{m_l} + (1-q) k_{m_R} \quad , \\ k_{m_l} &= y_1 k_1 + y_2 k_2 \quad , \\ \frac{1}{k_{m_R}} &= \frac{y_1}{k_1} + \frac{y_2}{k_2} \quad , \end{aligned} \tag{3}$$

IOP Conf. Series: Journal of Physics: Conf. Series **992** (2018) 012008 doi:10.1088/1742-6596/992/1/012008

where k_1 and k_2 are the heat conductivities of pure components 1 and 2 in W m^{-1} K^{-1}, y_1 and y_2 are the mole fractions of components 1 and 2, q is a function of the mole fraction of the lighter component given in [3]. The new simple method proposed in [4] is based on an iterative application of Brokaw's equation.

To determine the radial-time-resolved electron temperature, the following nonstationary heat conduction equation is numerically solved for the electron gas:

$$\eta C_e \frac{\partial T_e}{\partial t} = \frac{1}{r}\frac{\partial}{\partial r}\left(r k_e(T_e) \frac{\partial T_e}{\partial r} \right) + q_V(t) \quad , \tag{4}$$

where C_e and $k_e(T_e)$ are the heat capacity and heat conductivity of the electron gas, R is the discharge zone radius equal to 3.55 mm, T is the period between the excitation pulses equal to 50 μs at the pulse repetition frequency of 20 kHz, $q_V(t)$ is the electric power density deposited into the NPLD for electron heating, η is a coefficient specifying the discrepancy between the real heat capacity of the electron gas and the one predicted by the classical statistical mechanics, i.e. $C_e = \frac{3}{2} k_B N_e(t)$, where $N_e(t)$ is the time-dependent electron density and k_B is the Boltzmann constant.

3. Results and discussion

Figure 1 shows the heat conductivities as a function of gas temperature are for various 2-component gas mixtures. The heat conductivities calculated by equation (1) and (3) are plotted with solid and dash-dot lines, respectively. The discrepancy is less than 6 %, with the exception of the Ne-$H_{(2)}$ mixture, where using the Wassiljewa equation obviously does not account for the hydrogen molecule dissociation above 1600 K.

Figure 2 presents the heat conductivities as a function of the gas temperature for various 3-component gas mixtures. The heat conductivities calculated by equation (1) and (3) are plotted with solid and dash-dot lines, respectively. The discrepancy is less than 7 %, with the exception of the Ne-$H_{(2)}$-Cu mixture.

In our previous study [10] we the determined the electronic heat conductivity from the Wiedemann-Franz law $\frac{k_e}{\sigma} = c.T_e$ used in plasma physics, where σ is the specific electrical conductivity, $c = \frac{\pi^2}{3}\left(\frac{k_B}{e}\right)^2$ is a constant, k_B is Boltzmann constant, and e is the electron charge, assuming that σ is independent of the spatial coordinates and has a value of $\bar{\sigma}$. Our preliminary results showed that the electronic heat conductivity $k(T_e) = T_e$ derived from the Wiedemann-Franz law does not reflect the physical reality for the nonstationary electron temperature [10]. This is why we undertook a power index scanning for the electronic heat conductivity in an algometric form, i.e. $k_e = B.T_e^a$, as widely used for solving analytically and numerically the stationary heat conduction equation to determine the gas temperature.

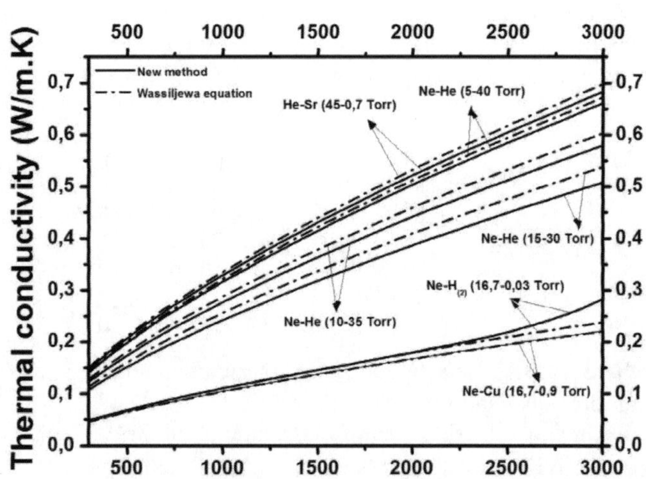

Figure 1. Thermal conductivities as a function of the gas temperature for various 2-component gas mixtures.

Equation (2) is numerically solved under the initial condition:

$$T_e(r,0) = T_e^{\,0}(r) = \left[T_e(R,0)^{1+a} + \frac{1+a}{4B} q_V(0)\left(R^2 - r^2\right)\right]^{\frac{1}{1+a}} ,$$

and the boundary conditions: (5)

$$\lim_{r\to 0} r k_e(T_e)\frac{\partial T_e}{\partial r} = 0 \text{ and } T_e(R,\,t) = 880\text{ K}.$$

The time-dependent electric power and electron concentration was experimentally determined in [11] by time-resolved measurement of electrical discharge parameters, such as tube voltage and discharge current. The $q_V(t)$ pulse is derived from the electric power pulse, taking into account that only 1 % of this power is deposited into heating of electrons [10], and is fitted by a Gaussian-type function:

$$q_V(t) = q_V^{\,0} + \frac{A}{w\sqrt{\frac{\pi}{2}}} e^{-2\left(\frac{t-t_0}{w}\right)^2} \quad (6)$$

while the electron density $N_e(t)$ is fitted by an exponential function:

$$N_e(t) = N_e^{\,0} + A.e^{\frac{t-t_0}{w}} \quad (7)$$

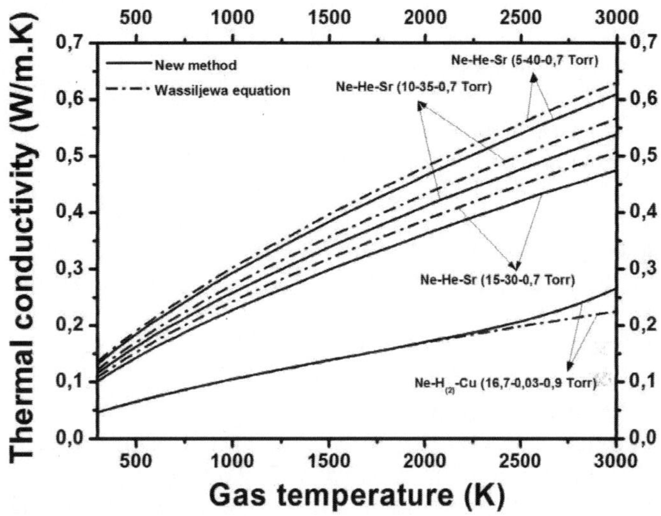

Figure 2. Thermal conductivities as a function of gas temperature for various 3-component gas mixtures.

The fitting parameters for both functions are given in table 1.

Table 1. Fitting parameters for the electric power density $q_V(t)$ and electron concentration $N_e(t)$.

Function	$q_V^{\,0}$ or $N_e^{\,0}$	A	w	t_0
$q_V(t)$	$1{,}61588.10^{8}$	1210,21118	$3{,}60545.10^{-8}$	$9{,}48298.10^{-8}$
$N_e(t)$	$-3{,}52795.10^{18}$	$1{,}06363.10^{19}$	$2{,}97375.10^{-8}$	$2{,}07038.10^{-8}$

Figure 3 illustrates the time-resolved electron temperature in an NPLD used for excitation of deep ultraviolet Cu^+ Ne-H_2-CuBr laser for different radial distances and $k_e(T_e) = T_e^{0{,}02}$.

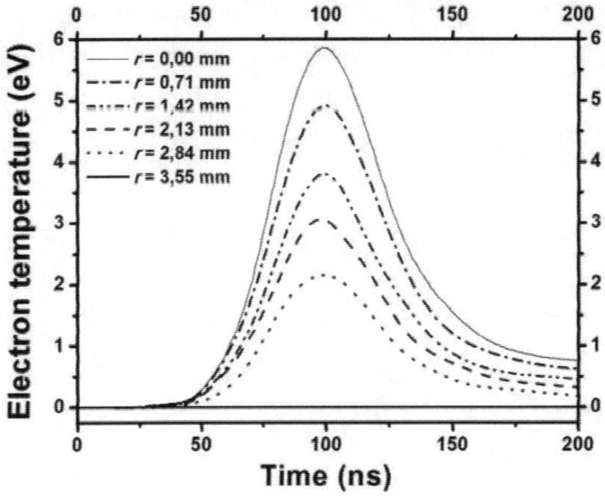

Figure 3. Temporal dependence for different radial distances of the electron temperature in an NPLD used for excitation of a deep ultraviolet Cu^+ Ne-H_2-CuBr laser.

4. Conclusions

In this paper, using Wassiljewa equation and a new simple method, the thermal conductivities of various 2- and 3-component gas mixtures are calculated and compared under gas-discharge conditions optimal for two prospective lasers excited in an NPLD. Using the new method for heat conductivity calculation is preferable when dissociation of molecules occurs in a gas discharge.

A 2D numerical model is also developed for determination of the radial and temporal dependences of the electron temperature $T_e(r, t)$ by solving numerically the nonstationary heat conduction equation for electrons. The result is in fair agreement with the time-dependent electron temperature obtained by the time-resolved measurement of electrical discharge parameters [11] and with the existing self-consistent models developed for Ne-Cu and Ne-CuBr lasers excited in an NPLD [12, 13].

Acknowledgements

T. P. Chernogorova is supported by the Bulgarian Science Fund for Sofia University Foundation under Grant No. 80.10-11/12.04.2017.

References

[1] Temelkov K A, Vuchkov N K, Freijo-Martin I and Ekov R P 2010 *J. Phys.* D*: Appl. Phys.* **43** 075206
[2] Wassiljewa A 1904 *Physik. Z.* **5** 737
[3] Reid R C and Sherwood T K 1966 *The properties of gases and liquids* (New York: McGraw-Hill Book Company)
[4] Temelkov K A, Slaveeva S A and Fedchenko Yu I 2016 *J. Phys.: Conf. Series* **700** 012005
[5] Mason E A and Saxena S C 1958 *Phys. Fluids* **1** 361
[6] Lindsay A L and Bromley L A 1950 *Ind. Eng. Chem.* **42** 1508
[7] Ely J F and Hanley H J M 1983 *Ind. Eng. Chem. Fundam.* **22** 90
[8] Chung T-H, Lee L L and Starling K E 1984 *Ind. Eng. Chem. Fundam.* **23** 8
[9] Brokaw R S 1955 *Ind. Eng. Chem.* **47** 2398
[10] Temelkov K A, Slaveeva S I and Chernogorova T P 2016 *J. Phys.: Conf. Series* **700** 012004
[11] Temelkov K A and Vuchkov N K 2014 *IEEE Trans.Plasma Sci.* **42**/12 3938-41
[12] Carman R J, Brown J W and Piper J A 1994 *IEEE J. Quantum Electronics* **30**/8 1876-94
[13] Vuchkov N K, Astadjov D N and Sabotinov N V 1994 *IEEE J. Quantum Electronics* **30**/3 750-58

DOUBLE code simulations of emissivities of fast neutrals for different plasma observation view-lines of neutral particle analyzers on the COMPASS tokamak

K Mitosinkova[1,2,4], M Tomes[1,2], J Stockel[1], J Varju[1] and M Stano[3]

[1]Institute of Plasma Physics of the CAS, Prague, Czech Republic
[2]Faculty of Mathematics and Physics, Charles University, Prague, Czech Republic
[3]Faculty of Mathematics, Physics and Informatics, Comenius University, Bratislava, Slovakia

Email: mitosinkova@ipp.cas.cz

Abstract. Neutral particle analyzers (NPA) measure line-integrated energy spectra of fast neutral atoms escaping the tokamak plasma, which are a product of charge-exchange (CX) collisions of plasma ions with background neutrals. They can observe variations in the ion temperature T_i of non-thermal fast ions created by additional plasma heating. However, the plasma column which a fast atom has to pass through must be sufficiently short in comparison with the fast atom's mean-free-path. Tokamak COMPASS is currently equipped with one NPA installed at a tangential mid-plane port. This orientation is optimal for observing non-thermal fast ions. However, in this configuration the signal at energies useful for T_i derivation is lost in noise due to the too long fast atoms' trajectories. Thus, a second NPA is planned to be connected for the purpose of measuring T_i. We analyzed different possible view-lines (perpendicular mid-plane, tangential mid-plane, and top view) for the second NPA using the DOUBLE Monte-Carlo code and compared the results with the performance of the present NPA with tangential orientation. The DOUBLE code provides fast-atoms' emissivity functions along the NPA view-line. The position of the median of these emissivity functions is related to the location from where the measured signal originates. Further, we compared the difference between the real central T_i used as a DOUBLE code input and the T_{iCX} derived from the exponential decay of simulated energy spectra. The advantages and disadvantages of each NPA location are discussed.

1. Introduction

The neutral particle analyzer (NPA) is a diagnostic tool measuring the energy spectra of fast neutral atoms escaping the plasma [1]. The COMPASS tokamak ($R = 0.56$ m, $a = 0.23$ m, $B_t = 0.9 - 2.1$ T, I_p up to 400 kA, line-integrated $n_e = 2 - 10\times10^{19}$ m^{-3}) [2] is equipped with one ACORD-24 NPA, which has 12 channels for hydrogen atoms (0.25 – 70 keV) and another 12 channels for deuterium (0.3 – 50 keV) and is installed at a mid-plane tangential port [3]. In the near future, a second, ACORD-5, NPA with 5 channels for detection of neutrals will be installed on the COMPASS tokamak. Although this NPA does not distinguish between different atomic masses, it can operate in a

[4] To whom any correspondence should be addressed.

sweeping mode, which allows one to measure several times as many points of an energy spectrum as the currently mounted detectors. It is possible to connect it at a perpendicular mid-plane port or on the top of the tokamak vessel with a view-line through the plasma center.

The fast neutral atoms escaping the plasma are formed by charge-exchange (CX) collisions between ions and electron donors. The latter are mostly neutral particles propagating from the edge towards the plasma center and undergoing multi-step CX collisions. Using the DOUBLE Monte-Carlo code [4, 5], developed in the Ioffe Institute, St Petersburg, we explored the influence of the plasma density n_e ($2 - 10\times10^{19}$ m^{-3}), the fast neutral atoms energy E_{FA} (300 – 10 000 keV) and the NPA view-line (tangential mid-plane, perpendicular mid-plane, and top center view) on the fast atoms emissivity profiles. The different view-lines are illustrated in figure 1.

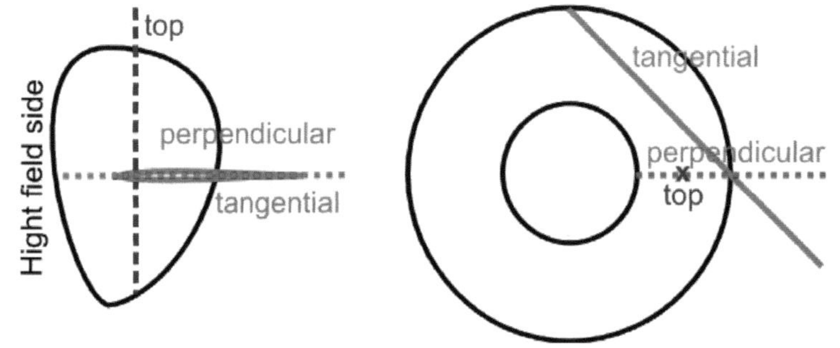

Figure 1. The different NPA view-lines. Left-hand side diagram – top-view; right-hand side diagram – chamber cross-section..

The data acquired by ACORD-24 are often processed assuming that the energetic fluxes measured originate from a region where the ion distribution is Maxwellian, i.e., can be characterized by one ion temperature [2]. This assumption allows one to calculate T_{iCX} from the slope of the linear part of the measured energy spectra when plotted in logarithmic scale. However, this assumption is usually not satisfied, as the fast atoms' region of origin has different ion temperatures. Figure 2 shows an example of a time-averaged signal measured by ACORD-24 during L-mode discharges with different n_e, but otherwise comparable.

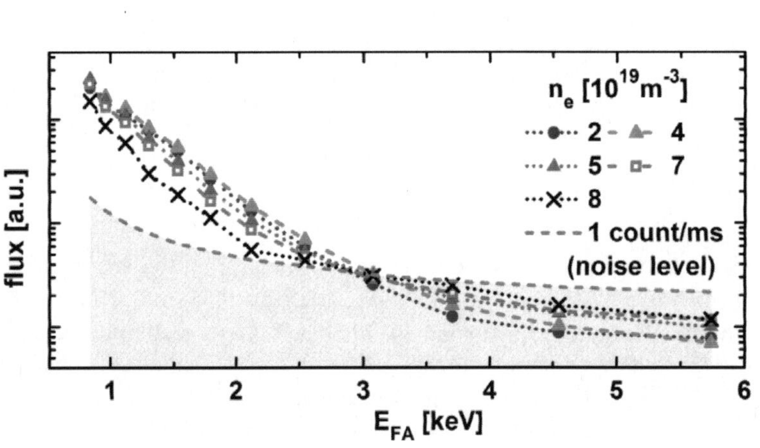

Figure 2. Time-averaged energy spectra measured during L-mode discharges for different electron densities n_e. The noise level is estimated to 1 count/ms.

The deviation from exponential decay at high energies is due to the poor signal-to-noise ratio. In the case of the lowest n_e, the data measured becomes comensurate with the noise level at the energy of about 3 keV; for the highest n_e this critical energy is shifted close to 2 keV. This means that in the case of tangential NPA orientation, useful signal can be obtained up to energies 2 – 3 keV only; however, as it is shown in the next section, the fluxes detected do not originate from the central plasma region. The situation can be improved by using a different NPA orientation. The distance between the plasma center and the NPA port is almost 0.4 m for a tangential view. For a perpendicular view at mid-plane, it is only 0.23 m. Thus, the signal from the plasma central part would be improved because of the shorter distance traversed by the fast neutrals.

A more detailed analysis of the spectra measured can be carried out by utilizing the DOUBLE Monte-Carlo code, as it simulates the energy spectra measured based on the plasma parameters provided, namely n_e, the T_e and T_i profiles, the magnetic field B_t map, the plasma rotation v_{rot}, the effective plasma charge Z_{eff}, the NPA orientation and the COMPASS vessel geometry. All inputs, except the T_i profile, can be derived from other diagnostic techniques. The T_i profile can be varied iteratively until the simulated and measured energy spectra of the escaping neutral atoms become sufficiently similar.

In this study, the DOUBLE inputs were fixed for the purpose of comparing the NPA performance for different view-lines at standard COMPASS L-mode discharges parameters ($B_t = 1.15$ T, $I_p = 180$ kA, elongation 1.8, $T_{i0} = 530$ eV, $T_{iedge} = 30$ eV, $n_{neutrals\_edge} = 10^{16}$ m^{-3}, $Z_{eff} = 1.55$, $v_{rot} = 0$ m/s). The edge neutral density, Z_{eff} and v_{rot} were not measured during this series of discharges, but were estimated for the purpose of the simulations. The n_e and T_e profiles were varied. Examples of n_e and T_e profiles used in the simulations are compared with real measured data obtained by Thomson-scattering (TS) [6] in figure 3. The T_i profile used is also presented. The value of T_{i0} in this profile is based on the exponential decay of the energy spectra measured by the NPA in the same series of discharges as the TS data.

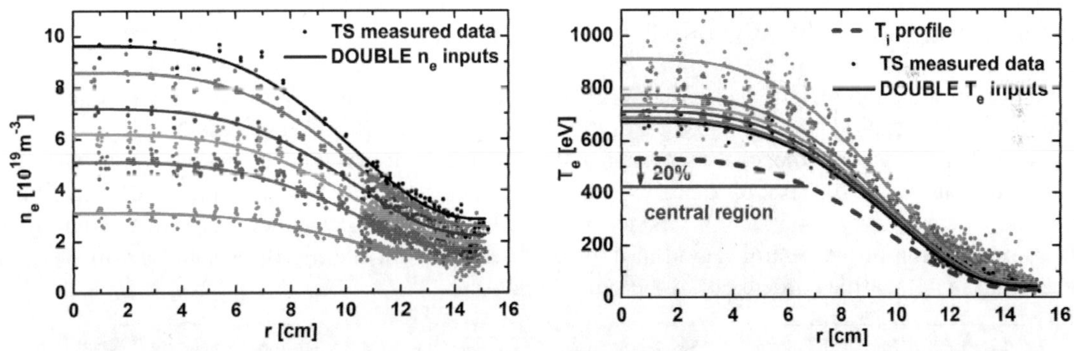

Figure 3. Example of n_e (left) and T_e and T_i (right) profiles used as input to DOUBLE. The measured data are presented by symbols; the data were obtained from several similar discharges for each n_e. The solid lines represent the n_e and T_e profiles calculated using formulas based on data fitting, with the central electron density n_{e0}. as input parameter. The input T_i profile is represented by a dashed line.

The curves were calculated using formulas with one varying parameter, namely, n_{e0} (central electron density) derived by fitting real plasma discharge profiles that have different requested n_e only. The T_e profiles are, naturally, different for different n_e. The colored region delineates the region of T_i where it is not lower than 80 % of T_{i0} and is referred to as the central plasma part further in the text.

2. Emissivity function of the detected fast neutral atoms

The fast neutrals flux detected is the integral of the emissivity along the NPA view line. The emissivity is influenced by the interplay of the probability of formation of fast neutrals and the probability of their re-ionization (attenuation of the fast neutrals flux) along the view-line. The formation of fast neutrals strongly depends on the CX cross-section, on the fast neutral energy E_{FA}, and on the local ion distribution function represented by the ion temperature profile and the donor density, i.e. the penetration depth of neutrals coming from the plasma edge. The re-ionization of fast neutrals depends mostly on n_e and E_{FA}. Examples of emissivity calculated by DOUBLE for different view-lines, central plasma densities n_{e0} and E_{FA} are shown in figure 4. The symbols represent the borders between quartiles (25 % of the signal entering the NPA originates in one quartile of its emissivity function); i.e. the middle symbols represent also the position of the median.

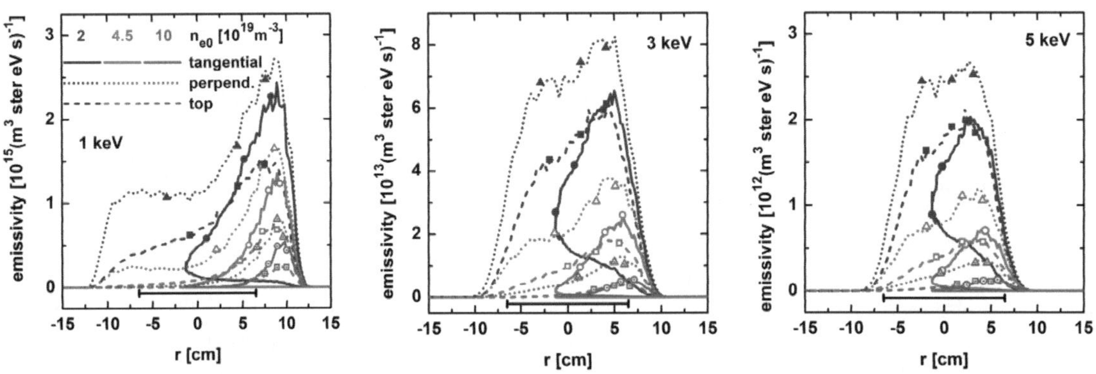

Figure 4. Emissivity for different central plasma densities n_{e0} [10^{19} m^{-3}], energies E_{FA} and view-lines remapped as a function of the minor plasma radius r. The location of the quartiles borders is highlighted by symbols. The colored area delineates the central plasma part.

The fluxes of less energetic fast atoms originate from closer to the edge due to the higher probability of their re-ionization, which prevents them from escaping the central plasma. For higher E_{FA} values, the signal originates mostly from the central part due to the longer mean-free-path of the fast atoms, and also due to the T_i drop towards the plasma edge, i.e. the lower high-energy ions population in the edge region. The reduction in the emissivity and the shift towards the edge as n_{e0} increases are also caused by the shortening of the escaping fast atoms' mean-free-path.

For higher n_{e0} values, the central plasma part cannot be observed in tangential view; e.g., for $E_{FA} = 1$ keV and $n_{e0} = 10^{20}$ m^{-3}, all quartile borders are out of the central region (as defined in figure 3), while the case of $n_{e0} = 4.5\times10^{19}$ m^{-3} is not much better. On the other hand, the perpendicular view allows observation of the central region in most of cases; only for the highest n_{e0} and the lowest E_{FA} are all emissivity quartile borders out of the central region.

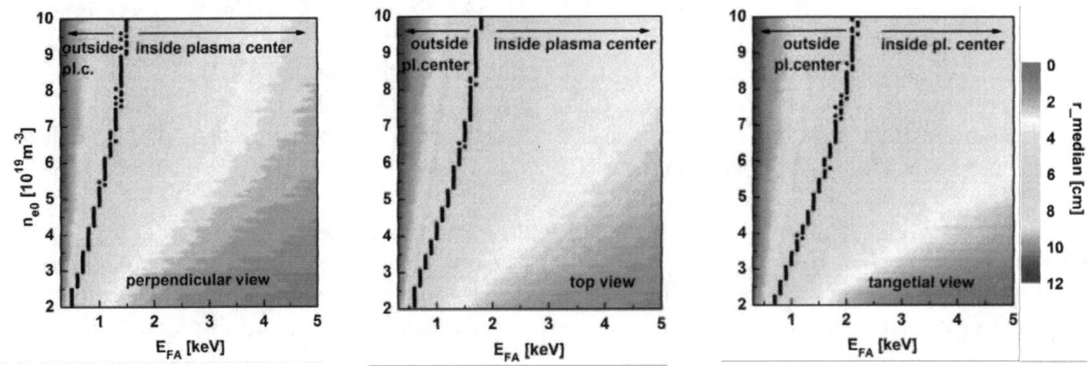

Figure 5. Dependence of the median radial position r_{med} of the emissivity function on the central plasma density n_{e0} and the fast neutrals energy E_{FA} for different NPA view-lines. The low-field side plasma center border $r = 6.7$ cm is denoted by a black line.

IOP Conf. Series: Journal of Physics: Conf. Series **992** (2018) 012009 doi:10.1088/1742-6596/992/1/012009

The variation of the median radial position r_{med} (the border between the second and third quartiles) of the fast neutral emissivity function with n_{e0} and E_{FA} is shown in figure 5 for different view-lines.

The median position r_{med} moves deeper into the plasma as E_{FA} increases and n_{e0} decreases. In the perpendicular view case, the neutrals still originating from the plasma center are detected above E_{FA} = 1.5 keV at the highest n_{e0}. This is much better than the tangential view case, where at the highest n_{e0} r_{med} crosses the plasma center border at E_{FA} above 2 keV. When we compare it with actually measured data (figure 2), the signal at higher n_e that is not buried under the noise does not in fact originate from the central region in the tangential view case, i.e. it is then impossible to determine the central ion temperature T_{i0}. In the perpendicular view case, the emissivity is higher (figure 4); therefore, the neutrals originate from the plasma center even for the highest n_{e0}.

3. Determination of the central ion temperature

As was mentioned in section 1, there exists a simple method of estimating T_{i0} directly from measured data using the exponential decay of the energy spectra T_{iCX}. Since one run of the DOUBLE code, which is normally used in an iterative loop for T_i profile derivation from measured data, takes about five minutes, we employed it to analyze T_{i0}. Figure 6 shows T_{iCX} thus derived using energy spectra simulated by DOUBLE. We also performed exponential decay fits for several different E_{FA} regions.

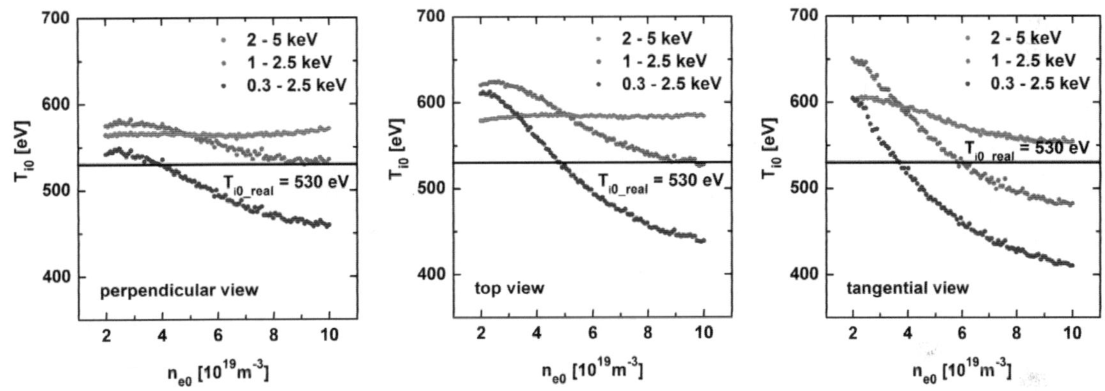

Figure 6. Central ion temperature T_{iCX} derived from simulated energy spectra using its exponential decay in different energy regions (2-5 keV, 1-2.5 keV and 0.3-2.5 keV) and for different NPA view-lines.

The deviation from the preselected T_{i0}, used as a DOUBLE code input, assumes its lowest value in perpendicular view. The T_{iCX} derived is independent of n_{e0} if one uses the 2-5 keV range for its evaluation in perpendicular and top views. Because the emissivity in the case of perpendicular view is higher (figure 4), there is also a possibility to use a higher E_{FA} range for T_{iCX} derivation than in the existing case of tangential view.

4. Conclusions

The analysis of different NPA views using the DOUBLE code with standard L-mode COMPASS parameters as input shows that:

Tangential view cannot be used for observation of the central plasma region at higher n_e ($> 6\times10^{19}$ m^{-3}), because the fast neutral atoms useful for determining T_i can only escape the central plasma and reach the NPA for the lowest n_e. If the energy spectra measured by the NPA are evaluated by a simple T_{iCX} analysis, the deviation from the real T_{i0} is at least + 15/ -17 %, depending on n_e;

The top view emissivity has similar or lower values as in tangential view, i.e. the signal intensity will be similar or lower. However, a larger part of the signal originates from the plasma center than in tangential view. T_{iCX} derived from exponential decay is not sensitive to n_e if derived in the E_{FA} range of 2 - 5 keV. This view-line provides better results than the tangential view;

Perpendicular view is the best of all tested possibilities. The escaping neutral atoms fluxes originate from the central plasma region with sufficient fast atom energies. Therefore, data acquired by an NPA installed with this view will be useful for measurements of T_i and T_{iCX} derived from energy spectra in the range of 2- 5 keV using the simple method that is independent of n_e; also, the systematic error (offset) is then the lowest, namely, ~ 35 eV, i.e. ~ 7%. Further, even when T_{iCX} is derived from the lower energy range, the deviation from T_{i0} is still under 10 %.

In summary, the second NPA (ACORD-5) should be tested in perpendicular view, where the escaping fast neutrals flux originates from the central plasma part; i.e., the NPA would be able to measure T_i in this region. There would also be possibility to complement the signal acquired by the second NPA with the data from the present one, which is oriented tangentially and observes a region closer to the edge. By combining these two NPA orientations, fast neutrals fluxes would be measured originating from a wide range of plasma radii.

Acknowledgements
This work was carried out within the framework of the EUROfusion Consortium and received funding from the Euratom research and training program 2014-2018 under grant agreement No 633053. The views and opinions expressed herein do not necessarily reflect those of the European Commission. The work was partly supported by the Ministry of Education, Youth, and Sports of the Czech Republic under projects Nos 8D15001 and LM2015045.

References
[1] Kislyakov A I, et al. *Plasma Phys. Rep.* **35**/7 535-51
[2] Panek R, et al 2016. *Plasma Phys. and Controlled Fusion* **58** 014015
[3] Mitosinkova K, et al. *Proc. 1st EPS Conf. Plasma Diagnostics (ECPD2015)*
[4] Afanasyev V, et al. *JET-R--00/04*
[5] Schlatter Ch, et al 2006. *Plasma Phys. and Controlled Fusion* **48**/12 1765
[6] Bilkova P, et al 2010. *Nuclear Instr.Methods in Phys. Res. A: Accelerators, Spectrometers, Deand Associated Equipment* **623.2** 656-9

IOP Conf. Series: Journal of Physics: Conf. Series **992** (2018) 012010 doi:10.1088/1742-6596/992/1/012010

Two-fluid model of the plasma-wall transition in the presence of warm ions

T Gyergyek[1,2,3] and J Kovačič[2]

[1]University of Ljubljana, Faculty of Electrical Engineering,
25 Tržaška, 1000 Ljubljana, Slovenia
[2]Jožef Stefan Institute, 39 Jamova, 1000 Ljubljana, Slovenia

E-mail: Tomaz.gyergyek@fe.uni-lj.si

Abstract. A one-dimensional two-fluid model of the plasma-wall transition in front of a large, negative, planar electrode is presented. The electrons are assumed to be isothermal, while for the ions the closure is made by the assumption that the ion heat flux is proportional to the negative gradient of the ion temperature. The introduction of the ion heat flow into the model changes the ion flow towards the electrode from mainly adiabatic into almost isothermal.

1. Introduction

Plasma-wall transition region is studied using a one-dimensional two-fluid model developed recently [1-3]. In this model the electrons are described by a continuity and momentum transfer equation, while the ions are described by the continuity, momentum transfer and energy transfer equation. The model is one-dimensional and steady state. For the electrons the closure is made by the assumption that the electrons are isothermal. For the ions the closure in the energy transfer equation is made by the assumption that the ion heat flux is proportional to the negative gradient of the ion temperature. Effects of variation of the heat conduction coefficient are studied in some detail.

2. Model

The basic equations of the one-dimensional two-fluid model obtained as moments of the Boltzmann equation in steady state are the following:

$$\frac{d}{dx}(n_i u_i) = S_i,\ \frac{d}{dx}(n_e u_e) = S_e, \tag{1}$$

$$m_i n_i u_i \frac{du_i}{dx} = -n_i e_0 \frac{d\Phi}{dx} - \frac{dp_i}{dx} + A_i - m_i u_i S_i, \tag{2}$$

$$m_e n_e u_e \frac{du_e}{dx} = n_e e_0 \frac{d\Phi}{dx} - \frac{dp_e}{dx} + A_e - m_e u_e S_e, \tag{3}$$

$$\frac{1}{2} u_i \frac{dp_i}{dx} + \frac{3}{2} p_i \frac{du_i}{dx} + \frac{dq_i}{dx} = M_i - u_i A_i + \frac{1}{2} m_i u_i^2 S_i. \tag{4}$$

[3] To whom any correspondence should be addressed.

IOP Conf. Series: Journal of Physics: Conf. Series **992** (2018) 012010 doi:10.1088/1742-6596/992/1/012010

The meaning of the symbols is the following: x is the space coordinate, m_i is the ion mass, m_e is the electron mass, n_i, and n_e are the respective particle densities, u_i, and u_e are the flow velocities of ions and electrons, e_0 is the elementary charge, S_i, and S_e are the source terms, discussed below, A_i and A_e are elastic collision terms and M_i is the energy transport term. In this type of models, the terms S_{ie}, A_{ie} and M_i are assumed known functions of x. For simplicity, the terms A_i, A_e and M_i are neglected in this work. The potential profile $\Phi(x)$ is determined by the Poisson equation:

$$\frac{d^2\Phi}{dx^2} = -\frac{e_0}{\varepsilon_0}\left(n_i(x) - n_e(x)\right). \tag{5}$$

Here ε_0 is the permittivity of the free space. The electron pressure p_e and electron temperature T_e are related by the ideal gas law $p_e = kn_eT_e$, where k is the Boltzmann constant. Because the electrons are isothermal, the electron pressure gradient term in (3) can be expressed as:

$$\frac{dp_e}{dx} = kT_e\frac{dn_e}{dx}. \tag{6}$$

Here k is the Boltzmann constant. The ion pressure p_i and ion temperature T_i are related by $p_i = kn_iT_i$, but in this case the ion pressure gradient term in (2) and (4) is expressed in the following way:

$$\frac{dp_i}{dx} = k\frac{dn_i}{dx}T_i + kn_i\frac{dT_i}{dx} = \kappa kT_i\frac{dn_i}{dx}. \tag{7}$$

Here κ is the polytropic function. It has been shown by Kuhn et al. [4] that the ratio between specific heats at constant pressure and at constant volume (the polytropic coefficient) in the plasma-wall transition region is not constant, but is space dependent. Thus, it is called a polytropic function instead of a polytropic coefficient. The definition of the polytropic function is read directly from (7):

$$\kappa = 1 + \frac{n_i}{T_i}\frac{\frac{dT_i}{dx}}{\frac{dn_i}{dx}} = 1 + \frac{n_i}{T_i}\frac{dT_i}{dn_i}. \tag{8}$$

The ion heat flux q_i is assumed to be proportional to the negative gradient of the ion temperature:

$$q_i = -K\frac{dT_i}{dx}. \tag{9}$$

Here K is the heat conduction coefficient.

Finally, the source terms S_i and S_e require some explanation. They give the difference between the number of created and annihilated charged particles of respective species per unit volume and per unit time as functions of x. Ionization and annihilation can have various physical mechanisms. If the main mechanism of ionization are ionizing collisions between neutrals and electrons, the number of created charged particles per unit volume and per unit time is proportional to the product n_nn_e, where n_n is the density of neutral particles. The number of recombined ions and electrons per unit time and per unit volume is on the other hand proportional to the product n_in_e. In addition, ions and electrons can be lost from the system via other mechanisms. These losses are given by the terms σ_i and σ_e. Thus, the source terms S_i and S_e could be written in the following form $S_{ie} = \gamma_1 n_n n_e - \gamma_2 n_i n_e - \sigma_{ie}$, where γ_1 and γ_2 are the respective proportionality constants. But, in order to keep the model as simple as possible, the source terms S_i and S_e are written simply as:

$$S_i = S_e = \frac{n_e}{\tau}. \tag{10}$$

Here τ should be understood as generalized ionization time, where creation and losses of charged particles are already taken into account.

The following variables are introduced:

$$\lambda_D = \sqrt{\frac{\varepsilon_0 k T_e}{n_0 e_0^2}},\ c_0 = \sqrt{\frac{kT_e}{m_i}},\ L = c_0\tau,\ \varepsilon = \frac{\lambda_D}{L},\ \mu = \frac{m_e}{m_i},\ \Psi = \frac{e_0\Phi}{kT_e}, \tag{11}$$

$$N_i = \frac{n_i}{n_0},\ N_e = \frac{n_e}{n_0},\ V_i = \frac{u_i}{c_0},\ V_e = \frac{u_e}{c_0},\ \Theta = \frac{T_i}{T_e},\ X = \frac{x}{\lambda_D}, \tag{12}$$

$$\chi = -\frac{d\Psi}{dX},\ P_i = \frac{kn_iT_i}{kn_0T_e} = N_i\Theta,\ K_2 = K\frac{e_0\sqrt{m_i}}{k^2T_e\sqrt{\varepsilon_0 n_0}}. \tag{13}$$

Here n_0 is the plasma density in the region far away from the electrode, where the plasma is not perturbed by its potential. By taking into account (6), (7) and (9) - (13), the set (1) - (5) is written in the following form:

$$\frac{d}{dX}(N_iV_i) = \varepsilon N_e,\ \frac{d}{dX}(N_eV_e) = \varepsilon N_e, \tag{14}$$

$$N_iV_i\frac{dV_i}{dX} = -N_i\frac{d\Psi}{dX} - \frac{dP_i}{dX} - \varepsilon V_iN_e, \tag{15}$$

$$N_eV_e\frac{dV_e}{dX} = \frac{N_e}{\mu}\frac{d\Psi}{dX} - \frac{1}{\mu}\frac{dN_e}{dX} - \varepsilon V_eN_e, \tag{16}$$

$$\frac{1}{2}V_i\frac{dP_i}{dX} + \frac{3}{2}P_i\frac{dV_i}{dX} - K_2\frac{d^2\Theta}{dX^2} = \varepsilon\frac{1}{2}N_eV_i^2, \tag{17}$$

$$\frac{d^2\Psi}{dX^2} = N_e(X) - N_i(X). \tag{18}$$

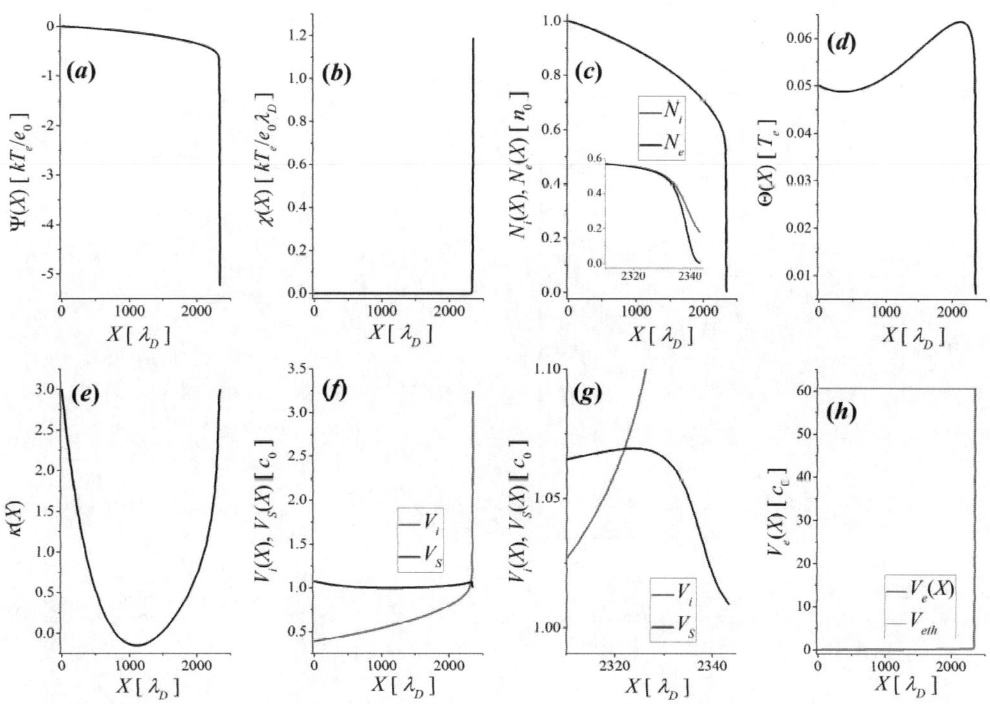

Figure 1. Solutions of the set (14) – (18) for $K_2 = 0$, $\mu = 2.724\times10^{-4}$, $\varepsilon = 10^{-4}$, $\Psi(0) = 0$, $N_i(0) = N_e(0) = 1$, $V_e(0) = 0$, $\Theta(0) = 0.05$, $V_i(0) = 0.387299$ and $\chi(0) = 7.74594\times10^{-5}$.

3. Results

We present two examples of solutions of set (14) - (18). First we assume that there is no ion heat flux in the plasma, $K_2 = 0$. In this case the only plasma parameter that has to be specified is the ion mass m_i. The Deuterium mass $m_i = 3.34\times10^{-27}$ kg is selected. This yields $\mu = 2.724\times10^{-4}$. In addition, $\varepsilon = 10^{-4}$ is selected. The set (14) - (18) is integrated from $X = 0$, where the plasma is unperturbed by the electrode, towards the electrode in the positive X direction. For a unique solution, seven boundary conditions defining the values of the unknown function at $X = 0$ must be specified. At $X = 0$ the plasma potential is set to zero and the plasma there is neutral, thus $\Psi(0) = 0$, $N_i(0) = N_e(0) = 1$ are a logical selection. As discussed in [2,3], the selection of the electron velocity $V_e(0)$ is rather arbitrary, so that $V_e(0) = 0$ is selected in this work. The ion temperature at $X = 0$ is selected as an independent boundary condition. In this work, $\Theta(0) = 0.05$ is selected. The ion velocity $V_i(0)$ and the electric field $\chi(0)$ are found in a self-consistent way, using the shooting method as explained in [2]. The values $V_i(0) = 0.387299$ and $\chi(0) = 7.74594\times10^{-5}$ are thus found. Figure 1 shows numerical solutions of the set (14) - (18).

In figure 2, the solutions of the system (14) - (18) are also presented, but with the ion heat flux being taken into account. In order to determine K_2, the plasma parameters have to be specified more precisely. The electron temperature and density that could correspond to the edge plasma in a small tokamak, such as COMPASS [5], are selected, or $kT_e = 5$ eV and $n_0 = 10^{18}$ m^{-3}. Together with $m_i = 3.34\times10^{-27}$ kg and $K = 1$ W/mK, this gives $K_2 = 281548$. The parameters $\mu = 2.724\times10^{-4}$ and $\varepsilon = 10^{-4}$ are the same as in figure 1. The boundary conditions are almost the same as in figure 1: $\Psi(0) = 0$, $N_i(0) = N_e(0) = 1$, $V_e(0) = 0$, $V_i(0) = 0.387298$ and $\chi(0) = 8.60663\times10^{-5}$. One additional boundary condition is required this time: $d\Theta/dX(0) = 0$. The corresponding numerical solutions are shown in figure 2.

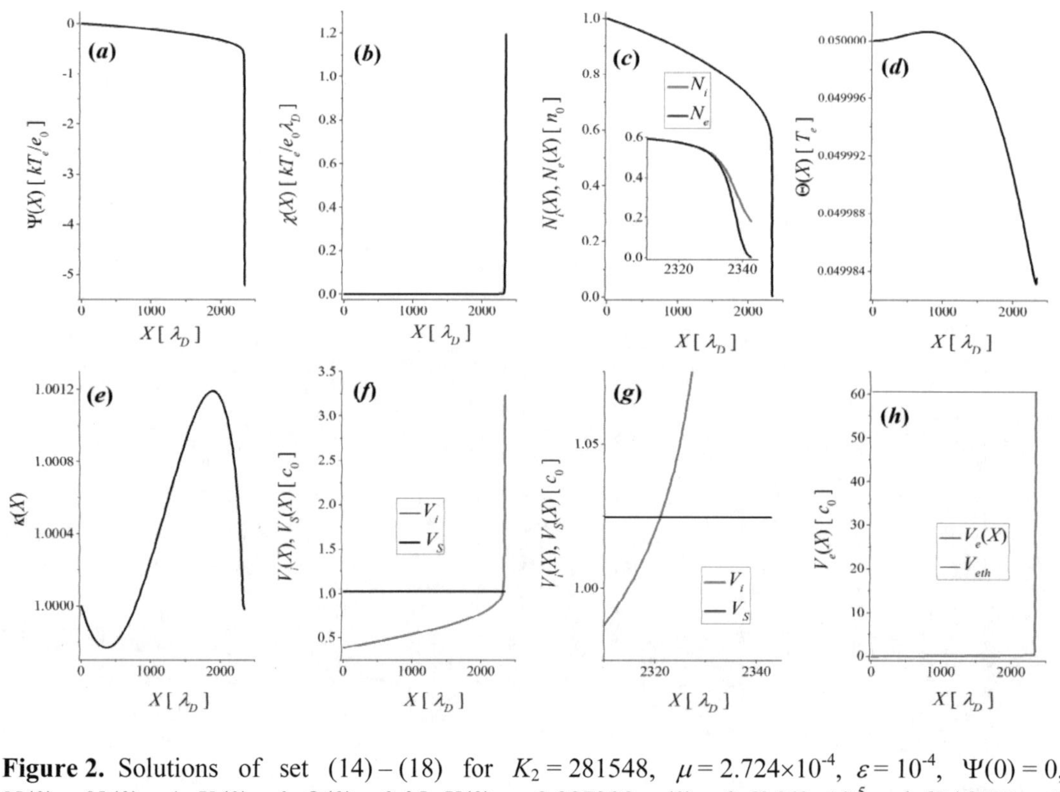

Figure 2. Solutions of set (14) – (18) for $K_2 = 281548$, $\mu = 2.724\times10^{-4}$, $\varepsilon = 10^{-4}$, $\Psi(0) = 0$, $N_i(0) = N_e(0) = 1$, $V_e(0) = 0$, $\Theta(0) = 0.05$, $V_i(0) = 0.387298$, $\chi(0) = 8.60663\times10^{-5}$ and $d\Theta/dX(0) = 0$.

At first glance, the solutions are very similar. In both cases the potential $\Psi(X)$ and the densities $N_i(X)$ and $N_e(X)$ are monotonically decreasing functions of X, while the electric field $\chi(X)$ and velocities $V_i(X)$ and $V_e(X)$ are increasing functions of X. But there is an obvious difference in the profiles of the ion temperature $\Theta(X)$, the polytropic function $\kappa(X)$ and the ion sound velocity $V_S(X)$, which is given by:

$$V_S(X)=\sqrt{\frac{1+\kappa(X)\Theta(X)}{1+\mu}}. \tag{19}$$

In the absence of an ion heat flux, the ion temperature decreases at first, then increases, reaches a maximum close to the sheath edge and then drops sharply in the sheath – figure 1 (*d*). The polytropic function - figure 1 (*e*) – indicates that at the entrance into the system the ion flow is adiabatic. Then ions are expanding and heating at the same time and $\kappa(X)$ becomes even negative. In the sheath, the ions are cooling adiabatically. In the presence of an ion heat flow, any ion temperature gradients are almost completely smoothed out and the ion temperature is (almost) constant – figure 2 (*d*). Also, the polytropic function - figure 2 (*e*) – confirms that the ion flow is isothermal since $\kappa(X)$ is (almost) equal to 1 everywhere in the system.

4. Conclusions

A two-fluid one-dimensional model of the plasma-wall transition in front of a planar presented earlier [1-3] is extended to include the ion heat flux in the energy exchange equation. If the heat conductivity is large, the ion flow towards the electrode becomes almost isothermal. The profiles of the potential, electric field, ion and electron density and also the ion and electron velocity remain almost unchanged when an ion heat flux term is introduced. This is valid also for the location of the singularity of set (14) – (18), which occurs when $V_e(X)$ reaches the electron thermal velocity $V_{eth} = \mu^{-1/2}$ (thin horizontal lines in both plots (*h*)) and also for the positon of the sheath edge (plots (*g*) in figures 1 and 2.

We conclude with the following remark. The model presented in Section 2 includes the first, the second, and in the case of ions, also the third moment of the velocity distribution function. The first moment (particle flux) is proportional to the electric current, which can be measured by, e.g., a Langmuir probe. The second and the third moment of the distribution function represent the pressure and heat flux gradients, which could be measured by force and calorimetric probes. Such advanced diagnostics of plasma sheaths are still in the early stages of development [6], but will be necessary to achieve a better understanding of the plasma-wall transition.

Acknowledgements

This work was partially supported by Grant No. P2- 0073 of the Slovenian Research Agency.

References

[1] Gyergyek T and Kovačič J 2016 *Phys. Plasmas* **23** 063510
[2] Gyergyek T and Kovačič J 2017 *Phys. Plasmas* **24** 063505
[3] Gyergyek T and Kovačič J 2017 *Phys. Plasmas* **24** 063506
[4] Kuhn S, Riemann K U, Jelić N, Tskhakaya D D Sr, Tskhakaya D Jr and Stanojević M 2006 *Phys. Plasmas* **13** 013503
[5] Popov Tsv, Dimitrova M, Pedrosa M A, López-Bruna D, Horacek J, Kovačič J, Dejarnac R, Stöckel J, Aftanas M, Böhm P, Bilkova, Hidalgo C and Panek R 2015 *Plasma Phys. Control. Fusion* **57** 115011
[6] Trottenberg T, Richter T and Kersten H 2015 *Eur. Phys. J.* D **69** 91

Thermophysical properties of binary mixtures containing oxygen and noble gases

M Damyanova[1,3], U Hohm[2] and E Balabanova[1]

[1]E. Djakov Institute of Electronics, Bulgarian Academy of Sciences, 72 Tzarigradsko Chaussee, 1784 Sofia, Bulgaria
[2]Institut für Physikalische und Theoretische Chemie der TU Braunschweig, 17 Gaußstr., D-38106 Braunschweig, Germany

E-mail: milenadam@gmail.com

Abstract. This work deals with the determination of reliable transport and equilibrium thermophysical properties of binary mixtures of O_2 and the noble gases Ar, Kr, and Xe. The second virial coefficients $B(T)$, the viscosities $\eta(T)$, and the binary diffusion coefficients $D_{12}(T)$ were calculated for each mixture in the temperature range between 200 K and 1000 K using a a temperature dependent (*n*-6) Lennard-Jones intermolecular interaction potential model. The properties of the mixtures were obtained by using the Lorentz-Berthelot and Hohm-Zarkova-Damyanova mixing rules. Our results were compared to other theoretical and experimental data and acceptable agreement was found in all cases. Fitting formulae for the calculation of the intermolecular potential parameters and equilibrium and transport properties of the examined mixtures are recommended proposed.

1. Introduction

Binary mixtures containing molecular oxygen and the noble gases Ar, Kr, and Xe are used in many industrial processes, e.g. Ar/O_2 is applied as a shielding gas in weldings [1]; Ar/O_2, Kr/O_2, and Xe/O_2 are used in magnetron sputtering [2]; Ar/O_2 or Kr/O_2 are applied in etching processes [3]. Recently, medical applications became popular; e.g., a Xe/O_2 mixture is used as a respiratory and anesthetic gas mixture [4-5] and Ar/O_2 [6] is used in tissue regeneration. In all applications, reliable information is needed on the equilibrium and transport properties at different concentrations and in a wide range of temperatures. The variety of possible constituents of mixtures, as well as various experimental difficulties, stimulate the calculation of thermophysical data by different models.

Our calculations were based on the (*n*-6) Lennard-Jones temperature dependent potential (LJTDP). By using the Lorentz-Berthelot [7] and Hohm-Zarkova-Damyanova [8] mixing rules, the intermolacular interaction potentials of the binary mixtures were obtained from the pure component data. Subsequently, these potentials were used to calculate the second virial coefficients (B_{12}, B_{mix}), the viscosities η_{12}, η_{mix} and the binary diffusion coefficients (D_{12}) for all three mixtures in the temperature range betwen 200 K and 1000 K. Our findings were compared to available theoretical and experimental data.

[3] To whom any correspondence should be addressed.

2. Theoretical background

All thermophysical properties of gas mixtures in the low pressure (under 1 atm) range depend on the binary intermolecular interaction energies between like and unlike constituents of the corresponding mixture. In our model, the intermolecular interactions between molecule 1 and 2 are described by the (n-6) Lennard-Jones potential with the characteristic potential parameters R_{m12} (equilibrium distance), ε_{12} (potential well depth), and n_{12} (repulsive parameter). In the case of molecules, we consider to be R_{m12} and ε_{12} as being functions of the temperature T, which leads us to using the Lennard-Jones temperature-dependent potential (LJTDP)[9].

The intermolecular interaction potential parameters R_{m11}, ε_{11}, and n_{11} for the pure gases have already been determined by Damyanova *et al.* and are given in Ref. [10] for pure O_2 and Ref. [11] for the noble gases.

The potential parameters R_{m12} and ε_{12} for unlike pairs are obtained via the Lorenz-Berthelot (LB) and Hohm-Zarkova-Damyanova (HZD) mixing rules; the latter is given in full detail in Ref. [8]. In both cases, the repulsive parameter $n_{12}=(n_1+n_2)/2$ does not depend on the temperature. The molecular properties used in this work are given in table 1.

Table 1. Pure gas values for mass, dipole polarizabilities, dispersion-interaction energy constants and zero temperature potential parameters used in this work; k_B is the Boltzmann constant.

Substance	Molar mass/ (g/mol)	$(\varepsilon_{11}(0)/k_B)$/ K	$10^{10}\,R_{m11}(0)$/ m	n_1	α/ a.u.[a]	C_6/ a.u.[a]
Ar	39.95	155.00 [11]	3.600 [11]	17.10 [11]	11.08 [12]	64.30 [14]
Kr	83.80	221.00 [11]	3.860 [11]	17.8 [11]	16.79 [12]	129.6 [14]
Xe	131.30	310.90 [11]	4.204 [11]	18.76 [11]	27.16 [12]	285.90 [14]
O_2	32.00	185.53 [10]	3.460 [10]	23.20 [10]	10.585 [12]	61.57 [13]

[a] Conversion factors are for α: 1 a.u.=$1.64878\cdot10^{-41}\,C^2m^2J^{-1}$ and C_6: 1 a.u. = $9.57345\cdot10^{-80}\,Jm^6$.

3. Results and discussion

The LJTDP model was used to calculate the thermophysical properties B_{12}, B_{mix}, η_{12}, η_{mix}, and D_{12} for all three O_2/X-mixtures (X=Ar, Kr, Xe) in the temperature range between 200 K and 1000 K. Both mixing rules, LB and HZD, were used. Our calculated results ("calc") were compared to other experimental and theoretical data in the literature ("other") (see the deviation plots given in figures 1 – 8). The absolute deviations $\Delta B=B_{other}-B_{calc}$ are displayed in the case of the second virial coefficient, whereas for η and D we show the relative deviations. The empty symbols are used for the results obtained by the LB mixing rule; the full symbols denote results obtained by the HZD mixing rule. We should first we note that most of the thermophysical properties of the three mixtures have not been well investigated. One can find measured B_{12} data [15,16] for the O_2/Ar mixture only. Calculated values for B_{mix}, η_{mix} and D_{12} of equimolar O_2/Ar, O_2/Kr, and O_2/Xe gas mixtures were published by Bzowski *et al.* [17]. Except for their calculations of η_{mix}, mixture viscosity measurements for different O_2 concentrations were reported by Hellemans *et al.* [18] for O_2/Ar and O_2/Kr mixtures; calculations were published by Katz *et al.* [19] for O_2/Xe. Measurements of the diffusion coefficients D_{12} were performed by Boardman *et al.* [20], Arrora *et al.* [21] and Trengove *et al.* [22] for O_2/Ar; Dunlop *et al.* [23] and Ivakin *et al.* [24] for O_2/Kr; and Dunlop *et al.* [23] for O_2/Xe. Calculated D_{12} data are given in Refs. [21, 25] for O_2/Ar, [24] for O_2/Kr, and [19] for O_2/Xe.

Figures 1-8 show that in the case of B_{mix} and D_{12}, the HZD mixing rule yields a slightly better agreement between the literature data and our findings, whereas the more popular LB mixing rule seems to work a little bit better in the case of B_{12} and η_{mix}.

IOP Conf. Series: Journal of Physics: Conf. Series **992** (2018) 012011 doi:10.1088/1742-6596/992/1/012011

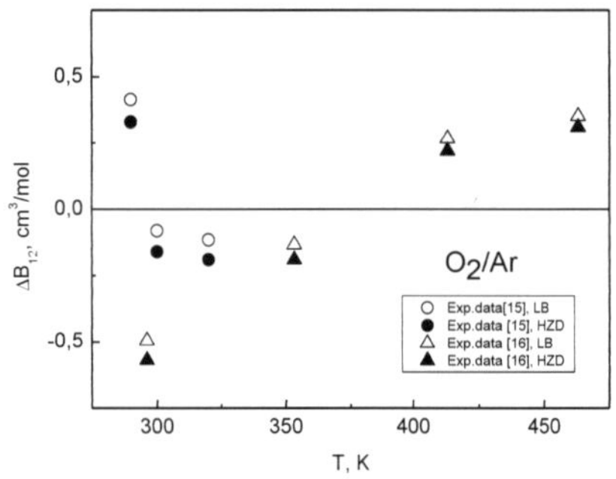

Figure 1. Absolute deviations ΔB_{12} of the interaction second virial coefficient, B_{12}, between Refs. [15,16] and our findings for O_2/Ar gas mixture.

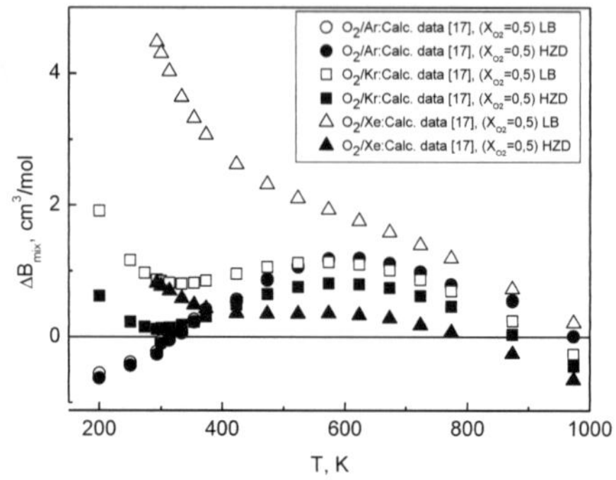

Figure 2. Absolute deviations ΔB_{mix} of the mixture second virial coefficient, B_{mix}, between Refs [17] and our findings for O_2/Ar, O_2/Kr, and O_2/Xe gas mixtures.

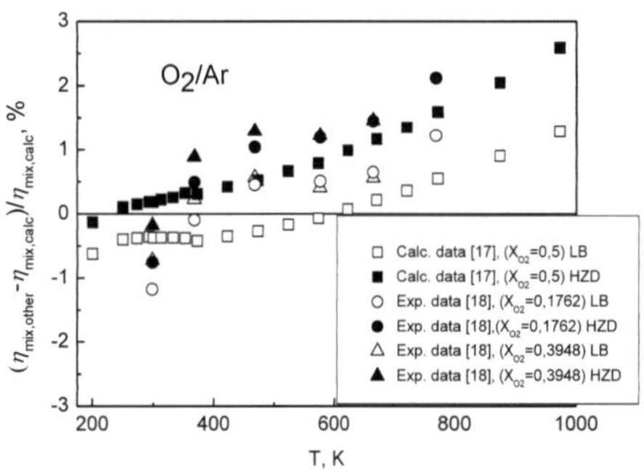

Figure 3. Relative deviations of the mixture viscosity, η_{mix}, between Refs. [17,18] and our findings for O_2/Ar gas mixture.

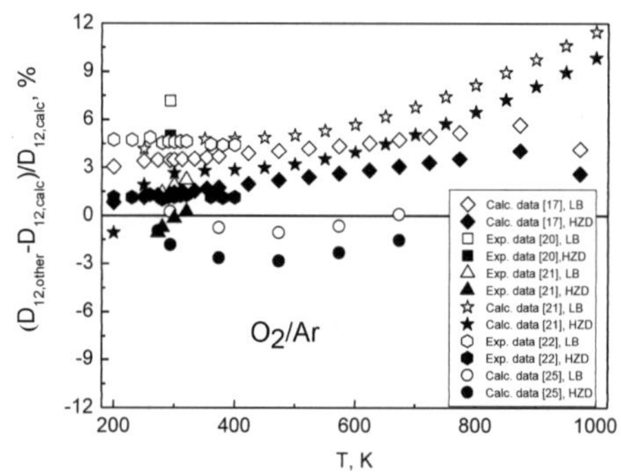

Figure 4. Relative deviations of the diffusion coefficient, D_{12}, between Refs. [17,20,21,22,25] and our findings for O_2/Ar gas mixture.

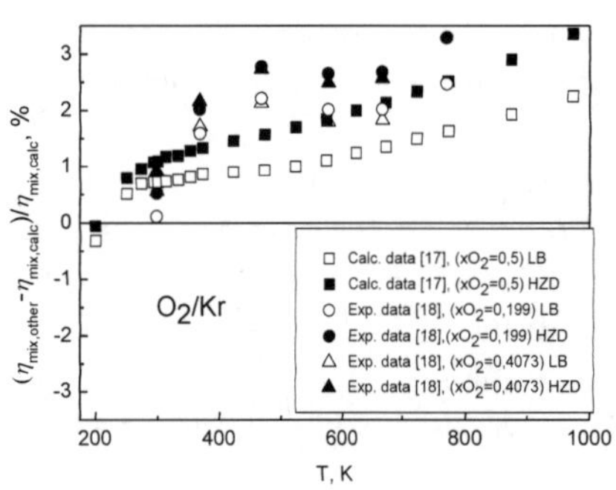

Figure 5. Relative deviations of the mixture viscosity, η_{mix}, between Refs. [17,18] and our findings for O_2/Kr gas mixture.

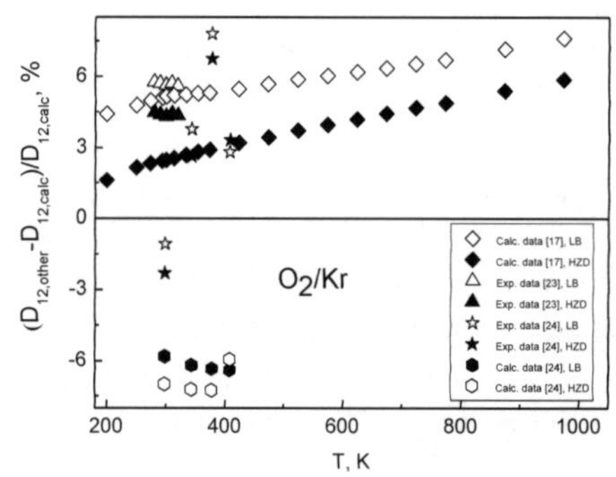

Figure 6. Relative deviations of the diffusion coefficient, D_{12}, between Refs. [17,23,24] and our findings for O_2/Kr gas mixture.

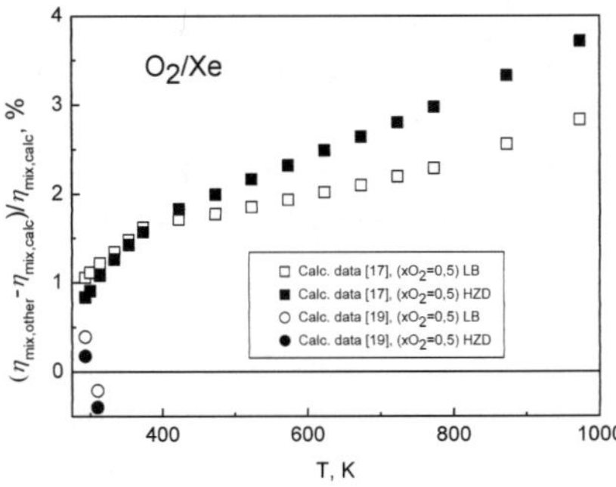

Figure 7. Relative deviations of the mixture viscosity, η_{mix}, between Refs. [17,19] and our findings for O_2/Xe gas mixture.

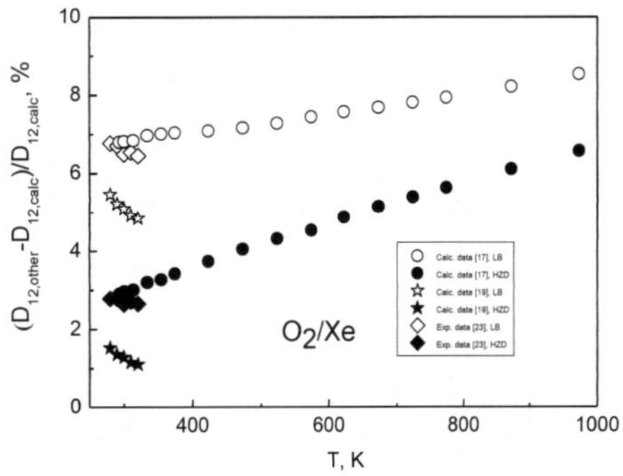

Figure 8. Relative deviations of the diffusion coefficient, D_{12}, between Refs. [17,19,23] and our findings for O_2/Xe gas mixture.

Our recommended values of $R_{m12}(T)$, $\varepsilon_{12}(T)$, $B_{12}(T)$, $\eta_{12}(T)$ and $D_{12}(T)$ in the temperature range between 200 K and 1000 K are presented in the form of fitting formulae. For the sake of consistency, all of them are based on the HZD mixing rule. The equilibrium distance $R_{m12}(T)$ is fitted to the dimensionless expression:

$$R_{m12}(T)/(10^{-10}\text{m})= R_{m12}(0)/(10^{-10}\text{m})+A_1\exp[-A_2/(T/\text{K})]+A_3\exp[-A_4/(T/\text{K})] . \quad (1)$$

Subsequently, the well depth $\varepsilon_{12}(T)$ can be calculated via $\varepsilon_{12}(T)= \varepsilon_{12}(0)[R_{m12}(0)/ R_{m12}(T)]^6$.

In the sequence O_2/Ar, O_2/Kr, and O_2/Xe, the zero temperature potential parameters $R_{m12}(0)/(10^{-10}\text{m})$ and $(\varepsilon_{12}(0)/k_B)$/K of the unlike interactions are given as 3.52987 and 169.368, 3.6630 and 198.875, 3.84464 and 225.440, respectively. For the thermophysical properties we have used the following fitting formulae.

The fitting constants are given in table 2. Here ΔP is the standard error of the fit.

$$B_{12}(T)/(\text{cm}^3/\text{mol}) = \sum_{k=1}^{5} A_k \left(\frac{T}{\text{K}}\right)^{1-k} \quad (2)$$

$$\eta_{12}(T)/(\mu\text{Pa}\cdot\text{s}) = \sum_{k=1}^{5} A_k \left(\frac{T}{\text{K}}\right)^{k-1} \quad (3)$$

$$D_{12}(T)/(10^{-6}\,\text{m}^2/\text{s}) = \sum_{k=1}^{5} A_k \left(\frac{T}{\text{K}}\right)^{k-1} \quad (4)$$

Table 2. Fitting constants of equations (1-4).

	Property	A_1	A_2	A_3	A_4	A_5	ΔP
O_2/Ar	$R_{m12}(T)$	0.076001	2275.20	0.075426	4920.25	-	$5.9\cdot10^{-7}$
	$B_{12}(T)$	33.439	$-1.00095\cdot10^{4}$	$-1.82492\cdot10^{6}$	$1.88479\cdot10^{8}$	$-1.47868\cdot10^{10}$	0.0034
	$\eta_{12}(T)$	1.07170	$9.65437\cdot10^{-2}$	$-8.53224\cdot10^{-5}$	$5.73430\cdot10^{-8}$	$-1.68199\cdot10^{-11}$	0.0055
	$D_{12}(T)$	-1.88266	$1.98262\cdot10^{-2}$	$1.91183\cdot10^{-4}$	$-7.65572\cdot10^{-8}$	$1.80548\cdot10^{-11}$	0.0083
O_2/Kr	$R_{m12}(T)$	0.075498	2274.87	0.074673	4911.94	-	$6.1\cdot10^{-7}$
	$B_{12}(T)$	38.310	$-1.41118\cdot10^{4}$	$-2.20146\cdot10^{6}$	$1.88790\cdot10^{8}$	$-1.72271\cdot10^{10}$	0.0040
	$\eta_{12}(T)$	-1.41488	$9.60991\cdot10^{-2}$	$-7.64399\cdot10^{-5}$	$4.70388\cdot10^{-8}$	$-1.28366\cdot10^{-11}$	0.0027
	$D_{12}(T)$	-1.30221	$1.23991\cdot10^{-2}$	$1.57433\cdot10^{-4}$	$-6.59022\cdot10^{-8}$	$1.67906\cdot10^{-11}$	0.0084
O_2/Xe	$R_{m12}(T)$	0.077214	2289.01	0.091152	5312.12	-	$3.8\cdot10^{-7}$
	$B_{12}(T)$	45.241	$-1.92441\cdot10^{4}$	$-2.75729\cdot10^{6}$	$1.95156\cdot10^{8}$	$-2.16278\cdot10^{10}$	0.0057
	$\eta_{12}(T)$	-1.28771	$8.54281\cdot10^{-2}$	$-5.89447\cdot10^{-5}$	$3.12228\cdot10^{-8}$	$-7.32284\cdot10^{-12}$	0.0050
	$D_{12}(T)$	-0.86444	$7.62039\cdot10^{-3}$	$1.37370\cdot10^{-4}$	$-5.92489\cdot10^{-8}$	$1.55300\cdot10^{-11}$	0.0077

IOP Conf. Series: Journal of Physics: Conf. Series **992** (2018) 012011 doi:10.1088/1742-6596/992/1/012011

4. Conclusions

Our recommended fitting formulae can be used for easy and reliable calculation of the HZD potential parameters and the thermophysical properties B_{12}, η_{12} and D_{12} of the binary mixtures O_2/Ar, O_2/Kr, and O_2/Xe in the temperature range between 200 K and 1000 K, covering the whole concentration range. The formulae for the potential parameters $R_{m12}(T)$ and $\varepsilon_{12}(T)$ might be useful for the calculation of other properties of oxygen/noble gas mixtutes as well.

References

[1] Mvola B, Kah P 2017 *Int. J. Adv. Manuf. Technol.* **88** 2369-87
[2] Bowes M and Bradley J W 2014 *Surf. Coat. Technol.* **250** 2-6
[3] Kumar A, Yerva S V and Barshilia H C 2016 *Sol. Energy Mater. Sol. Cells* **155** 184-93
[4] Bakhsheshi M F, Moradi H V, Stewart E E, Keenliside L and Lee T -Y 2015 *IEEE J. Transl. Eng. Health Med.* **3** 1500108
[5] Stoppe C, Ney J, Brenke M, Goetzenich A, Emontzpohl C, Schälte G, Grottke O, Moeller M, Rossaint R and Coburn M 2016 *Sports Med.* **46** 1753-66
[6] Kalmanov A S , Makarov E V, Shishkin A N, Pisarev A A and Bubeev Yu A 2016 *Aviakosmicheskaya i Ekologicheskaya Meditsina* **50** 48-54
[7] Hirschfelder J O, Curtiss C F and Bird R B 1954 *Molecular Theory of Gases and Liquids* (Wiley, London)
[8] Hohm U, Zarkova L and Damyanova M 2006 *Int. J. Thermophys.* **27** 1725-45
[9] Zarkova L and Hohm U 2002 *J. Phys. Chem. Ref. Data* **31** 183-216
[10] Damyanova M, Hohm U, Balabanova E and Zarkova L 2010 *J. Phys.: Conf. Ser.* **223** 012012
[11] Zarkova L, Hohm U and Damyanova M 2004 *Int. J.Thermophys.* **25** 1775-97
[12] Hohm U 2013 *J. Mol. Struct.* **1054-55** 282-92
[13] Kumar A, Meath J W, Bündgen P and Thakkar A J 1996 *J. Chem. Phys.* **105** 4927-37
[14] Tang K T and Toennies J P 2003 *J. Chem. Phys.* **118** 4976-83
[15] Martin M L, Trengove R D, Harris K R and Dunlop P J 1982 *Aust. J. Chem.* **35** 1525-9
[16] Vatter K, Schmidt H J, Elias E and Schramm B 1996 *Ber. Bunsenges. Phys. Chem.* **100** 73-9
[17] Bzowski J, Kestin J, Mason E A and Uribe F J 1990 *J. Phys. Chem. Ref. Data* **19** 1179-1232
[18] Hellemans J M, Kestin J and Ro S T 1973 *Physica* **65** 362-75
[19] Katz I, Gallibotte G, Martin A R and Arpentinier P 2011 *Medical Gas Res.* **1** 28-39
[20] Boardman L E and Wild N E 1937 *Proc. Roy. Soc.* **A162** 511-20
[21] Arrora P S, Robjohns H L, Bell T N and Dunlop P J 1980 *Aust. J. Chem.* **33** 1993-6
[22] Trengove R D, Harris K R, Robjohns H L and Dunlop P J 1987 *Physica* **144A** 254-5
[23] Dunlop P J and Bignel C M 1992 *Ber. Bunsenges. Phys. Chem.* **96** 1847-8
[24] Ivakin B A, Suetin P E and Plesovski W P 1967 *J. Techn. Phys.* **37** 1913-4
[25] *Handbook of Chemistry and Physics* 2006-2007 87th edition (CRC Press Taylor and Francis)

IOP Conf. Series: Journal of Physics: Conf. Series **992** (2018) 012012 doi:10.1088/1742-6596/992/1/012012

Numerical investigation of the effect of net charge injection on the electric field deviation in a TE CO_2 laser

Nahid Jahanianl[1,4], Majid Aram[2], Nader Morshedian[3] and Ahmad Mehramiz[1]

[1]Department of Physics, Faculty of Science,
Imam Khomeini International University, Qazvin, Iran
[2]Photonic and Quantum Technologies Research School,
Nuclear Science and Technology Research Institute,
P.O. Box 14155-1339, Tehran, Iran
[3]Plasma and Fusion Research School,
Nuclear Science and Technology Research Institute,
P.O. Box 14399-51113, Tehran, Iran

E-mail: nahid.jahanian@gmail.com

Abstract. In this report, the distribution of and deviation in the electric field were investigated in the active medium of a TE CO_2 laser. The variation in the electric field is due to injection of net electron and proton charges as a plasma generator. The charged-particles beam density is assumed to be Gaussian. The electric potential and electric field distribution were simulated by solving Poisson's equation using the SOR numerical method. The minimum deviation of the electric field obtained was about 2.2% and 6% for the electrons and protons beams, respectively, for a charged-particles beam-density of 10^6 cm^{-3}. This result was obtained for a system geometry ensuring a mean-free-path of the particles beam of 15 mm. It was also found that the field deviation increases for a the mean-free-path smaller than that or larger than 25 mm. Moreover, the electric field deviation decreases when the electrons beam density exceeds 10^6 cm^{-3}.

1. Introduction

The TE CO_2 laser was introduced in 1968. Initially, the output energy of these lasers was in the order of a joule, but this laser's technology began developing very quickly and the energy reached kilojoules, especially for the purposes of nuclear fusion [1], as a result of the large research effort concentrated on increasing the laser energy in the last four decades. Producing a high CO_2 laser pulse energy necessitates a large active medium with as uniform as possible excitation of the CO_2 molecules [2]. Apart from the electrical discharge as the main process of creating plasma, excited molecules and, subsequently, population inversion, complementary techniques, such as pre-ionization, net charge injection [3] and special electrodes geometry [4], have been considered in order to produce a uniform glow discharge, thereby achieving as high as possible population inversion in the active medium. In the present work, we focused our attention on the charged-particles injection technique. The

[4] To whom any correspondence should be addressed.

consecutive injection of pulsed electron or proton beams can generate a pre-plasma medium facilitating the development of the main discharge [9]. Generally, the method of charge injection can be considered as pertinent to switching or triggering a discharge system. For instance, when an analog switch is turned on and off, a small amount of charge can be capacitively coupled (injected) from the digital control line to the analog signal path [10].

In the work reported, using the numerical method of successive over relaxation (SOR), we studied the electric field distribution and uniformity in the active laser medium when net charge is supplied by injecting a charged particles beam.

2. Physical conditions

The charged particles beam is assumed to be injected perpendicularly to the laser optical axis in the front side of the discharge electrode. The cross section of the electrode, and the active medium geometry and boundaries are shown in the figure 1. The electrode edges are rounded with a radius of curvature of about 2 mm and all four external sides are grounded.

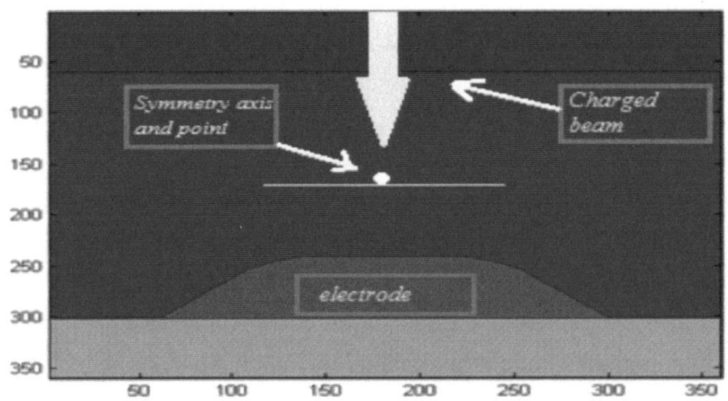

Figure 1. Schematic diagram of the system studied.

The distribution function of net charge beam injection is as follows;

$$\rho = en\,\boldsymbol{exp}(\frac{-x^2}{L^2}) \quad , \tag{1}$$

where, the ρ is the charge density, e is the electron charge, n is the charged-particles number density, L (mm) is the mean-free-path of the charges, or the distance where the charge density decays to 1/e of its initial value. If the value of L increases, then the space charge density in the active medium also increases. We solved numerically Poisson's equation (2) by using the SOR method for the space charge in the medium.

$$\nabla^2 V = -\frac{\rho}{\varepsilon_0} \quad , \tag{2}$$

where V is the potential and ε_0 is the permittivity of vacuum. The parameter ε_0 should be redefined for calculating the potential in a medium with different permittivity at the boundary conditions as shown schematically in figure 2. In the figure, h is the distance scale for the mesh construction. The solution of the potential equation at each point is based on the potential value obtained at that point in the previous stage, plus the average of the four potential values of the points that are in the vicinity of the point considered in the mesh construction.

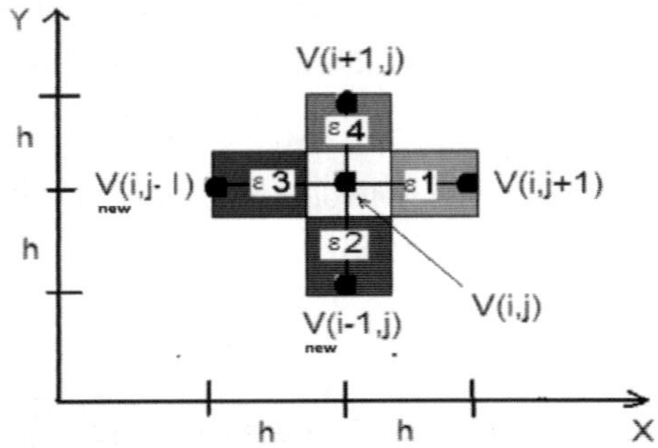

Figure 2. Potential of a point with four vicinity potentials and different permittivity at the boundaries.

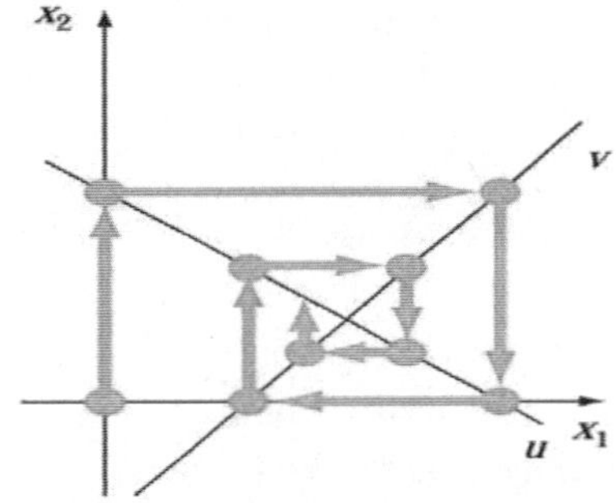

Figure 3. Path of the successive potential calculations for each point.

3. Numerical method

To simulate numerically the electric potential in the discharge medium, we chose to apply the SOR method. Generally, the memory of a personal computer may limit one in processing the large amount of data needed for mesh construction. On the other hand, if no limit is set on the processing time, the SOR method overcomes the memory restriction. Figure 3, illustrates the potential equation solution at each point as based on the potential value obtained according to relation (3) in the iteration process [6].

$$x_i^{(k)} = (1-w)x_i^{(k-1)} + \frac{w}{a_{ii}}\left\{ b_i - \sum_{j=1}^{i-1} a_{ij}x_j^{(k)} - \sum_{j=i+1}^{n} a_{ij}x_j^{(k-1)} \right\}, \tag{3}$$

where, $x^{(k)}$ is the response vector or scalar that refers to the potential quantity, $i = 1,2,\ldots n$, n being the number of iterations, w is the scaling factor (w = 0-2); if w = 1, the above expression is converted to the Gauss–Seidel equation. Further, a_{ij} is the matrix array of coefficients; b_i is the constant value of the equation and $x_j^{(k-1)}$ is the value of the potential obtained at the previous stage. Thus, the final value of the potential is derived by the iteration process provided that the solution converges by choosing a specific scaling factor (w) value. Rewriting relation (3) based on the SOR method, one arrives at relation (4) [6].

$$V_{new}(i,j) = (1-w)\times V(i,j) + \frac{w}{4}\left(\frac{\varepsilon_1\times V(i,j+1) + \varepsilon_3\times V_{new}(i,j-1) + \varepsilon_4\times V(i+1,j) + \varepsilon_2\times V_{new}(i-1,j)}{\varepsilon_1+\varepsilon_2+\varepsilon_3+\varepsilon_4}\right) + \frac{\rho h^2}{\varepsilon_1+\varepsilon_2+\varepsilon_3+\varepsilon_4}, \tag{4}$$

where, ε_1 and ε_2 are the permittivity at the boundary of the two media, $V(i\pm1, j\pm1)$ are the four potential values in the vicinity of the point considered, V_{new} $(i, j\text{-}1)$ is the potential value obtained at the previous stage, $\rho(i,j)$ is the density of the net charge injected and h is the distance scale of the mesh. The field deviation (FD) is defined as

$$FD = (\Delta E/E)\times 100; \qquad \Delta E = E - E_o, \tag{5}$$

IOP Conf. Series: Journal of Physics: Conf. Series **992** (2018) 012012 doi:10.1088/1742-6596/992/1/012012

where the E_o is a reference quantity defined as the electric field value at the central point, or the symmetry point. of the system, and ΔE is the difference between the electric field values at an arbitrary point in the mesh and the central point.

4. Results and discussion

The experimental conditions and system parameters were assumed as quoted by Maxwell Company (USA). Thus, the electrode potential is +75 kV and, as mentioned above, all external boundaries are at zero potential (grounded). The electrode voltage value is optimized in order to obtain a realistic reduced electric field (E/N) value, which for a CO_2 laser is about 4×10^{-16} V cm^2 [1]. The electron energy distribution has then a maximum at 1-2 eV when the mean-free-path is varied. The field deviation is a criterion for the electric field uniformity needed to achieve a proper glow discharge in the active medium, i.e., the generation is prevented of sparks and/or arcs in discharge. Table 1 summarizes the dielectric constant of different gas mixtures and the related field deviation in percents [8].

Typical simulated 3D images of the potential and electric field distributions are shown in figures 4 and 5 for Laplace equation, i.e. without any space charge.

Figures 6 and 7 illustrate the potential and electric field distributions obtained by solving the Poisson's equation when an electron beam is injected. The expected basic result becomes apparent, namely, the field deviation is reduced due to the injection of a net electron charge (compare with figures 4 and 5). The *FD* (%) decays rapidly vs. the charged particles density with a threshold of 10^7 cm^{-3} for electrons.

Table 1. Permittivity and field deviation values for different gas mixtures.

CO_2:N_2:He	K	$\varepsilon_{mix}(\times 10^{-18})\frac{C}{N.mm^2}$	*FD* (%)
1:1:1	1.0007	8.8565	0.9970
1:1:8	1.0006	8.8557	0.9970
1:1:3	1.0007	8.8560	0.9970

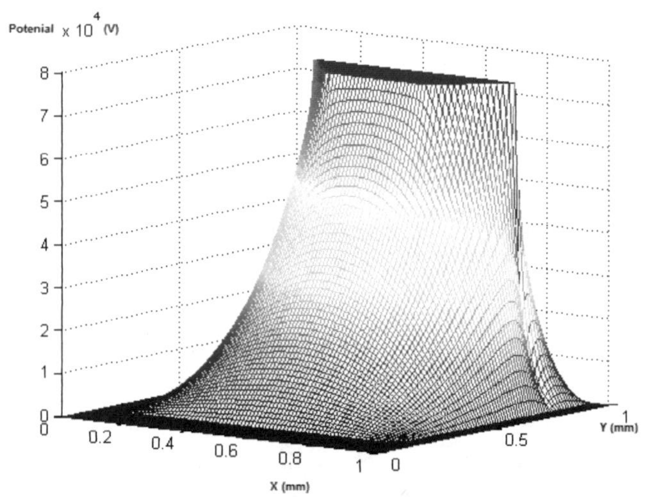

Figure 4. 3D image of the potential distribution without beam injection (electrode potential +75 kV).

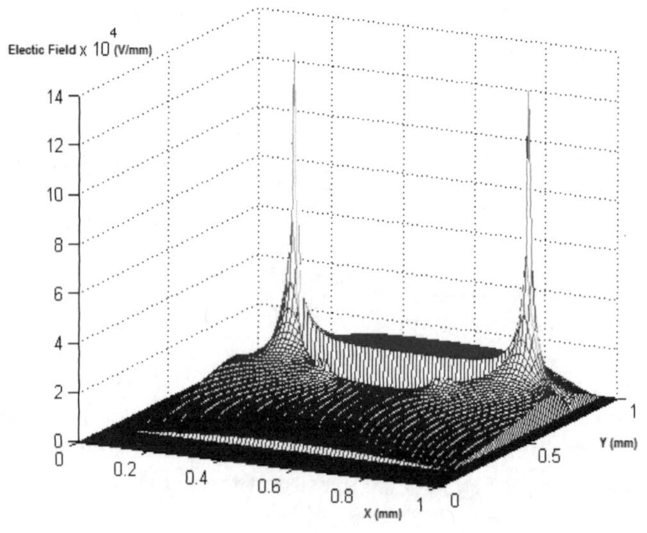

Figure 5. 3D image of the electric field distribution without beam injection (electric potential +75 kV).

However, for a beam of protons for a mean-free-path $L = 15$ mm this behavior is inverted, as shown in figure 8. Similarly, in the case of electron beam injection and $L = 25$ mm, *FD* starts decreasing, the threshold then being 10^6 cm^{-3} (figure 9). Moreover, *FD* as a function of the mean-free-path starts increasing at $L = 25$ mm for charged particles density of 10^8 cm^{-3}, as shown in figure 10.

Comparing the cases of electron or proton beam injection, one can be clearly see that the minimum *FD* is obtained using electron beam injection. There is a local minimum in *FD* vs. charged particles density using aproton beam, as shown in figures 8 and 9. In addition, for a proton beam there is generally an incremental *FD* dependence on *L*, as seen in figure 10. In contrast, the minimum of *FD* for an electron beam occurs at larger values of the charged particles density. This situation can be considered as a competition of the two parameters, i) the external electric field initiating the discharge and, ii) the electric field induced by the space charge due to the injection of a charged particles beam.

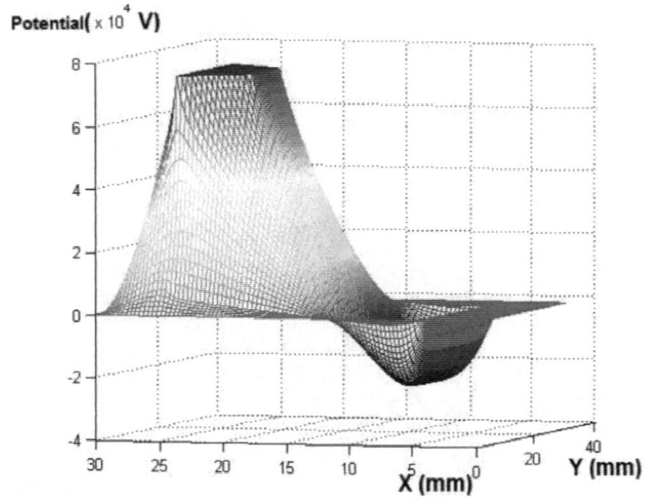

Figure 6. 3D image of the potential distribution of beam injection (electric potential +75 kV).

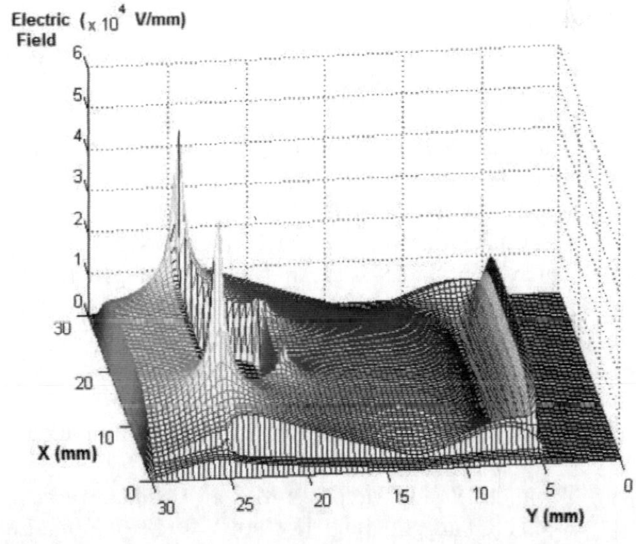

Figure 7. 3D image of the electric field distribution with beam injection (electric potential +75 kV).

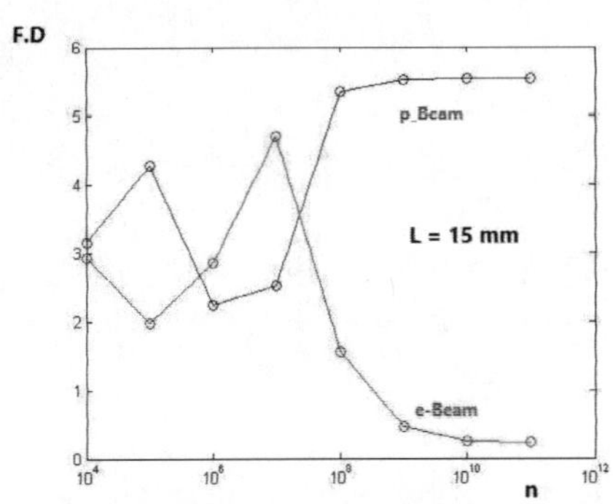

Figure 8. *FD* versus charged particles density for mean-free-path 15 mm.

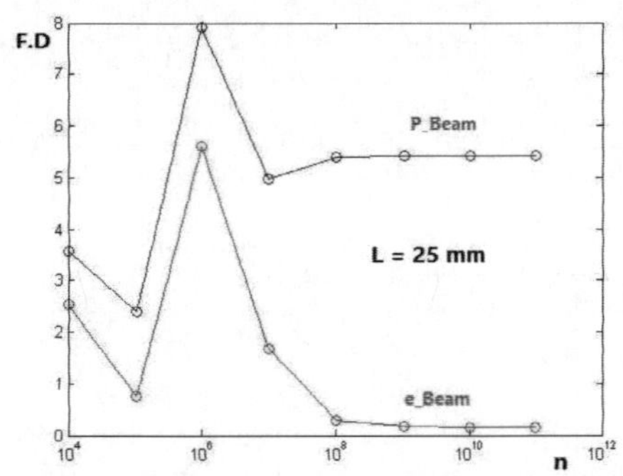

Figure 9. *FD* versus charged particles density for mean-free-path 25 mm.

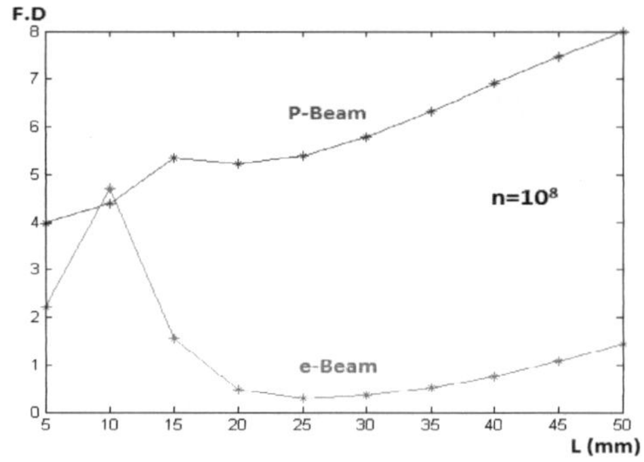

Figure 10. *FD* versus mean-free-path for charged beam density 10^8 cm^{-3}.

5. Conclusions

Pre-plasma generation is important for creating a uniform discharge medium and preventing arcs and/or sparks in the discharge medium. One of the methods of producing uniform conditions is injection of a charged beam in order to obtain a glow discharge and uniform pumping for optimal population inversion. We concluded that there are two basic parameters i.e., space charge density created by the particles beam and mean-free-path of the beam injected, both of which should be optimized in view of minimizing the electric field deviation in the active medium. We obtained the optimized parameters of the injected charged particles (electrons or protons) for a TE CO_2 laser by solving Poisson' equation using the SOR method. The results show that the *FD* is considerably less when an electron beam, rather than a proton beam.

Acknowledgements
The authors wish to thank the Technical Assistance Group and Heads of the two Research Schools.

References

[1] Witteman W H 1987 *The CO_2 Lasers* ch 6 (Springer-Verlag)
[2] Sourabh and Tripathi S B L 2012 *Calculation of Optical Gain as a Function of Pump-Power* (Research Scholar Dept of Physics CMJ University)
[3] Mesyats G, Osipov V V and Tarasenko V F 1995 *Pulsed Gas Lasers* (Nauka Moskow))
[4] Aghayari E, Sade F and Aram M 2015 Investigation of the electrodes cross sections for the TEA CO_2 laser *Proc. 3-th Engin. Conf. (KAM)* (2015 Iran)
[5] http://liberica.org
[6] Nagel J R 2012 *Solving the Generalized Poisson Equation Using the Finite-Difference Method (FDM)* (University of Utah)
[7] http://eng.utah.edu/~cfurse/NASA/User%20Guide/01%20Zo%20Catalog/Description%20of%20FDFD%20Method.pdf
[8] Schmidt J W and Moldover M R 2003 *Int. J. Thermophys.* **24**/2 375-403
[9] Sim H S, Park S, Kim T H, Choi Y K, Lee J S and Lee S H 2010 *Mater. Trans. JIM* **51** 1156
[10] Holloway M D and Nwaoha Ch 2012 *Dictionary of Industrial Terms* ISBN: 978-1-118-34457-6

Electron beam deflection control system of a welding and surface modification installation

E Koleva[1,2,3], V Dzharov[1], V Gerasimov[2], K Tsvetkov[2] and G Mladenov[1]

[1]E. Djakov Institute of Electronics, Bulgarian Academy of Sciences, 72 Tzarigradsko Chaussee, 1784 Sofia, Bulgaria
[2]University of Chemical Technology and Metallurgy, 8 St. Kliment Ohridski Blvd., 1756 Sofia, Bulgaria

E-mail: eligeorg@abv.bg

Abstract. In the present work, we examined the patterns of the electron beam motion when controlling the transverse with respect to the axis of the beam homogeneous magnetic field created by the coils of the deflection system the electron gun. During electron beam processes, the beam motion is determined the process type (welding, surface modification, etc.), the technological mode, the design dimensions of the electron gun and the shape of the processed samples. The electron beam motion is defined by the cumulative action of two cosine-like control signals generated by a functional generator. The signal control is related to changing the amplitudes, frequencies and phases (phase differences) of the generated voltages. We realized the motion control by applying a graphical user interface developed by us and an Arduino Uno programmable microcontroller. The signals generated were calibrated using experimental data from the available functional generator. The free and precise motion on arbitrary trajectories determines the possible applications of an electron beam process to carrying out various scientific research tasks in material processing.

1. Introduction

Electron beams find numerous applications in research and industry as a concentrated-energy heat source. The electron-beam deep-penetration welding of materials in vacuum is the most widely used method in non-conventional technologies of joining machine parts [1-3]. The reasons for this are: formation of deep and narrow welds with good physical-mechanical properties with minimal structural changes and heat deformations of the welded parts, possibility for high-speed welding in automated technological processes, elimination in many cases of the need for subsequent mechanical and heat treatment, possibility of welding near thermally-unstable structural elements, lack of necessity of special welding electrodes, protective gases or fluxes, improved use of materials, energy saving, especially when thick-wall structures are processed, etc.

In the present work, we developed a graphical user interface to generate control signals for the electron beam deflection system of an electron beam welding, surface modification and evaporation installation. The electron beam is moved in different patterns as determined by the magnetic field formed by the deflection system, depending on the process performed and the shape and location of

[3] To whom any correspondence should be addressed.

the samples to be processed. The software application developed by us provides motion control along different trajectories at different velocities.

2. Electron beam deflection system

Figure 1 presents a schematic diagram of an electron beam gun, where the electron beam is generated, accelerated, directed, focused and deflected.

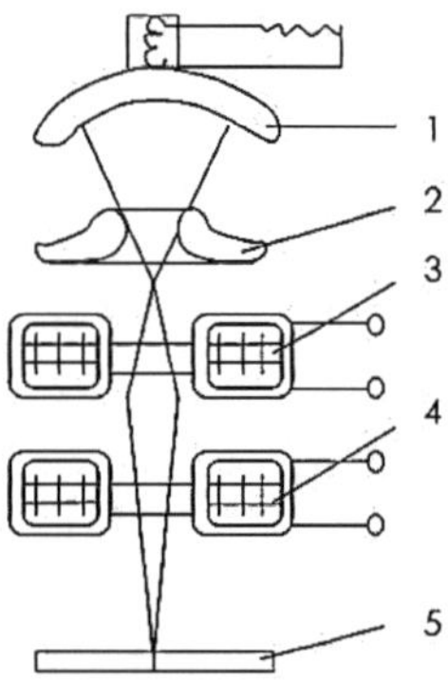

Figure 1. Schematic presentation of an electron beam gun: 1 - cathode, 2 - anode, 3 - focusing system, 4 - deflection system; 5 - sample.

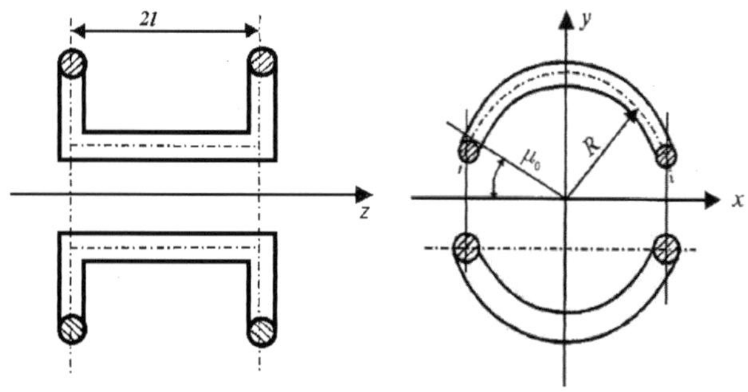

Figure 2. Schematic presentation of a saddle-shaped coil along axes a) *z* and b) *x*.

The deviation of the electron beam in two mutually perpendicular directions at angles of about 7-10 degrees is achieved by means of two pairs of deflecting coils (figure 2 [1]). It is assumed that each magnetic field is transversal to the beam axis and uniform within the beam deflection. The coils turns are usually wound in the form of a saddle. In order to obtain a uniform magnetic field, the main portion of the turns is wound at the ends of the coils.

The stationary model for the deflection distance is a particular solution of the dynamic equations of the electrons when the electric and magnetic field distributions are known. The relationship between the deflection distance along the axes Ox and Oy depends on the deflection coil currents i_{xdefl} and i_{ydefl} and at a constant value of the acceleration voltage U_a can be calculated by [4-6]:

$$\begin{cases} x_{defl} \approx \sqrt{\dfrac{e_o}{2m_e}}\dfrac{k_b \mu_o d_{defl} l_b}{a\sqrt{U_{ac}}} n i_{xdefl} = 3.7266\dfrac{k_b d_{defl} l_b}{a\sqrt{U_{ac}}} n i_{xdefl} \\ y_{defl} \approx \sqrt{\dfrac{e_o}{2m_e}}\dfrac{k_b \mu_o d_{defl} l_b}{a\sqrt{U_{ac}}} n i_{ydefl} = 3.7266\dfrac{k_b d_{defl} l_b}{a\sqrt{U_{ac}}} n i_{ydefl} \end{cases},$$

where e_0 is the elementary charge, m_e is the electrons stationary mass, μ_o is the vacuum permittivity, k_b is the deflection coil configuration coefficient, d_{defl} is the distance between the deflection coil and the processed material surface, l_b is the coil length, n is the number of turns, a is the coil dispersion. The dynamic models of the deflection system are [4-6]:

$$\begin{cases} L_s \dfrac{di_{xdefl}}{dt} + R_s i_{xdefl}(t) = u_{xdefl}(t) \\ L_s \dfrac{di_{ydefl}}{dt} + R_s i_{ydefl}(t) = u_{ydefl}(t) \end{cases},$$

where L_s is the coil equivalent inductivity and R_s is its equivalent resistivity.

The use of a magnetic conductor reduces about twice the required amperage, but it can cause field distortion, which is unacceptable. At high frequencies, the magnet may be a ferrite or not present at all and, in addition, the inductance and the capacity of the coils begin affecting the non-linearity of the deflection. Therefore, the required voltage (and power) increases and it is desirable to reduce the number of coil turns (n). It turns out that the product of the deflection angle and the frequency, at a constant power of the deflection source, is a good characteristic of the quality of the deflection system.

Typically, the magnetic deflection systems are outside the vacuum volume and the magnetic coils are separated from the vacuum space by a non-magnetic material – ceramics or glass, since the magnetic field frequencies are high. This prevents the generation of Foucault currents in the walls of the deflection systems' separation tube or cone, which can reduce the penetration of the magnetic field and can become an additional source of nonlinearities.

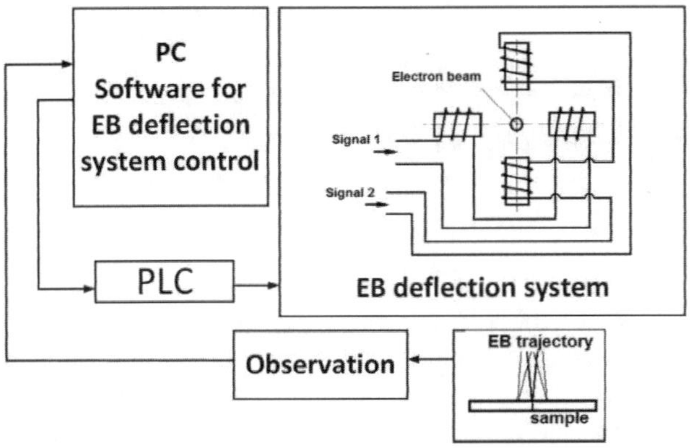

Figure 3. Electron beam deflection system control.

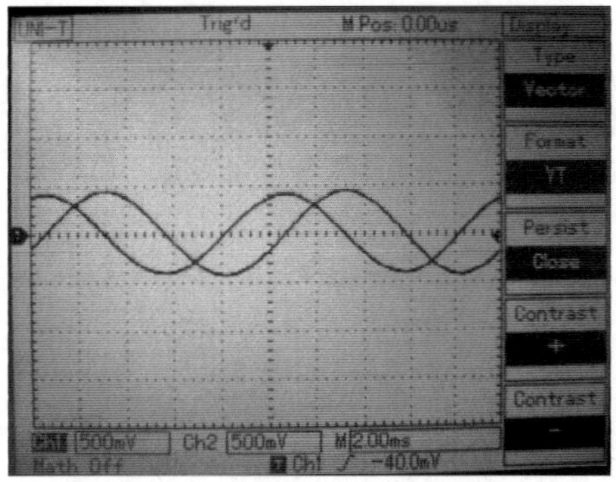

Figure 4. Deflection system control signals: amplitudes $A = B = 450$ mV, period $T = 10$ ms, frequency $\omega = 628.32$ Hz.

The deflection control system developed is presented schematically in figure 3. The available functional generator and amplifier are replaced by a computer and an Arduino Uno programmable microcontroller. The electron beam motion is controlled by varying the parameters of the two voltage signals (figure 4) – amplitudes, frequencies and phases (phase differences). Currently, verification and correction of the exact beam position is performed by observation; the implementation is considered of an image capturing system [5-7] for automated correction of the beam deflection.

3. Electron beam deflection control software and motion patterns

The electron beam motion is a result of the cumulative action of two sinusoidal control signals (voltage) fed to the deflection coils:

- Signal 1: $X(t) = A(t)\sin[\omega(t)t + \varphi_x(t)]$,
- Signal 2: $Y(t) = B(t)\sin[\omega(t)t + \varphi_y(t)]$,

where A is the amplitude of signal 1, B is the amplitude of the second signal, ω is the frequency of the oscillations, and the phase difference $\delta(t) = \varphi_y(t) - \varphi_x(t)$ is defined by the phases of the two signals (figure 4). The amplitudes, frequencies and phase differences of the two signals may be constant or may vary over time in a particular pattern. The period T (the time needed for one full turn of the electron beam) at a constant frequency is calculated as $T = 2\pi/\omega$, and if the working time is T_0, the number of turns is T_0/T. The characteristics of the two signals affect the magnetic field generated and, consequently, the electron beam path. The electron beam deflection is limited by the electron gun design characteristics, by the sample surface treated and by the type of technological process, together with the physical processes taking place.

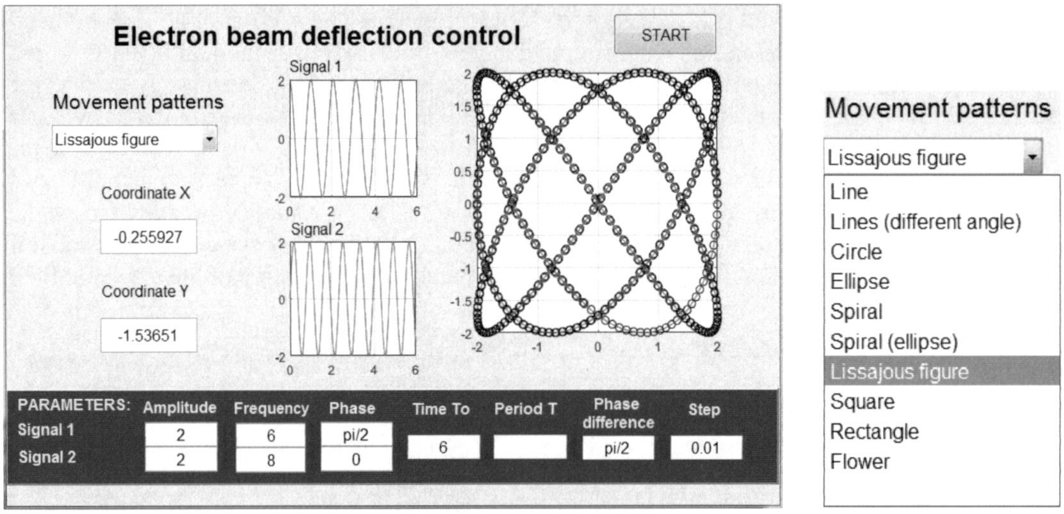

Figure 5. Electron beam deflection software.

Figure 5 presents the electron beam deflection control software developed. It allows one to select several motion patterns: line; several lines crossing at different angles, circle, ellipse spiral, spiral (ellipse), Lissajous figures, square rectangle and flower. It can be updated with other patterns. The parameters of the two signals can be set by the user, depending on the concrete electron beam process performed. The current coordinates of the overall effect of the two signals can be observed as values and a graph. The example shown represents a Lissajous figure with a frequency ratio of 6/8 and a phase difference $\delta = \pi/2$.

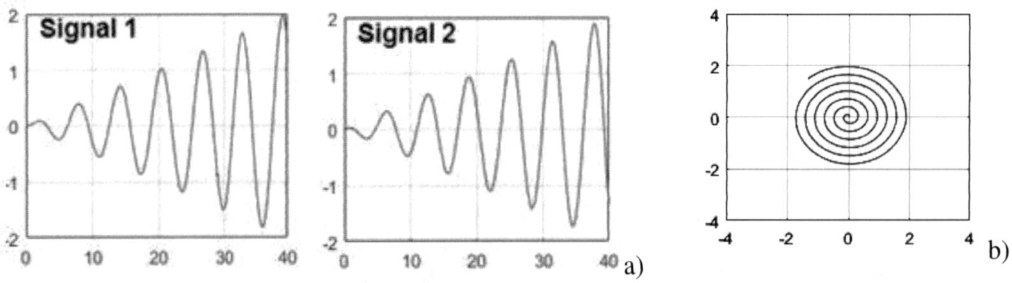

Figure 6. A spiral motion pattern –: a) Signal 1 and Signal 2 voltage as functions of time, b) voltage of Signal 1 vs. voltage of Signal 2.

Figure 6 presents another example of a motion pattern – spiral electron beam trajectory. The control signals parameters are: amplitude initial values $A_0 = 0.05$, $B_0 = 0.05$, increasing with at a step $s = 0.05$; frequency $\omega = 1$ Hz, working time $T_0 = 40$ s, period $T = 6.28$ s and phase difference $\delta = \pi/2$.

4. Conclusions

The electron beam welding is a complex process with many processing variables, which requires integration of interfaces for control of the processes on different levels through hierarchical equipment models, functional data and operation control models.

We developed software for electron beam deflection system control, implementing models for the characteristics of two control voltage signals — their amplitudes, frequencies, phases and phase differences. This is only one of the stages to realizing complete control of the electron beam settings, including its formation, characterization, focusing and motion. The beam pattern used in each case depends on the process (welding, surface modification, evaporation etc.) and the shape of the processed samples. The graphical user interface developed can be integrated with more complex shape trajectories in different modes, electron beam spot tracking, electron beam focusing system control etc. The operation of the deflection system must also be coordinated with the operation of the manipulators in the vacuum chamber, with which the work-pieces are moved.

Acknowledgements
This work was supported by the Bulgarian Academy of Sciences under contract DFNP-17-53 and by the Bulgarian National Science Fund under contract DNTS/Slovakia 01/1/2016 (project NTS/Slovakia 01/25-13).

References
[1] Mladenov 2009 *Electron and ion technologies* (Prof. Marin Drinov, Sofia, Bulgaria)
[2] Koleva E and Mladenov G 2011 *Practical Aspects and Applications of Electron Beam Irradiation* 95-133 ISBN: 978-81-7895-541-4
[3] Volker A, Uwe C, Dobeneck D, Krüssel T and Löwer T 2011 *Electron Beam Welding, The fundamentals of a fascinating technology* (Pro-beam AG & Co. KgaA)
[4] Chvertko A, Nazarenko O, Svyiatskiy A and Nekrasov A 1973 *Equipment for electron beam welding* (Naukova Dumka, Kiev)
[5] Oltean S, Grif, H and Duka A 2007 *Proc. Int. Conf. Interdisciplinarity in Eng. Sci.* (INTER-ENG) 1-6
[6] Oltean S and Abrudean M 2008 *J. Control Eng. and Appl. Informatics* **10**/1 40-8
[7] Oltean S and Dulau M 2011 *Proc. Int. Conf. Interdisciplinarity in Eng. Sci.* (INTER-ENG) 20
[8] Koleva E, Mladenov G, Vuchkov I, Velev K, Lamond B and Petrova D 2012. *Research and development of new materials on the base of recycling of reactive and refractory metals scrap through electron beam method* ed Vutova K (IE-BAS, Sofia, Bulgaria) 30-7
[9] Trushnikov D, Koleva E, Mladenov G and Belenkiy V 2013 *J. Material Proc. Technol.* **213**/9 1623– 34
[10] Koleva E and Dzharov V 2016 *Science, Engineering & Education* **1**/1

IOP Conf. Series: Journal of Physics: Conf. Series **992** (2018) 012014 doi:10.1088/1742-6596/992/1/012014

Integrated control system for electron beam processes

L Koleva[1], E Koleva [1,2,3], I Batchkova[1] and G Mladenov[2]

[1]University of Chemical Technology and Metallurgy,
8 Kliment Ohridski Blvd., 1756 Sofia, Bulgaria
[2]E. Djakov Institute of Electronics, Bulgarian Academy of Sciences,
72 Tsarigradsko Chaussee, 1784 Sofia, Bulgaria

E-mail: eligeorg@abv.bg

Abstract. The ISO/IEC 62264 standard is widely used for integration of the business systems of a manufacturer with the corresponding manufacturing control systems based on hierarchical equipment models, functional data and manufacturing operations activity models. In order to achieve the integration of control systems, formal object communication models must be developed, together with manufacturing operations activity models, which coordinate the integration between different levels of control. In this article, the development of integrated control system for electron beam welding process is presented as part of a fully integrated control system of an electron beam plant, including also other additional processes: surface modification, electron beam evaporation, selective melting and electron beam diagnostics.

1. Introduction

The application of the new electron-beam welding (EBW) technologies requires that the EBW equipment be formed into a complex plant, containing highly stabilized power sources and electronic blocks, a reliable and efficient vacuum system and a technological chamber with a high-precision 3D manipulator. They thus become truly software-controlled programmable systems with high efficiency and excellent reproducibility [1]. The highly sophisticated automated processes, including the vacuum pump system and pressure control, the cooling system control, the manipulator control, the processor-based high voltage and emission current control, the electron-beam movement control and its characterization and the PC-based automatic beam power distribution, need to be integrated based on using Manufacturing Operations Management (MOM) systems. The best practices show that MOM systems, when combined with Decision Support Systems (DSSs), can provide the organization with the necessary advanced process control solutions [2-4]. A way to integrate the DSSs (combined with Data Collection and Acquisition, Information Analysis and System Management Agents) with the control and monitoring systems during the operation of EBW is to develop and use integrated control system (ICS) based on MOM.

The ISO/IEC 62264 standard [5-9] supports the development and use of (MOM) systems, providing standard models and terminology for describing the activity models of MOM and interfaces between the business systems of an enterprise and its manufacturing operations and control systems, in order to integrate them. According to the ISO/IEC 62264, the activities in manufacturing operations management are divided into four main areas: production operation management, maintenance

[3] To whom any correspondence should be addressed.

IOP Conf. Series: Journal of Physics: Conf. Series **992** (2018) 012014 doi:10.1088/1742-6596/992/1/012014

operations management, quality operations management and inventory operations management. For all areas, a common activity model of eight activities (definition, planning, dispatching, resource provision, implementation, data collection, tracking and analysis) is suggested. To implement a cognitive system, it is necessary to apply an ontological approach for communication between different components and problem solving systems, which can be considered as part of an intelligent information system that integrates them. Ontologies can be used at all stages of the development and functioning of the cognitive system. UML environment will be used to build the cognitive management framework, as well as conceptual models and specifications.

2. Object models for manufacturing operation management

ISO/IEC 62264 [5-9] is a multi-part standard that defines the interfaces and provides standard models and terminology for enterprise (business) and manufacturing-control activities, functions and systems (figure 1 and figure 2). It defines a manufacturing hierarchical model and describes the manufacturing operations management domain and its activities, the interface content and the associated transactions within and between different levels (figure 2).

The basic information structural elements for enterprise application integration are product lifecycle management (PLM), enterprise resource planning (ERP), supply chain management (SCM), customer relationship management (CRM). The manufacturing operations are divided into four groups: production operations, maintenance operations, quality operations and inventory operations. Each operation is presented with an activity model, defined from a set of activities that are required for manufacturing: definition, scheduling dispatching, resource management, execution, data collection, performance analysis and tracking.

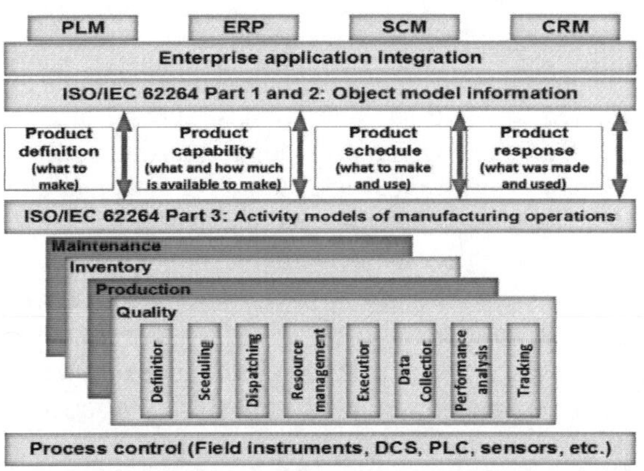

Figure 1. General structure of IEC 62264 standard [10].

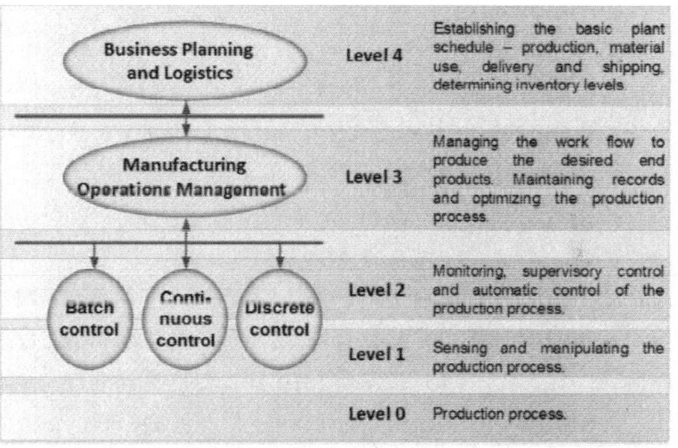

Figure 2. Functional hierarchy according to IEC 62264-3 [7].

3. Engineering support system architecture for EBW plant

From a system engineering point of view, the electron-beam plant is a complex system of processes such as electron-beam welding process, surface modification, electron-beam evaporation, selective melting and electron-beam diagnostics, which can be engineered to accomplish specific business and technological objectives.

Figure 3 shows the basic components of an Engineering Support System (ESS) architecture, presented as an UML "use-case" diagram (basic type of UML diagrams) in order to represent knowledge in a form that is either human or computer readable. It is integrated to all levels from the functional hierarchy of the enterprise (figure 4).

The basic use cases of the ESS for EBW plant are: Data Collection and Acquisition (DCA), Information Analysis (IA), Decision Support System (DSS) and System Management (SM).

The DCA subsystem has a direct connection to the EBW plant through the available sensors, which present the actual state of the plant, follow the system behavior and support the data acquisition and archiving.

The IA subsystem includes a component for analytical modeling of the different processes – providing the DSS user with calculated data (weld-geometry parameters, weld defects, temperature distribution, etc.); a component for statistical modeling, quality management and data analysis; components for information modeling; and components supporting optimization processes.

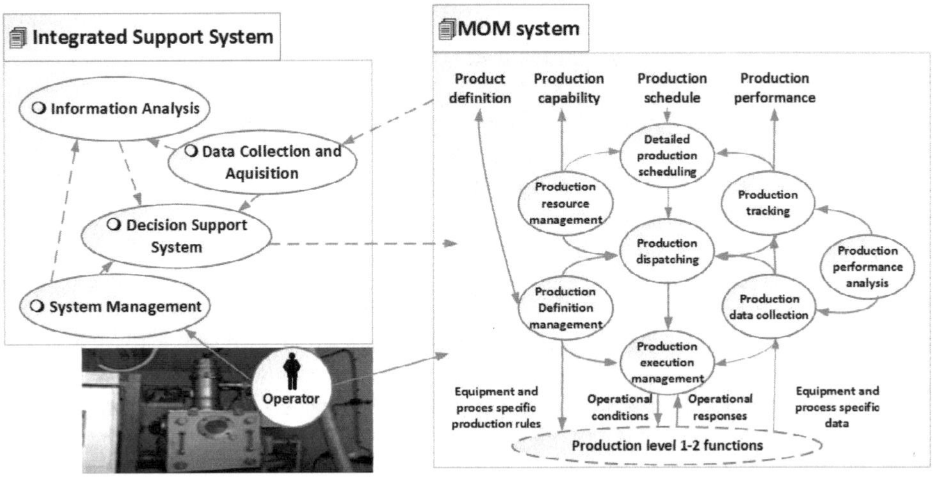

Figure 3. The architecture of ESS for EBW plant as "use-case" diagram.

The SM component involves the definition of requirements and parametric constraints. The DSS implements the results from the DCA, IA and SM components in order to provide the decision makers, the operators and the managers with key information that enables them to make more efficient and consistent decisions.

4. EBW plant domain model

The development of a MOM system is based on the application of an object-oriented approach and the Unified Modeling Language (UML) [11], and uses the ISO/IEC 62264 conforming meta-model, consisting of classes hierarchically composed in three categories of resources ("Personnel", "Equipment", and "Material"), "Process Segment" and five main interface models ("Product Definition", "Production Capability", "Production Schedule", "Process Segment Capability" and "Production Performance"). The classes defined in the domain model extend in detail the main meta-classes in "Material Model" and "Equipment Model" by introducing domain information. The domain model is integrated with the meta-model in such a way that all properties and relations, defined in the meta-model, are inherited in the domain model.

Figure 4 represents the electron-beam plant domain equipment model, which is based on the hierarchical equipment meta-model, suggested in ISO/IEC 62264. The EB_Plant represent the "Site" level of the standard hierarchical model. It contains four specific areas of processes: electron-beam welding (EBW), surface modification, electron-beam evaporation, selective melting and diagnostic area. In this paper, the EBW area is only covered. The "Equipment Model" of EBW contains four subclasses: "Manipulator", "Vacuum System", "Cooling System" and "Electron-Beam Gun". The "Manipulator" class has two working regimes (subclasses), depending on the type of manipulator used: "Rotary"- with velocity and rotation angle properties; and a "3D" manipulator, which has x, y and z coordinates. The "Vacuum System" contains three types of pumps as "Actuator_VS" subclasses: two rotary pumps (RP_VS), a turbo-molecular pump (TMP_VS) and a diffusion pump (DP_VS). All of them have one property of Boolean type, defining if the pump is on or off. Other subclasses of the

"Actuator" class in the "Vacuum System" are: an electromagnetic corner valve (EMC_VS) and an electro-pneumatic valve (EP_VS). A thermal binary sensor is placed in the "Vacuum Chamber" for pressure measurement. The "Cooling System" is another unit of the EBW area. It consists of diffusion pump cooling (DP_CS), turbo-molecular pump cooling (TMP_CS) and electron-beam gun cooling (EBG_CS) subclasses. Each of them has one attribute of Boolean type that sends the information on the cooling being on or off. The last unit of the "Equipment Model" is the "Electron-Beam gun" class. This class has his own properties: acceleration voltage, electron-beam current and cathode. Other electron beam properties are: electron-beam power, welding velocity, distance between the main surface of the magnetic lens of the electron gun and the beam-focusing plane, the distance between the main surface of the magnetic lens of the electron gun and the sample surface. The subclass "Deflection System" defines the movement of the electron-beam gun along the welded samples and has the following properties: amplitude, frequency and phase of Signal 1 and Signal 2. The second subclass is the "Focusing System" equipment with focusing-current property, which determines the position of the focus of the electron beam. The last subclass is the Sensor_EBG, based on gas discharge, which has one property for the level of the vacuum pressure in the zone of the electron-beam gun.

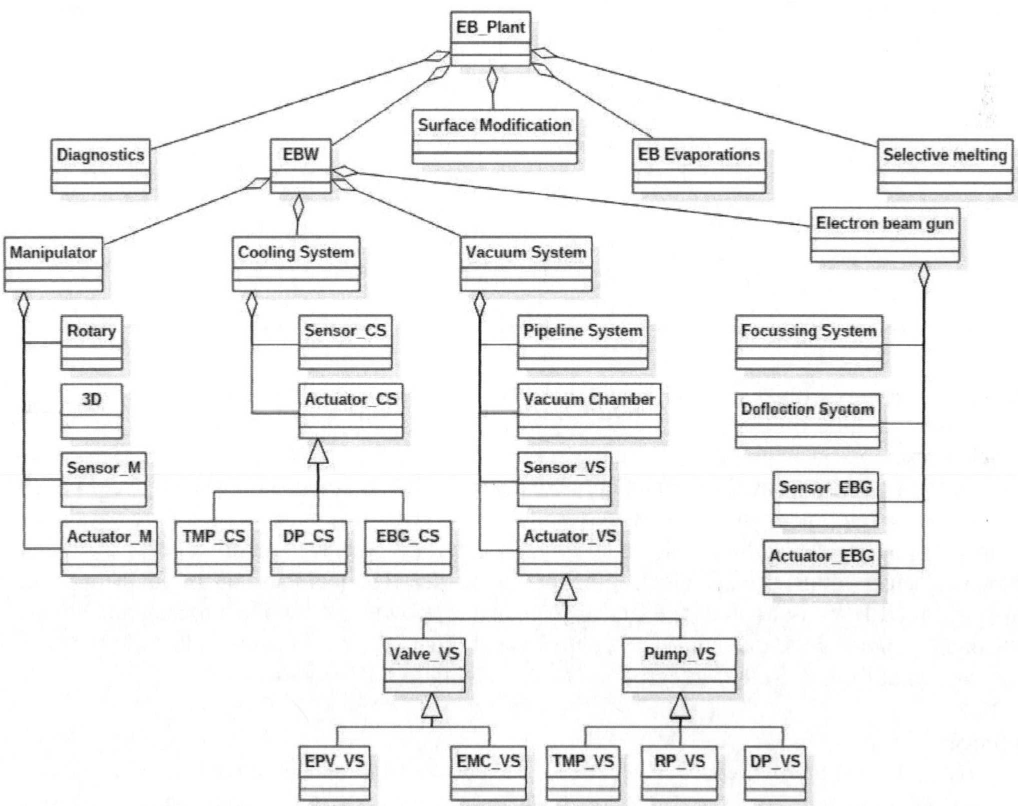

Figure 4. UML based EBW domain equipment model.

The "Material Model" contains the material characteristics of the samples that are welded by the EBW process. The welded joints can be from one type of material or of two different (dissimilar) materials – metals or alloys. Figure 5 shows the "Material model" as part of the meta-ontology, based on the ISO/IEC 62264 standard, using "Protégé". According to the ISO/IEC 62264 standard, the term "material" is a pooled concept for both raw materials, and intermediate and finished products. In general, the different material types and their characteristics are defined by the "Material Class" and "Material Class Property" classes. These can be, for example, welded parts dimensions and the requirements for zero defects and minimum spiking area in the weld root: weld depth, mean weld width, weld cross-section (area, form) and defects, etc. The specific material and its properties are given through the "Material Definition" and "Material Definition Property" classes. There, the specific raw materials are defined, as for example, the relative strengths of the steel grades, and the end products they produce, as well as their characteristics ,such as width, thickness, chemical composition, etc. The "Material Lot" and "Material Sublot" classes are introduced to describe the grouping of materials for scheduling purposes. The "QA Test Result" class includes quality analysis information for the joints formed by the EBW.

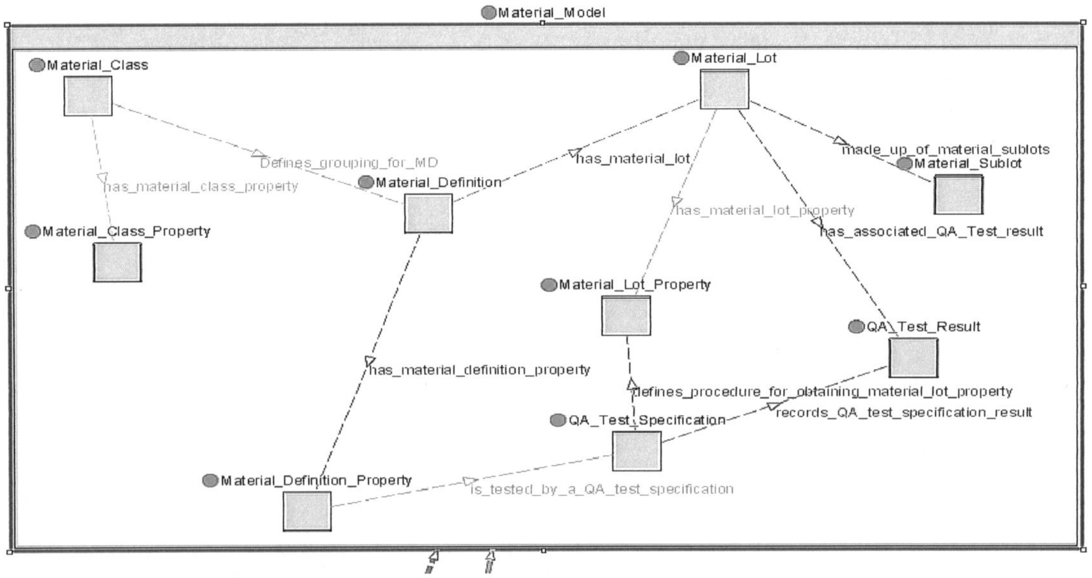

Figure 5. The "Material Model" as ontology in "Protege".

5. Conclusions

The development is presented of an integrated control system for electron-beam welding process based on the development of a MOM system, as part of a fully-integrated control system of an electron-beam plant, including also additional processes: surface modification, electron-beam evaporation, selective melting and electron-beam diagnostics. The main aim of the MOM for electron-beam welding (EBW) is to integrate and organize the knowledge for the process and to use this knowledge to improve the modeling and control capabilities, their efficiency, adaptability, flexibility and re-configurability, as well as integration of other technological processes.

References

[1] Koleva E G and Mladenov G M 2011 Practical aspects and applications of electron beam irradiation *Experience on electron beam welding* eds M R Nemtanu and M Brasoveanu; (Transworld Research Network) chapter 4 95-135

[2] Sheng-Tun L, Huang-Chin H and I-Wei S 2003 *Proc. Int. Conf. An ontology based knowledge management system for the metal industry* (Budapest 20-24 May 2003)

[3] Bowen J P and Hinchey M G 2005 Ten commandments revisited: a ten-year perspective on the industrial application of formal methods *Proc. Int. Conf. Formal Methods for Industrial Critical Systems FMICS'05* (Lisbon 5-6 Sept. 2005)

[4] Koleva E, Batchkova I and Velev K 2008 Engineering support system for modeling and control of the process EBMR of metals *Proc. Int. Conf. EBEAM 2008* (electronic) (Reno USA 26-28 Oct. 2008)

[5] *Int. Standard IEC 62264-1* 2013 Enterprise-control system integration Part 1: Models and terminology (Int. Electrotechnical Comm. *(IEC)* Press)

[6] *Int. Standard IEC 62264-2* 2013 Enterprise-control system integration Part 2: Object and attributes for enterprise-control system integration (Int. Electrotechnical Commission *(IEC)* Press)

[7] *Int. Standard IEC 62264-3* 2013 Enterprise-control system integration Part 3: Activity models of manufacturing operation management (Int. Electrotechnical Comm. *(IEC)* Press)

[8] *Int. Standard IEC 62264-4* 2015 Enterprise-control system integration Part 4: Object model attributes for manufacturing operation management (Int. Electrotechnical Comm. *(IEC)* Press)

[9] *Int. Standard IEC 62264-5* 2016 Enterprise-control system integration Part 5: Business to manufacturing transactions (Int. Electrotechnical Comm. *(IEC)* Press)

[10] Clark D 2010 *Invensys presentation standard ISA S95* Invensys operations management

[11] Fowler Martin 2003 *UML Distilled: A Brief Guide to the Standard Object Modeling Language (3rd Edition)* (Addison-Wesley Professional)

Empirical modeling of high-intensity electron beam interaction with materials

E Koleva[1,2,3], Ts Tsonevska[1] and G Mladenov[1]

[1]E. Djakov Institute of Electronics, Bulgarian Academy of Sciences, 72 Tzarigradsko Chaussee, 1784 Sofia, Bulgaria
[2]University of Chemical Technology and Metallurgy, 8 St. Kliment Ohridski Blvd., 1756 Sofia, Bulgaria

E-mail: eligeorg@abv.bg

Abstract. The paper proposes an empirical modeling approach to the prediction followed by optimization of the exact shape of the cross-section of a welded seam, as obtained by electron beam welding. The approach takes into account the electron beam welding process parameters, namely, electron beam power, welding speed, and distances from the magnetic lens of the electron gun to the focus position of the beam and to the surface of the samples treated. The results are verified by comparison with experimental results for type 1H18NT stainless steel samples. The ranges considered of the beam power and the welding speed are 4.2 – 8.4 kW and 3.333 – 13.333 mm/s, respectively.

1. Introduction

Electron beam welding (EBW) is a highly efficient and precise welding method increasingly used in manufacturing and of growing importance in different industrial sectors, such as the automobile and aerospace industries. The advantages of the method have to do with the possibility of obtaining welds with good physical-mechanical properties with minimal structural changes and thermal deformations of the individual welded parts, especially in cases of welding expensive structures and materials with different physical properties, reactive and refractory metals, or in precision welding of thin parts and micro-welding. In order to control the dimensions and shape of the welded seams cross-sections, as well as the heat-affected areas, adequate statistical and heat models of electron beam welding are required.

The EBW process depends on a large number of parameters in what concerns the material properties and the processing parameters of the electron beam installation. This is why one uses heat models to calculate the quasi-steady-state temperature distributions and the shapes of the melted zones as an approximation when simulating EB welding or melting and other electron beam processes [1-9]. They are based on different assumptions for the heat source – point, linear, exponential, combined etc. [6]. The solutions of the quasi-steady-state thermal physical models yield approximations for the shape of the temperature distributions taking into account some process and material parameters [6-9]. Due to the assumptions made in the physical models, the exact evaluation of the experimental results is difficult. It has been demonstrated that based on statistical analysis of the experimental and/or

[3] To whom any correspondence should be addressed.

IOP Conf. Series: Journal of Physics: Conf. Series **992** (2018) 012015 doi:10.1088/1742-6596/992/1/012015

calculated data one can predict the technological results, as well as choose appropriate processing parameters in view of obtaining desirable results when electron beam technologies are applied to the manufacturing of pure metals and alloys and joining of metal parts [10-12]. Additionally, regression models can be used to estimate the influence of the electron beam focus position, together with the relative position of the beam focus with respect to the surface of the processed samples [10].

This paper proposes an empirical modeling approach to predicting and optimizing the exact shape of the cross-sections of welded seams obtained by electron beam welding. The influence of the electron beam welding process parameters is taken into account, as follows: electron beam power, welding speed, the distances from the magnetic lens of the electron gun to the focus position of the beam and to the surface of the samples treated. The results are verified via comparison with the results of experiments with samples made of stainless steel type 1H18NT. The beam powers considered are in the range 4.2 – 8.4 kW, with the welding speed varying from 3.333 to 13.333 mm/s.

2. Experimental conditions and regression models

The experiment considered in this paper was electron beam welding of samples of austenitic stainless steel type 1H18NT. The following operating parameters were varied: electron beam power (P) – 4.2 kW, 6.3 kW and 8.4 kW; welding speed (v) – 80 cm/min, 20 cm/min and 40 cm/min, distance between the main surface of the magnetic lens of the electron gun and the beam focusing plane (z_o) – 176 mm, 226 mm and 276 mm; and different distances between the main surface of the magnetic lens of the electron gun and the sample surface (z_p) within 126 mm – 326 mm. The accelerating voltage was 70 kV. After polishing and etching, 81 experimental weld cross-sections were examined. The following quality characteristics were considered: weld depth H, mean weld width B and surface area of the weld cross-sections S. The geometry of the EBW experiment is presented in figure 1. The basic levels of the process parameters' intervals of variation during the experiments and their dimensions are presented in table 1.

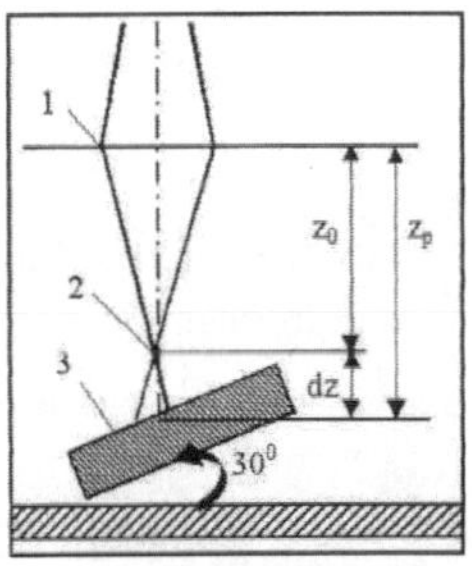

Figure 1. EBW experiment geometry: 1-magnetic lens; 2-electron beam; 3- welded sample.

Table 1. Basic levels and varying factors.

Parameter	Dimension	Coded	Lower level (Z_{imin})	Upper level (Z_{imax})
P-Z_1	kW	x_1	4.2	8.4
v-Z_2	cm/min	x_2	20	80
Z_o-Z_3	mm	x_3	176	276
Z_p-Z_4	mm	x_4	126	326

Table 2. Regression models for the geometry and the quality characteristics for stainless steel.

	Regression model	R^2 %	$R^2_{(adj)}$ %
$S_r =$	$35.665031+22.127596x_1-33.945939x_2+25.467878x_2^2-7.8344804x_2x_4-4.2384833x_3x_4+3.8790979x_1x_4\ 5.5226456x_1x_2x_4\ 6.1658247x_2^2x_3+8.5965428x_2^2x_4+5.2538552x_2x_3x_4-14.512189x_1x_2+6.8805937x_2x_3$	96.63	96.03
$H_r =$	$22.026299+5.9822119x_1-7.1080536x_2+3.8197955x_3-10.230833x_4-2.0413962x_1^2+3.7306182x_2^2-2.1965895x_3^2-15.797832x_4^2-2.8713398x_1x_2+12.976807x_3x_4-1.9193309x_1x_4$	89.66	88.01
$B_r =$	$1.9918913-0.62600498x_2-0.71935351x_3+2.3915622x_4+0.4291704x_2^2+0.26192114x_3^2+2.369174x_4^2+0.192194x_1x_3+0{,}18986534x_2x_3-2{,}2375194x_3x_4-0.25597695x_2x_4-0.58841498x_1^2x_4+0.28083044x_1x_2^2-0.67700299x_3x_4^2$	81.94	78.44

Using the response surface methodology [13], polynomial regression models were estimated (table 2) for the dependence of the weld depth, weld half-width and molten area of the weld cross-section for the process parameters investigated: electron beam power, welding speed, distances from the main surface of the magnetic lens of the electron gun to the beam focusing plane, as well as to the sample surface. The data processing was carried out in a coded scale, in order to avoid problems related to a possible multi-collinearity or other numerical problems in the evaluation of the coefficients of the regression models and bad prediction values of the quality indicator.

The process parameters, shown in table 1 and used in the regression equations in table 2, were coded in the range from -1 to 1, depending on their experimental ranges. Conversion from natural (z_i) to coded (x_i) values and vice versa was performed according to the formula:

$$x_i = \frac{2z_i - \left(z_{\max,i} + z_{\min,i}\right)}{z_{\max,i} - z_{\min,i}}.$$

For example, in order to evaluate the weld depth, weld half-width and molten area of the weld cross-section for the following process parameter values: $P = 8.4$ kW, $v = 20$ cm/min, $z_0 = 226$ mm and $z_p = 223$ mm, the dimensionless coded values that should be substituted in the corresponding equations from table 2 are: $x_1 = [2*8.4 – (8.4+4.2)]/(8.4–4.2) = 1$, $x_2 = [2*20 – (80+20)]/(80-20) = -1$, $x_3 = 1$, $x_4 = -0.1$. Table 2 presents the estimated regression models and the corresponding determination coefficients R^2 (the square of the multiple correlation coefficient) and the adjusted R^2_{adj}, which are measures for the model accuracy (the model is better if the coefficient is closer to 100%).

3. Empirical models for the welds shape

The Gaussian distribution is a good approximation of the energy distribution within the processed material in its interaction with a high-intensity electron beam. We proposed an approach to estimating empirical Gaussian models for the shape of the welds:

$$f_1(x) = \frac{Sk_1}{B\sqrt{2\pi}} e^{-\frac{x^2}{2\left(\frac{B}{k_1}\right)^2}} \text{ or } f_2(x) = He^{-\frac{x^2}{2\left(\frac{B}{k_2}\right)^2}} \text{ or } f_3(x) = He^{-\frac{x^2}{2\left(\frac{S}{H\sqrt{2\pi}}\right)^2}},$$

where H, B and S are correspondingly the weld depth, the weld half-width and the molten area of the weld cross-section. We estimated the coefficients k, k_1 and k_2 by minimizing the loss function Q_{res}, or the root-mean-square error ($RMSE$) criterion, namely:

$$Q_{res} = \sum_{i=1}^{n} (\hat{y}_i - y_i)^2 \to \min; \; RMSE = \sqrt{\frac{\sum_{i=1}^{n} (\hat{y}_i - y_i)^2}{n}} \to \min;$$

where $\hat{y}_i$ and y_i are the predicted and the experimental values of the parameters missing in the function – H, B or S, n being the number of data available. Table 3 presents the results obtained.

Table 3. Empirical models for the shape of the welds.

Function	$f_1(x)$	$f_2(x)$	$f_3(x)$
$\hat{y}_i$	$\hat{H} = f_1(x)_{\max}$	$\hat{S} = \frac{HB\sqrt{2\pi}}{k_2}$	$\hat{B} = \frac{Sk_3}{H\sqrt{2\pi}}$
y_i	H	S	B
Coefficient k_i	2.51	2.51	2.51
RSME	0.3323	0.7189	0.0296

One can see that for all models the value estimated for the coefficient k_i is 2.51. The *RSME* criterion has small values for all models, but it is the smallest for model $f_3(x)$.

Figure 2 presents the shape of the experimental and calculated by $f_3(x)$ weld cross-sections, obtained for different process parameter values for three sets of experimental condition, as follows: Experiment 1 – $P = 8.4$ kW, $v = 20$ cm/min, $z_0 = 226$ mm, $z_p = 223$ mm; Experiment 2 – $P = 6.3$ kW, $v = 40$ cm/min, $z_0 = 226$ mm, $z_p = 216$ mm; Experiment 3 – $P = 8.4$ kW, $v = 40$ cm/min, $z_0 = 276$ mm, $z_p = 308$ mm. The values of H and S estimated by model $f_3(x)$ for each process parameter set are calculated by the regression models in table 2 – H_r and S_r. The results obtained show a comparatively good accuracy of prediction of the welds shape.

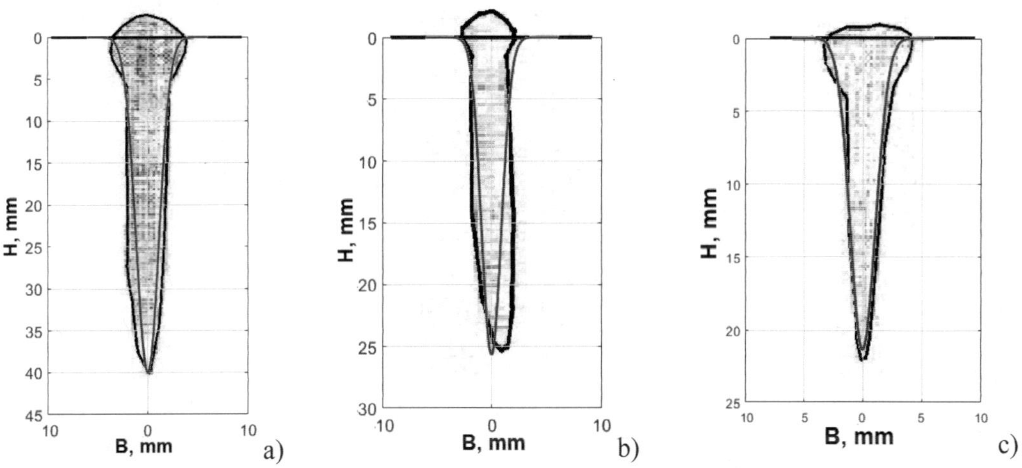

Figure 2. Experimental (black) and calculated by $f_3(x)$ (blue) shapes of the weld cross-sections, obtained for the process parameter values: a) Experiment 1 – $P = 8.4$ kW, $v = 20$ cm/min, $z_0 = 226$ mm, $z_p = 223$ mm, b) Experiment 2 – $P = 6.3$ kW, $v = 40$ cm/min, $z_0 = 226$ mm, $z_p = 216$ mm; c) Experiment 3 – $P = 8.4$ kW, $v = 40$ cm/min, $z_0 = 276$ mm, $z_p = 308$ mm.

Table 4. Weld geometry parameters – comparison between experimental results and results estimated by empirical models $f_1(x)$, $f_2(x)$, $f_3(x)$ and by regression models (table 2).

Model	*Exp.*	$f_1(x)$	*Abs. err.*	$f_2(x)$	*Abs. err.*	$f_3(x)$	*Abs. err.*	*Regr.*	*Abs. err.*
Param.	*H, mm*	$\hat{H}_1$*,mm*	$\lvert\varepsilon_i\rvert$*, mm*	$\hat{H}_2$*,mm*	$\lvert\varepsilon_i\rvert$*, mm*	$\hat{H}_3$*,mm*	$\lvert\varepsilon_i\rvert$*, mm*	H_r*, mm*	$\lvert\varepsilon_i\rvert$*, mm*
Exp. 1	39.9	40.119	0.2190	40.027	0.1270	40.027	0.1270	40.03	0.1300
Exp. 2	25.5	24.463	1.0370	25.675	0.1750	25.675	0.1750	25.68	0.1800
Exp. 3	23.4	27.33	3.9300	21.386	2.0140	21.387	2.0130	21.39	2.0100
Param.	*B, mm*	$\hat{B}_1$*, mm*	$\lvert\varepsilon_i\rvert$*, mm*	$\hat{B}_2$*, mm*	$\lvert\varepsilon_i\rvert$*, mm*	$\hat{B}_3$*, mm*	$\lvert\varepsilon_i\rvert$*, mm*	B_r*, mm*	$\lvert\varepsilon_i\rvert$*, mm*
Exp. 1	3.5	3.414	0.0860	3.414	0.0860	3.422	0.0780	3.27	0.2300
Exp. 2	1.9	2.008	0.1080	2.016	0.1160	1.924	0.0240	2.02	0.1200
Exp. 3	2.8	2.811	0.0110	2.736	0.0640	3.48	0.6800	2.80	0
Param.	S*,mm*^{2}	$\hat{S}_1$*,mm*^{2}	$\lvert\varepsilon_i\rvert$*,mm*^{2}	$\hat{S}_2$*,mm*^{2}	$\lvert\varepsilon_i\rvert$*,mm*^{2}	$\hat{S}_3$*,mm*^{2}	$\lvert\varepsilon_i\rvert$*,mm*^{2}	S_r*,mm*^{2}	$\lvert\varepsilon_i\rvert$*,mm*^{2}
Exp. 1	140.8	130.94	9.8600	130.64	10.1600	130.94	9.8600	130.94	9.8600
Exp. 2	48.2	49.45	1.2500	51.90	3.7000	49.45	1.2500	49.45	1.2500
Exp. 3	65.4	76.50	11.1000	59.86	5.5400	76.50	11.1000	76.50	11.1000

Table 4 summarizes the experimental values and those calculated by $f_1(x)$, $f_2(x)$, $f_3(x)$ and by regression models (table 2) for the weld depth *H*, mean weld half-width *B* and surface area *S* of the welds cross-sections. The absolute error is smaller for the depth and the half-width of the welds; larger differences are obtained between the values of the surface areas of the weld cross-sections due to the accuracy of the measurements, as well as to the precision of the estimated regression models.

Table 5. Comparison between mean absolute values and root-mean-square errors.

*Param.**Model*	*Error*	$f_1(x)$	$f_2(x)$	$f_3(x)$	*Regression*
H	ϵ_H	3.2559	2.4107	2.4107	2.4107
	$RMSE_H$	4.1705	3.0280	3.0280	3.0280
B	ϵ_B	0.3458	0.3463	0.4513	0.3139
	$RMSE_B$	0.4584	0.4541	0.6293	0.4174
S	ϵ_S	5.2677	5.2342	5.2631	5.2658
	$RMSE_S$	6.4933	6.7293	6.4847	6.4839

To compare the performance of the models (table 5), we used the mean absolute value of the error ϵ, calculated as the mean absolute difference between the predicted and the measured values of the weld geometry characteristics, as well as the root-mean-square error (*RMSE*). As seen, the three empirical models proposed $f_1(x)$, $f_2(x)$ and $f_3(x)$ give approximately the same results.

4. Conclusions

We proposed an empirical modeling approach to predicting the shape of the welded seam cross-sections, as obtained by electron beam welding. The methodology involves two steps: a) calculation of the weld depth, weld half-width and the molten area of the weld cross-section at given exact values of the process parameters: electron beam power, welding speed, distances from the main surface of the magnetic lens of the electron gun to the beam focusing plane, as well as to the sample surface, using estimated regression models, b) calculation of the shape of the EB welds by applying empirical models $f_1(x)$, $f_2(x)$ or $f_3(x)$. The calculated results were compared with the experimental data for samples from stainless steel type 1H18NT; a good prediction accuracy was found. The practical choice of the functions proposed will depend on the accuracy of the experimental measurements; some physical and metallurgical aspects of the specific process should also be considered, such as the appearance of spikes at the welds' roots and of various defects along the crater formed during the electron beam processing.

References

[1] Chiumenti M, Cervera M, Dialami N, Wu B, Jinwei L and Agelet de Saracibar C 2016 *Finite Elements in Analysis and Design* **121** 118-33
[2] Koleva E and Mladenov G 2000 *Bulgarian J. Phys.* **27**/4 83-96
[3] Tian Y, Wang Ch, Zhu D and Zhou Y 2008 *J. Mater. Processing Technol.* **199** 41–8
[4] Olshanskaya T 2016 *Elektrotechnica & Elektronica E+E* **51**/5-6 92-7
[5] Koleva E, Miadenov G and Vutova K 1999 *Vacuum* **53**/1-2 67-70
[6] Koleva E, Vutova K and Mladenov G 2001 *Vacuum* **62**/2-3 189-96
[7] Erokhin A 1976 To calculation of depth of pro-melting at electron beam welding *Svarochnoe proizvodstvo* **4** 4–6
[8] Kaidalov A 2004 *Electron beam welding and adjacent technologies,* 2-nd edition (Ekotekhnologiia, Kiev) 260
[9] Rykalin N, Uglov A and Zuev I 1978 *Bases of electron beam processing of materials* (Mashinostroenie Moscow) 239
[10] Koleva E 2001 *Vacuum* **62**/2-3 151-7

[11] Koleva E 2005 *Vacuum* **77** 413-421

[12] Trushnikov D, Koleva E, Mladenov G, Belenkiy V 2013 *Journal of Material Proc. Technology* **213** (9) 1623– 1634

[13] Vuchkov I, Boyadjieva L 2001 *Quality Improvement with Design of Experiments. A Response Surface Approach* Kluwer Academic Publishers

Study of residual stresses in CT test specimens welded by electron beam

I V Papushkin[1,4], D Kaisheva[2], G D Bokuchava[1], V Angelov[3] and P Petrov[2]

[1]Frank Laboratory of Neutron Physics, Joint Institute of Nuclear Research, 141980 Dubna, Moscow Region, Russia
[2]E. Djakov Institute of Electronics, Bulgarian Academy of Sciences, 72 Tzarigradsko Chaussee, 1784 Sofia, Bulgaria
[3]Institute for Nuclear Research and Nuclear Energy, Bulgarian Academy of Sciences, 72 Tzarigradsko Chaussee, 1784 Sofia, Bulgaria

E-mail: piv@nf.jinr.ru

Abstract. The paper reports result of residual stress distribution studies in CT specimens reconstituted by electron beam welding (EBW). The main aim of the study is evaluation of the applicability of the welding technique for CT specimens' reconstitution. Thus, the temperature distribution during electron beam welding of a CT specimen was calculated using Green`s functions and the residual stress distribution was determined experimentally using neutron diffraction. Time-of-flight neutron diffraction experiments were performed on a Fourier stress diffractometer at the IBR-2 fast pulsed reactor in FLNP JINR (Dubna, Russia). The neutron diffraction data estimates yielded a maximal stress level of ±180 MPa in the welded joint.

1. Introduction

One of the most important components of nuclear installation is the reactor pressure vessel (RPV). The safety of the RPV is ensured by mechanical testing of the surveillance Charpy or CT specimens. During reactor operation, the surveillance samples are exposed to the same thermal and radiation conditions as the RPV material.

In order to increase the number of irradiated samples of reactor steel, sample reconstruction by an electron beam welding (EBW) procedure is usually applied. The residual stresses and microstrains resulting from the EBW process are important parameters responsible for the service life of reconstituted Charpy or CT specimens [1, 2]. Over many years, residual stresses in materials have been studied using various non-destructive techniques: X-ray diffraction, ultrasonic scanning and various magnetic techniques (measurements of magnetic induction, permeability, anisotropy, Barkhausen effect, magnetoacoustics effects). The non-destructive techniques for stress determination seem to be the most informative, because they allow one to work with the original product and the real stress distribution. However, all these methods have strong limitations in the depth of penetration and materials used. Neutron diffraction enlarges and complements the capabilities of other methods for determining residual stresses. Neutron diffraction experiments are less common than X-ray diffraction due to the

[4] To whom any correspondence should be addressed.

difficulty in obtaining particle fluxes, but allow one to study a wider range of products without destroying them, due to the high penetrating ability of neutrons.

We present theoretical calculation of the temperature fields during electron beam welding, together with results of neutron diffraction measurement of the residual stresses.

2. Experiment and modeling

2.1. Experimental measurement of residual stress distribution

The material used in this study was 18MND5 low-alloyed steel with chemical composition (wt % C 0,18; Si 0,25, Mn 1,6; Cr 0,17; Ni 0,64; Mo 0,61; V 0,12; Cu 0,13). Figure 1 shows the welded CT specimen and a cross-section of the seam welded at a beam power of 3 kW and welding speed $V = 10$ mm/s. The insert from the test material (1) is welded to the holder (3) for the CT test. The insert had dimension 10×10×24 mm. The final dimensions of CT specimen were 26×24×10 mm.

The Fourier stress diffractometer (FSD) is a specialized high-resolution neutron diffractometer for residual stress studies [3, 4] located at channel 11A of the IBR-2 fast pulsed reactor in FLNP JINR. The correlation RTOF method used by the FSD allows a resolution of $\Delta d/d \approx 0.001 \div 0.004$ on a short flight base of ~5.5 m [5]. The instrument has two wide-aperture radial collimators with spatial resolution of 1.8 mm in front of each 90° detector. This allows the selection of a small scattering volume (gauge volume) inside the sample studied. Scanning the sample as installed on a precise Huber goniometer with a small gauge volume yields information on the residual strain distribution in the material. The typical accuracy of residual stress determination on the FSD diffractometer is 10-20 MPa for steel.

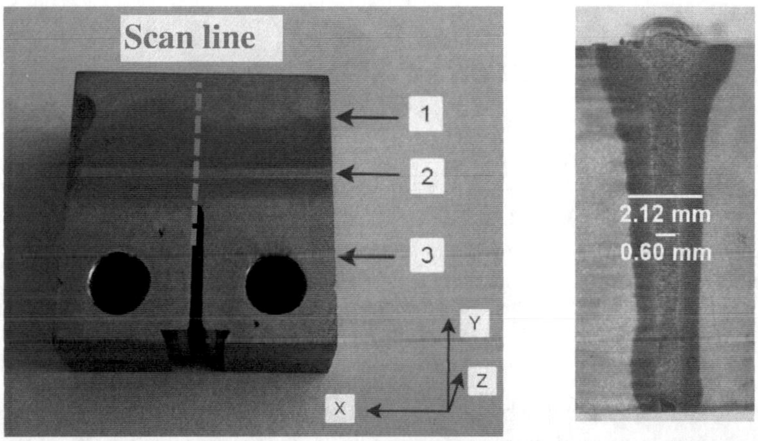

a) b)

Figure 1. a) CT welded specimen (1 – insert from the test material, 2 – weld seam, 3 – holder for CT test); b) cross-section of welding seam.

In the current study, the residual stresses were measured in the middle of the sample along the scan line (Y-axis) at 12-14 points. At each point investigated, three mutually orthogonal strain tensor components were measured with a gauge volume of 40 mm^3 (2×2×10 mm). The resulting stresses were averaged over the gauge volume. The scan step between two successive points was selected to be 2 mm.

2.2. Temperature fields modeling using Green`s functions

If the specific heat c_p, density ρ and heat conductivity λ of the material are temperature independent, the time and spatial temperature distribution $T(r,t)$ in a medium satisfies the transient conduction equation:

$$\frac{1}{\alpha}\frac{\partial T}{\partial t} - \nabla^2 T = \frac{\mathrm{f(r,t)}}{\lambda}, \tag{1}$$

where $\alpha = \lambda/\rho c_p$ is the heat diffusivity of the material and f(r,t) is the distribution of the source.

Equation (1) describes a three-dimensional non-homogeneous heat conduction problem, which has the following solution, expressed in terms of a three-dimensional Green`s function [6]:

$$\mathrm{T(r,t)} = \frac{\alpha}{\lambda}\int_{\tau=0}^{t} d\tau \int_{R} G(r,t|r',\tau) f(r',\tau) dV' + \int_{R} G(r,t|r',\tau)|_{\tau=0} F(r') dV', \tag{2}$$

where $F(r')$ is the initial temperature distribution.

The volume source beam distribution $f(r,t) = f_V(\xi,\eta,\zeta,\tau)$, where $\xi = x' = x - r\cos\theta$, $\eta = y' = y - r\sin\theta$ (r and θ are the polar coordinates) and ζ is the depth of the point of integration, can assume different forms depending on the type of the beam motion and the source shape chosen.

In case of welding along a straight line, the heat source function shape could be chosen to be a cone with a base radius r_0, height h with Gaussian distribution at any plane ζ and moving along a straight line. Then the integration over ζ is limited from $\zeta = 0$ to $\zeta = h$ and equation (2) is transformed to:

$$T_V(x,y,z,t) = T_0 + \int_0^t\int_0^h\int_{-\pi}^{\pi}\int_0^{\infty} \frac{\alpha}{\lambda}\frac{f_V(\xi,\eta,\zeta,\tau)}{\left(4\pi\alpha(t-\tau)\right)^{3/2}}\left[\exp\left(-\frac{r^2+(z-\zeta)^2}{4\alpha(t-\tau)}\right) + \exp\left(-\frac{r^2+(z+\zeta)^2}{4\alpha(t-\tau)}\right)\right] r dr d\theta d\zeta d\tau . \tag{3}$$

Bearing in mind that the cone radius in plane ζ is

$$\mathrm{r}_\zeta = \mathrm{r}_0\frac{h-\zeta}{h} \tag{4}$$

and the volume of the cone is smaller by a factor of three than the volume of a cylinder with the same base, which determines a normalizing coefficient over the cone volume, the cone heat source is defined as:

$$f_V(\xi,\eta,\zeta,\tau) = \frac{3}{h}\left(\frac{3Q}{\pi \mathrm{r}_0{}^2}\right)\exp\left(-3\frac{(\xi-v\tau)^2+\eta^2}{\mathrm{r}_\zeta{}^2}\right) = \frac{9Q}{\pi \mathrm{r}_0{}^2 h}\exp\left(-3\frac{h^2}{(h-\zeta)^2}\frac{(\xi-v\tau)^2+\eta^2}{\mathrm{r}_0{}^2}\right). \tag{5}$$

Equation (3) with the cone source (5) can be used for calculation of the temperature in case of welding.

We used the Wolfram Mathematica [9] numerical integration tool, which works adaptively, reaching a preliminary specified accuracy. Separating the numerical integration into many independent numerical integrations allows one to make use of the parallel computing implemented by Mathematica for multicore computers, thus achieving a significant acceleration of the calculations. The total temperature increase at a given point is calculated as the sum of the contribution of each integrand.

3. Results and discussion

The calculated temperature distribution for EBW of CT specimens presented in figure 2a exhibits high temperature gradient along the Y coordinate. At a lower source speed, heating to a higher temperature and a slower cooling rate are observed (see figure 1b and figure 2a). The closer a certain point to the weld seam, the higher the temperature. This directly affects the residual stress distribution after the welding process. In our case, the welding speed was 10 mm/sec, corresponding to the middle line in figure 2a.

The information on the hardness of the material in the weld region (figure 2b) was used as the initial experimental data. Neutron diffraction made it possible to measure directly the residual strains caused by the thermal action and then recalculate them into residual stresses. During the neutron diffraction experiments, the residual strain distributions were obtained in three perpendicular directions. Then, the residual stresses were calculated according to Hooke's law using a commonly applied approach [7]. The residual stress distributions obtained are shown in figure 3. Position Y = 0 mm corresponds to the weld center. The negative values of the Y coordinates belong to the insert of the witness sample material, the positive, to the CT test holder. It can be seen that all three components of the residual stress tensor have a quite low level (about 180 MPa) in the weld region and the stresses drop rapidly in the insert.

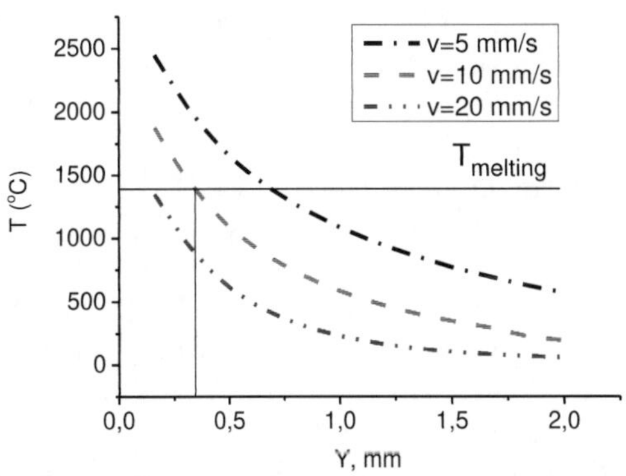

a)

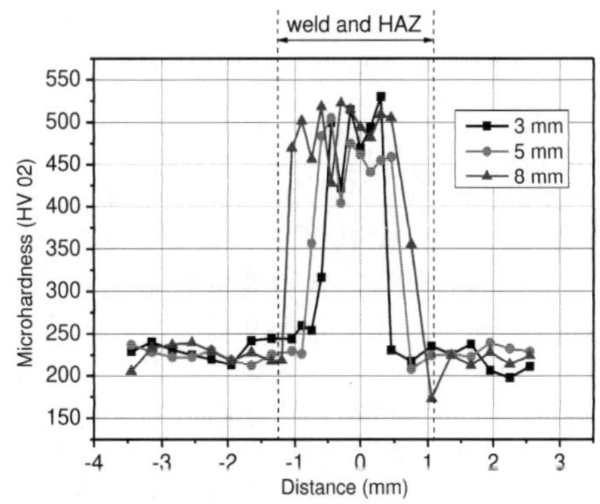

b)

Figure 2. a) Calculated temperature distribution for different welding speed and electron beam power 3 kW; b) Experimental microhardness of the weld and the heat-affected zone at beam power 3 kW and welding speed $V = 10$ mm/s.

The maximum values of the residual stresses of the X and Y components reach a level of 100 MPa in the near-weld region and the overheating section. At a depth of 3 mm from the welded seam, there are practically no effects from the heat-affected zone during welding, which is typical for electron-beam welding [8]. The tempering area ends in this zone and the residual stresses drop rapidly to almost zero. The divergence of the stresses in the holder, shown on the right-hand side of the graph, should be attributed to the peculiarities of the technology of its manufacture; it does not affect the results of the mechanical CT tests.

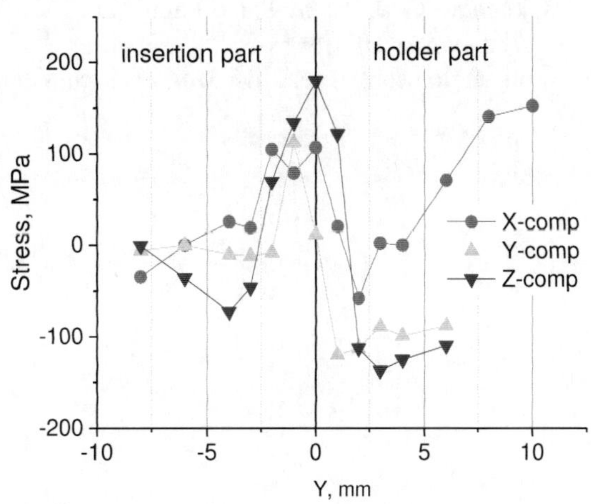

Figure 3. Measured residual stresses in the specimen.

The experimental results show a narrow region of thermal action for all three components of the strain tensor. This is in good agreement with the calculated temperature field data (figure 2a). The calculations show practically no effect of the zone of thermal action at a distance of ±2 mm from the weld at different welding speeds.

4. Conclusions

The temperature distribution during electron beam welding of alloyed steel show high gradients, which directly affects the residual stresses.

Using neutron diffraction study of the CT test sample, the residual stress distributions were determined in three directions in the material. All components of the strain tensor have a low level of residual stresses (~ 180 MPa) in the weld region.

The calculations width of the temperature-effect zone were confirmed by the neutron diffraction experiment. The width of the thermal exposure zone is several millimeters.

The low level of residual stresses and the narrow zone of thermal influence indicate a slight change in the structure of the material after welding. Therefore, the results obtained allow us to recommend the use of electron beam welding for manufacturing CT test samples.

Acknowledgements

The work was partially supported by the Russian Foundation for Basic Research (project No. 15-08-06418_a) and the Bulgarian National Scientific Fund (Contract No. DN 07/26).

References

[1] Masubuchi K 1980 *Analysis of welded structure* (Pergamon Press, New York) 642
[2] Mihailov V, Karhin V and Petrov P 2016 *Fundamentals of welding* (Polytechnic University Publishing St. Petersburg) 282
[3] Bokuchava G D, Aksenov V L, Balagurov A M, Kuzmin E S, Zhuravlev V V, Bulkin A P, Kudryashev V A and Trounov V A 2002 *Appl. Phys. A: Mater. Sci. & Processing* **74** 86-8
[4] Bokuchava G D, Balagurov A M, Sumin V V and Papushkin I V 2010 *J. Surf. Inv. X-ray Synchrotron and Neutron Techniques* **4**/6 879-90
[5] Balagurov A M, Bokuchava G D, Kuzmin E S, Tamonov A V and Zhuk V V 2006 *Zeitschrift für Kristallographie Suppl.* **23** 217-22
[6] Farrahi G H and Sistaninia M 2009 *IJE Trans. A: Basics* **22**/2 169-80
[7] Bokuchava G D, Papushkin I V, Tamonov A V and Kruglov A A 2016 *Romanian J. Phys.* **61**/3-4 491-505
[8] Bokuchava G D, Petrov P and Papushkin I V 2016 *J. Surf. Inv. X-ray, Synchrotron and Neutron Techniques* **10**/6 1143-53
[9] Wolfram Research Inc. 2015 *Mathematica* Version 10.3 (Champaign IL USA)

IOP Conf. Series: Journal of Physics: Conf. Series **992** (2018) 012017 doi:10.1088/1742-6596/992/1/012017

Analysis of high-efficiency widely-tunable N-resonances in Cs vapor

A Krasteva[1,4], S Gateva[1], S Tzvetkov[1], P N Ghosh[2], A Sargsyan[3] and S Cartaleva[1]

[1]Emil Djakov Institute of Electronics, Bulgarian Academy of Sciences, 72 Tsarigradsko Chaussee, 1784 Sofia, Bulgaria
[2]Department of Physics, University of Calcutta, 92 APC Road, 700009 Kolkata, India
[3]Institute for Physical Research, National Academy of Sciences of Armenia, Ashtarak-2, 0203 Armenia

E-mail: anna0kr0stz@gmail.com

Abstract. The narrow-band coherent N-type resonance, promising for the development of advanced atomic clocks, can be considered as a type of three-photon resonance, where a two-photon Raman excitation is combined with a resonant optical pumping field. In this communication, we present an experimental study and a theoretical analysis related to three-photon, bi-chromatic excitation of Cs atomic vapor contained in an 8-mm long cell with 20 Torr of neon. If a coupling laser is fixed at a frequency that is lower by several GHz than the position of the absorption profile of the $F_g = 4$ set of transitions, and a probe laser is tuned over the D_2 line (λ= 852 nm), a narrow high-contrast enhanced absorption N-resonance will be observed in the probe light profile, superimposed on the absorption profile of the $F_g = 4$ set of transitions. We present theoretical modeling aimed to clarify the processes behind the efficiency of the N-resonance preparation for different frequency positions of the coupling laser within the D_2 line of Cs.

1. Introduction

Coherent laser spectroscopy of alkali vapor contained in thermal optical cells is widely used in various applications; among them wavelength references, atomic clocks, precise optical magnetometers, slow and stored light, photonic sensors based on coherent population trapping (CPT), electromagnetically-induced transparency (EIT) resonances, etc. For these applications, various research groups are still exploring suitable approaches aimed at a profitable combination of the advantages of CPT/EIT with other processes. Recently, optical systems based on Rb atoms, in which narrow-band N-type resonances appear, have been extensively studied [1-4]. The N-resonance has been considered as a type of three-photon resonance, where a two-photon Raman excitation is combined with a resonant optical pumping field. Experimental works related to the three-photon, bi-chromatic excitation of Cs atomic vapor contained in an 8-mm-long cell with 20 Torr of neon have been reported in ref. [5-7].

In this communication, we report on an experimental study complemented by theoretical modeling of the related processes and analyze the N-resonance preparation for different frequency positions of the coupling laser within the D_2 line of Cs. A coupling laser was fixed at a frequency lower than the

[4] To whom any correspondence should be addressed.

IOP Conf. Series: Journal of Physics: Conf. Series **992** (2018) 012017 doi:10.1088/1742-6596/992/1/012017

position of the absorption profile of the $F_g = 4$ set of transitions, while the probe laser was tuned over the D_2 line ($\lambda = 852$ nm). A high-contrast narrow enhanced-absorption N-resonance was observed in the probe light profile, superimposed on the absorption profile of the $F_g = 4$ set of transitions. The theoretical model proposed, based on Bloch optical equations, is briefly described here in order to introduce the theoretical simulation concept. The theoretical model will be discussed in detail in a further communication. The simulation shows the formation of an enhanced absorption N-resonance superimposed on the $F_g = 4$ absorption profile at a frequency position depending on the coupling beam frequency position.

2. Experimental setup

Figure 1 shows a schematic diagram of the experimental setup, which is similar to that used in ref. [6, 7]. Two narrow-band distributed feedback (DFBs) lasers are used. The coupling laser has a fixed frequency ν_c, while the probe one with frequency ν_p is tuned over the D_2 line. Both laser beams have a 3×2-mm spot size and are carefully superimposed on a beam splitter BS and then directed at the 8-mm-long optical cell. The polarizations of the coupling and probe lasers are linear and mutually orthogonal. The optical cell is filled with atomic cesium with 20 Torr of neon gas added as a buffer. Part of the probe radiation is directed to an additional cell with a thickness $L = 6\lambda$ used as a frequency reference.

In the experiments, the transmission signals of the probe and coupling beams were registered with the probe beam scanned in a frequency interval covering the entire D_2 line absorption. The probe beam was separated from the coupling one by a polarizing beam splitter. The broad transmission spectrum of the probe laser beam was measured with a two-photon resonance superimposed on it, for different frequency positions of the coupling beam.

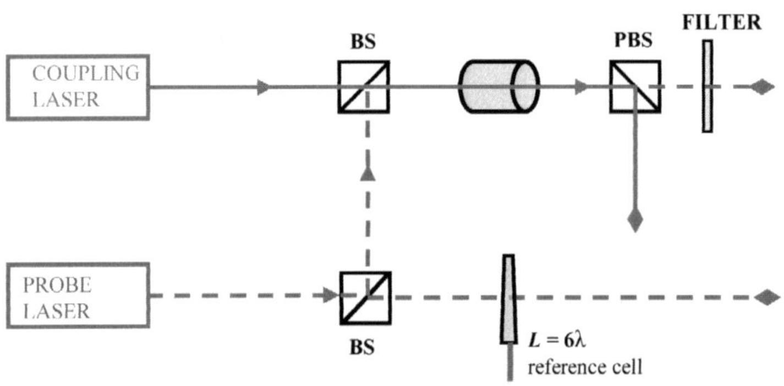

Figure 1. Schematic diagram of the experiment: DFB – distributed feedback diode-laser: coupling and probe, BS – beam splitter, PBS – polarizing beam splitter, interference filter at the wavelength of 852 nm. The $L = 6\lambda$ cell provides a reference spectrum.

3. N_p resonance within $F_g = 4$ absorption profile

In this section, we will briefly discuss the N-resonance superimposed on the absorption profile of the $F_g = 4$ set of transitions (also reported in ref. [5, 6]). Because the resonance is observed in the probe-beam transmission, hereinafter it is referred to as N_p resonance.

A schematic diagram of the relevant atomic levels and optical transitions is shown in figure 2. The lower levels used to form the required Λ system are the two ground $F_g = 3, 4$ levels, the upper level being $6P_{3/2}$, which consists of four hyperfine levels $F_e = 2, 3, 4, 5$. The probe laser frequency ν_p is tuned over the $F_g = 4 \rightarrow F_e = 3, 4, 5$ set of transitions, and the coupling laser frequency ν_c is fixed. In the probe-beam transmission spectrum, an N-type resonance involving a two-photon process was observed when the two-field frequency difference was $\nu_p - \nu_c = \Delta$, where Δ is the splitting ($\Delta = 9.2$ GHz) of the electronic ground state hyperfine levels. Figure 3 shows the experimentally observed N_p-resonances of enhanced absorption for two different frequency positions of the coupling beam. While the frequency of the coupling laser is fixed at two different positions within the lower-frequency wing of the $F_g = 4$ profile, the absorption of the probe laser is measured as a function of its frequency detuning. In both figures 3a and 3b, the N_p-resonance frequency position is shifted from that of the coupling laser by the electronic ground-state hyperfine level splitting $\Delta = 9.2$ GHz. Note that, although the coupling-beam frequency position is about 9.2 GHz lower than the $F_g = 4$ (and 18.4 GHz lower than the $F_g = 3$) set of transitions, the coupling demonstrates good-absorption Doppler-broadened profiles when the probe beam is in resonance with the $F_g = 4$ or $F_g = 3$ set of hyperfine transitions. Thus, the relatively strong laser intensity used, the high atomic source temperature and the buffer gas ensure efficient velocity-changing collisions and excitation transfer between Cs atoms.

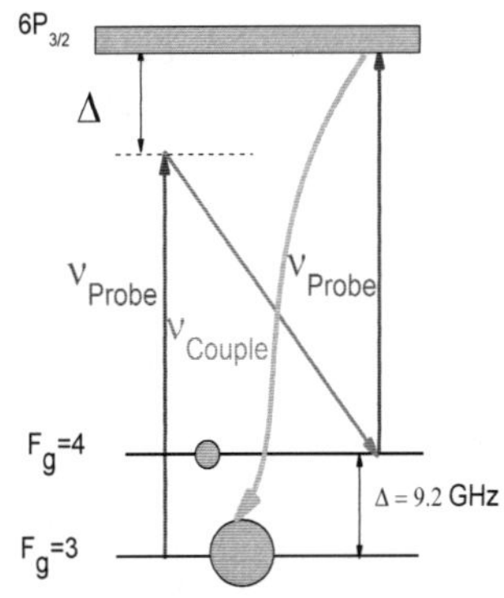

Figure 2. Probe frequency ν_P is scanned over the F_g=4→ F_e=3,4,5 transitions, while the coupling (ν_c) is fixed at a lower frequency [5, 6].

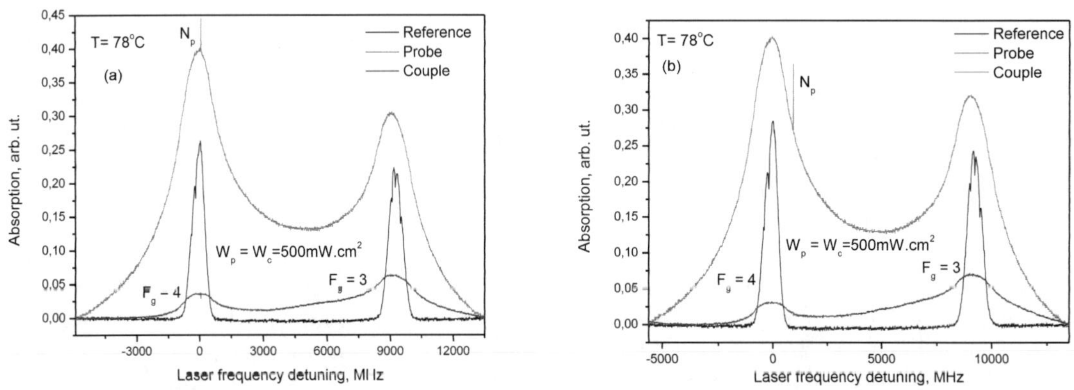

Figure 3. Probe-beam spectrum of the D_2 line involving the N_p resonance situated within the $F_g = 4$ absorption set and the fixed-frequency coupling-beam absorption, demonstrating two Doppler profiles. Fluorescence spectrum of the $L = 6\lambda$ cell used as a frequency reference. The coupling-beam frequency position is not shown, because it is well outside the probe frequency scan, in order to resolve the sub-Doppler features of the reference.

3. Theoretical simulation

To clarify the processes behind the N-resonance preparation and its dependence on the frequency positions of the coupling laser within the D_2 line of Cs, a theoretical modeling based on the Bloch optical equations is briefly described below.

Sttrong coupling and probe radiations are applied to a three-level atom (see figure 4a) with energy levels (a, b, c),. The coupling radiation has a fixed frequency ω_c corresponding to a transition from the lower level b to a hypothetical level h ($b \rightarrow h$), separated from the energy level c by a gap $\Delta = E_b - E_a$. The probe frequency ω_p is tuned over a broad range of frequencies ($a \rightarrow c$). When the probe frequency is such that the Raman condition $\omega_c - \omega_p = E_b - E_a$ is satisfied, a very sharp peak appears at the frequency $\omega_p = \omega_{bc}$.

The density matrix equations for the four-level system were written and the Bloch optical equations were solved numerically. As a results, we obtained a sharp resonance peak sitting on top of the $b \rightarrow c$

absorption transition from the $F_g = 4$ level, when the coupling frequency is tuned exactly to the $b \rightarrow h$ transition (figure 4b). This sharp peak moves away and weakens when the coupling frequency is moved away from the $b \rightarrow h$ transition to the $b \rightarrow c$ transition.

Comparing the theoretical simulations with the experimental results show that the theoretical modeling proposed is in good agreement with the experimental observations, as its reproduces well the N-resonance position and its amplitude, depending on the coupling-beam frequency.

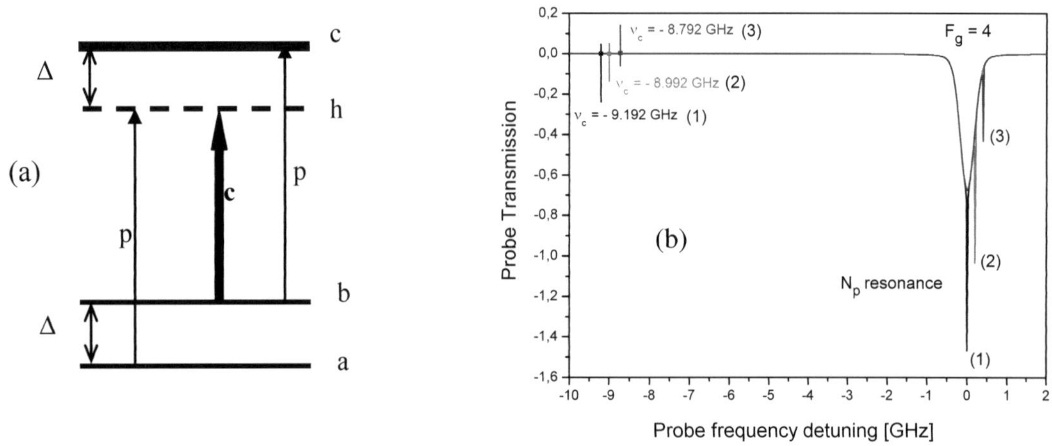

Figure 4. (a) Coupling and probe transitions for an N-type resonance observed in $F_g = 4$ absorption profile and (b) calculated sharp N_p resonance on the $F_g = 4$ absorption peak ($b \rightarrow c$) for different values of the coupling frequency detuning: (1) $\omega_c = -9.192$, (2) -8.992 and (3) -8.792 GHz.

4. Conclusions

We studied a narrow N_p-resonance superimposed on the $F_g = 4$ profile of the D_2 line of Cs atoms in a cell with a thickness of 8 mm in the presence of neon at a pressure of 20 Torr. The behavior of the N-resonance was demonstrated in the absorption of the probe beam measured as a function of its frequency detuning. A theoretical modeling based on the Bloch optical equations was proposed in view of studying the formation of the N_p-resonance in the $F_g = 4$ absorption profile for different frequency positions of the coupling beam.

The observation of a very good contrast and a sub-natural-width *N*-resonance and the simplicity of the experimental realization can be of interest for a large variety of fundamental and applied problems in optics and laser spectroscopy.

Acknowledgments

The work was supported by the Bulgarian National Science Fund under projects DNTS/Russia 01/5 23.06.2017 "Nonlinear spectroscopy of spatially restricted alkali vapor: methodology and applications" and DN 08-19/ 14.12.2016 "New coherent and cooperative effects in hot alkali vapor". ST and SG acknowledge the Young Scientists and PhD Students Support Program of the Bulgarian Academy of Sciences (Grant DFNP 17-76/2017).

References

[1] Zibrov A S, Ye C Y, Rostovtsev Y V, Matsko A B and Scully M O 2002 *Phys. Rev.* A **65** 043817/4

[2] Zibrov S, Novikova I, DPhillips D F, Taichenachev A V, Yudin V I, Walsworth R L and Zibrov A S 2005 *Phys. Rev.* A **72** 011801(R)/4

[3] Novikova I, Phillips D F, Zibrov A S, Walsworth R L, Taichenachev A V and Yudin V I 2006 *Opt. Lett.* **31** 2353-55

[4] Novikova I, Phillips D F, Zibrov A S, Walsworth R L, Taichenachev A V and Yudin V I 2006 *Opt. Lett.* **31** 622-4
[5] Slavov D, Sargsyan A, Sarkisyan D, Mirzoyan R, Krasteva A, Wilson-Gordon A D and Cartaleva S 2014 *J. Phys.* B*: At. Mol. Opt. Phys.* **47** 035001
[6] Krasteva A, Slavov D, Sargsyan A, Sarkisyan D, Wilson-Gordon A D and Cartaleva S 2015 *J. Phys.: Conf. Series* **594** 012009
[7] Krasteva A, Gateva S, Sargsyan A, Sarkisyan D and Cartaleva S 2017 *Rom. Rep. Phys.* **69** 202

IOP Conf. Series: Journal of Physics: Conf. Series **992** (2018) 012018 doi:10.1088/1742-6596/992/1/012018

Diffraction efficiency calculations of polarization diffraction gratings with surface relief

D Nazarova[1], P Sharlandjiev[1], N Berberova[1], B Blagoeva[1,4], E Stoykova[1,2] and L Nedelchev[1,3]

[1]Acad. J. Malinowski Institute of Optical Materials and Technologies, Bulgarian Academy of Sciences, Acad. G. Bonchev Str. Bl. 109, 1113 Sofia, Bulgaria
[2]VR/AR Research Center, Korea Electronics Technology Institute, Seoul, South Korea
[3]University of Telecommunications and Post, 1 Acad. Stefan Mladenov Str., 1700 Sofia, Bulgaria

E-mail: blagoeva.bl@gmail.com

Abstract. In this paper, we evaluate the optical response of a stack of two diffraction gratings of equal one-dimensional periodicity. The first one is a surface-relief grating structure; the second, a volume polarization grating. This model is based on our experimental results from polarization holographic recordings in azopolymer films. We used films of commercially available azopolymer (poly[1-[4-(3-carboxy-4-hydroxyphenylazo) benzenesulfonamido]-1,2-ethanediyl, sodium salt]), shortly denoted as PAZO. During the recording process, a polarization grating in the volume of the material and a relief grating on the film surface are formed simultaneously. In order to evaluate numerically the optical response of this "hybrid" diffraction structure, we used the rigorous coupled-wave approach (RCWA). It yields stable numerical solutions of Maxwell's vector equations using the algebraic eigenvalue method.

1. Introduction

Various techniques are used to produce diffractive optical elements (DOEs) that modulate and transform light in a predetermined way [1–8]. The DOEs, and surface-relief gratings in particular, are of major interest for applications in quantum electronics, integrated optics, spectroscopy, and holography. Applications include spectral filters, diffractive lenses, antireflection surfaces, beam splitters, beam combiners, beam deflectors, beam shaping elements, laser mirrors, polarization devices, couplers and switches, waveguide couplers, wavelength multiplexers and demultiplexers, etc. [9, 10]. The optical response of a DOE can be controlled by tuning various parameters, such as wavelength, polarization, ratio of the ridge-width to the pitch; the shape of the ridge, the refractive index of the material [11].

The optical response of high-frequency diffraction gratings has to be analyzed by rigorous diffraction approaches for solving the set of Maxwell's vector equations. Many approximate theories are available for the purpose, the most popular ones using the Kirchhoff approximation [12, 13]. These approaches are simple and easy to apply, but require certain conditions to be met in order to provide accurate results. For example, the wavelength must be much smaller than the grating period, and the

[4] To whom any correspondence should be addressed.

grating profile must be shallow. However, for a wide range of applications, these conditions are not fulfilled, and the approximations provide inaccurate results.

The analysis of the optical response of the diffraction element depends crucially on a high-precision rigorous modeling of the interaction between light and structure. The mathematical modeling requires the computation of the relation between the incoming and output waves. These quantities are described by a model based on the full set of Maxwell's vector equations [14, 15]. There exists a variety of different methods for obtaining a numerical solution of the resulting Helmholtz equation. One of the most popular is the rigorous coupled-wave approach (RCWA) [16-18]. Another one is the so-called C method [17]. Boundary value problems for elliptic partial differential equations could also be solved by the finite element method (FEM) [19 - 21].

In RCWA, the grating structure is divided into a number of thin layers (slices) parallel to the surface. Peng et al. [22] first proposed partitioning the grating into thin layers and approximating each layer by a rectangular profile. Each thin grating is then analyzed by using the state-variables method of solving the rigorous coupled-wave equations for that grating. As a result, the efficiencies of all propagating diffraction orders can be estimated.

In the present work, the rigorous coupled-wave approach is used to evaluate the diffraction efficiency of a structure consisting of a scalar surface-relief grating positioned on top of a volume polarization grating.

2. Diffraction structure model and computational implementation of RCWA

Figure 1 presents the "hybrid" diffraction structure considered. The incident plane wave propagates in the superstrate (air) and hits the modulated material (PAZO), which is deposited on the substrate (glass). Two diffraction regions are considered. The first one has a surface–relief grating (SRG) with topological sinusoidal profile *P* of period *d* along the *x*-axis; the second is a region with sinusoidal modulation of the refractive index. Both structures have the same period. It is assumed that there is no shifting between the two gratings.

The *y*-axis is perpendicular to the profile plane and the *z*-axis is the axis of invariance of the structure (Figure1). The electromagnetic properties of the PAZO material are represented by its complex refractive index [23].

RCWA that we used for evaluation of the diffracted field is based on the algebraic eigen system solution of the of the first-order coupled differential equation. This coupled-wave equation is derived from the full set of Maxwell's equations. The diffraction structure is split into a set of sublayers (slices), characterized by the material complex refractive index. Within the sublayer, the refractive index is assumed as independent of *y* (figure1), i.e. each slice is assumed to have rectangular pieces of modulation within one period of the grating. Using a truncated Fourier expansion of the refractive index, the general coupled-wave equation for the transverse electromagnetic field is solved explicitly for each slice. Solutions at the slice boundaries are evaluated by the continuity of the transverse field components.

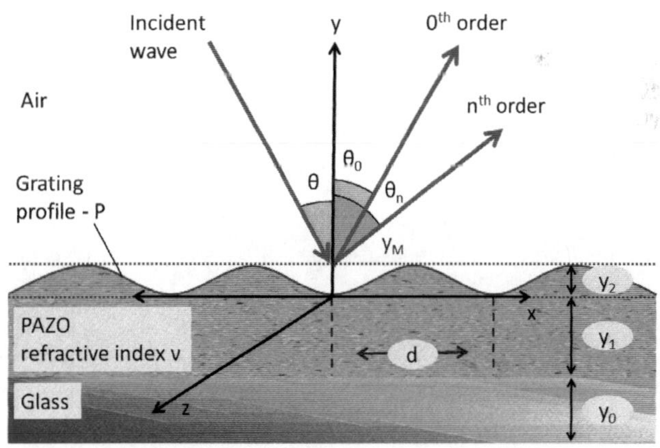

Figure 1. Diffraction grating notations.

Two parameters are of substantial importance: the number of diffraction orders taken in consideration and the number of sublayers used in the diffraction structure representation.

IOP Conf. Series: Journal of Physics: Conf. Series **992** (2018) 012018 doi:10.1088/1742-6596/992/1/012018

3. Results

We consider a recorded polarization grating in azopolymer thin films of commercially available azopolymer (poly[1-[4-(3-carboxy-4-hydroxyphenylazo) benzenesulfonamido]-1,2-ethanediyl, sodium salt]), – PAZO.

The film transmission and reflection are shown in figure 2. The optical constants of PAZO films are obtained using spectrophotometric data, as reported elsewhere [23].

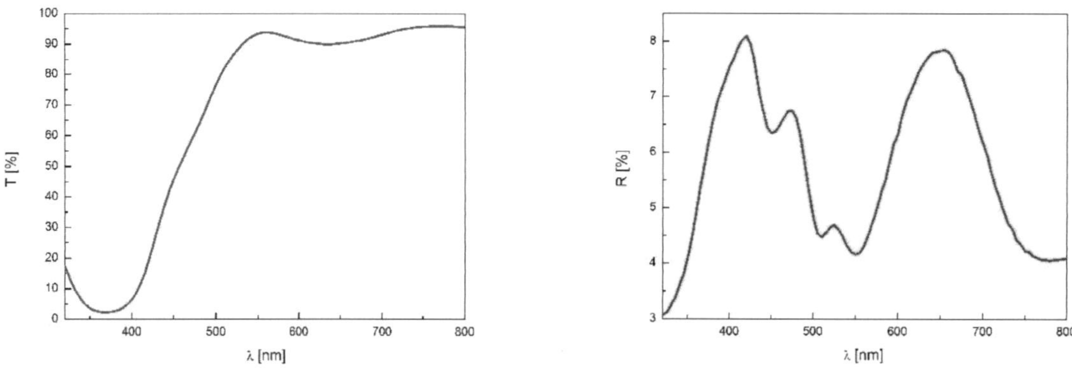

Figure 2. Transmission (left) and reflection (right) of a 300-nm-thick PAZO thin film layer.

During the recording, a polarization grating in the volume of the material and a relief grating on the film surface are formed simultaneously. To investigate how much of the diffraction efficiency is due to the SRG and how much, to the volume polarization grating, we first evaluate the optical response of the thin PAZO film. In this simulation the thickness is $y_1 = 300$ nm (figure 1). We then evaluate the diffraction efficiency of the SRG diffraction grating using the RCWA. The SRG is shown in Figure 1. The grating has a pitch $y_2 = 100$ nm, $d = 1000$ nm and is situated on top of a homogeneous thin film of thickness 300 nm (figure 3). The RCWA is applied to a stack of 12 sublayers with ± 5 diffraction orders considered. The homogeneous film is treated as one layer placed on a BK7 substrate.

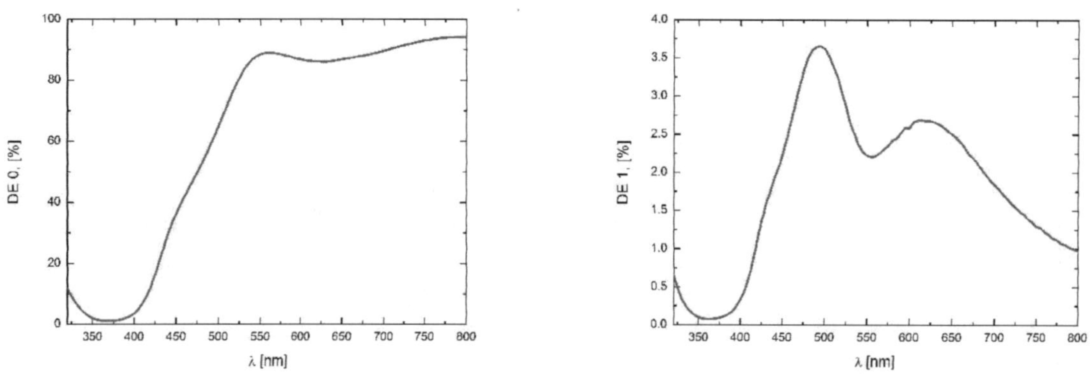

Figure 3. Wavelength dependence of the calculated transmission diffraction efficiency (DE) of SRG on top of a homogeneous thin film. Zero order – left and first order – right.

Next, we evaluate the diffraction efficiency of a structure consisting of an SRG grating with a pitch $y_2 = 100$ nm, and a 300-nm-thick volume grating with sinusoidal modulation. The average refractive index of the azopolymer model grating is set equal to 1.6 and several modulations of $\Delta n = [0.04, 0.06, 0.08]$ are considered. This diffraction structure model is closest to our experimental diffraction structure. The wavelength dependence of the calculated transmission diffraction efficiency for zero and +1 diffractive orders of this structure are shown in figure 4. In the RCWA evaluations, we used ± 5 diffraction orders, 12 sublayers for the SRG, 10 sublayers for the volume sinusoidal grating, and BK7 as a substrate.

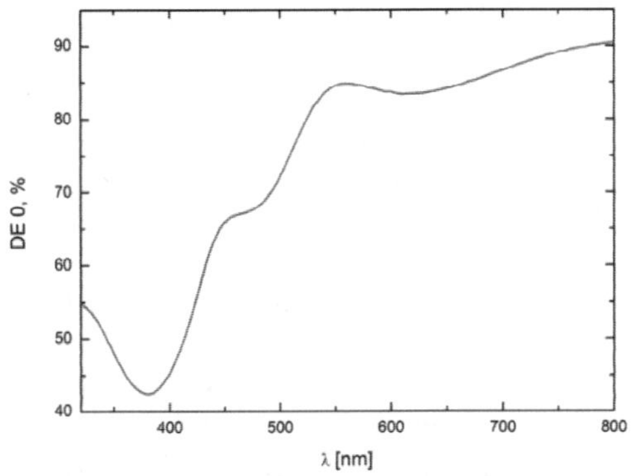

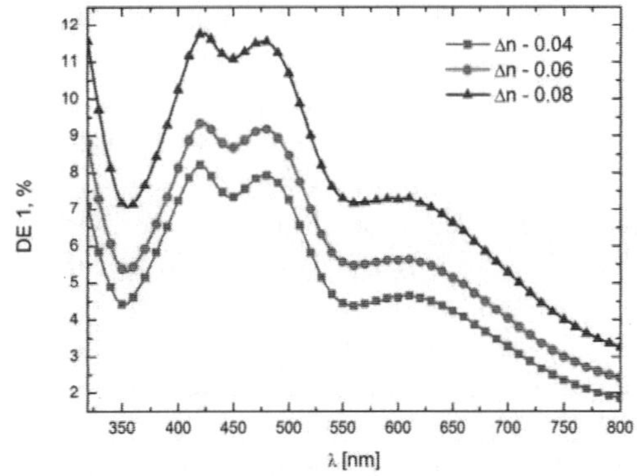

Figure 4. Wavelength dependence of the calculated transmission diffraction efficiency of an SRG diffraction grating on top of a volume sinusoidal grating.
Zero order – left and first order for three different modulations – right.

For our experimentally-recorded grating, at a wavelength of 633 nm we obtained a diffraction efficiency on transmission of first order of 6.1%. For the same wavelength, our model yields a diffraction efficiency of +1 transmission order of 5.45%, and a refractive index modulation slightly above $\Delta n = 0.06$, which agrees with independent measurements of Δn of PAZO [24].

4. Conclusions

We evaluated the optical response of a complex diffraction structure consisting of two diffraction gratings of equal one-dimensional periodicity: a thin-film surface-relief grating structure and a volume polarization grating.

The determination of the realistic models and their uncertainties are a challenge, which will require further experiments, mathematical analysis and computational work and post- data processing. As an important next step, we consider solving the inverse problem: estimation of the polarization grating characteristics from the optical response of the overall "hybrid" structure formed in the thin polymer film, having a sound model of the surface grating based on independent experimental data.

Acknowledgements
The authors are grateful for the financial support provided by Bulgarian National Science Fund under project DN 08/10.

References

[1] Paipulas D, Kudriašov V, Malinauskas M, Smilgevičius V and Sirutkaitis V 2011 *App. Phys.* A: *Mater. Sci. Process* **104** 769
[2] Nedelchev L, Todorov T, Nikolova L, Petrova Tz, Tomova N and Dragostinova V 2001 *Proc. SPIE* **4397** 338
[3] Pan C 2004 *J. Micromech. Microeng.* **14** 471
[4] Magnusson R and Shin D 2001 *Encyclopedia of physical science and technology* Third ed Meyers R ed (San Diego: Academic Press) p 421
[5] Martinez-Ponce G, Petrova T, Tomova N, Dragostinova V, Todorov T and Nikolova L 2004 *Opt. Lett.* **29** 1001
[6] Ramanujam P S, Dam-Hansen C, Berg R. H, Hvilsted S and Nikolova L 2006 *Optics and Lasers in Eng.* **44** 912

IOP Conf. Series: Journal of Physics: Conf. Series **992** (2018) 012018 doi:10.1088/1742-6596/992/1/012018

[7] Sabat R G 2014 *Proc. SPIE* **9006** 900606
[8] Todorov T, Nikolova L, Stoilov G and Hristov B 2007 *Appl. Opt.* **46** 6662
[9] Martin S, Akbari H, Keshri S, Bade D, Naydenova I, Murphy K and Toal V 2017 Holographically recorded low spatial frequency volume Bragg gratings and holographic optical elements *Holographic Materials and Optical Systems* eds. I Naydenova, D Nazarova and T Babeva (InTech)
[10] Tamulevičius T, Šeperys R, Morkūnas V, Kopustinskas V, Andrulevičius M and Tamulevičius S 2010 Diffraction grating based beamsplitters: calculations and implementation *Radiation interaction with material & its uses in technologies* ed Grigonis A (Technologija) p 357
[11] Bartelt H, Glaser T and Schroter S 2001 *Optik* **112** 283
[12] Gaylord T K and Moharam M G 1985 *Proc. IEEE* **73** 894
[13] Beckmann P and Spizzichino A 1987 *Scattering of Electromagnetic Waves from Rough Surfaces* (Artech House Publishers, New York)
[14] Petit R 1980 *Electromagnetic theory of gratings. Topics in current Physics* (Springer-Verlag, Berlin)
[15] Chen X and Friedmann A 1991 *Trans. Amer. Math. Soc.* **323** 465
[16] Moharam M G and Gaylord T K 1981 *J. Opt. Soc. Amer.***71** 811
[17] Moharam M G, Grann E B, Pommet D A and Gaylord T K 1995 *JOSA* A **12** 1068
[18] Li L 1997 *JOSA* A **14** 2758
[19] Bao G 1995 *SIAM J. Numer. Anal.* **32** 1155
[20] www.gsolver.com
[21] www.JCMwave.com
[22] Peng S T, Tamir T and Bertoni H L 1975 *IEEE Trans. Microwave Theory Tech. MTT*-**23** 123
[23] Berberova N, Sharlandjiev P, Stoilova A, Nedelchev L, Nazarova D, Blagoeva B 2017 Optical constants of azopolymer thin films in the spectral range 320 - 800 nm *submitted to J. Phys.: Conf. Series*
[24] Nedelchev L, Nazarova D, Mateev G, Berberova N 2015 *Proc. SPIE* **2004** 9447

Optical constants of azopolymer PAZO thin films in the spectral range 320 – 800 nm

N Berberova[1], P Sharlandjiev[1], A Stoilova[2], L Nedelchev[1,3], D Nazarova[1] and B Blagoeva[1,4]

[1]Acad. J. Malinowski Institute of Optical Materials and Technologies,
Bulgarian Academy of Sciences, Acad. G. Bonchev Str. Bl. 109, 1113 Sofia, Bulgaria
[2]University of Chemical Technology and Metallurgy,
8 Kliment Ohridski Blvd., 1756 Sofia, Bulgaria
[3]University of Telecommunications and Post,
1 Acad. Stefan Mladenov Str., 1700 Sofia, Bulgaria

E-mail: blagoeva.bl@gmail.com

Abstract. Azobenzene containing polymers (azopolymers) have been investigated for a number of applications – holographic data storage, polarization optical elements, diffractive gratings, etc. Herein we report on the optical characteristics of films obtained from the commercially available water soluble azopolymer PAZO (poly[1-[4-(3-carboxy-4-hydroxyphenylazo) benzenesulfonamido]-1,2-ethanediyl, sodium salt). Thin films were deposited by spin coating and their thickness was determined by a Talystep precision profilometer. The FTIR spectra of PAZO were measured between 3000 cm^{-1} and 500 cm^{-1}. For the first time, to the best of our knowledge, we evaluated the PAZO complex refractive index continuously in a broad spectral range (320 – 800 nm) based on transmission and reflection spectrophotometric data. The wavelength dependence of the complex refractive index of organic dyes layers cannot be readily described with dispersion models used for crystals, metals or semiconductors. This is why we used a derivative approach on a wavelength-by-wavelength basis to solve a set of two nonlinear equations with two unknowns – the real and imaginary parts of the refractive index. The determination of the optical constants of PAZO films is effective and robust.

1. Introduction

Azobenzene containing polymers (azopolymers) have been investigated for a number of applications – holographic data storage, polarization optical elements, diffractive gratings etc. [1-6]. All these applications are based on an important property of the azocompounds – the reversible trans-cis photoisomerization observed in them.

In many cases it is essential to know the complex refractive index (RI) of the azopolymer. For example, in the case of thickness measurement using thin-film interference, the measurement precision can be increased significantly if the thin film RI is provided for the entire spectral range used.

[4] To whom any correspondence should be addressed.

However, data about the RI of azocompounds are given rarely and even in such a case, usually for fixed wavelengths [7,8].

Herein we report on the optical characteristics of films obtained from a commercially available azopolymer PAZO (poly[1-[4-(3-carboxy-4-hydroxyphenylazo) benzenesulfonamido]-1,2-ethanediyl, sodium salt). The spectrum of the complex refractive index of PAZO in the range 320 – 800 nm was calculated from transmission and reflection spectrophotometric data at normal incidence.

Our approach was based on the use of incoherent partial waves. The inverse problem was solved on a wavelength-by-wavelength basis. Further, the numeric procedure was based on the so-called "trust-region-dogleg" algorithm that is specially designed to solve nonlinear equations. The calculated transmission and reflection spectra with the evaluated optical constants of PAZO film match the experimental results to a negligible uncertainty.

2. Experimental procedures, model and computational approach

The chemical structure of PAZO is shown in figure 1. This azopolymer is commercially available from Sigma Aldrich. The main advantage of PAZO is its water solubility.

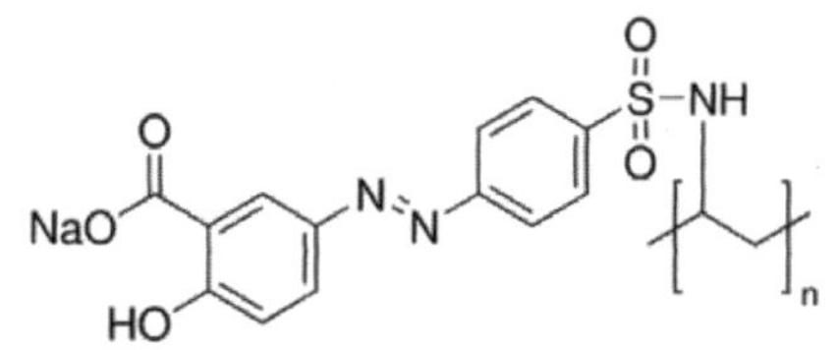

Figure 1. Chemical structure of the azopolymer PAZO.

Thin layers were prepared by spin coating for 30 s at 1500 rpm of a PAZO water solution with concentration 100 mg/ml on 1.96-mm-thick quartz substrates. The film thickness was measured by a high-resolution Talystep profilometer. The experimental data reported below were acquired from a 440-nm-thick PAZO sample.

The transmittance and reflectance spectra of azo-films were measured by a UV-Vis-NIR Cary 5E spectrophotometer (Varian) at normal incidence. The spectral range of evaluation covered 320 nm to 800 nm. The results are presented in figure 2.

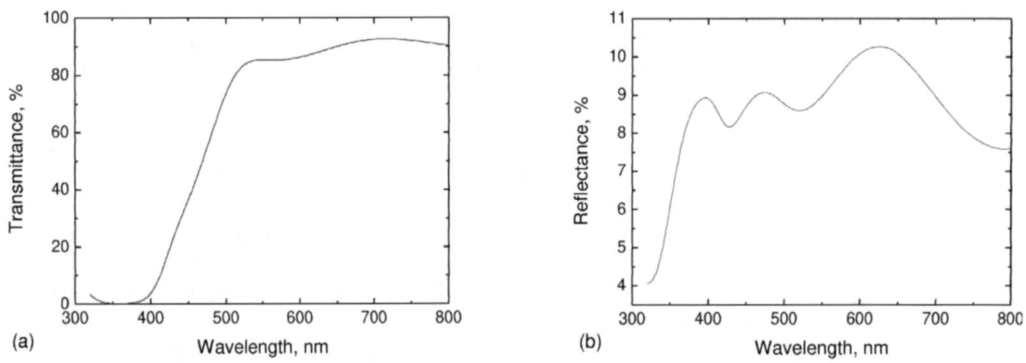

Figure 2. The transmittance (a) and reflectance (b) spectra of the 440-nm-thick PAZO film.

The powdered PAZO polymer was studied by a Varian 600 Series FTIR Spectrometer with a spectral resolution of 0.07 cm^{-1} (figure 3) using a method described earlier [9]. The strong absorbance at 1750 cm^{-1} in the IR spectrum could be explained with the presence of a carbonyl group (C=O). The peak at 1523 cm^{-1} could be related to stretching vibrations of azo compounds bearing the functional group R^1–N=N–R^2, namely to phenyl rings linked by an N=N double bond. The peak at 1195 cm^{-1} shows the presence of an S=O double bond. The absorbance at 867 cm^{-1} and 761 cm^{-1} could be related to deformation vibrations of aromatic rings with two neighboring hydrogen atoms [10-13]. This deformation suggests the presence of certain inhomogeneities in the deposited thin film that could lead to light scattering or functional dependence of the layer's refractive index on the film depth. For this reason, we will treat the complex refractive index evaluated as an effective index of equivalent homogeneous isotropic non-corrugated film with a predetermined physical thickness.

Our goal was to determine the wavelength dependence of the PAZO layer's real refractive index N and extinction coefficient K from experimental data by mathematical minimization techniques. Usually, the estimation of the unknown parameters N and K is regarded as a heavily over-determined inverse problem, i.e. 6 to 20 unknowns (describing N and K) in a set of 300 to 500 wavelength-dependent equations. Besides, the spectral dependence of the complex refractive index of layers containing organic dyes, as the azo-polymer PAZO, cannot be readily described with dispersion models used for crystals, metals, semiconductors: Drude, Lorentz, Farouhy and Blooner, Lorentz and Tauc models, to name a few. This is why we used a derivative approach to solve a set of two nonlinear equations, related to the measured transmission (T) and reflectance (R), on a wavelength-by-wavelength basis. The procedure was based on the so-called 'trust-region-dogleg' algorithm [14]. The evaluation was focused on computing a trial step to a global minimum. This step was constructed from a step along the steepest descent direction and a step in the Gauss-Newton direction [14]. The leading idea of this method is to define a region in the parametric N-K space (trust-region), where the optical response can be approximated by a simpler function.

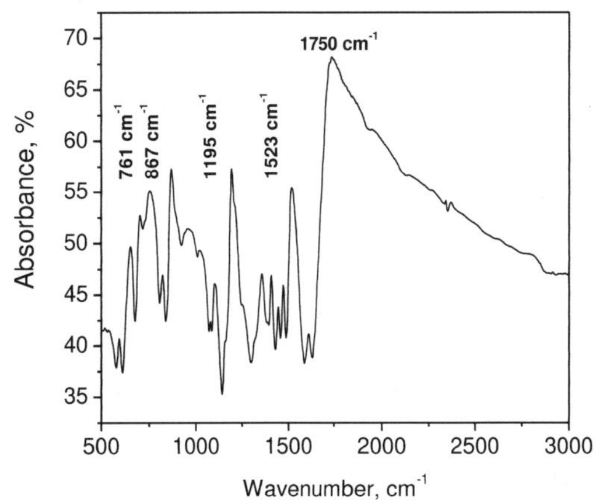

Figure 3. FTIR characterization of PAZO polymer.

3. Results and discussion

We tried two different procedures for the construction of the minimization goal function. In the first one, the Abelès characteristic matrix approach, the complex-amplitude reflection and transmission coefficients were calculated from 2×2 matrix elements. The physical picture behind this is the coherent superposition of multiple reflected partial waves within the film. This is the "classical" way for thin-films optical response estimation. If, for some reasons, the coherence of the partial waves is destroyed, the optical response is estimated by multiple reflections of partial wave's intensities, not complex field amplitudes. This was the second method, namely the incoherent approach [15]. The overall layer reflectance and transmittance were calculated by the elements of the product of three 2×2 matrices, whose elements themselves are functions of the complex refractive index, physical thickness, angle of incidence and wavelength. All this is easily generalized for a multilayer stack. Details can be found in ref. 15. The results obtained showed that the incoherent approach is by far more appropriate for our problem. As a measure for the goodness of the estimation, we calculated the root-mean-square uncertainty (RMSU), normalized by the number of degrees of freedom. Thus, RMSU gives the error per one point of the observations. It is a statistical measure of the discrepancy between the measured spectra of T and R, and the estimated transmission and reflectance. RMSU is compared to the experimental uncertainty.

First, we estimated the optical constants of the uncoated substrate. Then we considered the stack PAZO layer on the substrate and constructed its optical response. Then, with the "trust-region-dogleg" algorithm, we calculated the optical constants of the 440-nm-thick PAZO film as functions of the wavelength. The estimation of the optical constants N and K by means of the proposed model and the described minimization technique resulted in RMSU ~ 0^{-9}. This is much lower than the experimental error, which was ~ 10^{-4} in T and ~ 5×10^{-4} in R. The results are presented in figure 4.

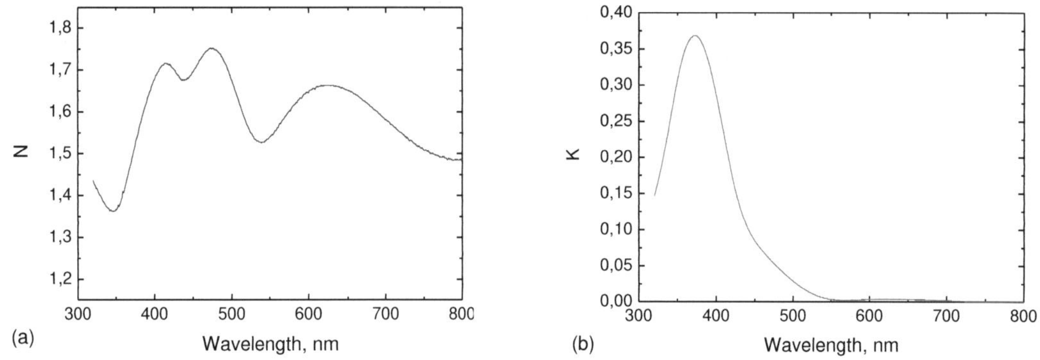

Figure 4. Wavelength dependence of the calculated optical constants. (a) polymer refractive index *N* and (b) extinction coefficient *K*.

4. Conclusions

The complex refractive index of a PAZO layer was evaluated from transmission and reflection spectrophotometric data at normal incidence between 320 nm and 800 nm. For determination of the spectral dependence of the refractive index of azopolymer thin films, we proposed a minimization procedure based on solving a set of two nonlinear equations on a wavelength-by-wavelength basis. The equations were related to experimental data under the assumption of multiple incoherent wave propagations within the thin layer. A solution of the inverse optical problem was obtained by the so-called “trust-region-dogleg” algorithm. The evaluation of the optical constants of PAZO films is effective and robust. Although there is a significant number of publications on optical properties of azopolymers, to the best of our knowledge, this is the first time that spectra of complex refractive index are reported. We regard the proposed approach as sound and reliable that could be applied to other azopolymer layers as well, even in the case of a photo-induced anisotropic optical response. This approach could promote the research on organic-dye sensitized polymers and the design of new types of optical devices.

Acknowledgement

A. Stoilova, L. Nedelchev, D. Nazarova and B. Blagoeva are grateful for the financial support provided by the Bulgarian Science Fund under project DN 08/10. N. Berberova acknowledges the financial support of “For Women in Science” (Fondation L’OREAL).

References

[1] Todorov T, Nikolova L and Tomova N 1984 *Appl. Opt.* **23** 4309

[2] Natansohn A and Rochon P 2002 *Chem. Rev.* **102** 4139

[3] Nikolova L and Ramanujam P S 2009 *Polarization Holography* (New York: Cambridge University Press)

[4] Matharu A, Jeeva S and Ramanujam P S 2007 *Chem. Soc. Rev.* **36** 1868

[5] Gindre D, Boeglin A, Fort A, Mager L and Dorkenoo KD 2006 *Opt. Express* **14** 9896

[6] Forcen P, Oriol L, Sanchez C, Rodriguez F, Alcala R, Hvisted S and Jankova K 2007 *Europ. Pol. J.* **43** 3292

[7] Chen C, Yang Z, Qiu F, Cao T, Cao G, Guan Y and Yang D 2016 *Z. Phys. Chem.* **230** 211

[8] Martinez-Ponce G, Petrova T, Tomova N, Dragostinova V, Todorov T and Nikolova L 2004 *J. Opt.* A*: Pure Appl. Opt.* **6** 324

[9] Georgiev A, Bubev E, Dimov D, Yancheva D, Zhivkov I, Krajcovic J, Vala M, Weiter M and Machkova M 2017 *Spectrochim. Acta Mol. Biomol. Spectrosc.* **175** 76

[10] Tolstoy V P, Chernyshova I V and Skryshevsky V A 2003 *Handbook of infrared spectroscopy of ultrathin films* (Hoboken: Wiley–Interscience) chapter 3 p 140

[11] Gregoriou V G and Rodman S E 2002 Vibrational spectroscopy of thin organic films *Handbook of Vibrational Spectroscopy* vol. 4 (New York: Wiley–Interscience) pp 3010-33
[12] Pretsch E, Buhlmann P and Badertscher M 2009 *Structure determination of organic compounds* (Berlin Heidelberg: Springer-Verlag) 4th Completely Revised Edition
[13] Georgiev A, Dimov D, Spassova E, Assa J and Danev G 2014 *J. Mol. Struct.* **1074** 100
[14] Powell M J D 1983 *Variable Metric Methods for Constrained Optimization, Mathematical Programming: The State of the Art* eds A Bachem, M Grotschel and B Korte (Springer Verlag) pp 288–311
[15] Harbeke B 1986 *Appl.Phys.* B **39** 165

IOP Conf. Series: Journal of Physics: Conf. Series **992** (2018) 012020 doi:10.1088/1742-6596/992/1/012020

Monte-Carlo simulation of the birefringence kinetics in azobenzene-containing materials

E Stoykova[1,2,4], G Mateev[1], B Blagoeva[1], L Nedelchev[1,3] and D Nazarova[1]

[1]Acad. J. Malinowski Institute of Optical Materials and Technologies, Bulgarian Academy of Sciences, Acad. G. Bonchev Str., Bl. 109, 1113 Sofia, Bulgaria
[2]VR/AR Research Center, Korea Electronics Technology Institute, 8 Floor, 11 World cup buk-ro 54-gil, Seoul, South Korea
[3]University of Telecommunications and Post, 1 Acad. St. Mladenov, 1700 Sofia, Bulgaria

E-mail: estoykova@iomt.bas.bg

Abstract. The photoinduced optical anisotropy (birefringence and dichroism) in azobenzene-containing materials gives rise to a variety of applications in photonics by way of controlling the anisotropic properties at a nanoscale level. These applications require modeling the photoisomerization process kinetics. This is a complex and stochastic task that depends on many parameters. The paper presents a computationally effective Monte-Carlo simulation of recording, relaxation and erasure of the photoinduced anisotropy for a large number of 3D-oriented molecules at small angular increments between the neighboring orientations. The modeling is in good agreement with the experiment for measuring the birefringence in the PAZO polymer.

1. Introduction

The phenomenon of photoinduced optical anisotropy (POA) in azobenzene-containing materials, manifested as the appearance of birefringence and dichroism under irradiation with polarized light [1], finds various applications in photonics, optoelectronics and holographic data storage. Controlling the anisotropic properties at a nanoscale level is of vital importance for these applications and requires modeling the kinetics of the photo-reorientation of azochromophores during all typical processes occurring in the material, as e.g. recording, relaxation and erasure of birefringence [2-4].

The polarized light induces angular reorientation of the molecules via a sequence of reversible *trans-cis-trans* photoisomerization cycles. The *trans* isomers are axial molecules with a rod-like shape, so that the probability of their excitation depends on the angle between the polarization direction and the molecule dipole moment oriented along the long molecular axis. Thus, the governing mechanism of creating birefringence is the angular hole burning effect expressed as an excess of inactive azochromophores aligned perpendicularly to the pump light polarization direction. Alignment occurs through *trans-cis-trans* photoisomerization and remains preserved in its larger part after the end of the recording process.

Describing the kinetics of the angular reorientation of molecules while taking into account the spontaneous relaxation of the *cis*-state and thermal effects, such as rotational diffusion, is a complex

[4] To whom any correspondence should be addressed.

IOP Conf. Series: Journal of Physics: Conf. Series **992** (2018) 012020 doi:10.1088/1742-6596/992/1/012020

and stochastic in nature task characterized by many parameters. The Monte-Carlo method is a powerful approach for solving such a task due to its flexibility and the option for tracking the possible impact of any of the parameters involved [5,6]. The Monte-Carlo determination of the time-evolution of angular distributions of *trans* and *cis* isomers is based on a set of kinetic equations [2,7-10]. The statistical accuracy of the modeling depends on the number of processed molecules and a three-dimensional (3D) description of their orientations [1-14]. This fact entails using simplifications to decrease the computational burden of the simulation. Some solutions considered the problem as two-dimensional [7] and restricted the number of simulated orientations [7,11], excluded the thermal effects [11] or took a small number of molecules as a consequence of the limited power of the available computers [11,13].

This paper presents a Monte-Carlo simulation of recording, relaxation and erasure of the POA for a large number of 3D-oriented molecules at small angular increments between the neighboring orientations. We consider also the relaxation processes in which some of the reoriented *trans* molecules are randomized by thermal motion. The results are compared to an experiment performed using thin films of the azopolymer (poly[1-[4-(3-carboxy-4-hydroxyphenylazo) benzenesulfonamido]-1,2-ethanediyl, sodium salt]), shortly denoted as PAZO.

2. Description of the simulation approach

To simulate the photoinduced anisotropy kinetics, we estimated the signal at the output of an optical set-up consisting of two crossed polarizers and a thin film of an azobenzene-containing material placed between them [7]. In such an experiment, the power transmitted through the second polarizer is usually measured by a probe laser beam. The simulation includes three stages: i) recording stage under the action of linearly polarized light with a polarization direction at 45° to the transmission axis of the second polarizer by using a pump laser emitting in the absorption band of the azochromophore; ii) relaxation or reading stage, when the recording beam is turned off; iii) erasure stage under the action of a circularly polarized light or heating. The output signal is given by

$$S(t) \propto N_T(\theta,t)\sin\left[2\left(\theta-\frac{\pi}{4}\right)\right], \tag{1}$$

where $N_T(\theta,t)$ is the number of *trans* molecules oriented at an angle θ between the polarization direction and the transition dipole moment of the molecules at moment t. As seen, the molecules aligned normally to the polarization axis of the pumping light give the main contribution into the signal.

We consider a photoisomerization cycle which includes the following steps: i) resonant excitation of a *trans* isomer followed by transition to the meta-stable ground state of the *cis* isomer; ii) resonant excitation of a *cis* isomer followed by transition to the *trans* ground state; iii) spontaneous relaxation of the metastable *cis* form to the stable *trans* form at a given lifetime of the *cis* state. The excited states of both *trans* and *cis* isomers are too short-lived with respect to the time scales of the POA kinetics to be taken into account. The photoisomerization and spontaneous relaxation are accompanied by reorientation of the molecules. The other reorientation mechanism is the thermally-induced rotational diffusion. The polarization direction of the pump light introduces symmetry in the model; the orientation of a molecule is given by $\Omega=(\theta,\varphi)$ where φ is the azimuthal angle in the plane normal to the polarization direction.

Similarly to the previously reported models, our Monte-Carlo model follows the photoisomerization process as given at any moment t by the kinetic equations for the population densities $T(\Omega)$ and $C(\Omega)$ of *trans* and *cis* isomers with orientation $\Omega=(\theta,\varphi)$ [10]:

$$\frac{dT(\Omega)}{dt}=-\eta P_T(\Omega)T(\Omega)+\eta'\int R_{CT}^{P}(\Omega'\to\Omega)P_C(\Omega')C(\Omega')d\Omega'+\frac{1}{\tau_C}\int R_{CT}^{S}(\Omega'\to\Omega)C(\Omega')d\Omega'+\left(\frac{dT(\Omega)}{dt}\right)_{th}, \tag{2}$$

IOP Conf. Series: Journal of Physics: Conf. Series **992** (2018) 012020 doi:10.1088/1742-6596/992/1/012020

$$\frac{dC(\Omega)}{dt} = \eta \int R_{TC}^{P}(\Omega' \to \Omega) P_T(\Omega') T(\Omega') d\Omega' - \eta' P_C(\Omega) C(\Omega) - \frac{1}{\tau_C} C(\Omega) + \left(\frac{dC(\Omega)}{dt}\right)_{th}, \quad (3)$$

where $P_{T,C}(\Omega)$ and η, η' are the probabilities and quantum yields of *trans*→ *cis* and *cis*→ *trans* transitions, τ_C is the lifetime of the *cis* state, $R_{CT}^P, R_{TC}^P, R_{CT}^S$ are the probabilities for reorientation $\Omega' \to \Omega$ at *cis*→*trans* transition, *trans*→*cis* transition and spontaneous relaxation of the *cis* state respectively. As mentioned in [10], equations (2) and (3) are written in a formal form in view of the fact that the *cis* configuration resembles more a spherical than an axial molecule. The last terms on the right-hand side of both equations give the kinetics of reorientation due to thermal diffusion. The excitation probabilities $P_{T,C}(\Omega)$ differ for the three simulated stages as follows: i) $P_T(\Omega) = \eta I \cos^2\theta$ and $P_C(\Omega) = \eta' I$ during the recording with I being the intensity of the pump light; ii) $P_T(\Omega) = 0$ and $P_C(\Omega) = 0$ during relaxation at turning off the pump light; iii) $P_T(\Omega) = \eta I_{circ}$ and $P_C(\Omega) = \eta' I_{circ}$ during erasure with a circularly polarized light with intensity I_{circ}. We assume uniform distribution of the probabilities $R_{CT}^P, R_{TC}^P, R_{CT}^S$ with an option to introduce correlation between the new and the old molecule orientations.

In simulation, the quantities $dT(\Omega)$ and $dC(\Omega)$ were found at a single Monte-Carlo step, i.e. $dt = 1$. At each step, we simulated all stochastic processes observed for the molecules in the *trans* and *cis* states according to their probabilities. A probability check was performed for each molecule, and a molecule involved in a *cis*→*trans* transition was excluded from the spontaneous relaxation. To describe the 3D orientation of molecules, we represented the angular distributions $T(\Omega)$ and $C(\Omega)$ at a moment t as 2D arrays of size $N_\theta \times N_\varphi$ where N_θ was the number of the simulated discrete values of the angle θ and N_φ was the number of φ values. Thus, a single cell $(k,l), k = 1..N_\theta, l = 1..N_\varphi$ in these arrays gives the number of *trans* or *cis* isomers oriented at $\Omega = (\theta = k\delta\theta, \varphi = l\delta\varphi)$. A third dimension was added as $T(k,l,n) \equiv T(k\delta\theta, l\delta\varphi, ndt)$ and $C(k,l,n) \equiv C(k\delta\theta, l\delta\varphi, n\Delta t)$ to follow the photoisomerization process in time according to the scheme:

$$T(k,l,n) = T(k,l,n-1) - DT(k,l) + RPC(k,l) + RSC(k,l), \quad (4)$$

$$C(k,l,n) = C(k,l,n-1) - DC(k,l) + RPT(k,l) - DSC(k,l), \quad (5)$$

where $n = 1..N$ indicates the current simulation step. The 2D arrays introduced in the right-hand side of equations (4) and (5) account for the reordering of the molecules orientations as readdressing the molecule which has undergone a certain change of state and orientation as a result of transition or relaxation to a new cell. Accordingly, the *DT* and *DC* arrays give the number of molecules that have undergone *trans*→*cis* and *cis*→*trans* transitions; *DSC*, the number of spontaneous relaxations of *cis* molecules; *RPC* and *RSC*, the number of reoriented *trans* isomers at *cis*→*trans* photoisomerization and spontaneous relaxation; *RPT*, the number of reoriented *cis* isomers at *trans*→*cis* photoisomerization. The randomizing effect of the rotational diffusion was simulated at substantial correlation between the new and old orientations. Usage of arrays allows for easy simulation of the reorientation by generating random integer numbers. As initial conditions, we assume that all molecules are in a *trans* state *T* and have an isotropic distribution.

3. Results and discussion

The algorithm developed was implemented as a MatLab code. Figure 1 shows the time evolution of the output signal $S(t)$. The signal is given in arbitrary units. Initially, 50 molecules were added to each cell of the array *T* at $t = 0$ or $n = 1$. The angular increments were $\delta\theta = 2°$ and $\delta\varphi = 4°$, so that *T* and *C* arrays were of size 91×90 and contained a total of 409 500 molecules. The lifetime of the *cis* state was 20 steps, $\eta I = 0.2$, $\eta' I = 0.4$, $\eta I_{circ} = 0.03$, $\eta' I_{circ} = 0.06$. The simulation was conducted for the case when the probability for excitation of the *cis* level is higher than that for the *trans* level. The curve obtained

IOP Conf. Series: Journal of Physics: Conf. Series **992** (2018) 012020 doi:10.1088/1742-6596/992/1/012020

corresponds to recording the photoinduced anisotropy from $n = 1$ to $n = 600$, to the relaxation process without illumination – from $n = 601$ to 1200, and to erasure from $n = 1201$ to 1400. The smoothness of the curve in figure 1 is due to the large number of processed molecules. The algorithm allows for observation of the angular distribution of the *trans* molecules at different moments. Four such distributions corresponding to the numbers in figure 1 are shown in figure 2, where figure 2 (a) reflects the early stage of recording when the *trans* isomers start grouping around the direction $\theta = \pi/2$; in figure 2 (b), the *trans* isomers are practically grouped around this direction. The relaxation process, when the orientations of trans isomers are slightly randomized due to spontaneous relaxation of the *cis* state and rotational diffusion, is shown in figure 2(c). The alignment of *trans* isomers normally to the polarization direction is destroyed under the action of circularly polarized light, as seen in figure 2(d).

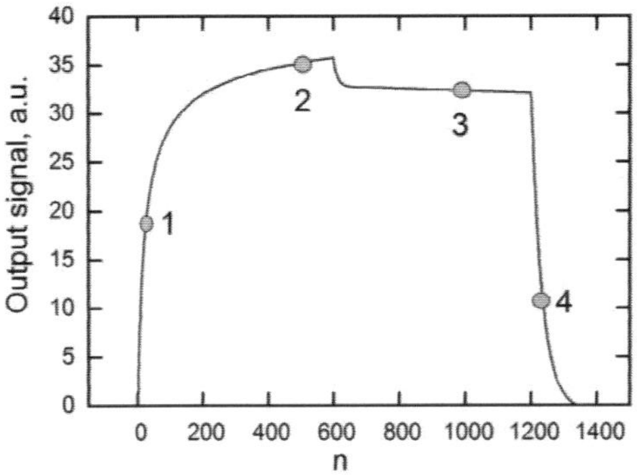

Figure 1. Time evolution of the output signal at i) illumination with a linearly polarized light from $n = 1$ to 600 (recording), ii) turning off the light from $n = 601$ to 1200 (relaxation or reading), iii) illumination with a circularly polarized light from $n = 1201$ to 1400 (erasure); n is the number of Monte-Carlo steps. The numbers correspond to the distributions in figure 2.

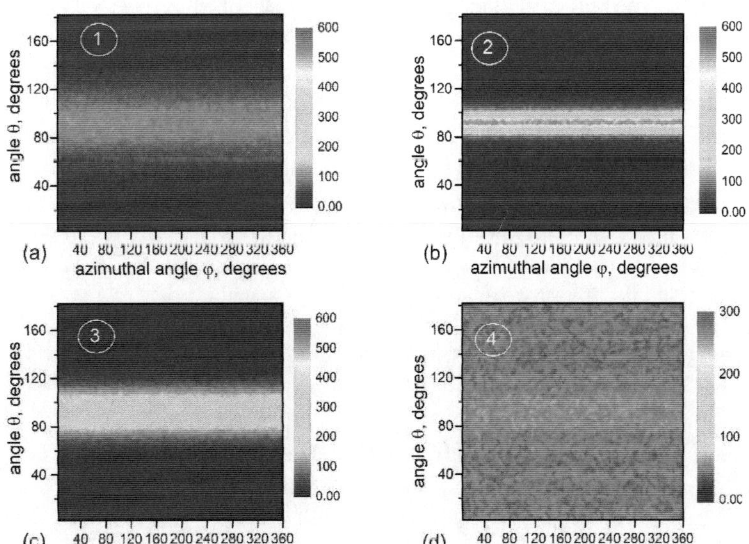

Figure 2. Angular distributions of *trans* molecules at (a,b) recording, (c) relaxation and (d) erasure.

For optical testing, the PAZO polymer was dissolved in water and films were prepared by spin coating at 1500 rpm on glass substrates, thus forming 500-nm uniform thin films. The film thickness was measured by a high-precision Talystep profilometer (Taylor Hobson). To record and probe the photoinduced birefringence, two diode-pumped solid-state lasers were used: a recording laser with $\lambda_{rec} = 444$ nm, power of 35 mW; and a reading (probe) laser with $\lambda_{probe} = 635$ nm and power below

2 mW. The reading laser beam passed through a polarizer, oriented at 45°, while the recording beam passed through a vertical polarizer. After passing through the azopolymer sample, the reading beam entered a polarimeter (PAX5710 Polarization Analysing System, Thorlabs), which monitors the Stokes parameters of light in real time. In this case, the birefringence is given by the expression:

$$\Delta n = \frac{\lambda_{probe}}{2\pi d}\arctan\left(\frac{S_3}{S_2}\right), \quad (6)$$

where d is the film thickness, and S_2 and S_3 are the Stokes parameters [15,16]. During the recording, the light from the laser at 444 nm passed through a vertical polarizer, while for erasure, a quarter-wave plate was inserted in the beam path to produce circularly polarized light which randomized the orientation of the azochromophores and erased the birefringence. The result is shown in figure 3.

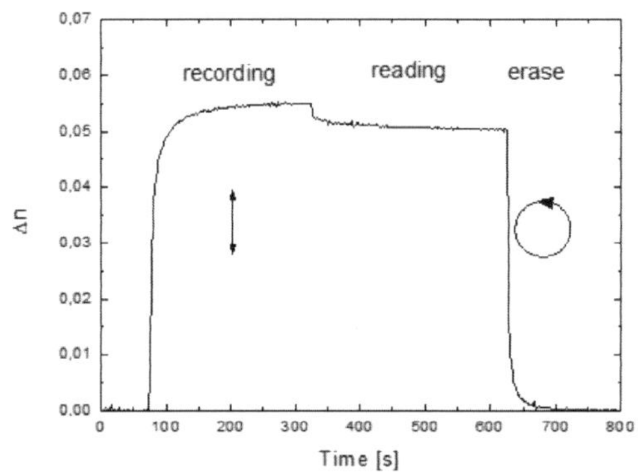

Figure 3. Evolution of the birefringence measured for PAZO polymer at illumination with a linearly polarized light (recording), turning off the light (reading) and erasure with a circularly polarized light.

4. Conclusions

In summary, we proposed and implemented an algorithm for Monte-Carlo simulation of the photoinduced anisotropy kinetics in azobenzene containing materials under resonant excitation with a polarized light. The algorithm is computationally effective and allows one to include a large number of 3D oriented molecules at fine sampling of the orientations in space. The algorithm allows one to simulate recording, relaxation and erasure processes. The simulation result obtained for 409500 molecules oriented at 2° sampling of orientation in the plane passing through the polarization direction and the transition dipole moment of the *trans* molecules and 4°-sampling in the plane normal to the polarization direction are in good agreement with the birefringence curve measured in an experiment we performed with a PAZO polymer. Synthetic experiments with four times as many molecules and increased angular resolution were also performed.

Acknowledgements
The work was supported by the Bulgarian National Science Fund under project DN 08/10.

References

[1] Todorov T, Nikolova L, Stoilov G and Hristov B 2007 *Appl. Opt.* **46** 6662
[2] Kiselev A 2002 *J. Phys.: Cond. Matter.* **14**/49 13417-28
[3] Raschellà R, Marino I, Razzetti C, Bersani D and Lottici P 2007 *JOSA* B **24**/3 504-9
[4] Toshchevikov V, Ilnytskyi J and Saphiannikova M 2017 *J. Phys. Chem. Lett.* **8**/5 1094-98
[5] Cantatore V, Granucci G and Persico M 2012 *J. Comput. Chem.* **33**/10 1015-22
[6] Battaile C 2008 *Computer Methods in Appl. Mechanics and Eng.* **197**/41 3386-98
[7] Natansohn A, Rochon P, Gosselin J and Xie S 1992 *Macromolecules* **25**/8 2268-73
[8] Dumont M, Sekkat Z, Loucif-Saibi R, Nakatani R and Delaire J 1993 *Nonlinear Opt.* **5** 395
[9] Dumont M and Osman A 1999 *Chem. Phys.* **245**/1 437-62
[10] Dumont M 2011 *JOSA* B **28**/8 1855-65
[11] Pawlik G, Mitus A, Miniewicz A and Kajzar F 2003 *J. Chem. Phys.* **119**/13 6789-6801.
[12] Pawlik G, Mitus A, Miniewicz A and Kajzar F 2004 *J. Nonlinear Optical Phys. & Mater.* **13**/03 481-9

[13] Pawlik G, Kordas W, Mitus A, Sahraoui B, Czaplicki R and Kajzar F 2007 Optical materials in defence systems technology *Proc. SPIE* IV 6740**07**
[14] Tavarone R, Charbonneau P and Stark H 2016 *J. Chem. Phys.* **144**/10 104703
[15] Nedelchev L, Matharu A S, Hvilsted S and Ramanujam P S 2003 *Appl. Opt.* **42** 5918
[16] Berberova N, Nazarova D, Nedelchev L, Blagoeva B, Kostadinova D, Marinova V and Stoykova E 2016 *J. Phys.: Conf. Series* **700** 012032

IOP Conf. Series: Journal of Physics: Conf. Series **992** (2018) 012021 doi:10.1088/1742-6596/992/1/012021

Optimum efficiency lidar sensing of multilayer hydrometeors through a turbid atmosphere

Ts T Evgenieva[1] and L L Gurdev

E. Djakov Institute of Electronics, Bulgarian Academy of Sciences,
72 Tzarigradsko Chaussee, 1784 Sofia, Bulgaria

E-mail: tsevgenieva@ie.bas.bg

Abstract. The detected lidar return power is a basic factor determining the brightness of the detected lidar images and the signal-to-noise ratio (SNR) of a given measurement. At equal other characteristics, the laser radiation wavelength should influence the lidar return signal and assume an optimum value depending on the specificity of the objects investigated. As such a problem had not been considered systematically, we recently began developing a modeling approach to solving it, based on evaluating the mean and the noisy lidar profiles and the SNR profile of the measurement along the lidar line of sight by using the lidar equation and well known realistic models of the atmospheric objects and background. The main purpose of the present work is to estimate by numerical modeling the detectability of the lidar return from different distances and multilayer cirrus clouds, depending on the laser radiation wavelengths. The results obtained confirm the expectations that at a higher atmospheric turbidity, a relatively higher sensing efficiency (return power) is achievable by longer-wavelength laser radiation, within the NIR range.

1. Introduction

The lidar remote sensing of the atmosphere is an effective, non-disturbing and large-scale approach to studying the physical and chemical structure, state and dynamics of the atmosphere [1].

The detected lidar-return power is a basic factor determining the brightness and detectability of the "lidar images", as well as the measurement signal-to-noise ratio (SNR), conditioning in turn the minimum detectable contrast. The corresponding lidar signal depends on the characteristics of the lidar system and the media investigated. At equal other characteristics, the laser radiation wavelength should influence the lidar return signal strength and have some optimum value in this sense, depending on the specificity of the investigated objects. As such a question had been investigated mainly for some concrete specific wavelength pairs [2, 3], rather than considered systematically, we began recently [4] developing a modeling approach to the problem, based on evaluating the mean and the noisy lidar profiles and the SNR profile of the measurement along the lidar line of sight (LOS).

A high atmospheric turbidity is unfavorable in view of express and vast-scale ground-based lidar sensing because of the strong light attenuation. At the same time, the sensing of specific contrasting objects under high-turbidity conditions may be of special research interest. Therefore, the main purpose of this work is to estimate by numerical modeling the possibility of successful sensing of such

[1] To whom any correspondence shoul be addressed.

objects, depending on the wavelength of the laser radiation within the UV, VIS and NIR spectral ranges.

2. Basic expressions, approaches and models

2.1. Lidar equation

The optical thickness of the atmospheric layers of interest in this work may exceed unity in some cases. Nevertheless, we shall hereinafter employ the single-scattering lidar equation, assuming corrections performed for the multiple scattering [5], or not so extreme particulate-scattering-due turbidity and a sufficiently small angle of acceptance of the receiving optical system.

The single-scattering elastic lidar equation is convenient to be written in the form [4]

$$N_s(\lambda,z)=(A/z^2)\eta(z)C_L\beta(\lambda,z)\exp\left\{-2\int_0^z \mu_e(\lambda,z')dz'\right\}, \tag{1}$$

where λ is the laser radiation wavelength; $z = ct/2$ is the coordinate of the scattering volume along the LOS, and t is the time period after the pulse emission; A is the receiving aperture area; $\eta(z) = \Omega/(\Omega+A/z^2)$ is the receiving efficiency function, where Ω is the solid angle of view of the receiving optical system; $C_L = cE_0S\tau_e/2e^-$, c is the speed of light, E_0 is the laser pulse energy, S [A/W] is current-to-(light)power photodetection sensitivity, τ_e is the integration (response) time of the receiving electronics, and e^- is the electron charge; $\beta(\lambda, z)$ and $\mu_e(\lambda, z)$ are the LOS profiles of the atmospheric backscattering and extinction coefficients, respectively; and $N_s(\lambda, z)$ is the number of photoelectrons produced in a photon detector during a τ_e-long interval. In the numerical calculations we assume that $A = \pi 10^{-2}$ m^2, $\Omega = \pi 10^{-6}$ sr, $E_0 = 750$ mJ, $S = 0.5$ A/W, and $\tau_e = 5$ ns.

2.2. Signal-to-Noise Ratio (SNR)

The SNR of measuring lidar signal profiles may be written as [1]

$$SNR(\lambda,z)=N_s(\lambda,z)\left\{\left[N_s(\lambda,z)+N_b(\lambda)+N_d\right]\right\}^{-1/2}, \tag{2}$$

where $N_b(\lambda)$ and N_d are the mean numbers of background-due photoelectrons and dark current-due electrons, respectively. Equation (2) is derived [1], assuming Poisson statistics of the photoelectron and dark electron fluctuations. At a dark current of ~ 10^{-10} A [6, 7], we obtain that $N_d = 3$ for $\tau_e = 5$ ns. Also, using experimental data [8], at a 2-nm interference filter, realistic values are determined of $N_b = 302$, 437, 221 and 72 for wavelengths $\lambda = 337.1$, 514.5, 694.3 and 1060 nm, respectively.

2.3. Models of the extinction and backscattering coefficient profiles $\mu_e(\lambda, z)$ and $\beta(\lambda, z)$

The profiles of $\mu_e(\lambda, z)$ and $\beta(\lambda, z)$ consist of several components corresponding to the atmospheric molecular and aerosol constituents and strongly scattering compact objects, like clouds, etc. Each of these components has its own extinction and backscattering coefficient profiles, respectively, $\mu_{em}(\lambda, z)$ and $\beta_m(\lambda, z)$, $\mu_{ea}(\lambda, z)$ and $\beta_a(\lambda, z)$, and $\mu_{ec}(\lambda, z)$ and $\beta_c(\lambda, z)$. In turn, $\mu_{em}(\lambda, z)$, $\mu_{ea}(\lambda, z)$ and $\mu_{ec}(\lambda, z)$ are sums of the corresponding absorption (index ab) and scattering (index s) coefficients: $\mu_{em}(\lambda, z) = \mu_{abm}(\lambda,z)+\mu_{sm}(\lambda, z)$, $\mu_{ea}(\lambda, z) = \mu_{aba}(\lambda,z)+\mu_{sa}(\lambda, z)$, and $\mu_{ec}(\lambda, z) = \mu_{abc}(\lambda,z)+\mu_{sc}(\lambda, z)$. The backscattering and extinction (scattering) coefficients are connected by the relations:

$$\beta_m(\lambda,z)=b_m(\lambda,z)\mu_{sm}(\lambda,z),\ \beta_a(\lambda,z)=b_a(\lambda,z)\mu_{ea}(\lambda,z),\ \beta_c(\lambda,z)=b_c(\lambda,z)\mu_{ec}(\lambda,z), \tag{3}$$

where $b_a(\lambda, z)$ and $b_c(\lambda, z)$ are the corresponding lidar ratios, and $b_m(\lambda, z) = 3/8\pi \approx 0.119$ sr^{-1}. The ratio $b_a(\lambda, z)$ is considered here as constant along the vertical, i.e., $b_a(\lambda, z) \equiv b_a(\lambda)$. It is estimated at visibilities S_m of 23 km (clear atmosphere) and 5 km (hazy atmosphere), using an empiric procedure [9]. The values obtained for clear atmosphere and $\lambda = 337.1$, 514.5, 694.3 and 1060 nm are 0.023 sr^{-1}, 0.032 sr^{-1}, 0.041 sr^{-1}, and 0.056 sr^{-1}, respectively. The corresponding values for hazy atmosphere are

IOP Conf. Series: Journal of Physics: Conf. Series **992** (2018) 012021 doi:10.1088/1742-6596/992/1/012021

0.012 sr^{-1}, 0.016 sr^{-1}, 0.021 sr^{-1}, and 0.028 sr^{-1} (see also in [10, 11]). For cirrus clouds we assume that $b_c(\lambda, z) = \text{const} = 0.1\ sr^{-1}$ [1, 5, 12]. The overall optical thickness is

$$Th(\lambda, z) = \int_0^z \mu_e(\lambda, z')dz' = \int_0^z \mu_{em}(\lambda, z')dz' + \int_0^z \mu_{ea}(\lambda, z')dz' + \int_0^z \mu_{ec}(\lambda, z')dz'. \quad (4)$$

2.3.1. Clear and hazy atmosphere. For an atmosphere without clouds we have $\mu_e(\lambda, z) = \mu_{em}(\lambda, z)+\mu_{ea}(\lambda, z)$ and $\beta(\lambda, z) = \beta_m(\lambda, z)+\beta_a(\lambda, z)$, where the profiles $\mu_{em}(\lambda, z)$, $\mu_{sm}(\lambda, z)$ and $\mu_{ea}(\lambda, z)$ are basic for the numerical modeling and analysis. Realistic data concerning these profiles are taken from a report of McClatchey *et al.* [11], where the vertical sampling interval is 1 km and more. The cases of midlatitude clear and hazy atmospheres are considered corresponding to visibilities of 23 and 5 km [11]. For convenience, the data profiles are approximated here by simple analytical expressions:

$$\mu_{pq}(\lambda, z) = A_{pq}(\lambda) / \left\{1+\exp\left[(z - z_{0,pq}(\lambda)) / w_{pq}(\lambda)\right]\right\}, \quad (5)$$

where p ≡ e or s when q = m, and p ≡ e when q = a; A_{pq} [m^{-1}], $z_{0,pq}$ [m], and w_{pq} [m] are best-fit least-squares approximation parameters. The values obtained of the parameters A_{em}, $z_{0,em}$, w_{em}, A_{sm}, $z_{0,sm}$, w_{sm}, A_{ea}, $z_{0,ea}$ and w_{ea}, are given in table 1. The model data [11] for $\mu_{em}(\lambda, z) = \mu_{sm}(\lambda, z)$ and $\mu_{ea}(\lambda, z)$ along with the corresponding approximating curves in the case of hazy atmosphere and wavelength 514.5 nm are illustrated in figures 1a and 1b, respectively.

Table 1. Parameters of the curves approximating the vertical extinction profiles in clear and hazy atmosphere.

Medium	λ [nm]	A_m [m^{-1}] / A_{sm} [m^{-1}][a]	$z_{0,m}$ [m] / $z_{0,sm}$ [m][a]	w_m [m] / w_{sm} [m][a]	A_a [m^{-1}]	z_{0a} [m]	w_a [m]
	337.1	1.1391E-4	4768.6191	6015.5248	3.20757E-4	901.34287	876.44783
Clear		1.05644E-4	5653.13	5515.10			
Atmoshpere	514.5	3.11541E-5[b]	-377.72112[b]	6654.69124[b]	2.29643E-4	912.98584	895.10918
23km	694.4	1.09915E-4	849.40119	1533.79435	1.76006E-4	890.48116	903.22858
visibility		8.72105E-6	139.1284	6555.60679			
	1060	1.64867E-6	46.13384	6577.02863	1.21505E-4	878.65179	906.27013
	337.1	1.1391E-4	4768.6191	6015.5248	0.00148	886.55158	736.65616
Hazy		1.05644E-4	5653.13	5515.10			
Atmoshpere	514.5	3.11541E-5	-377.72112	6654.69124	0.00107	876.50855	738.06736
5km	694.4	1.09915E-4	849.40119	1533.79435	9.17684E-4	611.12836	819.11022
visibility		8.72105E-6	139.1284	6555.60679			
	1060	1.64867E-6	46.13384	6577.02863	9.39205E-4	-202.60498	1169.59388

[a] Parameters of the curves approximating only the molecular scattering profiles.
[b] $A_m = A_{sm}$, $z_{0,m} = z_{0,sm}$, and $w_m = w_{sm}$ at λ = 514.5 and 1060 nm.

2.3.2. Clouds in the atmosphere. In the presence of clouds we have $\mu_e(\lambda, z) = \mu_{em}(\lambda, z)+\mu_{ea}(\lambda, z)+\mu_{ec}(\lambda, z)$ and $\beta(\lambda, z) = \beta_m(\lambda, z)+\beta_a(\lambda, z)+\beta_c(\lambda, z)$. The additional terms $\mu_{ec}(\lambda, z)$ and $\beta_c(\lambda, z)$ describing the contribution of clouds may be modelled as superpositions, $\mu_{ec}(\lambda, z) = \sum_i \mu_{ec,i}(\lambda, z)$ and $\beta_c(\lambda, z) = \sum_i \beta_{c,i}(\lambda, z)$, of bell-shaped modes

$$\mu_{ec,i}(\lambda, z) / \mu_{ec,i}(\lambda, z_{0,i}) = \beta_{c,i}(\lambda, z) / \beta_{c,i}(\lambda, z_{0,i}) = [1+(z - z_{0,i})^p / w_{c,i}^p]^{-1} \quad (6)$$

that are symmetric with respect to the corresponding positions $z_{0,i}$ of their peak values $\mu_{ec}(\lambda, z_{0,i})$ and $\beta_c(\lambda, z_{0,i})$ and have characteristic widths $w_{c,i}$; p is some integer. We have considered here such a threefold cirrus cloud model with mode parameters: $z_{0,1} = 7500$ m, $\mu_{ec,1}(\lambda, z_{0,1}) = \mu_{ec,1}(z_{0,1}) = 0.5\times10^{-3}$ m^{-1}, and $w_{c,1} = 400$ m; $z_{0,2} = 8500$ m, $\mu_{ec,2}(\lambda, z_{0,2}) = \mu_{ec,2}(z_{0,2}) = 1\times10^{-3}$ m^{-1}, and $w_{c,2} = 250$ m; and $z_{0,3} = 9500$ m, $\mu_{ec,3}(\lambda, z_{0,3}) = \mu_{ec,3}(z_{0,3}) = 0.35\times10^{-3}$ m^{-1}, and $w_{c,3} = 150$ m. The value chosen of p is 4. When considering cirrus clouds, we assume that their extinction and backscattering coefficients do not depend on the laser wavelength within the UV, VIS and NIR spectral ranges [12]. Physically, this is due to the much larger size of the scattering ice particles compared to the wavelength.

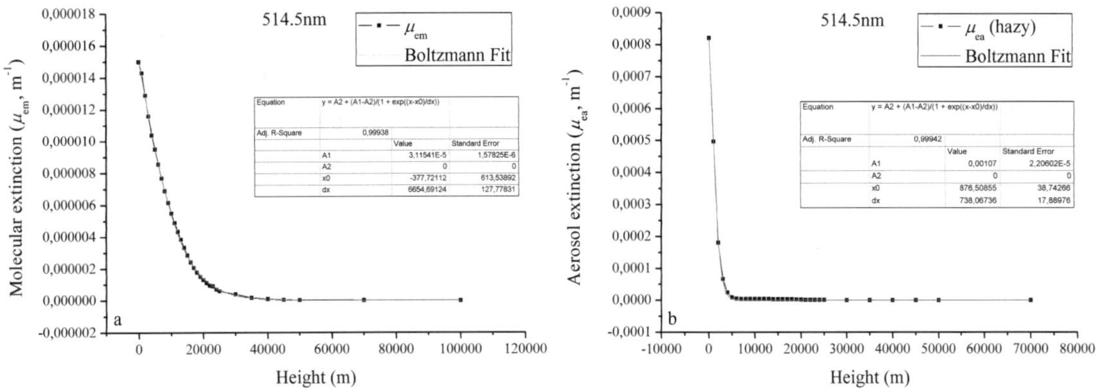

Figure 1. Model data and approximating curves for the molecular (a) and aerosol (b) extinction coefficients in hazy atmosphere at $\lambda = 514.5$ nm.

3. Numerical results and discussion of N_s and *SNR* profiles obtained under different atmospheric conditions and laser wavelengths

All the models and approximations, equations, and concrete values of characteristic parameters described in section 2, as well as some expressions obtained in [4], were employed to evaluate and model profiles of the lidar return signal $N_s(\lambda, z)$ and the corresponding $SNR(\lambda, z)$ from clear and hazy atmospheres containing multilayer cirrus cloud structures. In figure 2, one can see the graphs of the lidar signal profiles from clear and hazy atmospheres with a three-layer cirrus cloud. In figure 2a it is seen that the signal from the space below the cloud is determined mainly by the process of backscattering, because at relatively low turbidity the influence of the attenuation is not so essential. Then, the signal is stronger at shorter wavelengths. The signal from the cloud zone however is stronger at longer wavelengths because of the lower attenuation at these wavelengths. At wavelengths λ=514.5

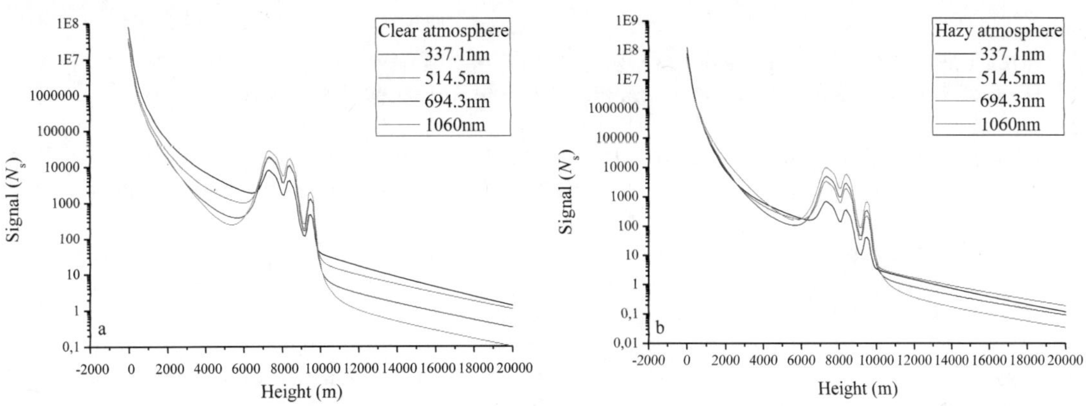

Figure 2. Lidar profiles from clear (a) and hazy (b) atmosphere containing a three-layer cloud field.

IOP Conf. Series: Journal of Physics: Conf. Series **992** (2018) 012021 doi:10.1088/1742-6596/992/1/012021

and 694.3 nm, the signals are comparable because of the comparable attenuation. Above the cloud zone, the backscattering again becomes determinant and the shorter sensing wavelengths provide the stronger signals. If the analog detection threshold is $N_s = N_{s,th} = 1000$ (32 nA photocathode current), the aerosol stratification is detectable in analog mode up to $z = 4$ km.

The main fragments of the cloud are well detectable at $\lambda = 1060$ nm, and hardly, at $\lambda = 514.5$ and 694.3 nm. The detection of clouds through hazy atmosphere is illustrated in figure 2b. It is seen that the signals from the clouds at longer wavelengths are many times higher than those at shorter wavelengths. Besides, large intervals of altitudes appear, below the cloud field, from where the longer wavelengths (694.3 and especially 1060 nm) provide the highest return signals. These are the intervals where the aerosol optical thickness is mainly formed (see also in [3]).

The SNR profiles for clear and hazy atmospheres have a similar run as those of $N_s(z)$. A minimum of about 3% signal variations is detectable, through clear atmosphere, from all the cloud layers at $\lambda = 1060$, 694.3 and 514.5 nm. In a hazy atmosphere, such variations are detectable only at $\lambda = 1060$ nm. Thus, the NIR wavelengths are advantageous for sensing cirrus clouds through a hazy atmosphere.

4. Conclusions

Using an approach we formulated recently, we investigated the efficiency of detecting multilayer cirrus clouds by elastic lidar, through clear (23 km visibility) and hazy (5 km visibility) atmospheres, depending on the sensing lidar radiation wavelength. The approach consists in modeling realistically the lidar signal and SNR profiles at wavelengths in the UV, VIS and NIR spectral ranges. Thus, we determined the optimum sensing wavelengths that ensure maximum signal and SNR levels and, respectively, maximum measuring and imaging sensitivity, accuracy and contrast. We showed that below the cloud base in a clear atmosphere, the shorter wavelengths are advantageous in the above sense because of the stronger (back)scattering and the weak attenuation. Because of the lower attenuation, however, the cloud fields are better detectable by the longer wavelengths. As the atmospheric turbidity increases in a hazy atmosphere, the advantages of the shorter wavelengths decrease and even vanish below the cloud base, because of the increased attenuation. In this case, the cloud layers are still relatively better imaged by the longer wavelengths.

The investigations will be continued with computer simulations of noisy profiles and statistical data processing, illustrating the wavelength influence under various atmospheric conditions.

Acknowledgments

This research is funded by the Bulgarian Academy of Sciences Program for supporting young scientists and PhD students in BAS – 2017, under the project "Optimal effective elastic lidar sensing of aerosol atmosphere containing strongly scattering compact objects" (DFNP-17-4/24.07.2017).

References

[1] Measures R M 1984 *Laser remote sensing* (New York: Wiley)
[2] Russell P B, Swissler T J and McCormick M P 1979 *Appl. Opt.* **18**/22 3783
[3] Ansellet G M, Menzies R T and Tratt D M 1988 *Appl. Opt.* **27**/23 4907
[4] Evgenieva Ts T and Gurdev L L 2017 *Proc. SPIE* **10226** 102260Z
[5] Sakai T, Orikasa N, Nagai T. Murikami M, Kusinoki K, Mori K, Hashimoto A, Matsumura T and Shibata T 2006 *J. Atmos. Sci.* **63** 2156
[6] Ultrafast photodetectors, UPD series, new models: faster rise times and wider wavelength ranges www.aphalas.com
[7] Photomultiplier tubes: Basics and applications 2007 www.hamamatsu.com/resources/pdf/etd/PMT_handbook_v3aE.pdf
[8] Takeuchi N 2005 *Laser remote sensing* ed T Fujii and T Fukuchi (Boca Raton London NY Singapore: CRC Taylor and Francis Group)
[9] Samokhvalov I V (ed) 1985 *Correlation methods for laser location measurements of wind*

velocity (Novosibirsk: Nauka - Siberian branch)

[10] Papayannis A, Mamouri R E, Amiridis V, Remoundaki E, Tsaknakis G, Kokalis P, Veselovskii I, Kolgotin A, Nenes A and Fountoukis C 2012 *Atmos. Chem. and Phys.* **12** 4011

[11] McClatchey R A et al 1972 Optical properties of the atmosphere *Environmental Research Papers* (Air Force Cambridge Res. Lab.) 411

[12] Wright M L, Proctor E K, Gasiorek L S and Liston E M 1975 A preliminary study of air-pollution measurement by active remote sensing techniques NASA CR-132724

Modeling the natural UV irradiation and comparative UV measurements at Moussala BEO (BG)

N Tyutyundzhiev[1], Ch Angelov[2,3], K Lovchinov[1], Hr Nitchev[1], M Petrov[1] and T Arsov[2]

[1]Central Laboratory of Solar Energy and New Energy Sources, Bulgarian Academy of Sciences, 72 Tzarigradsko Chaussee, 1784 Sofia, Bulgaria
[2]Moussala BEO, Institute for Nuclear Research and Nuclear Energy, Bulgarian Academy of Sciences, 72 Tzarigradsko Chaussee, 1784 Sofia, Bulgaria

E-mails: pv-jet@phys.bas.bg, hangelov@inrne.bas.bg

Abstract. Studies of and modeling the impact of natural UV irradiation on the human population are of significant importance for human activity and economics. The sharp increase of environmental problems – extraordinary temperature changes, solar irradiation abnormalities, icy rains – raises the question of developing novel means of assessing and predicting potential UV effects. In this paper, we discuss new UV irradiation modeling based on recent real-time measurements at Moussala Basic Environmental Observatory (BEO) on Moussala Peak (2925 m ASL) in Rila Mountain, Bulgaria, and highlight the development and initial validation of portable embedded devices for UV-A, UV-B monitoring using open-source software architecture, narrow bandpass UV sensors, and the popular Arduino controllers. Despite the high temporal resolution of the VIS and UV irradiation measurements, the results obtained reveal the need of new assumptions in order to minimize the discrepancy with available databases.

1. Introduction

Overexposure to solar radiation, especially in the ultraviolet (UV) range (wavelengths 280–400 nm) of the solar spectrum, is the predominant risk factor for the development of erythema (sunburn) and all forms of skin cancer in humans. The UV radiation is also of great importance in the degradation of construction and household materials. The depletion of the ozone layer and the resulting changes in the UV radiation exposure reflect negatively on the conditions of the atmosphere and the oceans, and on all human activities. Therefore, studies on the variability of UV irradiation and its impact acquire increased urgency.

The UV irradiation's biological impact (positive or negative) varies enormously with its wavelength, so that it is subdivided into three zones: UV-A (400–320 nm), UV-B (320–280 nm) and UV-C (280–200 nm). McKinlay and Diffey [1], found that erythema is caused by irradiation within 295–320 nm, the effect peaking near 305 nm. Both the quality (spectrum) and quantity (intensity) of terrestrial UV irradiation vary with several factors, including the solar zenith angle (SZA), the altitude, and the absorption and scattering by gas molecules (e.g., ozone, nitrogen oxides), water vapor or cloud

[3] To whom any correspondence should be addressed.

formations in the atmosphere. In fact, the solar atmosphere is not in thermodynamic equilibrium, so that the solar UV spectrum cannot be described correctly by Plank's blackbody spectrum (~5800 K). The calculated solar irradiance outside the atmosphere, but at the Earth's mean distance from the sun, is termed as "solar constant" and is approximately 1367 W/m^2. However, similar values of the global horizontal solar irradiance (GHI) have been measured at ground level. The maximal solar UV irradiance and solar GHI (of 1500 W/m^2) have been estimated at high elevations near Cuzco (over 6000 m), Peru, under the assumption of full snow cover and 20% enhancement due to scattered clouds, as reported by Liley and McKenzie [2]. Figure 1 presents a model of the terrestrial solar spectra at sea level, including the combined effect of 33 simulations of the atmospheric gas absorption bands and the harmful solar UV bands.

The solar irradiation variability is predetermined by the space configuration, namely, the planets' elliptical orbits and the Earth's rotational axis inclination with respect to its orbital plane. In addition, the solar energy is not constant but varies with the 27-day apparent solar rotation and an 11-year cycle of sunspot activity. The variability depends also on the scattering and optical absorption of mostly those wavelengths ($\lambda < 290$ nm) that are sensitive to molecular vibrations of atmospheric gases [3]. Solar radiation is not just scattered, some radiation is absorbed by molecules and particles and re-emitted back as thermal infrared (IR) radiation.

The UV irradiation influence on human health is being investigated intensively [4, 5]. There are positive and negative effects. The DNA damage induced by UV radiation is wavelength-dependent because of the strong absorption by DNA chains at wavelengths below 320 nm. The health benefits of UV-A radiation are principally derived from the synthesis of vitamin D in the skin, which supports bone health. UV-B rays are extremely energetic and harmful to the skin; they cause skin damage such as erythema and are responsible for 65 % of skin tumors.

The UV index (UVI) is a measure of the intensity of solar UV radiation at the Earth's surface, weighted erythemally. According to the World Health Organization recommendations (2002) [6], the UV Index values can be grouped into five exposure categories (UVI: 0-12 units).

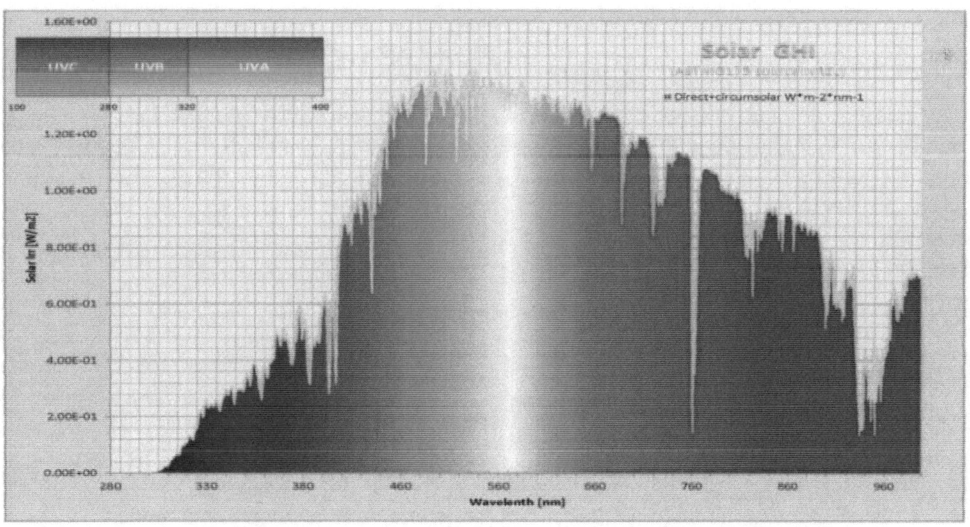

Figure 1. Adopted standard solar terrestrial spectra (ASTM G173) including harmful solar UV bands.

At present, the majority of countries have adopted four skin types:

(white -> yellow/white -> yellow/black -> black),

as a function of the tanning capacity and the UV dose [J/m^2] limits [7].

The low irradiation in wintertime results in diseases associated with vitamin D deficiency. The threshold for insufficient vitamin D production is about 0.7 kJ/m^2 per day, the erythemally-weighted

UV dose is approximately 0.55 kJ/m^2 per day, with a corresponding peak UVI = 1.3 at noon. In contrast, in bright summertime, if one considers a person without sun-protection having the most sensitive skin type I (UV dose required to induce erythema of 250 J/m^2) and UVI = 12, then the skin damage time of exposure is 13.9 min. Such short exposure times are usually underestimated in climatic UV measurements and solar irradiation forecasting.

2. Measurements and modeling

2.1. Instruments and measurements

In the work presented, we developed specific UV sensors and portable data-loggers following the open-source software and open-source hardware approach. The purpose of our UV sensor design based on open-source architecture [8] was to explore the UV irradiation with a higher temporal resolution (< 2 min).

The Arduino platform provided additional flexibility, a higher temporal resolution and automation to the on-field measurement techniques and procedures, thus demonstrating a successful new low-cost approach to the open-source fabrication of scientific tools. The novel lightweight sensors for UV solar irradiation are based on the properties of wide-bandgap semiconductor materials (in particular diamond, GaN, SiC and Al-Ga-nitrides), and are characterized by visible solar blindness, radiation hardness and ageing robustness.

In this study, we made use of the latest generation of digital sensors (VEML6070) to measure the UV irradiance and to calculate the UV Index. A narrow-band Fabry-Perot interference filter (patented FILTRON thin-film technology and CMOS fabrication process) was attached directly to the photodetector. A VEML 6070 sensor breakout (manufacturer specified accuracy of 10%) was employed to measure the global UV-A and UV-B irradiance between the wavelengths of 305 nm and 375 nm. It provides in-factory sensor calibration and 16-bit analog to digital circuitry on the same chip. The spectral response of the sensor is shown in figure 2. The UVI calibration procedure was performed under laboratory conditions. The corrected voltages were converted into UV indexes following the calibration equation below:

$$\boldsymbol{UVI} = 0.04(\int S(\lambda)Ery(\lambda)), \quad (1)$$

where $SS(\lambda)$ is the solar irradiance spectrum and $Ery(\lambda)$ is the McKinlay-Diffey erythemal action spectrum defined in [1]. Under the assumption that the total UV radiation ($UVTotal$) is the UV irradiation measured by the sensor at the fixed ratio (defined by the sensor manufacturer) UVA = 94.1% of $UVTotal$, and UVB = 5.9% of $UVTotal$, then

,

$$UVI = 0.04UVTota(0.941k1 + 0.059k2), \quad (2)$$

where $k1 = 0.00070$, $k2 = 0.05221$ are the averaged erythemal factors for UV-A and UV-B, respectively.

Our field campaign was conducted between 20 June and 10 July 2017, at the Moussala BEO (2925 m ASL) in Rila Mountain. The measurements were performed under cloudy- and near clear-sky conditions. The instrumental setup consisted of several commonly-used and specially-designed sensing instruments: a broadband Kipp-Zonen pyranometer for solar GHI irradiation, UV-A, UV-B selective sensors for UV irradiation measurements, pressure and humidity sensors and a MICROTOPS II Ozone Monitor-Sun-Photometer.

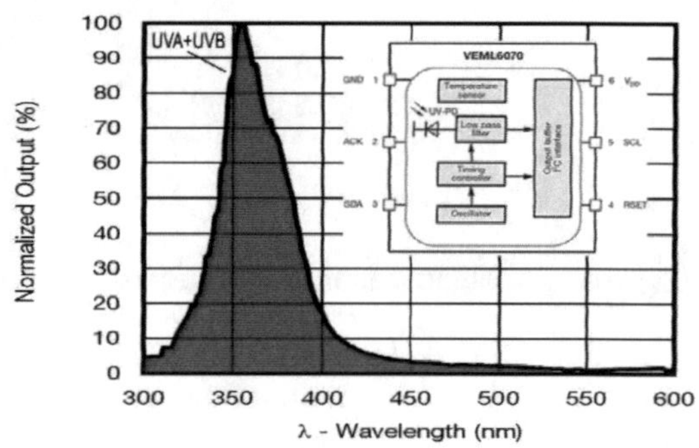

Figure 2. UVA + UVB sensor sensitivity.

The UV sensors readings were acquired and recorded as 2-min averages by an open-source hardware platform (Arduino-based Adafruit Feather M0 data logger). Open-source integrated development environment (IDE) programing tools and a computer code based on C++ were used to develop in-home flexible control software. The Observatory roof and the location of the UV sensors are shown in figure 3.

The light-weight, low-cost UV sensors were chosen due to the open-hardware architecture, easy replacement and protection against the high risks of lightning and atmospheric electrostatic discharges at the high mountain area during UV-active seasons.

Figure 3. UV sensor equipment and bio-container at Moussala BEO.

Comparisons of the UV instruments performed by other authors [9] characterize the accuracy of the professional and low-cost UV instruments to about 0.5 UVI and 1 UVI, respectively. The variability under these thresholds is not significant. The UV measuring equipment based on photon UV sensors is a good low-cost instrumentation able to evaluate UV health risks and provide recommendations to the general public.

2.2. Modeling

Besides measurement activities, modeling the radiative transfer of UV radiation is deemed an important factor in assessing the UV irradiation effects. We considered the UV reduction due to clouds on the basis of daily doses.

In the model used, the daily values of the clear-sky global solar radiation ($GHI_{clearsky}$) were calculated using an empirical relation [10]. A fitted expression as a function of the cosine of the zenith angle (μ) was derived from clear-sky radiometer readings in The Netherlands, Germany, and Greece, and measurements at EU sites which are validated to cover the intra-regions, such as the Bulgarian site of interest.

This, the daily UV-reduction factor (*GDF*) vs. solar *GHI* (day-to-day basis) was assumed to be

$$GHI_{clearsky} = 1120(0.15 + \mu)\exp(-0.213 R_{\mathrm{eff}}) \qquad \text{den Outer, [1998],} \qquad (3)$$

where R_{eff} is solar radiation's relative path length through a curved atmosphere; R_{eff} varies from 1 for overhead sun ($\mu = 1$) to ~ 15 at sunset ($\mu = 0$).

Based on 125 ground stations, Matthijsen et al. [2000] defined:

$$GDF = a(GHI/GHI_{\mathrm{clearsky}})^b = UV/UV_{clearsky} \qquad \text{Bodeker and McKenzie, [1996]),} \qquad (4)$$

where $a = 1.05$ and $b = 0.84$.

The daily clear-sky UV doses ($UV_{clearsky}$) were calculated at sea level using their dependence on the ozone column TOZ and on the solar zenith angle SZA. The daily mean ozone column data were obtained from the satellite total ozone mapping database.

The comparison with the ground-based data obtained by the MICROTOPS II Ozone Monitor revealed differences of more than 30%; these may be attributed to the wide grid of the satellite datasets, the non-relevant cloudiness index and the lack of a parameter describing the altitude (atmospheric pressure) effect.

3. Results and discussions

The largest variability and the maximum amplitude were observed in early summer during the June–July transition. As can be seen in figure 4, the highest UVI value of 9.32 was registered at 12:12 h on 27 June 2017 (the hottest day of the campaign). The large amplitude variability (more than 90% in a time span of 4 min) occurred under mixed (partly) cloudy conditions due to the strong reflection from high cloud edges and the increased vertical transmissivity inside the clouds enhancing the diffuse UV component.

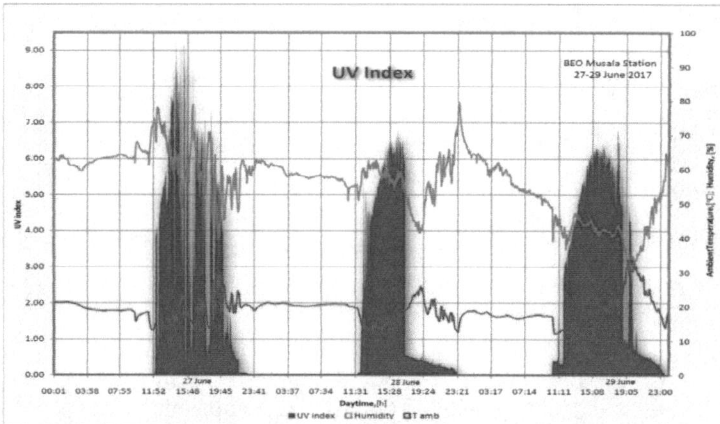

Figure 4. Measurement of the daily solar UV irradiation, environmental temperature and humidity at a high mountain area (Moussala Peak) during the hottest summer week of 2017.

We consider the “deep gaps” (figure 5) observed in the UV index in comparison with the solar GHI during the late afternoon at high SZA to be of paramount significance. Typically, the UV irradiation is calculated as a percentage of the averaged solar GHI values – with similar peaks and flat or slow variations. In the case reported here, the UV short-term variations are of quite different behavior as to be attributed to water drops in low clouds. It is more reasonable to assume that under partly cloudy conditions (40–60%) with sharp peaks (as on 27 June 2017), more than 80% the UV absorption should be attributed to tropospheric clouds of aerosols or pollution gasses. We did not measure critical values

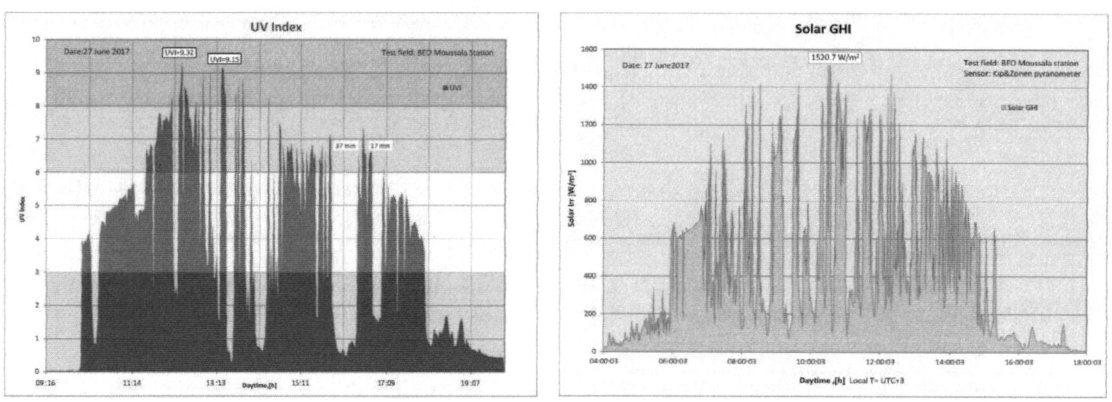

Figure 5. Daily distribution of the UV index, solar GHI and the corresponding risk zones (27 June 2017).

of UVI = 10-12. The average daily ozone column in the same day was 316.5 DU. One study [Erlick and Frederick, 1998] has demonstrated that the attenuation due to aerosols is generally stronger in the UV region than at longer wavelengths. Regardless of their origin, such huge fluctuations in the UV irradiation will cause harmful effects on materials, solar photovoltaic (PV) installations and human skin. Obviously, UV irradiation measurements with a higher temporal resolution can reveal new interactions in the atmospheric UV absorption. Further studies of the tropospheric vertical UV profile could be performed by UV sensors mounted on drones or balloons.

A comparison of our daily UV and solar GHI measurements and online available databases [11] can be extracted from the data presented in table 1. The erythemal UV values of the SoDa database are closer to the ground measurements of the UV daily dose, namely, the same trend and equal overestimation of 34–38% is seen. In what concerns the accuracy of the available solar GHI databases, Meteotest data are closer than other databases to our ground measurements when the cloud cover was less than 20% (29 June), the tolerance being less than 0.1%. On that standard summer day, the daily total solar GHI radiation was close to 8992 Wh/m^2 day, while the daily maximum approached 1032 W/m^2.

Table 1. Comparison between daily parameters from solar UV measurement, simulation and available databases.

Parameter: Daily mean Daily sum		27 June' 17	28 June'17	29 June'17
measured UV dose		0.879 [W.h/m^2.day]	0.640 [W.h/m^2.day]	0.916 [W.h/m^2.day]
calculated UV irradiation		0.639 [W.h/m^2.day]	0.622 [W.h/m^2.day]	0.681 [W.h/m^2.day]
SoDa database:	UV-A	17.69 [W/m^2]	15.53 [W/m^2]	20.05 [W/m^2]
		424.65 [W.h/m^2.day]	372.86 [W.h/m^2.day]	481.29 [W.h/m^2.day]
	UV-B	0.55 [W/m^2]	0.49 [W/m^2]	0.61 [W/m^2]
		13.21 [W.h/m^2.day]	11.77 [W.h/m^2.day]	14.70 [W.h/m^2.day]
	Erythemal UV	0.055 [W/m^2]	0.049 [W/m^2]	0.062 [W/m^2]
		1.336 [W.h/m^2.day]	1.190 [W.h/m^2.day]	1.488 [W.h/m^2.day]
Meteotest database:	UV-A	20.09 [W/m^2]	16.69 [W/m^2]	20.52 [W/m^2]
		482.16 [W.h/m^2.day]	400.56 [W.h/m^2.day]	492.48 [W.h/m^2.day]
	UV-B	0.84 [W/m^2]	0.70 [W/m^2]	0.85 [W/m^2]
		20.16 [W.h/m^2.day]	16.80 [W.h/m^2.day]	20.40 [W.h/m^2.day]
	Erythemal UV	0.079 [W/m^2]	0.079 [W/m^2]	0.080 [W/m^2]
		1.896 [W.h/m^2.day	1.896 [W.h/m^2.day]	1.920 [W.h/m^2.day]
Temis database: erythemal UVI		9.7	9.4	10.3
Temis database: erythemal UV		5.5 [kJ/m^2]	5.4 [kJ/m^2]	5.9 [kJ/m^2]
measured solar GHI		5650.9[Wh/m^2.day]	8004.9[Wh/m^2.day]	8992.2[Wh/m^2.day]
calculated solar GHI		6187 [Wh/m^2.day]	6187 [Wh/m^2.day]	6242 [Wh/m^2.day]
SoDa database:	solar GHI	231 [W/m^2]	359 [W/m^2]	266 [W/m^2]
		5544 [Wh/m^2.day]	8616 [Wh/m^2.day]	6384 [Wh/m^2.day]
Meteotest database:	solar GHI	369 [W/m^2]	370 [W/m^2]	374 [W/m^2]
		8856 [Wh/m^2.day]	8880 [Wh/m^2.day]	8976 [Wh/m^2.day]
NASA database:	solar GHI	255 [W/m^2]	269 [W/m^2]	315 [W/m^2]
		6120 [Wh/m^2.day]	6470 [Wh/m^2.day]	7570 [Wh/m^2.day]

In contrast, under partly cloudy condition (27 June), the daily maximum of solar GHI = 1520 W/m^2 is an extremely high value for Moussala Peak (latitude 42°10'45"N, longitude 23°35'07"E, altitude 2925 m ASL), as is the peak ambient temperature exceeding 30 ^{o}C. Such a "record" value registered in early summer exceeds even the extra-terrestrial solar irradiation and must be taken into account in the planning of future solar PV systems, outdoor construction works and safety measures against forest fires and equipment malfunctions.

IOP Conf. Series: Journal of Physics: Conf. Series **992** (2018) 012022 doi:10.1088/1742-6596/992/1/012022

4. Conclusions

Our new open-source UV monitoring system is currently under evaluation. The preliminary results show a good agreement between our UV sensor design and high-end state-of-the solar GHI instruments. We thus demonstrated the benefits of coupling heterogeneous sensors and detectors for measuring solar irradiation. The open-source software and hardware approach is a convincing argument that the cost of scientific equipment can be radically reduced by applying open-source designs in combination with Arduino electronics. The entire process of rapid design accelerates the development of scientific tools and innovations.

The insufficient quality of the numerical UV prediction models seem to be the main origin of the gap between the results of modeling and measuring the UV irradiation. The cloud cover is a complex object for parametrization which needs further efforts. Further analyses are ongoing.

Acknowledgements
This work was financed in part by the Bulgarian National Science Fund under project DN 04/1/13.12.2016.

References

[1] McKinlay A and Diffey B 1987 *A reference action spectrum for ultra-violet induced erythema in human skin* eds Passchier W F and Bosnajakovic B F M (Elsevier, Amsterdam) 83-7
[2] Liley B and McKenzie R 2006 *Reprint NIWA*, New Zeeland
[3] Diffey B L 2002 *Elsevier Science, Methods* **28** 4–13
[4] Milon A, Sottas P E, Bulliard J L and Vernez D 2007 *J. Expo Sci. Environ Epidemiol.* **17** 58–68
[5] Rivas M, Rojas E, Araya M C and Calaf G M 2015 *Oncol. Lett.* **10**/4 2259-64
[6] World Health Organization 2002 *Global Solar UV index: A Practical Guide*
[7] Estupian J, Raman S, Crescenti H,Streicher J and Barnard W 1996 *J. Geophysical Res.* **101** /D1 16807-16
[8] Amini N, Matthews J E, Dabiri F, Vahdatpour A, Noshadi H and Sarrafzadeh M 2009 *Proc. Int. Conf. Biomedical Electronics and Devices* (14-17 Jan. 2009 Porto Portugal)
[9] Jegou F, Godin-Beekman S, Correa M P, Brogniez C, Auriol F, Peuch V H, Haeffelin M, Pazmino A, Saiag P, Goutail F and Mahe E 2011 *Atmos. Chem. Phys.* **11** 13377–94
[10] Matthijsen J, Slaper H and Reinen H A J M 2000 *J. Geophysical Res.* **105**/D4 5069–80
[11] Pagola I, Gaston M, Brnados A and Fernandez-Peruchena C 2014 *Energy Procedia* **57** 1037 – 43

IOP Conf. Series: Journal of Physics: Conf. Series **992** (2018) 012023 doi:10.1088/1742-6596/992/1/012023

Temperature and excitation power influence on the velocity-selective optical pumping resonances of ^{133}Cs atoms confined in an extremely thin cell

T Vartanyan[1,4], V Polishchuk[1], A Sargsyan[2], A Krasteva[3], St Cartaleva[3] and G Todorov[3]

[1]ITMO University, 49 Kronverkskiy Pr., 197101 St. Petersburg, Russian Federation
[2]Institute for Physical Research, National Academy of Sciences of Armenia, Ashtarak-2, Armenia
[3]Emil Djakov Institute of Electronics, Bulgarian Academy of Sciences, 72 Tsarigradsko Chaussee, 1784 Sofia, Bulgaria

E-mail: Tigran.Vartanyan@mail.ru

Abstract. Linear and nonlinear absorption spectra of ^{133}Cs vapor confined in an extremely thin cell were computed via iterations with respect to the resonance radiation intensity. When the incident radiation intensity is low, the transient polarization of the atoms that undergo frequent collisions with the cell walls leads to sub-Doppler features in the absorption spectra. Higher incident radiation intensities result in the appearance of velocity-selective optical pumping resonances. The theory developed agrees quantitatively with the experimental findings.

1. Introduction

The extremely thin layers of atomic vapors possess unique spectral properties due to the frequent collisions of the atoms with the walls of the optical cell containing the alkali vapor. The development of extremely thin cells (ETC) in 2001 [1] considerably widened the sub-Doppler spectroscopy capabilities. The main features of an atomic vapor confined in an ETC and subjected to resonant electromagnetic excitation are the strong anisotropy of the gas phase, the frequency selection of the atomic velocities and the dominant role of the polarization transients. As the width of the ETC is less than the mean-free-path, the atoms do not reach steady-state polarization. Acting together, these mechanisms lead to the appearance of narrow resonances in the resonance fluorescence excitation spectra, as well as in absorption and reflection. Such narrow resonances are of increasing interest from both fundamental and practical points of view. Many properties of these resonances and their dependence on the excitation conditions have been the object of numerous studies [e.g. 2-5].

In the present paper we present a comparative study of the influence of the temperature and the excitation power density on the absorption resonances of Cs in an ETC. The main new goal was to separate the effects of these experimental factors in order to suggest the possible mechanisms of interaction in the atomic layer and on the cell walls. In the experiments, the power was varied in a very wide range – from 1.25 mW/cm^2 to 125 mW/cm^2 for three values of the temperature, namely, 85 °C,

[4] To whom any correspondence should be addressed.

99 °C and 120 °C. The proposed theoretical description based on an iteration approach is only adequate for power densities in the order of threshold values of the saturation power not exceeding 3 mW/cm^2. This constraint notwithstanding, by varying the relaxation parameters in the numerical experiments, in the third-order solutions we were able to obtain good quantitative agreement with the experimental spectra for values of the power exceeding several times the saturation power density for this particular transition.

2. Theory

Our approach to the theoretical description of an ETC interacting with resonant laser radiation was detailed in [6]. The linear and nonlinear polarizations of atomic vapor sandwiched between transparent dielectric walls are computed for arbitrary values of the atomic levels' full angular momenta. The master equations are written in the irreducible tensor operator (ITO) representation and solved while accounting for the transients induced by the atomic collisions with the ETC walls [4,7-11]. Also, the coherence and population transfer via spontaneous emission is accounted for [6]. The analytical solutions for the components of the density matrix are obtained up to terms of third order with respect to the laser field. The components of the density matrix in the ITO representation ρ_q^{κ} ($\rho = f, \varphi, \xi$), have the following meaning: (f_q^{κ}) characterizes the upper level population and alignment; (φ_q^{κ}) does the same for the lower level; while (ξ_q^{κ}) is the optical coherence directly connected with the polarization of the medium.

We derived solutions for the tensor components of the optical coherence ${}^{(s)}\xi_0^1(x,v) = {}^{(s)}\xi_0^1(x,v>0) + {}^{(s)}\xi_0^1(x,v<0)$ in the first (s = 1) and third (s = 3) order of the iteration procedure and calculated the respective amplitudes of the transmitted field ${}^{(s)}E_{tr}^1$:

$$ {}^{(s)}E_{tr}^1 = (-1)^{F_f - F_\varphi}(2F_\varphi + 1)^{-1/2} 2\pi ikd \int_{-\infty}^{+\infty} dv \left\langle W(v) \int_0^l \exp[-ikx']\, {}^{(s)}\xi_0^1(x',v)dx' \right\rangle; \quad s = 1, 3. \tag{1} $$

We should emphasize that the numerical integration of the analytical solutions was performed by assuming that the relaxation parameters are different γ_ρ^k ($\rho = f, \varphi, \xi$) and by varying their values.

Designating the results of the averaging over the Maxwellian velocity distribution as follows (${}^{(s)}J^+ \equiv \int_0^{\infty} du\langle ...\rangle$, ${}^{(s)}J^- \equiv \int_{-\infty}^{0} du\langle ...\rangle$) $u = v / v_T$, $v_T = \sqrt{2RT/\mu}$), the first order amplitudes may be written in the following form ${}^{(1)}E_{tr}^1 / E_0 = K(F_\rho, \gamma_\rho, kv_T) \left[{}^{(1)}J_1^+ + {}^{(1)}J_1^- \right]$ with

$$ K(F_\rho, \gamma_\rho, kv_T) = \frac{2\pi^{1/2}}{3} \left(\frac{d_{F_f F_\varphi}^2}{\hbar(kv_t)}\right)\left(\frac{\varphi_0^0}{2F_\varphi + 1} - \frac{f_0^0}{2F_f + 1}\right). \tag{2} $$

In the third-order expression for the transmitted field, the last term describes the contribution of the coherence and population transfer via the spontaneous emission process:

$$ {}^{(3)}E_{tr}^1 = \beta_0 . K(F_\rho, \gamma_\rho, kv_T) E_0 . \sum_{k=0,2} \left\langle (S_{000}^{1k} C_{0\,00}^{k1})_{F_\varphi}^{F_f}\, {}^{(3)}J_{f^\kappa}^1 + R_{000}^{1k} \left[B_{000}^{k1}\, {}^{(3)}J_{\varphi^\kappa}^1 - C_{000}^{k1} \left(\frac{\Gamma_{F_f F_\varphi}^{\kappa}}{kv_T}\right) {}^{(3)}J_{sp.tr}^1 \right] \right\rangle, $$

$$ \beta_0 = C_0^k(F_f, F_\varphi) \Omega_{Rabi}^2 (\Gamma_f^{k_2})^2 \quad \Gamma_s^{k_i} = \gamma_s^{k_i} / kv_t, \quad s = f, \xi, \varphi. \tag{3} $$

IOP Conf. Series: Journal of Physics: Conf. Series **992** (2018) 012023 doi:10.1088/1742-6596/992/1/012023

Describing correctly the spontaneous emission is critical in establishing the connection between the nonlinear resonance sign-reversal on the $F_\varphi = 4 \rightarrow F_f = 5$ closed transition of the D_2 line with the analogous sign reversal of the ground-state longitudinal alignment that depends on the transition principal quantum numbers, and on the cell thickness and temperature.

The transmitted field intensity is determined by the squared modulus of the sum of the incident field and the polarization-induced fields ${}^{(1)}E^1_{tr}$ and ${}^{(3)}E^k_{tr}$, and can be written as:

$$ {}^{(\Sigma)}I_{tr} \propto |E_0|^2 \left[1 + 2\,\mathrm{Re}\left(\frac{{}^{(1)}E^1_{tr}}{E_0} + \frac{{}^{(3)}E^{\kappa}_{tr}}{E_0} \right) \right]. \quad (4) $$

3. Numerical modeling and comparison with the experimental findings

Figure 1 plots the results of the numerical modeling and demonstrates the influence of the field intensity and the atomic source temperature. Only the region of the Cs D_2 line spectrum is shown that involves the $F_\varphi = 4 \rightarrow F_f = 4,5$ hyperfine transitions (figure 1a).

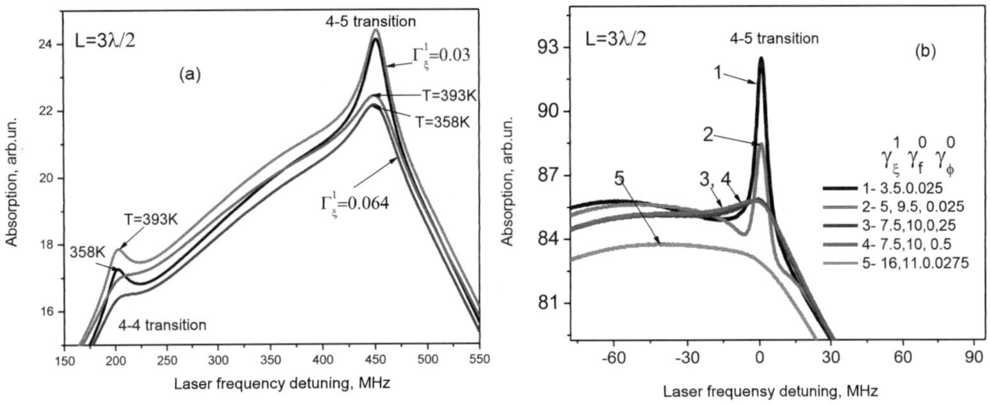

Figure 1. Absorption of an ETC filled with Cs vapor. (a) demonstrates the temperature dependence of the absorption in a 3λ/2-thick ETC, where λ is the resonance wavelength, while (b) shows the 4-5 transition absorption dependence on the relaxation parameters.

As figure 1a shows, a rise of about 10 % of the atomic source temperature leads to a slight but noticeable rise of the resonance amplitude. On the other hand, a twofold rise of the optical coherence relaxation rate Γ^1_ξ reduces the resonance amplitude considerably. The calculation of the nonlinear absorption leads to the conclusion that a fivefold increase of Γ^1_ξ leads to the disappearance of the resonance on the $F_\varphi = 4 \rightarrow F_f = 5$ transition. In addition, we found that raising the relaxation parameters can be used to simulate the power broadening. Indeed, in a strong laser field, the role of the spontaneous transitions diminishes and the resonance absorption follows the same trend (figure 1 b).

In the case of an ETC of thickness $L = 3\lambda/2$, we compared the experimentally-observed nonlinear narrow-width absorption resonances with the respective theoretical simulations (figure 2), for two different atomic source temperatures: 85°C and 120°C. It can be seen that for the lower laser beam intensity (figure 2a), the experimentally-observed nonlinear resonance is very well reproduced by an optical coherence relaxation rate $\Gamma^1_\xi = 0.03$. For the same two temperatures and following the theoretical approach developed, the laser-light-intensity enhancement is reproduced by means of a higher optical coherence relaxation rate, namely $\Gamma^1_\xi = 0.064$ (figure 2b). Thus, the nonlinear resonance broadening could be related to optical coherence relaxation. The computed absorption resonances agree well with the experimental results.

IOP Conf. Series: Journal of Physics: Conf. Series **992** (2018) 012023 doi:10.1088/1742-6596/992/1/012023

We further considered the case of a cell thickness equal to an integer number of optical wavelengths [6] (figure 3a, b). For $L = \lambda$, 3λ, narrow resonances with reduced absorption are observed for all $F_\varphi = 4 \rightarrow F_f = 3,4,5$ transitions. Figure 3 compares the absorption resonances for two different light intensities and atomic source temperature $T = 120\ ^\circ$C. It is important to note that even for incident intensity as high as 15 mW/cm^2, qualitative agreement with the experimental spectra is achieved (figure 3b).

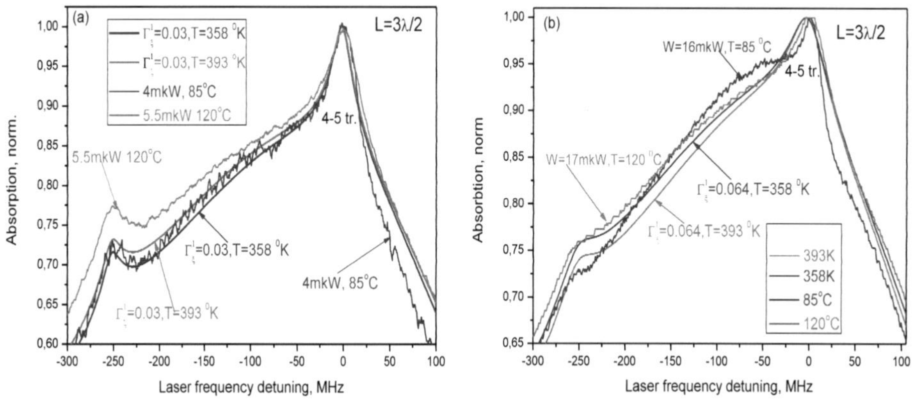

Figure 2. Temperature dependence of the absorption of an ETC filled with Cs vapor:

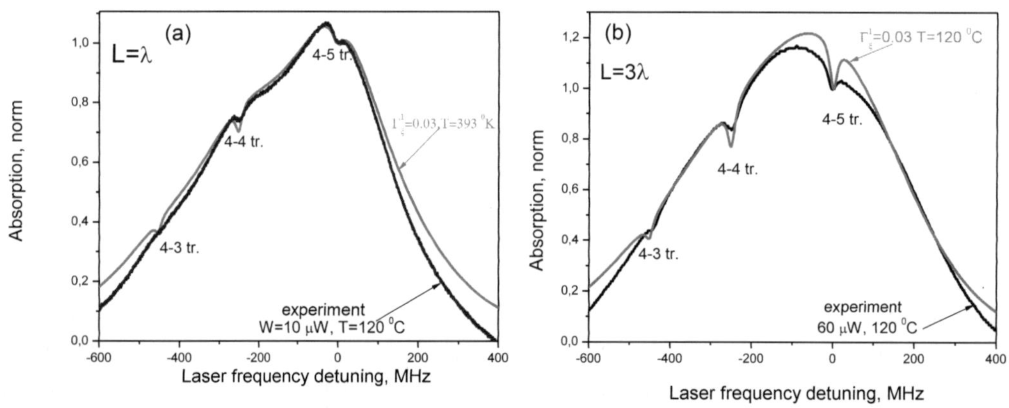

Figure 3. Field intensity dependence of the absorption of an ETC filled with Cs vapor with thickness $L = \lambda$ (a) and $L = 3\lambda$ (b).

4. Conclusions

We performed comparative studies of the roles played by the cell temperature and the incident light intensity in the formation of sub-Doppler resonances in an ETC filled with Cs vapor. In agreement with the experimental findings, the computed absorption spectra demonstrated a reduction in the amplitudes of the sub-Doppler resonances as the optical coherence relaxation rate $\Gamma_\xi^1 = \gamma_\xi^1 / kv_T$ increased. On the other hand, the amplitude of the resonance rose with the rise of the cell temperature. The simulation proposed of the power broadening via an artificial increase of the relaxation parameters $\Gamma_\rho^\kappa = \gamma_\rho^\kappa / kv_T$, $\rho = f, \varphi, \xi$ in the third-order terms of the perturbation series resulted in a good qualitative agreement with the experimental spectra, even for incident intensities several times larger than the saturation threshold.

Acknowledgements
This work was supported by the National Science Fund of Bulgaria under contract DN 08-19/14.12.2016 and by the bilateral project "Nonlinear spectroscopy of spatially restricted alkali vapor: methodology and applications" between Republic of Bulgaria and the Russian Federation (RFBR grant No 17-52-18037, grant 074-U01) and state order No 3.4903.2017/6.7.

References

[1] Sarkisyan D, Bloch D, Papoyan A and Ducloy M 2001 Sub-Doppler spectroscopy by sub-micron thin Cs vapor layer *Opt. Comm.* **200** 201
[2] Izmailov A 1992 Manifestations of sub-Doppler structure of spectral-lines of gas particles in the emission of traveling monochromatic wave of pumping *Laser Phys.* **2** 76
[3] Izmailov A 1993 *Opt. Spectrosc.* **74** 25
[4] Briaudeau S, Bloch D and Ducloy M 1996 Detection of slow atoms in laser spectroscopy of a thin vapor film *Europhys. Lett.* **35** 337
[5] Vartanyan T A and Lin D L 1995 Enhanced selective reflection from a thin layer of a dilute gaseous medium *Phys. Rev.* A **51**
[6] Dutier G, Yarovitski A, Saltiel S, Papoyan A, Sarkisyan D, Bloch D and Ducloy M 2003 Collapse and revival of a Dicke-type coherent narrowing in a sub-micron thick vapor cell transmission spectroscopy *Europhys. Lett.* **63** 35
[7] Todorov G, Polishchuk V, Krasteva A, Sargsyan A D, Cartaleva S and Vartanyan T A 2016 *J. Opt. Tech.* **83** 659-66
[8] Decomps B, Dumont M and Ducloy M 1976 Linear and nonlinear phenomena in laser optical pumping; Laser spectroscopy of atoms and molecules *Topics in Appl. Phys.* **2** (Springer-Verlag 283-347
[9] D'yakonov M I and Perel V I 1966 On the theory of the gas laser in a magnetic field *Opt. Spectrosc.* **20** 472-80
[10] Vartanyan T A and Lin D L 1998 Nonlinear diffraction due to the transient polarization in a thin film of atomic gases *Eur. Phys. J.* D **1** 217-21
[11] Vartanyan T A 1985 Resonant reflection of intense optical radiation from a low density gaseous medium *Sov. Phys JETP* **61** 674–7

IOP Conf. Series: Journal of Physics: Conf. Series **992** (2018) 012024 doi:10.1088/1742-6596/992/1/012024

Preparation and characterisation of carbon-free Cu(111) films on sapphire for graphene synthesis

J. Lehnert, D. Spemann, S. Surjuse, M. Mensing, C. Grüner, P. With, P. Schumacher, A. Finzel, D. Hirsch, B. Rauschenbach

Leibniz Institute of Surface Engineering (IOM), Permoserstr. 15, 04318 Leipzig, Germany

E-mail: daniel.spemann@iom-leipzig.de

Abstract. This work presents an investigation of carbon formed on polycrystalline Cu(111) thin films prepared by ion beam sputtering at room temperature on c-plane Al_2O_3 after thermal treatment in a temperature range between 300 and 1020 °C. The crystallinity of the Cu films was studied by XRD and RBS/channeling and the surface was characterised by Raman spectroscopy, XPS and AFM for each annealing temperature. RBS measurements revealed the diffusion of the Cu into the Al_2O_3 substrate at high temperatures of > 700 °C. Furthermore, a cleaning procedure using UV ozone treatment is presented to remove the carbon from the surface which yields essentially carbon-free Cu films that open the possibility to synthesize graphene of well-controlled thickness (layer number).

1. Introduction

In recent years, a large research effort has been invested on graphene because of its unique physical, mechanical, and electronic properties [1, 2]. Graphene is attractive for a wide variety of applications including semiconductor and high frequency devices [3, 4], optoelectronics [5], spintronics [6] or even catalysis [7] or sensor applications [8]. There are many synthesis routes for graphene [9], some of them employ metallic substrates, for example chemical vapour deposition (CVD) [10, 11] or carbon ion implantation. Indeed, several studies of graphene synthesis using carbon ion implantation can be found in the literature that employ either polycrystalline Ni [12, 13, 14, 15] or Cu foils [16, 17, 18] as substrate for graphene growth. The latter are advantageous due to the large Cu grain sizes in the range of mm achieved after thermal treatment [19]. Note that the Cu foils in these studies had a Cu(200) orientation which results in a lattice mismatch of 19.9% between graphene and the underlying Cu substrate. In contrast, Cu(111) substrates would allow for a lattice mismatch of only 3.8% [20, 21, 22, 23] and are therefore desirable for growth of high quality graphene.

It is essential to start with a carbon-free substrate for the synthesis, in order to have full control on the layer number of the synthesized graphene and therefore on its physical properties, e.g. the behaviour as semiconductor or conductor [24, 25]. Any additional contribution from intrinsic carbon that may have been incorporated into the Cu film during its preparation process has to be avoided. This carbon mostly comes from the residual gas within the deposition chamber. Note that the Cu-foils used in graphene synthesis so far may also not be free of carbon [17].

The aim of this study is the preparation of Cu(111) thin films that are essentially free of carbon and can therefore serve as substrates for high quality graphene growth. For this purpose, Cu(111) thin films sputtered on Al_2O_3 substrates were thermally annealed and the formation of carbon on the surface of the Cu films studied as a function of annealing temperature. The origin of the carbon is discussed and an UV ozone treatment described that allows obtaining carbon-free Cu substrates for subsequent graphene synthesis.

2. Experimental details

The Cu films with a thickness of 200 nm and 450 nm were prepared on 1 cm^2 *c*-plane Al_2O_3 at room temperature by magnetron and ion beam sputtering, respectively (see section 1 in supplementary material). The 200 nm films were annealed in vacuum ($\approx 5 \times 10^{-7}$ mbar) at 300 °C, 500 °C, 700 °C, 750 °C, 800 °C, 850 °C, 900 °C for 1 h, whereas the 450 nm films were annealed at 1020 °C for 20 min.

The chemical composition of the surface of the deposited Cu films on sapphire was determined after annealing with X-ray photoelectron spectroscopy (XPS). Monochromatic Al $K\alpha$ X-rays with 1486.6 eV energy were used for photoelectron emission. Quantification was done with the help of the program UNIFIT (Unifit, Leipzig) using the experimentally estimated analyser transmission functions. For further investigations, Raman measurements were performed with a 473 nm laser excitation to reduce the fluorescence background from the Cu substrate. The laser was coupled on a LabRam HR Evolution spectrometer with a resolution of 1.3 cm^{-1} (600 l mm^{-1} grid) and an Olympus MPlan N 100x/0.90 objective.

The crystallinity was studied by X-ray diffraction (XRD) using a diffractometer Ultima IV equipped with a Cu anode X-ray tube, a graded multilayer mirror to obtain a parallel beam, a theta-theta goniometer, a parallel slit analyser and a scintillation detector.

Rutherford backscattering spectrometry (RBS) and ion channeling were performed to study the composition and crystalline quality of the Cu films as well as interdiffusion of Cu film and Al_2O_3 substrate. For this purpose, a 2.07 MeV He^+ ion beam with a diameter of ≈ 0.8 mm was used. The film composition was determined from RBS data using XRUMP [26].

The surface topography was measured by atomic force microscopy (AFM) with a large sample scanning force microscope. The resolution was 1024 × 1024 pixels for each measurement. The root mean square (RMS_{sq}) roughness was calculated from the processed images.

The carbon formed on the Cu surface was removed by UV-ozone treatment using a Xe_2^* excimer lamp 172/630 Z lamp system, applying a VUV power of up to 30 mW cm^{-2} and a distance from the sample of 2 cm. The excimer lamp has a continuous emission in the VUV wavelength range from λ = 160 to 195 nm (photon energy 7.75 eV $> E_{Ph} >$ 6.36 eV; 614 kJ $mol^{-1} < \Delta E <$ 748 kJ mol^{-1}) with the maximum at 172 nm (E_{Ph} = 7.21 eV; 696 kJ mol^{-1}). Irradiation was carried out in a chamber flushed with nitrogen containing 0.25 vol.-% oxygen to minimize VUV absorption by molecular oxygen and ozone formation on one hand [27] and maintain oxidative conditions on the other, as published for the conversion of perhydropolysilazane to silica [28].

3. Results and discussion

3.1. Crystallinity

The XRD measurements reveal a Cu(111) orientation of the as-prepared Cu films (see fig. 1(a)). Note that no other crystal orientations were found. In contrast to the literature [23] the films were not single-crystalline but consisted of twinned Cu(111) crystals for all annealing treatments and film thicknesses studied here (see section 4 of supplementary material). The peaks at 41.7 ° and 90.8° shown in fig. 1(a) originate from the Al_2O_3 substrate.

In order to investigate the crystallinity of the Cu films RBS-channeling measurements were performed. Figure 1(b) shows two spectra for a sample annealed at 700 °C for 1 h with random

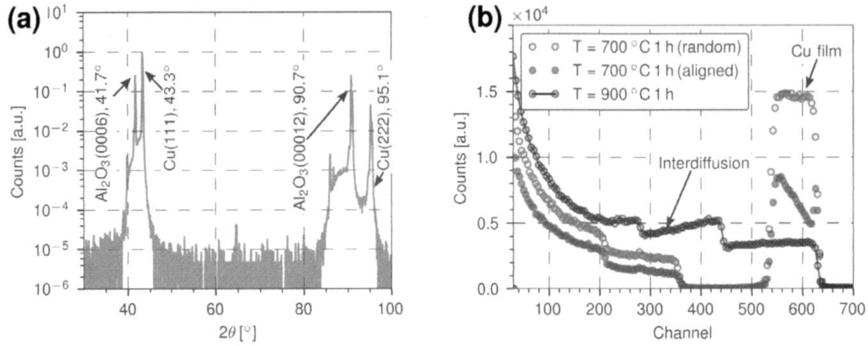

Figure 1. (a) Typical XRD spectrum showing the Cu(111) crystal orientation and (b) RBS spectra of a sample (200 nm thickness) annealed at 700 °C for 1 h for random and aligned crystal orientation as well as a spectrum of a Cu film annealed at 900 °C (200 nm thickness) for 1 h showing strong interdiffusion of the Cu and the Al_2O_3 substrate.

orientation (red circles) and after alignment of the Cu ⟨111⟩ axis with respect to the ion beam (green full circles) indicating ion channelling of the He^+ ions, respectively. For the normalised minimum yield χ_{min} values down to 0.22 were obtained indicating Cu(111) orientation of the film over macroscopic lateral extensions. Channeling into the ⟨111⟩ axis is also possible when crystals are twinned around this axis.

Furthermore, an RBS spectrum of a sample annealed at 900 °C for 1 h (circle-line) is shown in fig. 1(b), which demonstrates the interdiffusion of Cu film and Al_2O_3 substrate at this high temperature and long annealing time. Whereas the Cu film of the sample annealed at 700 °C for 1 h still consists of pure Cu, the composition after 900 °C for 1 h has changed to $Cu_{0.12}Al_{0.36}O_{0.52}$ (see section 2 of supplementary material for details on RBS analysis). Due to this diffusion process the high temperature annealing steps above 1000 °C were performed for the 450 nm thick Cu films and for 20 min only.

3.2. Surface topography

Prior thermal treatment the Cu film show grain sizes of around 50 nm (see fig. 2(a)). Annealing leads to an increase of the grain sizes to several μm. Figure 2(b) shows the RMS_{sq} roughness of the Cu films after the different annealing procedures. The RMS_{sq} roughness of Cu obtained after sputter deposition is drastically lower than the one measured for Cu foils of any quality used for graphene synthesis [16, 19]. This low roughness is beneficial for 2D graphene growth. The large roughness values for temperatures above 800 °C originate from the presence of etch holes (see figure 2(c)). The etch holes are of hexagonal shape with a depth given by the thickness of the Cu film.

3.3. Characterisation of carbon films

In order to investigate the formation of carbon on the surface of the Cu films and thus the diffusion of the impurity carbon atoms from the bulk of the film to its surface, Raman (range $1000\,cm^{-1}$ to $3000\,cm^{-1}$ to observe the graphite/graphene-related D-, G- and 2D-bands) and XPS measurements were performed. Directly after the deposition of the Cu on the Al_2O_3 no carbon was found on the surface. After annealing within the 300 °C to 700 °C temperature range Raman spectra were obtained which indicate the presence of amorphous carbon at the surface (see fig. 3). This is supported by XPS measurements where an increase of the carbon content at the surface from 30 at.-% after the 300 °C anneal to 67 at.-% after the 700 °C anneal

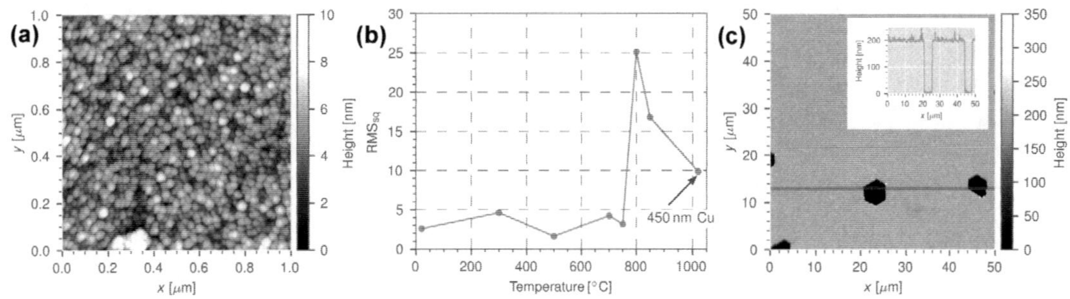

Figure 2. (a) Typical AFM image of a non-annealed sample showing Cu grains of 50 nm size. (b) RMS_{sq} roughness of the differently annealed Cu films and (c) an AFM scan and line plot of a Cu film annealed at 800 °C for 1 h showing etch holes of hexagonal shape and a depth of 200 nm.

has been observed. This carbon is believed to originate from surface adsorbates (see section 3 in supplementary information for more details). The samples annealed at 750 °C, however, showed no Raman bands, despite a carbon content of 56 at.-% as obtained by XPS studies. At temperatures above 800 °C the 2D-band which is indicative for graphene appears in the Raman spectra. This graphite/graphene is probably formed by carbon atoms that start to diffuse at this temperature from the bulk of the Cu film to the surface. From XPS analysis, the sp^2-content of the carbon was determined to 12.5% for an annealing temperature of 1020 °C. For the lower temperatures, the sp^2-content could not be unambiguously determined using XPS.

3.4. UV-ozone cleaning

The carbon formed at the surface during the thermal annealing could be removed by a UV ozone treatment for 30 min, which represents a radiant exposure of 33 J cm^{-2} at the sample height. During this process, the temperature of the sample increased to approximately 50 °C. After this UV ozone cleaning and an additional annealing step at 1020 °C no carbon signal was observed on the surface of the 450 nm Cu film demonstrating the purification of the Cu film with regard to carbon contamination.

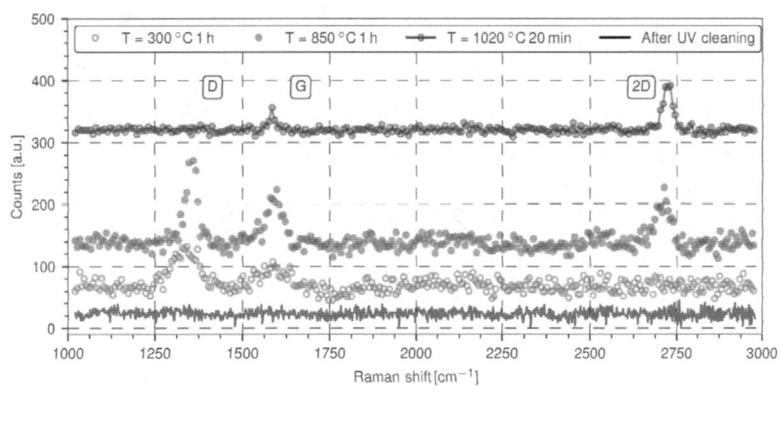

Figure 3. Raman spectra of samples annealed at 300 °C and 850 °C for 1 h each as well as 1020 °C for 20 min. Furthermore, the removal of the carbon film after UV ozone cleaning is indicated by the missing Raman features of the grey curve.

4. Conclusion

In this study, a pathway is shown to obtain carbon-free Cu(111) polycrystalline substrates for the synthesis of graphene. The substrate purity achieved with regard to carbon will enable better control of the layer number of the graphene by the synthesis process. It is also shown that a thermal anneal at temperatures above 850 °C is required for a sufficient mobility of the carbon atoms within the Cu film to allow for purification. In initial experiments, a diffusion of the Cu into the Al_2O_3 was observed for temperatures above 700 °C, indicating that annealing times should be carefully chosen at high temperatures to preserve the metallic Cu film for further synthesis.

Acknowledgments

The use of the LIPSION facility of the Group of Nuclear Solid State Physics at the University Leipzig for the RBS measurements is gratefully acknowledged.

References

[1] Geim A K and Novoselov K S 2007 *Nature Materials* **6** 183–191
[2] Geim A K 2009 *Science* **324** 1530–1534 ISSN 0036-8075
[3] Wang X, Zhi L and Müllen K 2008 *Nano Letters* **8** 323–327
[4] Schuler S, Schall D, Neumaier D, Dobusch L, Bethge O, Schwarz B, Krall M and Mueller T 2016 *Nano Letters* **16** 7107–7112
[5] Chang D W and Baek J B 2016 *Materials Research Express* **3** 044001
[6] Kheirabadi N, Shafiekhani A and Fathipour M 2014 *Superlattices and Microstructures* **74** 123–145
[7] Higgins D, Zamani P, Yu A and Chen Z 2016 *Energy and Environmental Science* **9** 357–390
[8] Shao Y, Wang J, Wu H, Liu J, Aksay I A and Lin Y 2010 *Electroanalysis* **22** 1027–1036
[9] Shams S S, Zhang R and Zhu J 2015 *Materials Science-Poland* **33** 566–578
[10] Zhang Y, Zhang L and Zhou C 2013 *Accounts of Chemical Research* **46** 2329–2339
[11] Munoz R and Gomez-Aleixandre C 2013 *Chemical Vapor Deposition* **19** 297–322 ISSN 1521-3862
[12] Garaj S, Hubbard W and Golovchenko J A 2010 *Applied Physics Letters* **97**
[13] Kim J, Lee G and Kim J 2015 *Applied Physics Letters* **107** 033104
[14] Baraton L, He Z B, Lee C S, Cojocaru C S, Chtelet M, Maurice J L, Lee Y H and Pribat D 2011 *EPL (Europhysics Letters)* **96** 46003
[15] Wang G, Ding G, Zhu Y, Chen D, Ye L, Zheng L, Zhang M, Di Z and Liu S 2013 *Materials Letters* **107** 170 – 173 ISSN 0167-577X
[16] Lehnert J, Spemann D, Hatahet M H, Mändl S, Mensing M, Finzel A, Varga A and Rauschenbach B 2017 *Applied Physics Letters* **110** 233114
[17] Lee J S, Jang C W, Kim J M, Shin D H, Kim S, Choi S H, Belay K and Elliman R 2014 *Carbon* **66** 267 – 271 ISSN 0008-6223
[18] Hui L, Yan-Xia S, Zao-Di Z, Ze-Song W, Rui Z and De-Jun F 2015 *Chinese Physics B* **24** 018502
[19] Huang J, Zhang J, Cuevas A and Tu K 1997 *Materials Chemistry and Physics* **49** 33 – 41 ISSN 0254-0584
[20] Shin H J, Yoon S M, Choi W M, Park S, Lee D, Song I Y, Woo Y S and Choi J Y 2013 *Applied Physics Letters* **102** 163102
[21] Kim D W, Kim S J, Kim J S, Shin M, Kim G T and Jung H T 2015 *ChemPhysChem* **16** 1165–1171 ISSN 1439-7641
[22] Gao L, Guest J R and Guisinger N P 2010 *Nano Letters* **10** 3512–3516
[23] Verguts K, Vermeulen B, Vrancken N, Schouteden K, Van Haesendonck C, Huyghebaert C, Heyns M, De Gendt S and Brems S 2016 *The Journal of Physical Chemistry C* **120** 297–304
[24] Oostinga J B, Heersche H B, Liu X, Morpurgo A F and Vandersypen L M K 2007 *Nature Materials* **7** 151–157
[25] Castro E V, Novoselov K S, Morozov S V, Peres N M R, dos Santos J M B L, Nilsson J, Guinea F, Geim A K and Neto A H C 2007 *Phys. Rev. Lett.* **99**(21) 216802
[26] Doolittle L 1985 *Nucl. Instr. Meth. B* **9** 344–351
[27] Prager L, Dierdorf A, Liebe H, Naumov S, Stojanovic S, Heller R, Wennrich L and Buchmeiser M 2007 *Chemistry A European Journal* **13** 8522–8529 ISSN 1521-3765
[28] Laszlo Z, Peintler G I and Dombi A 1998 *Ozone: Science & Engineering* **20** 421–432

Magnetic-particles-composed wire structures produced by pulsed laser deposition in a magnetic field

Ru Nikov[1], A Dikovska, N Nedyalkov and P Atanasov

E. Djakov Institute of Electronics, Bulgarian Academy of Sciences,
72 Tzarigradsko Chaussee, 1784 Sofia, Bulgaria

E-mail: rumen_nikov24@abv.bg

Abstract. We demonstrate the possibility to fabricate wire structures composed by arranged magnetic particles using pulsed laser deposition (PLD) in the presence of a magnetic field. Ablation of Ni and Co targets was performed in air by nanosecond laser pulses delivered by a Nd:YAG laser system oscillating at 355 nm. Due to the high density of the ambient, particles and clusters were formed by condensation in the plasma plume close to the target. The strong deceleration of the ablated material under these conditions further benefited the efficiency of applying a magnetic field to the plume. We also studied the effect of the target-to-substrate distance and the ambient pressure on the morphology of the deposited structures.

1. Introduction

Magnetic micro- and nanostructures continue to sustain the interest of researchers due to their potential use in areas such as high-density data storage, biomedical applications, drug delivery, magnetic separation and sensors [1–7]. The properties of these structures differ from those of their bulk counterparts because of their high surface-to-volume ratios. Moreover, such structures could also exhibit novel properties influenced by their arrangement and the ability to form complex systems. When an external magnetic field is applied to an ensemble of micro- or nanoparticle, alignment of the particles' magnetic moments can take place and, as they interact, mask-free assemblies of 1D, 2D or 3D systems can be produced [8]. The possibility of additional control of the system orientation on a macroscale by the external magnetic field can result in these structures acquiring unique tunable directional-dependent electric, magnetic and optical properties [9].

The magnetic structures' unique characteristics led to the development of a wide range of fabrication methods, including microemulsions, sol-gel synthesis, sonochemical reactions, hydrothermal reactions, hydrolysis and thermolysis of precursors, flow injection synthesis, and electrospray synthesis [10–16]. However, these chemically-based techniques usually involve several steps and often require toxic reagents. Even in the well-developed chemical, physical and combined methods, the cost-efficient, easy-to-implement and time-saving fabrication of 2D and 3D magnetic micro- or nanostructures is still a challenge.

As a flexible method, the pulsed laser deposition (PLD) has proven its capability to produce complex structures of a wide variety of materials. The technology includes ablation of a target and deposition of the ejected material on a substrate. The morphology of the deposited structure can be

[1] To whom any correspondence should be addressed.

controlled via different experimental parameters: laser processing characteristics, deposition geometry, substrate temperature and environment parameters. Besides the conventional parameters, the application of an external magnetic field to the plasma plume has also been established as being an important parameter in the PLD technique. It was demonstrated that the absence of a magnetic field during laser ablation of gold in water leads to the formation of nanowires with a length of ~100 nm, while single nanoparticles and shorter nanowires (~10–20 nm) are formed when a magnetic field is applied [17]. Recently, ordered magnetic nanostructures were produced by femtosecond laser ablation of steel in air and in the presence of an external magnetic field [18]. The specific dynamics of the ejected material under these conditions leads to the formation of nanoparticle wires with lengths of tens of micrometers on the substrate. Despite the great possibilities of the method, the need for a femtosecond laser system limits the flexibility of the method, which hampers its efficient application in the industry. Furthermore, fabricating magnetic structures with a desired orientation by nanosecond laser ablation in the presence of an external magnetic field still presents considerable challenges.

In this paper, we present results on the nanosecond laser ablation of Co and Ni when a magnetic field is applied. We found that the deposition process produces wire structures composed of arranged particles. We studied and discussed the effect of varying the target-to-substrate distance on the morphology of the formed structures, and how the ambient pressure affected the arrangement of the deposited material.

2. Experimental

The fabrication of wire structures was implemented by applying the PLD process in the presence of external magnetic field. The targets of ferromagnetic metal (Co–99.99 % or Ni–99.99 %) were mounted on a rotating holder and ablated by nanosecond laser pulses delivered by a Nd:YAG (Lotis LS-2147) laser system operating at the wavelength of 355 nm. The pulse duration was 15 ns at a repetition rate of 10 Hz. The angle between the incident radiation and the target surface was about 30°. The deposition process was performed in a standard on-axis geometry (figure 1) with laser fluences of 20 J/cm^2 or 45 J/cm^2 and deposition times of 2 min and 5 min. The ablated material was deposited on a quartz substrate fixed at 5 mm or 3 mm from the target. A permanent magnet producing a magnetic field of $B = 0.2$ T was attached to the back side of the substrate. The depositions were conducted in air at atmospheric pressure or in He at pressure of 500 Torr. The composition of the material deposited on the substrate was analyzed by EDX (Quantax 200 Bruker). The morphology of the structures produced was studied by SEM (LYRA Tescan).

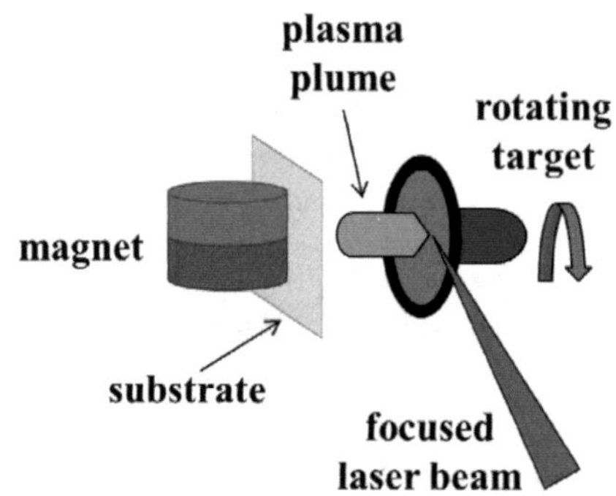

Figure 1. A scheme of PLD process in the presence of an external magnetic field.

3. Results and discussion

Figure 2 shows SEM images of the material deposited after ablation of a Co target in open air, without application of a magnetic field (a) and in the presence of a magnetic field (b). The ablation was performed at a laser fluence of 20 J/cm^2 with a deposition time of 5 min in both cases. The distance between the target and substrate was 5 mm. As seen, deposition without an external magnetic field applied leads to the formation of a structure consisting of randomly distributed nanoparticle aggregates and free nanoparticles with diameters up to 100 nm. The origin of these nanostructures is related to the high density of the ambient atmosphere. In the case of deposition in air at atmospheric pressure, the nanoparticles and nanoparticle aggregates are formed by condensation close to the target [19]. When the magnetic field is applied during the deposition process, distinct wire structures with lengths of several tens of microns are formed on the substrate (figure 2b). The orientation of the wires is predominantly parallel to the magnetic field applied. It should be mentioned that the spatial confinement of the plasma plume and the drop in the particles' velocity further benefit the efficiency of application of the magnetic field on the flying material. A high-magnification SEM image of this sample reveals the presence of free nanoparticles (inset in figure 2b). Similar to the case without a magnetic field, the nanoparticle substructure has a random distribution over the substrate. The lack of arrangement of these particles is related to their small size and the high temperatures they acquire in the plasma plume.

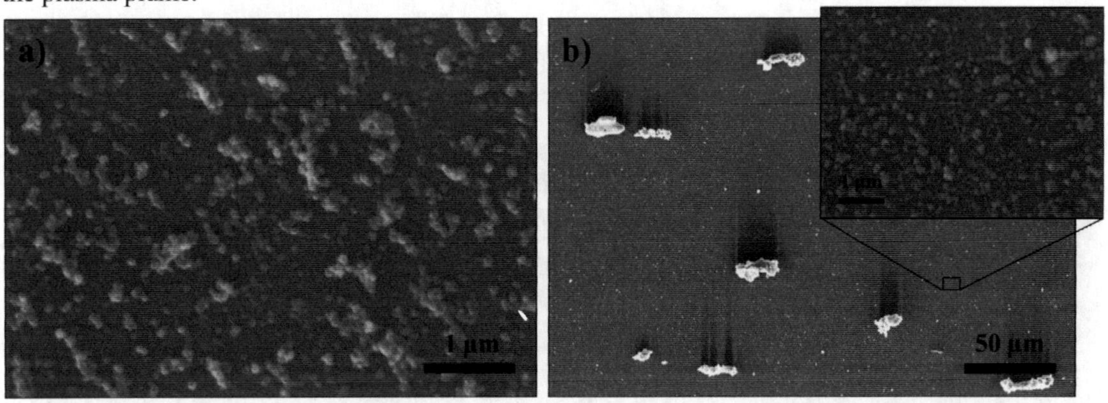

Figure 2. SEM images of Co structures produced by PLD in open air, without application of an external magnetic field (a) and in the presence of an external magnetic field (b). The samples are produced at a laser fluence of 20 J/cm^2, deposition time of 5 min and target-to-substrate distance of 5 mm. The inset in (b) is a high-magnification image of the sample.

The composition of the structures produced by PLD of Co in open air was revealed by EDX analysis. Figure 3 is an EDX spectrum taken from the structure shown on figure 2b. It should be noted that besides Co, a peak corresponding to oxygen is also observed in the EDX spectrum. The presence of oxygen in the sample is related to oxidation of the top layer of the Co wires due to the high concentration of surface defects.

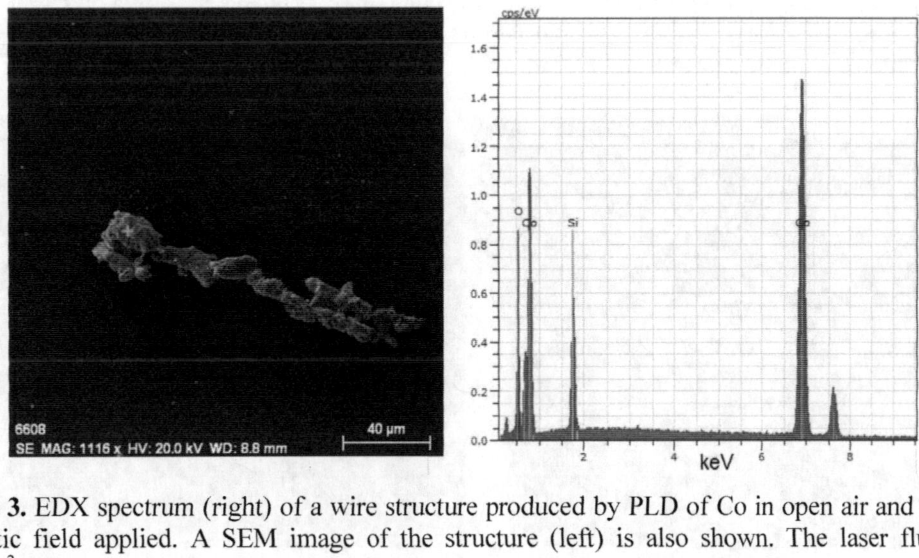

Figure 3. EDX spectrum (right) of a wire structure produced by PLD of Co in open air and external magnetic field applied. A SEM image of the structure (left) is also shown. The laser fluence is 20 J/cm^2, the deposition time is 5 min, and the target to substrate distance is 5 mm.

Further, we explored the influence of the process parameters on the morphology of the structures deposited in the presence of an external magnetic field. Figure 4 shows SEM images of samples deposited at different target-to-substrate distances – 5 mm (a) and 3 mm (b). The depositions were performed in open air at a fluence of 45 J/cm^2 and a deposition time of 5 min. Clearly separated wires oriented along the external magnetic field are observed in both structures. However, the Co wires morphologies in the two samples differ significantly. When the target to substrate distance is 5 mm, the deposition process results in elongated features with a length ranging from 25 to 45 μm and a width of 4-10 μm. In the case of deposition at 3 mm, fine wires with lengths in the range from 8 to 35 μm and a width of 2-3 μm are formed. The formation of thicker structures in the sample deposited at 5 mm could be attributed to the higher degree of aggregation of the ablated material due to the longer impact of the magnetic field on the plasma plume compared to the deposition at 3 mm.

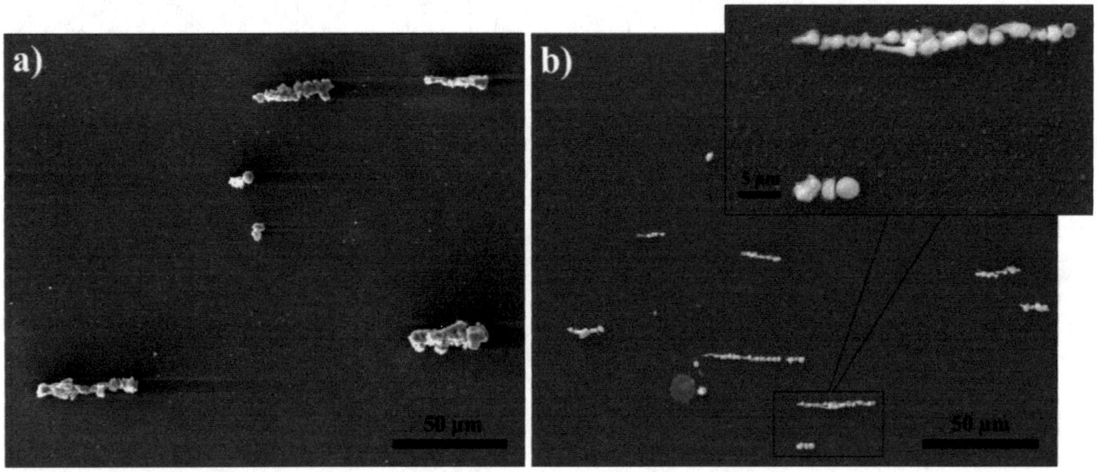

Figure 4. SEM images of Co wire structures produced by PLD in open air and an external magnetic field applied. The target to substrate distance is (a) 5 mm and (b) 3 mm. The laser fluence is 45 J/cm2 and the deposition time is 5 min. The inset in (b) shows a high-magnification image of the structure.

In order to examine the capabilities of the method utilized, we also performed deposition of Ni in a magnetic field. Figure 5 presents SEM images of the material after the ablation of a Ni target in air at atmospheric pressure (a) and in He at 500 Torr (b). The experiments were carried out under the same conditions as the structure shown in figure 4a. A shorter ablation time was used for the sample

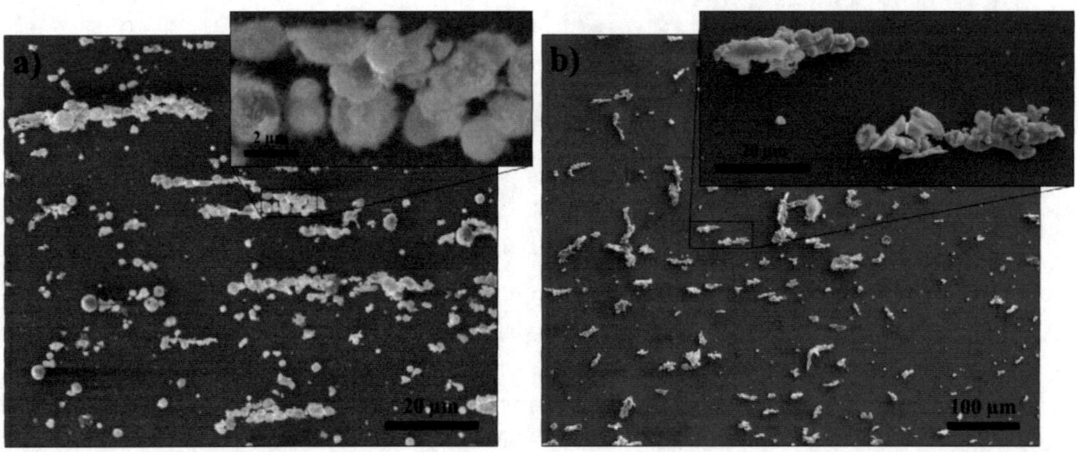

Figure 5. SEM images of Ni structures produced in the presence of an external magnetic field by PLD (a) in air at atmospheric pressure for a deposition time of 5 min, and (b) in He at pressure of 500 Torr for a deposition time of 2 min. The laser fluence is 45 J/cm^2, and the target to substrate distance is 5 mm. The insets show high-magnification images of the structures.

produced at 500 Torr due to the more efficient deposition of material compared to the case of ablation at atmospheric pressure. As seen, the application of an external magnetic field during PLD of Ni also affects the morphology of the deposited material. The deposition at atmospheric pressure results in wires composed of particles arranged along the magnetic field, as well as in free particles (figure 5a). The particle wires formed have lengths from 10 to 40 μm and widths of 2-4 μm. Similar elongated features are observed in the case of deposition in He at the lower pressure, but they are significantly less magnetically-oriented and rather randomly distributed (figure 5b). The lower degree of orientation is related to the pressure of the ambient medium. A lower ambient pressure will lead to a higher velocity of the ablated material, which will limit the time of the external magnetic field acting on the plasma plume, so that the efficiency of the orientation process will be reduced.

4. Conclusions

We described a technique for fabrication of 2D wire structures composed of arranged magnetic particles. The method is based on the implementation of the PLD process in the presence of an external magnetic field. The ablation of Co under these conditions leads to the formation of distinct wire structures with lengths of several tens of microns. The wires on the substrate are predominantly oriented along the direction of the magnetic field applied. When no magnetic field is applied during the deposition process, randomly distributed nanoparticle aggregates and free nanoparticles are formed on the substrate. We found that the size of the Co wires strongly depends on the target-to-substrate distance. When this distance is 5 mm, thick elongated structures with lengths in the range from 25 to 45 μm and widths of 4-10 μm are produced. The deposition at a 3-mm distance results in fine wires with lengths ranging from 8 to 35 μm and width of 2-3 μm. The ablation of Ni in a magnetic field also leads to the creation of oriented wire structures. The Ni-wires length ranges from 10 to 40 μm, the widths being 2-4 μm. The technique presented could find applications as an inexpensive and flexible way for fabrication of new materials and structures.

Acknowledgements

The authors acknowledge the financial support of project DFNP-17-116 “Fabrication of ordered nanostructures by pulsed laser deposition in magnetic field” under the Assistance for Young Scientists Program of the Bulgarian Academy of Sciences.

References

[1] Frey N A, Peng S, Cheng K and Sun S 2009 *Chem. Soc. Rev.* **38** 2532
[2] Weller D and Doerner M F 2000 *Annu. Rev. Mater. Sci.* **30** 611
[3] Colombo M, Carregal-Romero S, Casula M F, Gutierez L, Morales M P, Bohm I B, Heverhagen J T, Prosperi D and Parak W J 2012 *Chem. Soc. Rev.* **41** 4306
[4] Kami D, Takeda S, Itakura Y, Gojo S, Watanabe M and Toyoda M 2011 *Int. J. Mol. Sci.* **12** 3705
[5] Arruebo M, Fernandez-Pacheco R, Ibarra M and Santamaria J 2007 *Nanotoday* **2** 22
[6] Liang Y, Liu P, Xiao J, Li H, Wang Ch and Yang G 2013 *Sci. Rep.* **3** 3051
[7] Koh I and Josephson L 2009 *Sensors* **9** 8130
[8] Hu L, Zhang R and Chen Q 2014 *Nanoscale* **6** 14064
[9] Richter H J 1996 *J. Magn. Magn. Mater.* **154** 263
[10] Chin A B and Yaacob I I 2007 *J. Mater. Process. Technol.* **191** 235
[11] Albornoz C and Jacobo S E 2006 *J. Magn. Magn. Mater.* **305** 12
[12] Kim E H, Lee H S, Kwak B K and Kim B K 2005 *J. Magn. Magn. Mater.* **289** 328
[13] Wan J, Chen X, Wang Z, Yang X and Qian Y 2005 *J. Cryst. Growth* **276** 571
[14] Kimata M, Nakagawa D and Hasegawa M 2003 *Powder Technol.* **132** 112
[15] Alvarez G S, Muhammed M and Zagorodni A A 2006 *Chem. Eng. Sci.* **61** 4625
[16] Basak S, Chen D R and Biswas P 2007 *Chem. Eng. Sci.* **62** 1263
[17] Musaev O R, Sutter E A, Wrobel J M and Kruger M B 2016 *Appl. Phys.* A **122** 95
[18] Nedyalkov N, Nakajima Y and Terakawa M 2016 *Appl. Phys. Lett.* **108** 043107
[19] Girault M, Hallo L, Lavisse L, Lucas M C M, Hébert D, Potin V and Jouvard J M 2012 *Appl. Surf. Sci.* **258** 9461

Influence of loading QCMs with electrochemically-deposited ZnO on their NO_2-sensing properties

B Georgieva[1,4] , H Nichev[2], M Petrov[2], T Koutzarova[1], V Georgieva[3] and D Dimova-Malinovska[2]

[1]E. Djakov Institute of Electronics, Bulgarian Academy of Sciences, 72 Tzarigradsko Chaussee, 1784 Sofia, Bulgaria
[2] Central Laboratory of Solar Energy and New Energy Sources, Bulgarian Academy of Sciences, 72 Tzarigradsko Chaussee, 1784 Sofia, Bulgaria
[3] G. Nadjakov Institute of Solid State Physics, Bulgarian Academy of Sciences, 72 Tzarigradsko Chaussee, 1784 Sofia, Bulgaria

E-mail: b.georgiewa@abv.bg

Abstract. This paper reports on ZnO layers' sensitivity to NO_2 exposure. ZnO layers were grown by electrochemical deposition on the surface of quartz crystal microbalances (QCMs) with Au electrodes; the sensitivity was estimated by the frequency-time characteristics (FTCs) of the QCM, namely, its resonance-frequency-shift response. The sorption process was investigated in NO_2 test gas. The behavior was studied of three different sensors with ZnO layers deposited for different times – 30, 35 and 60 min. The change in the frequency, ΔF, of the QCM as a function of the loaded mass of NO_2 was detected in different NO_2 concentrations in the range of 250 – 5000 ppm and the value of the sorbed mass was calculated, together with the rate of the NO_2 sorption and desorption. As the time of ZnO layers deposition was increased, the sorbed NO_2 mass increased for all concentrations used in the experiment. This can be explained by changes in the ZnO layers' structure with the time of deposition.

1. Introduction

ZnO has shown growing potential for applications in new optoelectronic and electronic devices [1-4]. It has high chemical stability, low dielectric constant and high optical transmittance. Additionally, ZnO films doped with different elements have demonstrated sensitivity to different gases [3-5]. As a gas-sensing material, it is one of the earliest discovered and most widely applied gas-sensing oxide materials because of its suitable electrical and chemical properties [6].

The quartz crystal microbalance (QCM) is among the extremely sensitive mass devices [3]. It can be used as a gas sensor [5-7] or a biosensor [8, 9]. The QCM sensor's properties, as sensitivity, selectivity and time response, are strongly influenced by the sensing film's properties. These sensors are highly sensitive to mass changes in the presence of the test gas. In the present work, we studied the sensing ability to NO_2 in air of QCMs coated with electrochemically-deposited nanostructured (NS) ZnO films. The sensitivity was estimated by following the frequency-time characteristics (FTCs) of

[4] To whom any correspondence should be addressed.

IOP Conf. Series: Journal of Physics: Conf. Series **992** (2018) 012026 doi:10.1088/1742-6596/992/1/012026

the QCM, particularly, its resonance-frequency-shift response, ΔF. The behavior was monitored of three different sensors with ZnO layers deposited for different time - 30, 35 and 60 min.

2. Experimental

We used 16-MHz QCMs made of AT-cut quartz substrates with 60-nm Au electrodes thermally evaporated on both sides of the quartz substrates and a Cr underlayer improving the adhesion of Au to the quartz plate. The diameters of the substrates and electrodes were 8 mm and 4 mm, respectively. The quality of a QCM was characterized by its series resonance frequency (f_s), equivalent dynamic resistance (R_s) and quality factor (Q) [10]. These were the parameters that we measured and controlled before and after modifying the QCMs by depositing thin ZnO nanostructured (NS) layers, thus fabricating gas sensors. The layers were formed electrochemically from a slightly acid aqueous solution of $ZnCl_2$ (5×10^{-3} M) and KCl (0.5 M) with pH 4.0 at 80 ºC and –1000 mV (*vs.* SCE) using a three-electrode electrochemical cell [11]. The electrolyte was agitated by a magnetic stirrer. The quartz crystal resonator with a Au electrode was used as a cathode and a spectrally-pure-graphite plate was the anode. The deposition was carried out while controlling the redox potential of the system. The total oxygen content in the solution was controlled by a DO&T meter (Hanna Instruments 9146). The duration of the NS ZnO layers deposition was 30, 35 and 60 minutes. The study of the structural properties of the NS ZnO layers was performed by scanning electron microscopy (SEM) using a Philips 515 SEM apparatus.

The gas-sensing properties of the QCM-NS-ZnO system were investigated at different NO_2 concentrations by using an experimental laboratory setup consisting of a gas module, a gas mixing and control module, a test chamber, a thermostat, an oscillator and a frequency counter, and a computer system. The setup was described in detail in [12]. The QCM-NS-ZnO structures investigated were placed in the test chamber near a Pt heat-sensor, so that the temperature of the process was maintained constant and strictly controlled. All samples were measured at the temperature of 24°C ± 0.2°C. The setup was operating in a dynamic mode – in a continuous gas flow. The testing gas consisted of 10 000 ppm NO_2 diluted in high-purity synthetic air as a carrier gas. The flow rates of the two gases were measured and controlled by mass flow controllers (FC-260 and FC-280), with the desired NO_2 concentration obtained by varying the ratio of the two gas flows. The QCM frequency was registered by a Hameg 8123 frequency counter, and the data was saved on a computer.

The frequency changes as a function of the deposition time (FTCs) for each QCM-NS-ZnO structure were measured at NO_2 concentrations ranging from 250 ppm to 5000 ppm. The measurement process can be divided into the following stages:

- with the QCM-NS-ZnO structure placed, the test chamber was purged by the main flow of carrier gas until the frequency reached a constant value;
- the test gas NO_2 was included to the main flow; the frequency value decreased due to the loading of the structure. The process continued until a constant frequency value, i.e., saturation, was reached;
- the gas flow was turned off and the frequency started increasing due to unloading of the NS ZnO layers. The process continued until the recovery was complete and the initial frequency was reached.

At this step, the measurements were stopped and the FTC was completed.

The sensitivity of the QCM-NS-ZnO system with different NO_2 loads was determined by measuring the FTCs. The sorbed NO_2 mass (Δm) was calculated from the QCM resonant frequency-shift measured, in accordance with Sauerbrey's equation [13].

3. Results and discussion

Figure 1 shows SEM micrographs of NS ZnO layers deposited for 30 and 60 min. The structure consists of planar hexagonal ZnO nano-whiskers with thickness of about 50 – 100 nm in the sample deposited for 30 min, and 200 – 300 nm in that deposited for 60 min, with a height of about 0.5 µm and 1–1.2 µm, resepectively. It is seen that the size of the whiskers increases with the electrochemical deposition duration. Increasing the nano-whiskers' height leads to an increase of the effective surface of the NS ZnO layers. The structure of the sample deposited for 35 min is very similar to that of the NS ZnO layer deposited for 30 min.

(a) (b)

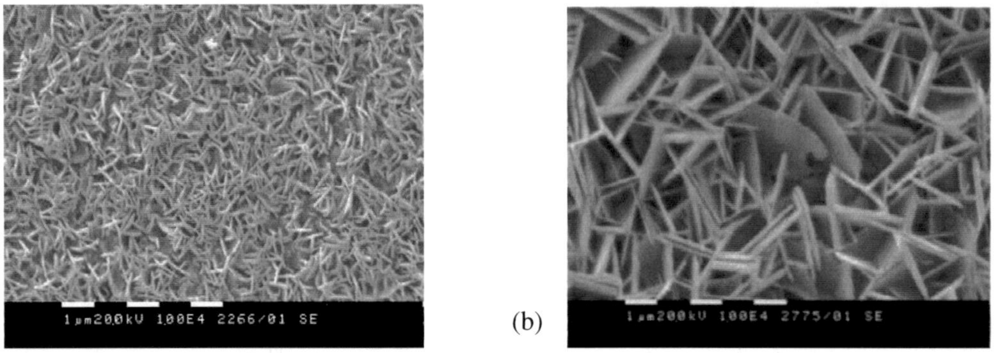

Figure 1. SEM micrographs of ZnO nanostructured layers deposited electrochemically on quartz resonators for different time: (a) 30 min, (b) 60 min. The markers correspond to 1 μm.

Figure 2 demonstrates the change of the frequency (the frequency-shift, ΔF) of the QCM after ZnO layers deposition as a function of the time. After the ZnO deposition, the QCM frequency was reduced by 25.05 kHz, 30.29 kHz and 76.24 kHz for deposition times of 30, 35 and 60 min, respectively. According to Sauerbrey's equation [13], the frequency-shift measured after ZnO deposition is:

$$\Delta F = -\frac{2.26\times10^{-6} f^2 \Delta m}{A},$$

where ΔF is the frequency-shift measured after deposition of NS ZnO layers (Hz), f is the frequency of the QCM before deposition of NS ZnO layers (MHz), Δm (g) is the mass of the NS ZnO layers deposited on the QCM surface and $A = 0.1256$ cm^2 is the surface area of the electrode.

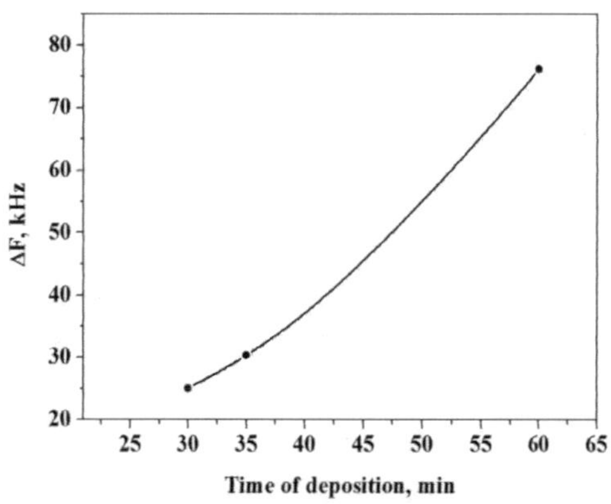

Figure 2. Frequency-shift of the QCMs, ΔF, with the time of deposition of NS ZnO films.

The deposition mass of ZnO was calculated to be 5.4 μg, 6.7 μg and 16.7 μg, for deposition times of 30, 35 and 60 min, respectively. The other QCM parameters (R_s, Q) were also measured and calculated before and after ZnO deposition. After the ZnO deposition, the R_s value increased slightly, while that of Q decreased slightly. This slight deterioration of the QCM parameters did not have a negative influence on the sensors' properties.

Figures 3 (a), (b) and (c) illustrate the FTCs curves of the three QCMs with NS ZnO layers deposited for the three periods of time considered (i.e., with the three different masses of ZnO deposited as quoted above) measured at a 500-ppm concentration of the NO_2 test gas. Thus, the time of sorption decreases – 276 s, 252 s and 248 s, as the mass of the NS ZnO films deposited increases, i.e., the rate of sorption increases – 0.123 Hz/s, 0.146 Hz/s and 0.177 Hz/s. The time of desorption at this NO_2 concentration increases with the time of ZnO deposition – 104 s, 140 s and 308 s, i.e., the speed of desorption decreases – 0.32 Hz/s, 0.26 Hz/s, 0.14 Hz/s. Or, as it is only to be expected, the mass of ZnO deposited affects the FTCs of the QCMs and the NS ZnO structures' gas sensing characteristics, since the mass of sorbed NO_2 increases with the ZnO mass increase.

Further, we measured the FTCs of QCMs with different mass of NS ZnO films at NO_2 concentrations of 250 ppm, 500 ppm, 1000 ppm, 2500 ppm and 5000 ppm. From each FTC, the frequency-shift was calculated as the difference between the initial frequency value and the frequency value at saturation. Based on the frequency-shift measured, the maximal adsorbed NO_2 mass was calculated using the Sauerbrey's equation.

Figures 4 (a) and (b) illustrate the frequency-shift and the NO_2 sorbed mass calculated, respectively, as a function of the NO_2 concentration for the QCMs with NS ZnO layers deposited for different times. As seen in the figures, the behavior of the dependences obtained

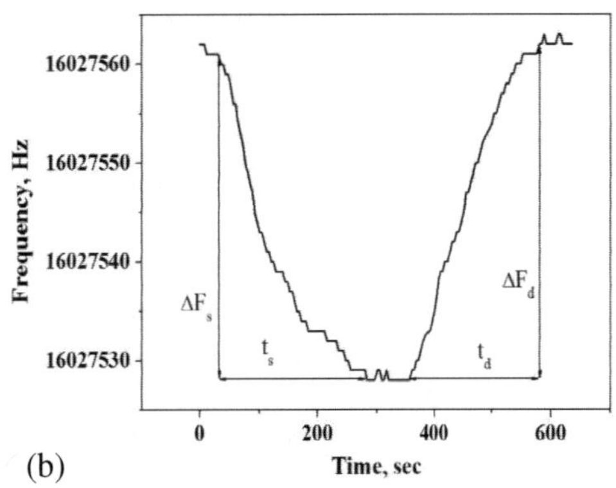

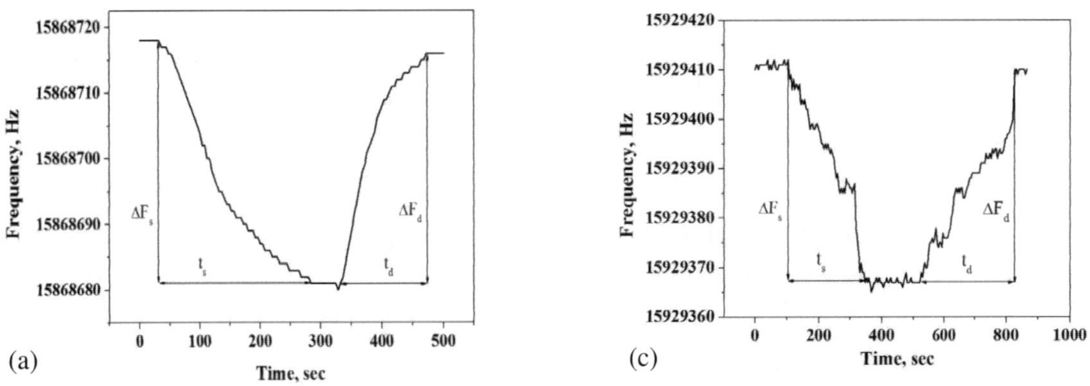

Figure 3. Frequency-time characteristics of QCMs with NS ZnO layers deposited for different times 30 (a), 35 (b) and 60 (c) min at 500 ppm NO_2 concentrations.

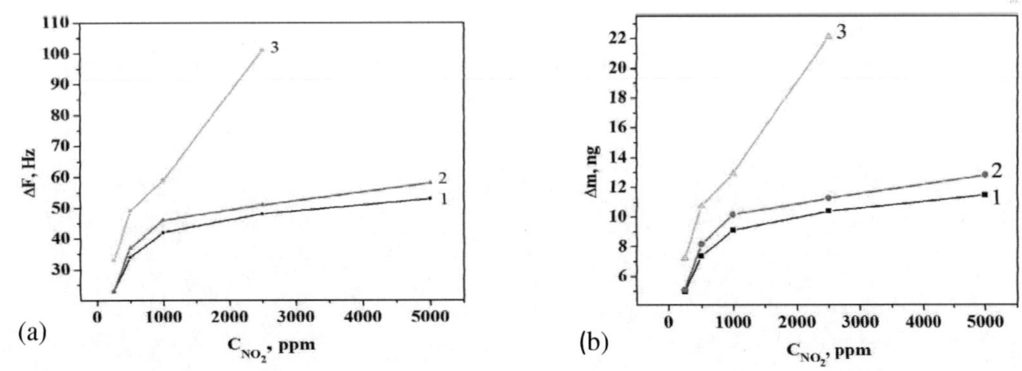

Figure 4. Dependences of the frequency-shift, ΔF, (a) and calculated NO_2 sorbed mass, Δm, (b) on the NO_2 concentration in the gas flow, C_{NO_2}, for QCMs with NS ZnO layers deposited for different time: 1 – 30 min, 2 – 35 min and 3 – 60 min.

can be divided into two parts regarding the variation of the NO_2 concentration, namely, from 250 ppm to 1000 ppm, and from 1000 ppm to 5000 ppm. In the cases of the lower deposition times of NS ZnO layers (30 and 35 min) (lines 1 and 2 in figure 4) and, respectively, of low ZnO mass, the absorption of the gas reaches saturation at about 1000 ppm. In the case of the longest deposition time (60 min), the mass of the NO_2 sorbed increases almost linearly with C_{NO2} up to $C_{NO2} = 2500$ ppm; at higher C_{NO2}, the frequency, f, and the change in ΔF could not be measured. The changes in the behavior of the QCMs sensitivity with the NS ZnO layers deposited for different time is related to the increase of the deposited ZnO mass and of the size of the ZnO nano-whiskers, as seen in the SEM micrographs.

4. Conclusions

We studied the sensing characteristic to NO_2 of 16-MHz QCMs with Au electrodes coated electrochemically by nanostructured ZnO films. The SEM studies revealed that the NS ZnO layers' surface had developed morphology and high effective surface area consisting of nano-whiskers; the size of the nano-whiskers and the mass of the ZnO grown increased with the time of deposition.

We measured the response and recovery times, the maximal frequency-shift and the mass of gas sorbed on the QCMs with nanostructured ZnO films at different NO_2 concentrations. The QCMs showed good sensitivity to NO_2 at room temperature, making them suitable for detection of NO_2 at concentrations over 250 ppm. The QCM-NS-ZnO system's sensitivity increased with the ZnO deposition duration, which is explained by the increased ZnO mass and size of the ZnO nano-whiskers, or correspondingly, the increased effective surface area of the NS ZnO layers on a QCM sensor.

References

[1] Wagh M S, Jain G H, Patil D R, Patil S A and Patil L A 2006 *Sensors and Actuators* B **115** 128
[2] Navale S C, Ravi V, Mulla I S, Gosavi S W and Kulkarni S K 2007 *Sensors and Actuators* B **126** 382
[3] King W H 1964 *Anal. Chem.* **36** 1735
[4] Nakamura K, Nakamoto T and Moriizumi T 2000 *Mater. Sci. Eng.* C **12** 3
[5] Sun L-X and Okada T 2000 *Anal. Chim. Acta* **421** 83
[6] Seiyama T, Kato A, Fujiishi K and Nagatani M 1962 *Anal. Chem.* **34** 1502
[7] Zhang J, Hu J Q, Zhu F R, Gong H and O'Shea S J 2002 *Sensors and Actuators* B **87** 159
[8] Bunde R L, Jarv E J and Rosentreter J J 1998 *Talanta* **46** 1223
[9] Guibault G G, Hock B and R. Schmid R 1992 *Biosens. Bioelect.* **7** 411
[10] Manolov S and Tihchev H 1982 *Generators* (Sofia: Technika)
[11] Dimova-Malinovska D, Andreev P, Sendova-Vassileva M, Nichev H and Starbova K 2010 *Energy Procedia* **2** 55
[12] Georgieva V, Stefanov P, Spassov L, Raicheva Z, Atanassov M, Tincheva T, Manolov E and Vergov L 2009 *J. Optoelectr. Adv. Mater.* **11** 1363
[13] Sauerbrey G 1959 *Zeitschrift für Physik* **155** 206

Investigation of the structural, optical and piezoelectric properties of ALD ZnO films on PEN substrates

B S Blagoev[1, 5], M Aleksandrova[2], P Terziyska[1], P Tzvetkov[3], D Kovacheva[3], G Kolev [2], V Mehandzhiev[1], K Denishev[2] and D Dimitrov[1, 4]

[1]G. Nadjakov Institute of Solid State Physics, Bulgarian Academy of Sciences, 72 Tzarigradsko Chaussee, 1784 Sofia, Bulgaria
[2]Technical University of Sofia, Department of Microelectronics, 8 Kliment Ohridski Blvd, 1000 Sofia, Bulgaria
[3]Institute of General and Inorganic Chemistry, Bulgarian Academy of Sciences, Acad. G. Bonchev Str., Bl. 11, 1113 Sofia, Bulgaria
[4]Acad. J. Malinowski Institute of Optical Materials and Technologies, Bulgarian Academy of Sciences, Acad. G. Bonchev Str. Bl. 109, 1113 Sofia, Bulgaria

E-mail: blago_sb@yahoo.com, blago@issp.bas.bg

Abstract. We present the results of studies on the structural, optical and piezoelectric properties of ZnO thin films deposited by ALD on flexible polyethylene naphthalate (PEN) substrates. Changes were observed in the optical transmission and crystal structures as the deposition temperature was varied. The electromechanical behavior, dielectric losses and voltage generated from ZnO flexible devices were investigated and discussed, in order to estimate their suitability for potential application as microgenerators activated by human motion.

1. Introduction

Flexible microelectronics is a fast-growing modern technology with a variety of attractive applications: energy harvesting, micro-/nano-electromechanical systems (sensors and actuators), organic light-emitting displays (OLEDs), flexible smartphones, electronic paper, wearable microelectronics and biological applications. Recently, mechanical energy scavenging from human motions has attracted great attention due to its potential application in the bioelectronics and medicine [1, 2]. These motions take place at low frequencies (up to 50 Hz), compared to other ambient sources of vibration. In addition, there exist many other sources of low-frequency vibrations in the environment, such as pumps, compressors, diesel engines, wind and fans, that could also contribute to electrical energy generation [3]. Therefore, the study is of particular interest of piezoelectric microgenerators in the low-frequency range. When used for self-sufficient power supply of biosensors, these elements must be small-sized, ultrathin, lightweight and possibly flexible [4].

An increasing number of biocompatible components would benefit from the use of such structures. Replacing the metal or metal-oxide electrodes with conductive polymers is a good starting point in the process leading to the fabrication of fully-compatible microgenerators [5, 6]. Furthermore, it results in

[5] To whom any correspondence should be addressed.

more stable flexible devices that are not affected by multiple mechanical loading due to the high elasticity of the electrode [7]. Replacing the piezoelectric functional film with a polymer is still an obstacle restricting the device efficiency due to the relatively low piezoelectric coefficient d_{ij} of this class of materials [8]. This is the reason why conventional inorganic piezoelectric materials, such as piezoceramics and their composites, are still a basic component of layered microgenerators, locked between polymer-containing electrodes [9]. All this provoked the interest in zinc oxide (ZnO), which exhibits good piezoelectric properties along the [0001]-direction due to its non-centrosymmetric structure. Along with its other properties, such as high transmittance in the visible range and low DC resistivity, ZnO is the most promising environmentally-friendly (lead-free) material for energy harvesting applications, transparent electronics, solar cells and optical devices. A variety of chemical and physical deposition techniques have been reported for preparation of ZnO films, nanocoatings and nanostructures (nanowires, nanorods, etc.) [10, 11]. To obtain flexible devices, low-temperature deposition processes are needed, e.g., atomic layer deposition (ALD). The advantages of ALD over other deposition techniques are: large area thickness uniformity, atomically flat and smooth coatings, perfect 3D conformability, uniform covering on high-aspect-ratio features and possibility to control the thickness on a nanometer scale [11-13].

In this paper, we report on the structural, optical and piezoelectric properties of ZnO thin films deposited by ALD on flexible polyethylene naphthalate (PEN) substrates. We present the electromechanical behavior, dielectric losses and voltage generated by a flexible device with a piezoelectric nanocoating of atomic-layer-deposited ZnO films. Such a device is very promising for potential application as a microgenerator activated by human motion. To the best of our knowledge, this is the first report on a piezoelectric energy-harvesting flexible device with ALD-grown ZnO films.

2. Experimental

We discuss several samples based on ALD ZnO films. The films were grown directly on polyethylene naphthalate (PEN) and p-Si (100) substrates; another set of ZnO films were deposited on the top of three different contacts: Al, Au and PEDOT:PSS/Au. Thus, PEN/ZnO, Si/ZnO, PEN/Al/ZnO, PEN/Au/ZnO and PEN/PEDOT:PSS/Au/ZnO structures were obtained. The ZnO film in the PEN/Al/ZnO and PEN/Au/ZnO structures were grown at 100 °C, the rest were grown at 80, 100 and 120 °C.

2.1. Bottom contact deposition

The PEN substrates were cleaned by isopropanol in a supersonic cleaner. The bottom electrodes were prepared as follows: 1) for the polymer-containing ones, a poly(3,4-ethylenedioxythiophene) polystyrene sulfonate (PEDOT:PSS) film with a thickness of 35 nm was spin-coated on the PEN surface at a spinning speed of 1500 rpm for 30 s; a gold (Au) ultrathin film (15 nm) was then grown by DC sputtering; 2) for the bottom Al electrode, an Al film with a thickness of 230 nm was prepared by thermal evaporation of a wire with 99.998 % purity at a basic pressure of 10^{-5} Torr.

2.2. ALD of ZnO films

Thin ZnO films were prepared in a Beneq TFS 200 ALD system on p-type Si (100), PEN, PEN/Al, PEN/Au and PEN/PEDOT:PSS/Au substrates. The deposition process was conducted at 80, 100 and 120 °C and a pressure of 2 mbar. To obtain ZnO films, alternating deposition from DEZ and H_2O precursors was carried out. The pulse and purging durations were the same for both precursors – 200 ms and 1 s, respectively. A pure N_2 gas flow at 600 sccm was maintained during the precursor introduction and the purging steps [11]. For each temperature, 550 cycles were performed.

2.3. XRD Analysis

The XRD patterns of all samples were recorded on a Bruker D8 Advance powder diffractometer equipped with a copper tube (CuKα) and a LynxEye position-sensitive detector. The patterns were recorded in the angular range 5.5 – 80.0 °2θ, with a step of 0.04 °2θ and counting statistics of 0.1

second per detector strip (17.5 s totally). More detailed patterns were recorded in the range 30 – 38 °2θ, a step of 0.04 °2θ and counting statistics of 1 second per strip (175 s totally). The phase identification of the collected data was performed by using the ICDD PDF-2 (2014) database. According to the reference PDF file number 00-036-1451 of synthetic ZnO (Zincite), the first three lines of this compound with indices (100), (002) and (101) should appear in the range 30 – 38 °2θ.

2.4. Ellipsometric measurements

The ellipsometry measurements were performed using a Woollam M2000D rotating compensator spectroscopic ellipsometer with a wavelength range from 193 nm to 1000 nm in reflection mode. The spectroscopic ellipsometry data of Ψ and Δ were taken at room temperature at angles of incidence of 65°, 70° and 75°. The data acquisition and analysis software were CompleteEASE 5.10 J. A. Woollam Co., Inc. The ZnO thickness was extracted from the ellipsometric data of the films grown on Si substrates. Transmission intensity measurements on the PEN/ZnO samples were performed.

2.5. Piezoelectric measurements

To acquire information on the processes in the ALD ZnO grown on a polymer-covered PEN substrate, the top electrode was made of aluminium. Using top electrode same as the bottom one would result in a weaker signal due to its lower conductivity compared to aluminium. The top electrodes were cut from aluminium tape with one side covered by a conductive adhesive, forming active areas of 2 mm^2. The set-up allowed us to apply a low-frequency bending force (up to 52 Hz controlled by the rotation speed of an electric motor), similar to the targeted mechanical sources. All samples were preliminary poled in an electric field with strength of 5 V/cm^2 for one min. The samples were fixed to the bendable beam of the stand to avoid non-uniform distribution of the mechanical wave across the structure and the appearance of additional resonance modes. A reference force sensor measured the mechanical load with the data shown on the stand's display. The shape of the resulting voltage produced by the cyclic loading of the samples was recorded by a DQ2042CN digital storage oscilloscope. The piezoelectric response was excited by a mechanical stimulus of about 5 kg mass load. Additionally, the resistance, capacitance and dielectric losses were measured at a standard bias voltage of 1 V and a frequency of 1 kHz by an Instek 816 RLC meter. This measurement was carried out before and after polarization of the samples to determine how the film is affected by the polling voltage, i.e., to assess its piezoelectric susceptibility. All measurements were conducted with no electric load (open circuit mode) [6].

3. Results and discussions

The thicknesses of ZnO films grown on Si substrates at temperatures 80, 100 and 120 °C were 50, 70 and 86 nm, respectively. The same behavior, i.e., a growth rate rise with the temperature, was also seen for ZnO grown on glass and polycarbonate substrates; we thus assumed that this applies to all PEN substrates, since this temperature range is on the front edge of the ALD window [14]. The transmission of PEN/ZnO samples and pure PEN substrate is presented in figure 1a.

For the sample grown at 120 °C, the transmission was measured at two different points of the substrate. While all samples with ZnO films were less transparent than pure PEN substrate, the thickness dependence of transparency was more complex (figure 1). The sample of middle thickness (PEN/ZnO grown at 100 °C) showed the lowest transmission in the wavelength range 380 – 1000 nm (which may be due to neutral oxygen vacancies and singly ionized zinc vacancies [15]). In the shorter wavelength range of 380 – 525 nm the thickest ZnO film (PEN/ZnO grown at 120 °C) had higher transmission, whereas in the longer wavelength range (525 – 1000 nm), the thinnest ZnO film (PEN/ZnO grown at 80 °C) displayed a higher transmission probably due to singly ionized zinc vacancies [15].

All samples discussed exhibited a polycrystalline hexagonal ZnO phase with a wurtzite-type structure (figure 1 b and figure 2). The analysis of the first three peaks of ZnO between 30 – 38 °2θ showed that at a temperature of 120 °C, the predominant ZnO peaks were (100) and (101), while (002) was with low intensity. As the temperature decreased, the intensity of the (100) and (101) peaks

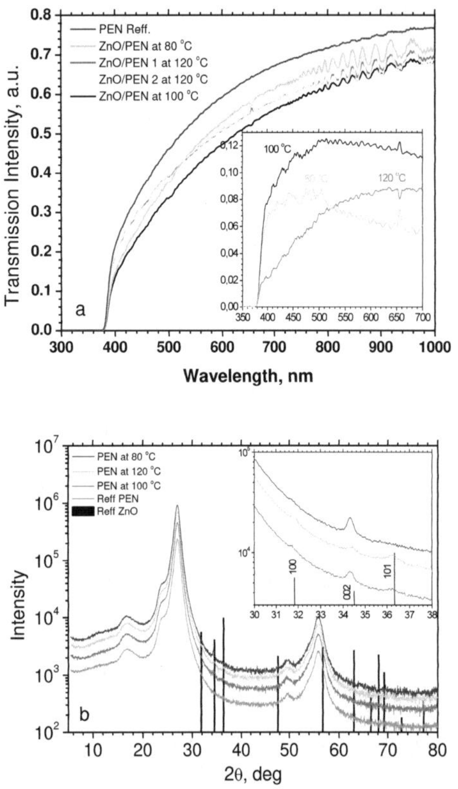

Figure 1. Optical transmission a) and X-ray 2θ intensity b) for ZnO films on PEN substrates at 80, 100 and 120 ºC deposition temperatures; inset in figure 1a – transmission with subtracted substrate; inset in figure 1b – detailed patterns in the range 30 – 38°2θ.

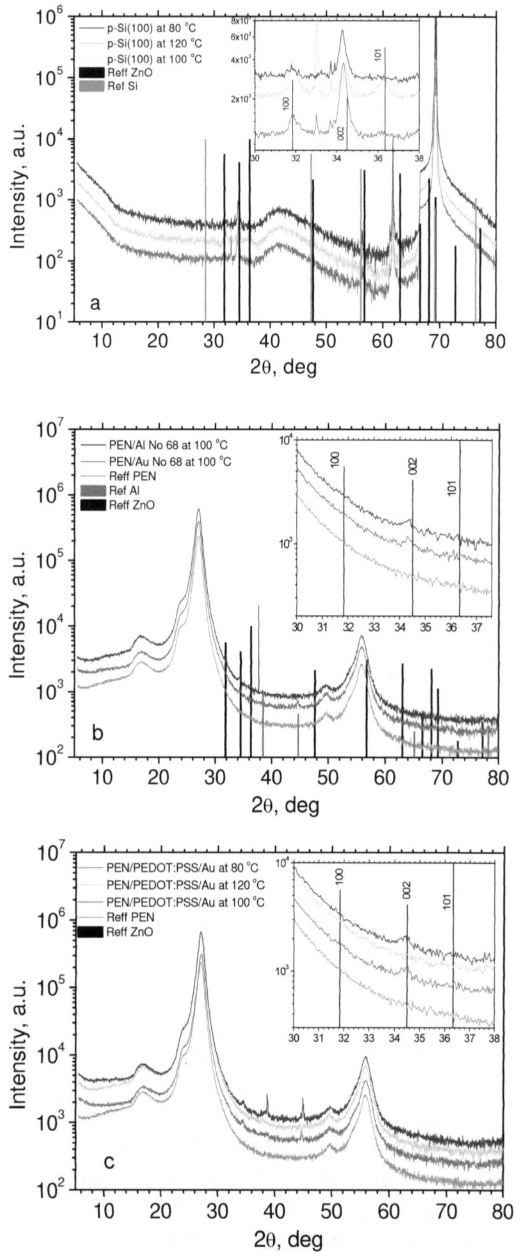

Figure 2. X-ray 2θ intensity for ZnO films on a) p-Si(100), b) PEN/Al and PEN/Au and c) PEN/PEDOT:PSS/Au substrates at 80, 100 and 120 ºC deposition temperatures. The insets show more detailed patterns in the range 30 – 38°2θ.

decreased, while that of (002) increased. Similar results have been presented in other works [12, 16]. The film grown at the middle temperature (PEN/ZnO at 100 ºC) possessed more crystallites of both orientations and, probably, of smaller sizes. This could explain its lowest transparency. To prove this, or other propositions, additional investigations are needed.

Next, we demonstrated the ability of the flexible device with ALD ZnO to generate measurable piezoelectric voltage with parameters (periodicity, effective value, distortions) suitable for application as an energy harvesting microgenerator. As is shown in figure 3 a-d, periodically bending and releasing the structures resulted in the generation of an alternating voltage that in most of the cases strictly followed the frequency of the stimulating vibration. In figure 3, we present the piezoelectric characterization of PEN/PEDOT:PSS/Au/ZnO samples (figure 3 a-c) grown at differen temperatures, and a PEN/Al/ZnO sample (figure 3 d) grown at 100 ºC. The output voltage response for a representative flexible microgenerator was measured under continuous actuation at a frequency of 52 Hz (deflection of the beam and sample of ~3–4 mm). The extracted voltages were 29.7 mV, 86.16 mV and 11.48 mV, corresponding to ZnO grown at temperatures of 80, 100 and 120 °C, respectively. The lateral positioning of the top electrodes for the structure with an Al bottom electrode also gave promising results regarding measurable piezoelectric voltage distinguishable from the noise (figure 4).

We further found that the electrical output was strongly affected by the mechanical load frequency. At frequencies higher than 52 Hz, noise could only be registered, most probably due to high dielectric losses and inability of the dipoles to follow the mechanical wave direction change. This observation is in good agreement with the results for the RC time constant of each device, varying between 29.6 ms

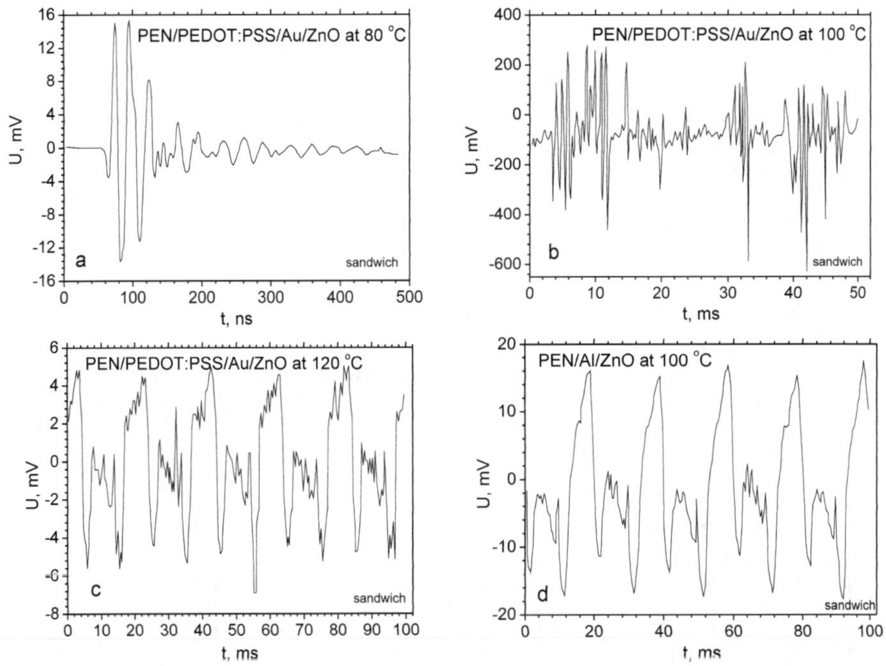

Figure 3. Piezoelectric response of flexible microgenerators with ALD ZnO films grown on PEN/PEDOT:PSS/Au substrates at temperatures a) 80 °C; b) 100 °C; c) 120 °C; and on a PEN/Al substrate grown at 100 °C d).

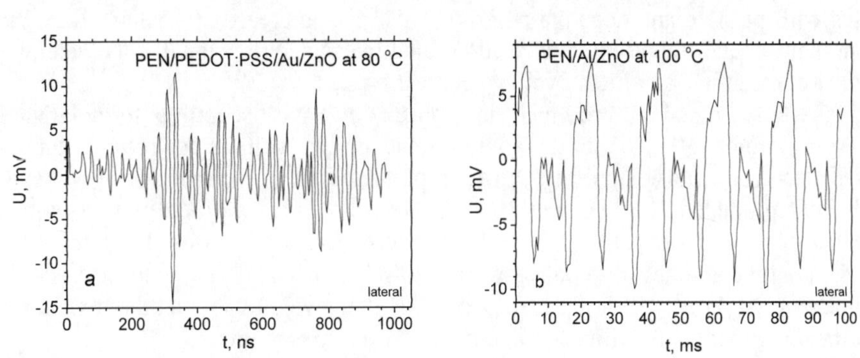

Figure 4. Piezoelectric response of (a) PEN/PEDOT:PSS/Au/ZnO/Al and (b) PEN/Al/ZnO/Al devices.

IOP Conf. Series: Journal of Physics: Conf. Series **992** (2018) 012027 doi:10.1088/1742-6596/992/1/012027

and 420 ms depending on the ZnO-growth temperature and the sample design. One period of a sinusoidal vibration with a frequency of 52 Hz is 19 ms, meaning that none of the prepared structures generating periodical signals could follow a mechanical load of higher frequency.

Moreover, we found that, due to the rising losses, the magnitude of the piezoelectric voltage was maximal in the frequency range 20-52 Hz. The sign of the measured capacitance (table 1) at stretching the samples along the poling axis was positive, which means that the voltage generated had a polarity coinciding with the poling electric field direction. This was the reason for the positive pulses seen on the oscilloscope; similarly, after discontinuing the stress, negative voltage peaks were seen.

Figure 3 a shows the lack of periodicity in the signal. Random pulses with a maximal amplitude of 29.7 mV were observed, but neither a frequency nor an effective value could be defined. This is to be expected, because the RC parameters did not change significantly after polarization. Otherwise, the resistance and capacity should be changed at least by an order of magnitude [17]. Keeping the same order of magnitudes (or less) means no response to poling or weak susceptibility of the dipoles to re-orientation, resulting in a negligible piezoelectric reaction. When the measurement was conducted between two adjacent electrodes (a lateral structure, instead of a sandwich-type one), a periodic signal was recorded (figure 4a), but its effective value was only 5.4 mV. The dielectric losses were high, as can be seen in table 1.

For the device whose signal is shown in figure 3 b, the effective value measured of the generated voltage was 86.16 mV. The pulses look random, but the oscilloscope picked up repeating pulses and measured a frequency of 20,83 Hz which is lower by a factor of 2.6 than that of the exciting vibration, probably meaning that the dipoles had a relatively low mobility. One can notice a symmetry with respect to the pulses' rise and fall times – 24 ms for pulses with positive and negative polarity. This gives us reason to believe that they did not have a random nature and represented a well-defined piezoelectric signal. The lateral electrode configuration did not result in a significant voltage generation.

Table 1. Resistance, capacitance and dielectric losses of samples for different electrode configurations and ZnO-growth temperatures.

Device	R, C and tgδ before poling	R, C and tgδ after poling
PEN/PEDOT:PSS/Au/ZnO/Al, **ZnO at 80 °C**	C = 37 μF R = 80 Ω tgδ = 0.105	C = 18 μF R = 44.5 Ω tgδ = 0.099
PEN/PEDOT:PSS/Au/ZnO/Al **ZnO at 100 °C**	C = 1 nF R = 24 Ω tgδ = 0.098	C = 35 μF R = 12 kΩ tgδ = 0.094
PEN/PEDOT:PSS/Au/ZnO/Al **ZnO at 120 °C**	C = 119 nF R = 57 Ω tgδ = 0.091	C = 15 μF R = 53 Ω tgδ = 0.068
PEN/Al/ZnO/Al **ZnO at 100 °C**	C = 109 nF R = 420 Ω tgδ = 0.140	C = 2.2 μF R = 100 Ω tgδ = 0.010
PEN/Au/ZnO/Al **ZnO at 100 °C**	C = 934 nF R = 1.3 kΩ tgδ = 0.068	C = 623 nF R = 1.2 kΩ tgδ = 0.041

The structure whose signal is presented in figure 3 c was susceptible to polarization, with capacitance varying from 119 nF prior to applying the poling voltage to 15 μF after poling. Its dielectric losses decreased most significantly after polarization, compared to the other measured structures with ZnO produced at 80 °C and 100 °C. The signal was periodic with a high degree of symmetry of the positive and negative pulses. The effective value was 11.48 mV, the largest amplitudes of the positive and negative pulses were 21.9 mV and –32.4 mV, respectively. The pulses' rise and fall times were 54 ms, with a frequency of 25 Hz, or twice as low as that of the mechanical load, probably meaning that the mobility of the dipoles was relatively low.

The sandwich structure with configuration illustrated in figure 3 d produced a periodic signal with a well-expressed amplitude of 15.3 mV and an effective (rms) value of 10.85 mV. It also demonstrated the lowest value of dielectric losses of all structures prepared. The pulses' frequency was 52 Hz (exactly the mechanical stimulation frequency); therefore, the dipoles were re-orienting smoothly without delay following the mechanical wave. The pulses' rise/fall times was 10.8 ms. The lateral structure with adjacent upper electrodes (figure 4 b) with the signal measured between them was also functioning, although with twice as low amplitude of the piezoelectric voltage generated – 8.82 mV, and steeper pulses – 1.6 ms.

We also prepared a structure with all Au electrodes, but the results were not satisfactory, as a significant piezoelectric reaction was lacking. Such behavior is expected, since the capacitance after polarization remains small and even decreases, while the resistance of the polarizable medium is in the order of kiloohms.

Regarding the usability of the film for piezoelectric devices, the increased (002) peak suggests a predominant *c*-axis orientation of the crystallites [18], which coincides with the applied force direction for the sandwich structures and may be the reason for the highest achieved effective value of the generated voltage (see figure 3b). Moreover, it has been proved that samples with high crystallinity at (002) orientation are more conductive, which facilitates the electrical poles formation [19].

Compared to previous reports for thin film generators with ZnO nanostructures (nanowires, nanorods, etc.), the ALD-ZnO based one showed promising results in terms of the voltage generated. The reason is probably in the more reliable contact between the ALD film and the electrodes. In contrast, the deformation of the above-mentioned nano-formations leads to unstable interfaces between the needle-type piezoelectric film and the electrode, thus resulting in a lower yield [20].

Although the values of all measured voltages were in the order of millivolts, the current flowing through the structure was still rather small – in the range of 500 nA to 16 μA, resulting in microwatts power.

4. Conclusions

ALD-ZnO nanofilms were deposited on flexible PEN substrates and their structural, optical and piezoelectric properties were investigated in dependence of the deposition temperature and contact type. Some of the structures discussed possessed high transparency and flexibility, which could make them useful for applications in flexible optoelectronic. Other structures showed promising piezoelectric properties and are potential candidates for flexible microgenerators activated by human motion.

Our future work will be related to increasing the structures' conductivity, thus raising the electrical power density. Long-term stability studies after cyclic repeated bends should also be carried out on piezoelectric microgenerators with ALD ZnO sandwiched between symmetrical types of electrodes, especially polymer-based ones.

Acknowledgement

This work was supported by the Bulgarian National Science Fund under Grant DH 07/13 and by Project DCOST 01/12. The authors appreciate Prof. V. Videkov's (Technical University, Sofia) assistance in processing the oscilloscope data. The authors also acknowledge PhD student Yordanka Vucheva's technical support with the measurement stand.

References

[1] Diao L-W, Zheng J, Pan X-D, Zhang W, Wang L-F and Sun L-Z 2014 *J. Thorac. Dis.* **6**/9 1300
[2] Zheng Q, Shi B, Li Z and Wang Z L 2017 *Adv. Sci. (Weinh)* **4**/7 1700029
[3] Leventhall G 2003 *A Review of Published Research on Low Frequency Noise and its Effects* (Nobel House, London)
[4] Ani Melfa Roji M, Jiji G and Ajith Bosco Raj T 2017 *RSC Adv.* **7** 33642
[5] Cheng T, Zhang Y-Z, Zhang J-D, Lai W-Y and Huang W 2016 *J. Mater. Chem.* A **4** 10493
[6] Kolev G, Aleksandrova M, Dobrikov G, Pathan H, Fartunkov M and Denishev K 2017 Piezoelectric energy harvesting device with nanobranched ZnO on Polymer/Metal/Polymer coated flexible substrate *Proc. 15th IEEE Int. Conf. Electrical Machines, Drives and Power Systems* (Sofia, Bulgaria, 1-3 June 2017) pp 320-4
[7] Aleksandrova M, Kurtev N, Videkov V, Tzanova S and Schintke S 2015 *Microelectr. Eng.* **145** 112

[8] Ramadan K S, Sameoto D and Evoy S 2014 *Smart Mater. Struct.* **23** 033001
[9] Ponnammar D, Chamakh M M, Deshmukh K, Ahamed M B, Erturk A, Sharma P and Al-Maadeed M A-A 2017 Ceramic-based polymer nanocomposites as piezoelectric materials *Springer Series on Polymer and Composite Materials* pp 77-93
[10] Özgür Ü, Alivov Ya I, Liu C, Teke A, Reshchikov M A, Doğan S, Avrutin V, Cho S-J and Morkoç H 2005 *J. Appl. Phys.* **98** 041301
[11] Blagoev B S, Dimitrov D Z, Mehandzhiev V B, Kovacheva D, Terziyska P, Pavlic J, Lovchinov K, Mateev E, Leclercq J and Sveshtarov P 2016 *J. Phys.: Conf. Ser.* **700** 012040
[12] Kumar A, Kumar P, Kumar K, Singh T, Singh R, Asokan K and Kanjilal D 2015 *J. All. Comp.* **649** 1205
[13] Kowalik I A, Guziewicz E, Kopalko K, Yatsunenko S, Wójcik-Głodowska A, Godlewski M, Dłużewski P, Łusakowska E and Paszkowicz W 2009 *J. Cryst. Growth* **311** 1096
[14] Guziewicz E, Kowalik I A, Godlewski M, Kopalko K, Osinniy V, Wójcik A, Yatsunenko S, Łusakowska E, Paszkowicz W and Guziewicz M 2008 *J. Appl. Phys.* **103** 033515
[15] Krajewski T A, Terziyska P, Luka G, Lusakowska E, Jakiela R, Vlakhov E S and Guziewicz E 2017 Diversity of contributions leading to the nominally n–type behavior of ZnO films obtained by low temperature Atomic Layer Deposition, *J. Alloys and Compounds* In Press Accepted Manuscript DOI 10.1016/j.jallcom.2017.08.206
[16] Malm J, Sahramo E, Perälä J, Sajavaara T and Karppinen M 2011 *Thin Solid Films* **519** 5319
[17] Chalioris C E, Karayannis C G, Angeli G M, Papadopoulos N A, Favvata M J and Providakis C P 2016 *Case Studies in Construction Materials* **5** 1
[18] Huang W 2012 C-axis oriented ZnO piezoelectric thin films prepared by RF magnetron sputtering for saw filters *Proc. 2nd Int. Conf. Computer Appl. and System Modeling* (2012 Paris France) pp 2017-20
[19] Shariffudin S S, Zakaria N Z, Herman S H and Rusop M 2011 Effect of deposition temperature on the characteristics of Zinc Oxide nanoparticles thin films deposited by thermal chemical vapor deposition *Proc. Int. Conf. Electronic Devices, Systems and Applications* (25-27 April 2011 Kuala Lumpur Malaysia)
[20] Li T, Li Y T, Qin W W, Zhang P P, Chen X Q, Hu X F and Zhang W 2015 *Nanoscale Res. Lett.* **10** 394

Electrochemical oxidation of sulfites by DWCNTs, MWCNTs, higher fullerenes and manganese

Dzhamal Uzun[1], George Pchelarov, Ognian Dimitrov, Sasho Vassilev, Willi Obretenov and Konstantin Petrov

Acad. Evgeni Budevski Institute of Electrochemistry and Energy Systems, Bulgarian Academy of Sciences, Acad. G. Bonchev Str., Bl. 10, 1113 Sofia, Bulgaria

E-mail: dzhamal.uzun@gmail.com

Abstract. Different electrocatalysts were tested for oxidation of sulfites to sulfates, namely, manganese thin films deposited on fullerenes and carbon nanotubes. The results presented clearly show that electrodes containing HFs (higher fullerenes), DWCNTs (double-wall carbon nanotubes) and manganese acetate are effective catalysts in S/O_2 fuel cells. HFs and DWCNTs have high catalytic activity and can be employed as standalone catalysts. Manganese was deposited on DWCNTs, HFs and fullerenes C_{60}/C_{70} by a thermal process. The electrocatalysts were characterized by scanning electron microscopy (SEM) and X-ray diffraction (XRD). The electrochemical testing was carried out by plotting the E/V polarization curve. The polarization curves of the electrodes composed of pristine DWCNTs showed the lowest overpotentials.

1. Introduction

The present work considers the possibility of utilizing electrocatalysts, such as fullerenes and carbon nanotubes. The use of fullerenes and nanohorns as catalysts in fuel cells has been discussed in the past [1]. In this study we used 'higher fullerenes' (also termed 'small-bandgap fullerenes') as prepared by Diener and Alford by applying the carbon-arc method [2]. These were purchased together with manganese acetate from the Sigma Aldrich company. The fullerenes C60/C70, DWCNTs (double-wall carbon nanotubes) and MWCNTs (multiwall carbon nanotubes) were bought from SES Research, Huston, Texas. Vulcan XC-72 has been utilized as an electrocatalyst support in the past; this was acquired from the Cabot Corporation [3]. The first application of nanotubes in electrochemistry was reported by Britto [4], who, using bromoform as a binder, packed a paste of nanotubes into a glass tube in order to study the redox reaction of dopamine. The process was found to occur at a much faster rate than that observed using other catalytic surfaces. Later, Britto [5] and co-authors were the first to show that MWCNTs can be used for the electrocatalysis of oxygen. In the work reported here, our studies on DWCNTs found that they have superior electrochemical properties as compared to other fullerene-type materials. The role of carbon in fuel cells was studied in the past by Dicks [6] and recently reported in [7].

[1] To whom any correspondence should be addressed.

Our aim was to develop a fuel cell for simultaneous purification of sulfates and nitrites. We studied novel catalysts incorporated in the electrode material and used half cells to oxidize SO_3 to SO_4.

The current work is part of a project entitled "New fuel cells based on chemical and microbial processes", whose objective is exploring the possibility to decontaminate compounds containing sulfur and nitrogen.

2. Experimental

2.1. Experimental Setup

The catalytic properties of a set of anode electrodes were determined with the aid of a three-electrode cell (figure 1) while studying the process of oxidation of sulfites to sulfates in solution

$$SO_3^{2-} + 0.5O_2 \leftrightarrow SO_4^{2-}.$$

The solution was 1 M Na_2SO_3 + 18 g l^{-1} NaCl as an additive electrolyte. The cell volume was 50 ml. The counter electrode was platinum foil and the reference electrode, a "Gaskatel" hydrogen electrode.

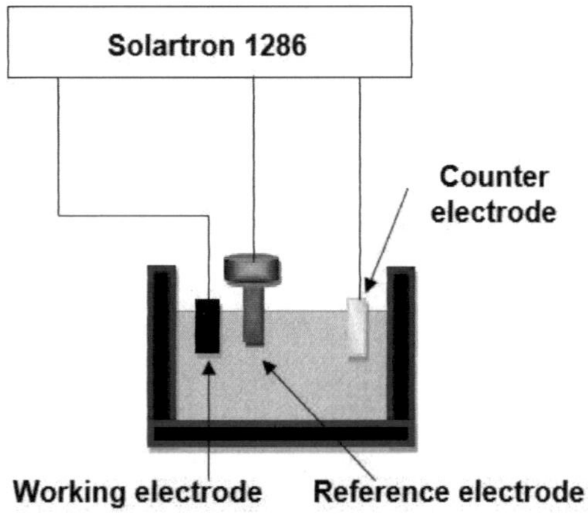

Figure 1. A drawing of the three-electrode cell used in the experiments.

2.2. Measurements

2.2.1. XRD Measurements. The X-ray diffraction (XRD) patterns were recorded by a Philips diffractometer using CuK_α radiation (λ = 1.54178 Å, 40 kV, 30 mA) at a scanning rate of 2° min^{-1}. The intense diffraction peaks at 36.92°, 42.42 and 55.96° were referred to the characteristic peaks of γ-MnO_2, while the peaks at 17.94°, 28.78°, 49.0° and 60.44° were ascribed to α-MnO_2 [8]. The XRD of manganese acetate as purchased from Sigma Aldrich and scanned in our lab is presented in figure 2. The XRD of the sample of manganese acetate with added polypyrrole is presented in figure 3. The XRD of MWCNTs as purchased from SES Research and scanned in our lab is presented in figure 4. The XRD of DWCNTs as purchased from SES Research and scanned in our lab is presented in figure 5. Shown in figure 6 is the XRD of higher fullerenes, DWCNTs and manganese acetate. Shown in figure 7 is the XRD of HFs and manganese acetate. Presented in figure 8 is the XRD of the d space for the sample comprised of HFs. Shown in figure 9 is the XRD of pristine HFs.

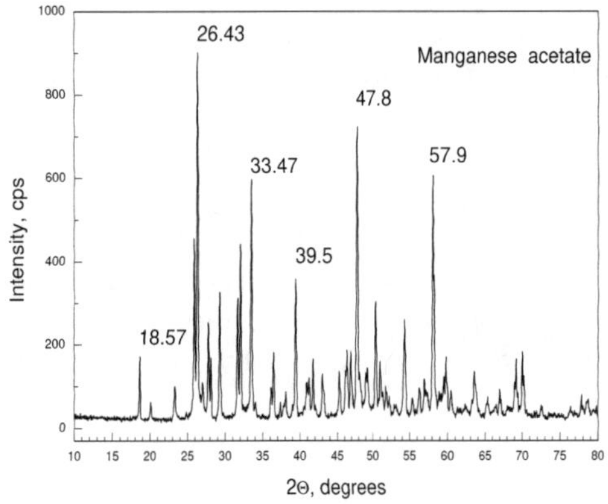

Figure 2. XRD of manganese acetate.

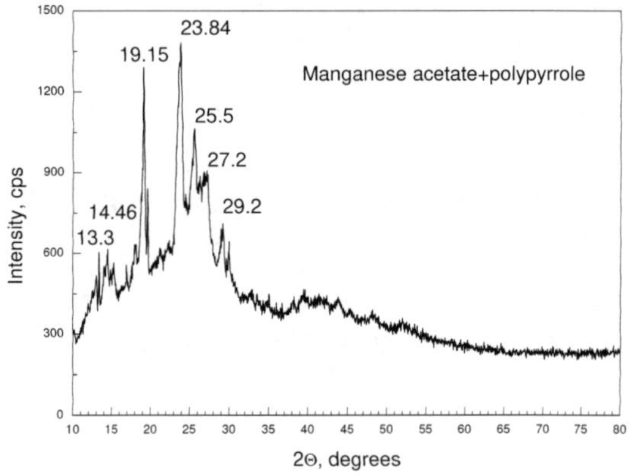

Figure 3. XRD of manganese acetate and polypyrrole.

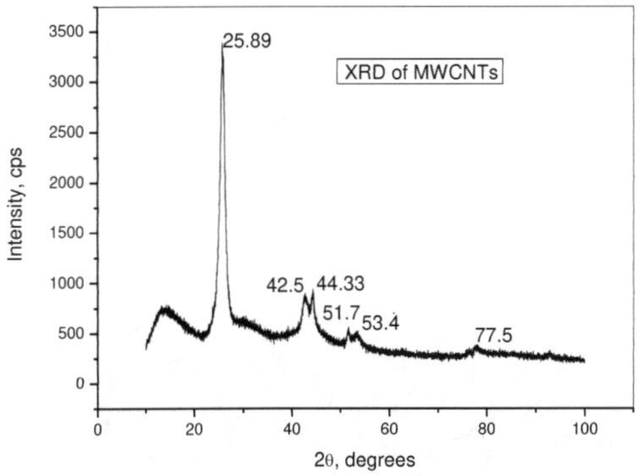

Figure 4. XRD of MWCNTs.

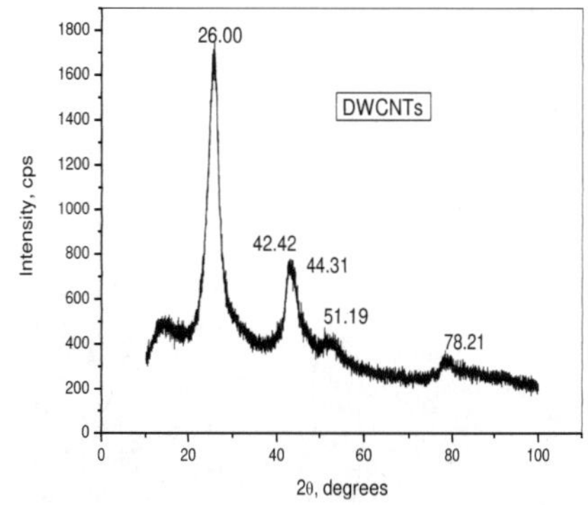

Figure 5. XRD of DWCNTs.

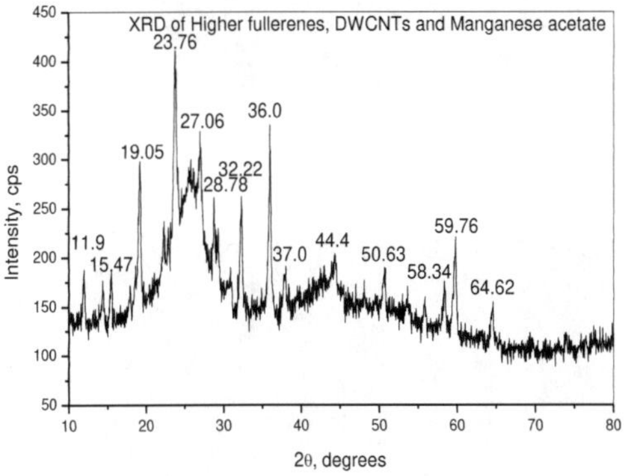

Figure 6. XRD of HFs, DWCNTs and manganese acetate.

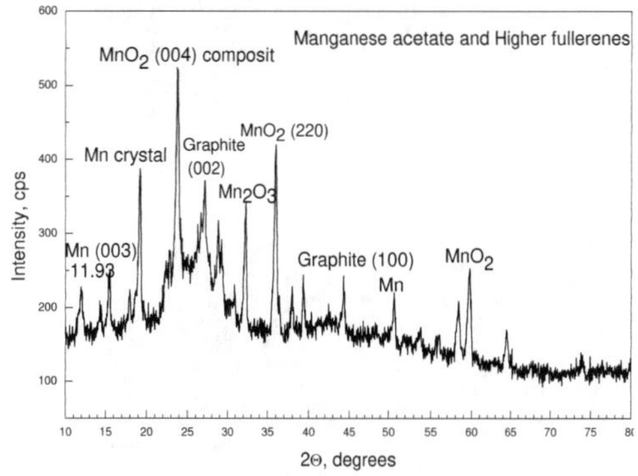

Figure7. XRD of HFs and manganese acetate.

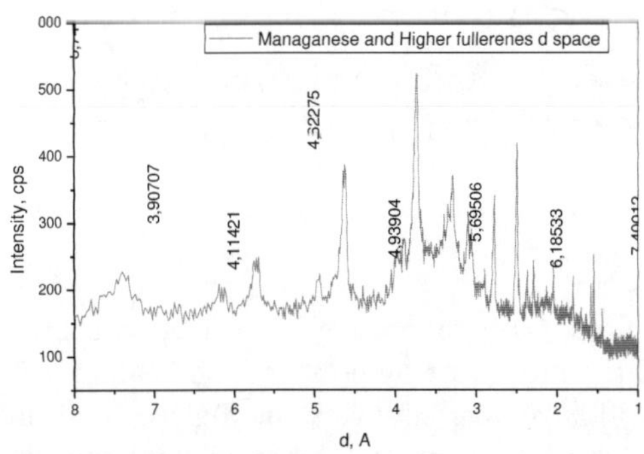

Figure 8. XRD study of the d space for the sample composed of higher fullerenes and manganese acetate.

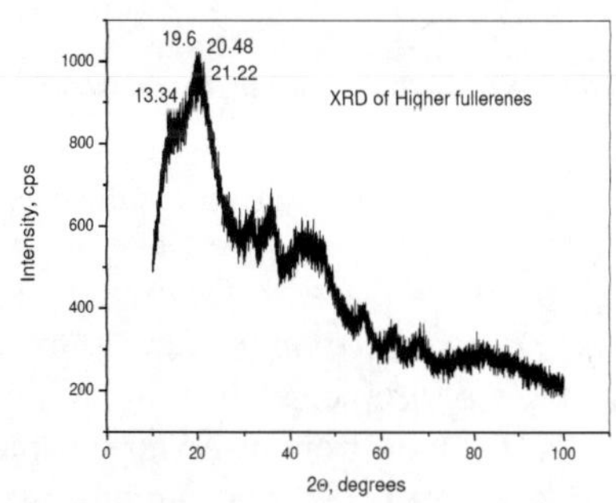

Figure 9. XRD of pristine HFs.

2.2.2. Catalyst synthesis. Small quantities of MWCNTs and of DWCNTs were each dispersed in 6 ml of distilled water in a sonic bath for 15 min. Next, 40 mg of manganese acetate and 60 mg of polypyrrole were slowly added to the aqueous suspensions. These ternary components were then baked at 180 °C for 12 h in a Teflon autoclave. Thus, MnO_2 particles were deposited on the carbon nanotube lattice structures with polypyrrole binding. The DWCNTs used clearly maximize the electro-active surface area of the catalyst. The porous structure of the catalyst consisting of DWCNTs and manganese was observed by SEM and is shown in figure 10. The catalyst consisting of MWCNTs and manganese was observed by a SEM and is shown in figure 11.

IOP Conf. Series: Journal of Physics: Conf. Series **992** (2018) 012028 doi:10.1088/1742-6596/992/1/012028

Several partial polarization curve measurements were performed to characterize the DWCNTs, MWCNTs and higher fullerenes with deposited MnO_2 particles and polypyrrole binder in different ratios.

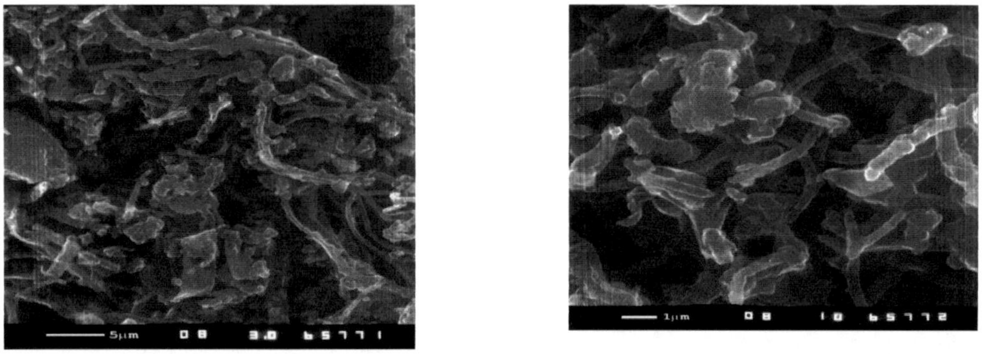

Figure 10. SEM images of the catalyst consisting of DWCNTs and manganese acetate.

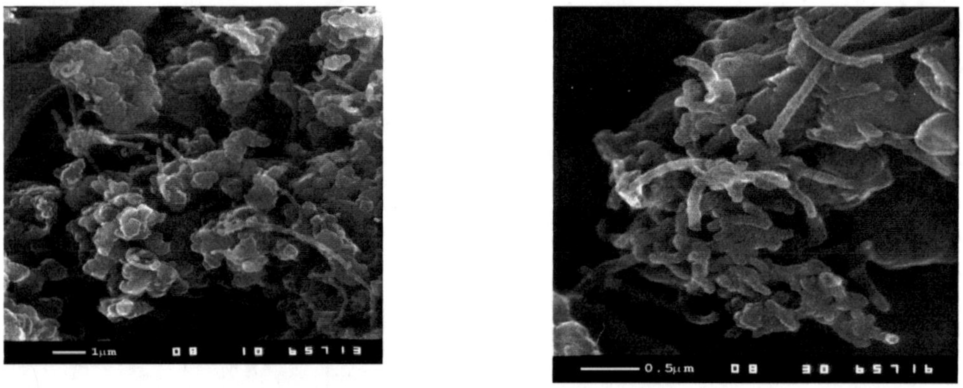

Figure 11. SEM images of the catalyst consisting of MWCNTs and manganese acetate.

2.2.3 Electrode preparation. Several sample electrodes with geometrical area of 1 cm^2 were prepared by depositing the sample catalyst on both sides of the electrode followed by pressing and heating. All electrodes were made from a mixture of the catalyst and teflonized carbon black [60 mg cm^{-2} Vulcan XC-72 (35% Teflon)] [9]. The mixture was pressed at 300 kg cm^{-2} on both sides of a stainless steel current collector at 150 °C. The catalysts studied were as follows: a NORIT & manganese oxide catalyst; HFs, DWCNTs and manganese oxide; HFs and manganese oxide. We used higher fullerenes and DWCNTs to maximize the electro-active surface area of the catalyst.

In the course of the experiment, several electrodes from each catalytic mass were prepared. The volt/ampere characteristic of each electrode was measured several times and the average values calculated.

The electrodes contained the following catalysts: 1 – 100 mg DWCNTs; 2 – 60 mg manganese acetate; 3 – 100 mg higher fullerenes; 4 – 100 mg MWCNTs; 5 – 120 mg fullerenes C_{60}/C_{70}; 6 – 20 mg C_{60}/C_{70} + 60 mg manganese acetate + 40 mg polypyrrole; 7 – 3 mg MWCNTs + 37 mg manganese acetate + 57 mg polypyrrole; 8 – 3 mg DWCNTs + 37 mg manganese acetate + 57 mg polypyrrole.

IOP Conf. Series: Journal of Physics: Conf. Series **992** (2018) 012028 doi:10.1088/1742-6596/992/1/012028

2.2.4. Polarization curves. As it is seen, the polarization curves of the pristine DWCNTs samples as presented in figures 12 and 13 are the most advantageous, yielding the lowest overpotentials at the respective current densities. The DWCNTs electrode as compared to the MWCNTs electrode shows a 300-mV decrease in overpotential at $i = 10$ mA cm^{-2}.

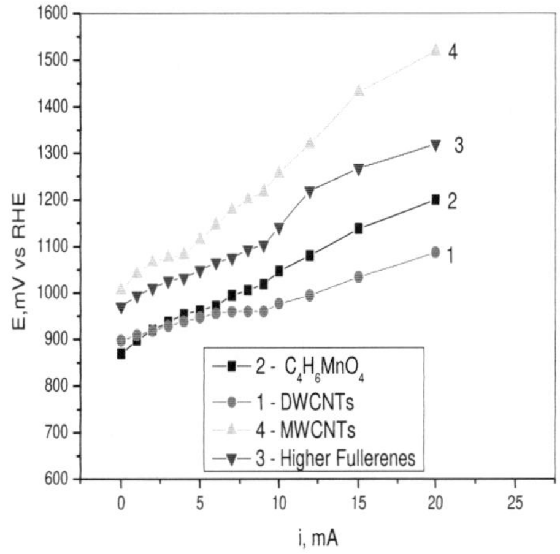

Figure 12. Polarization curves of sulfites oxidation by electrodes fabricated with different catalysts: 1 – 100 mg DWCNTs; 2 – 60 mg manganese acetate; 3 – 100 mg higher fullerenes; 4 – 100 mg WCNTs. The electrolyte was 1 M Na_2SO_3 + 18 g l^{-1} NaCl; $t = 20$ °C.

3. Results and discussions

The multiwall carbon nanotubes (MWCNTs) comprise many graphene layers decreasing in diameter for each successive layer. The double-wall carbon nano tubes (DWCNTs), on the other hand, consist of only a few layers that are hollow and larger in diameter. It was argued in the past that the inner tube formation of the DWCNTs structure could be very defective, containing pentagons, heptagons and octagons [10]. In catalysis, a large number of active sites are sought in order to optimize the process. The rolled-up carbon networks of DWCNTs clearly would play a major role in the synthesis of more effective catalysts. We found that the electrode catalyst sample composed of DWCNTs was the most effective, yielding the lowest overpotentials at the respective current densities for the polarization curves. Therefore, it can be expected that this catalyst will exhibit the highest mass activity in the process of oxidation of sulfites to sulfates. Our recent study [11] showed that it is possible to implement electrochemically reduction of nitrates and oxidation of sulfite by means of electrodes containing DWCNTs and HFs.

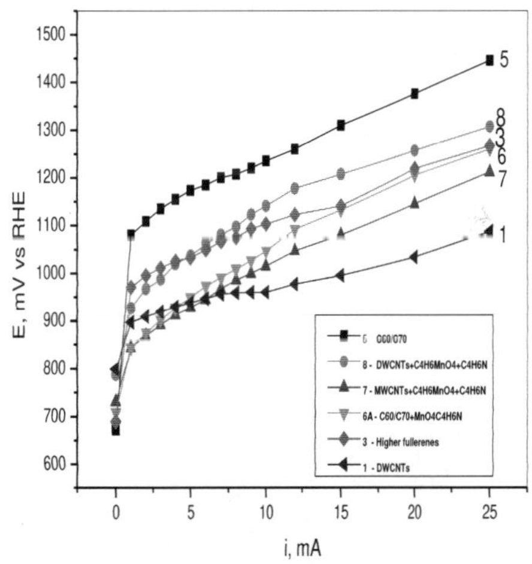

Figure 13. Polarization curves of sulfites oxidation by electrodes fabricated with different catalysts: 1 – 100 mg DWCNTs; 3 – 100 mg higher fullerenes; 5 – 120 mg fullerenes C_{60}/C_{70}; 6 – 20 mg C60/C70 + 60 mg manganese acetate + 40 mg polypyrrole; 7 – 3 mg MWCNTs + 37 mg manganese acetate + 57 mg polypyrrole; 8 – 3 mg DWCNTs + 37 mg manganese acetate + 57 mg polypyrrole. The electrolyte was 1 M Na_2SO_3 + 18 g l^{-1} NaCl; $t = 20$ °C.

4. Conclusions

The results presented show that higher fullerenes and DWCNTs have high catalytic activity on their own, i.e. without any deposited traditional catalysts such as manganese, and can be employed as stand-alone catalysts. Due to their large cation exchange capacity, strong adsorption ability, huge surface area and suitable electronic properties, DWCNTs exhibit high catalytic activity, as clearly evidenced by the polarization curve plots.

Acknowledgements
This work was part of project "New fuel cells based on chemical and microbial processes, through the Ministry of Education and Science and the Bulgarian National Science Research Fund, contract DFNI E02/15.

References

[1] Koprinarov N, Konstantinova M, Pchelarov G and Marinov M 2005 Fullerene structures suitable for fuel cells electrodes 2004 *Proc. NATO Adv. Res. Workshop Fuel Cell Technol.* (6-10 June 2004 Kyiv Ukraine) p 81

Sammers N et al eds 2004 Fuel Cell Technologies: State and Perspectives *NATO Sci. Series II – Math. Phys. and Chem.* **202** 91-5

[2] Diener M D and Alford J M 1998 Isolation and properties of small-gap fullerenes *Nature* **393** 668-71

[3] Ralph T R and Hogarth M P 2002 Catalysis for low-temperature fuel cells – part 1: The cathode challenges *Platinum Metals Rev.* **46**/1 4-5

[4] Britto P J, Santhanam K S V and Ajayan P M 1996 *Bioelectrochem. and Bioenergetics* **41** 121

[5] Britto P J, Santhanam K S V, Alonso V, Rubio A P and Ajayan M 1999 *Adv. Mater.* **11** 154

[6] Dicks A L 2006 The role of carbon in fuel cells *J. of Power* **156** 2

[7] You P Y and Kamarudin S K 2017 Recent progress of carbonaceous materials in fuel cell applications: An overview *Chem. Eng. J.* **309** 1

[8] Dai J X, Li S F Y, Siow K S and Gao Z 2000 *Electrochimica Acta* **45** 2211

[9] Uzun D, Razkazova–Velkov E, Petrov K and Beschkov V 2015 Electrochemical method for energy production from hydrogen sulfide in the Black sea waters in sulfide-driven fuel cell *Bulgarian Chem. Communic.* **47** 929

[10] Hernández E, Meunier V, Smith B W, Rurali R, Terrones H and Buongiorno M 2003 Fullerene coalescence in nanopeapods: a path to novel tubular carbon *Nano Lett.* **3**/8 1037-42

[11] Pchelarov G, Uzun D, Dimitrov O, Vassilev S and Petrov K 2017 Electrocatalysts for sulphur ions oxidation based on DWCNTs, MWCNTs, higher fullerenes and manganese *Bulgarian Chem. Communic.* **49** 218

IOP Conf. Series: Journal of Physics: Conf. Series **992** (2018) 012029 doi:10.1088/1742-6596/992/1/012029

Yellow emitting Iridium (III) phenyl-benzothiazole complexes with different β-diketone ancillary ligands as dopants in white organic light-emitting diodes

P Ivanov, P Petrova and R Tomova[1]

Acad. J. Malinowski Institute of Optical Materials and Technologies, Bulgarian Academy of Sciences, Acad. G. Bonchev Str., Bl. 109, 1113 Sofia, Bulgaria

E-mail: reni@iomt.bas.bg

Abstract. We discuss the influence of the type of β-diketone ancillary ligand in Iridium (III) bis phenyl-benzothiazole complexes ($(bt)_2Ir$(β-diketone)) on their photophysical and electroluminescent properties when they are used as dopants in white organic light-emitting diodes (WOLED). For this purpose, we investigated four novel yellow cyclometalated complexes: $(bt)_2Ir$(dbm), $(bt)_2Ir$(fmtdbm), $(bt)_2Ir$(tta) and $(bt)_2Ir$(bsm), where dbm = 1,3-diphenylpropane-1,3-dionate; fmtdbm = 1-(4-fluorophenyl)-3-(4-methoxyphenyl)propane-1,3-dionate; tta = 4,4,4-trifluoro-1-(thiophene-2-yl)butane-1,3-dionate; and bsm = 1-phenylicosane-1,3-dionate). To obtain white light by mixing emissions of two complementary colors (yellow emitted by the dopant and blue, by another emitter), we chose the following OLED structure: ITO/doped HTL/EIL/ ETL/M, where ITO was a transparent anode of In_2O_3:SnO_2; M, a metallic Al cathode; HTL, 4,4'-Bis(9H-carbazol-9-yl)biphenyl (CBP) involved in a poly(N-vinylcarbazole) (PVK) matrix; EIL, an electroluminescent layer of aluminum(III)bis(2-methyl-8-quninolinato)-4-phenylphenolate (BAlq); and ETL, an electron-transporting layer of zinc(II)bis(2-2-hydroxyphenyl)benzothiazole. We found that all complexes are suitable candidates for fabrication of WOLED. The best results were demonstrated by the device doped with 2 wt % of $(bt)_2Ir$(bsm), which had twice as high luminescence (1100 cd/m^2) and one-and-a-half as high current efficiency (5 cd/A) as the device doped with 1.25 wt % of the known $(bt)_2Ir$(acac), with its 580 cd/m^2 and 3.4 cd/A at approximately the same CIE (Commission Internationale de L'Eclairage) (*x*/*y*) coordinates of the warm white light emitted by the two devices.

1. Introduction

Phosphorescent organic light-emitting diodes offer an intriguing future for the next generation of flat-panel displays and solid-state lighting sources because of their high quantum efficiency in comparison with fluorescent OLEDs [1-3]. Owing to the spin-statistics, the singlet/triplet ratio of the electro-generated excitons is 25/75 %. As it is known, only singlet excitons decay radiatively, which limits the external quantum efficiency to 25 %. This constraint can be removed by using electrophosphorescent complexes of transition metals, as Ir, Pt and Os [4–6]. These complexes demonstrate efficient intersystem crossing due to the metals with d6 electronic configurations allowing

[1] To whom any correspondence should be addressed.

radiative relaxation of excited triplet states. Thus, the light emitted by both type of excitons can be collected, and the external quantum efficiency can be raised up to 100 % [7, 8].

Among all complexes of transition metals, the cyclometalated Iridium(III) complexes are the most promising because of their high thermal stability, short triplet state lifetimes, high photoluminescent quantum yields and quasi-octahedral geometry providing opportunities for controlled entry of appropriate ligands. To improve the color purity and the quantum efficiency of Ir complexes, special attention should be paid to synthesizing compounds with new, or modified, main [4] [8-11] or ancillary [5, 6] ligands. Typically, phosphorescent complexes are used as emitting guests in hole-transporting hosts of carbazole derivatives, as CBP (4,4'-Bis(9H-carbazol-9-yl)biphenyl) and PVK [poly(9-vinylcarbazole)] [7, 12-14] or polyfluorenes [15, 16].

The excited triplet state of the guest molecule is formed by direct recombination of an electron and a hole on the guest molecule or by energy transfer from the host followed by recombination on the guest [17–18]. Generally, the dopant concentration is in the range 3 – 10 wt %. At higher dopant concentrations, undesirable side effects appear in the host-guest system, such as concentration quenching and triplet-triplet annihilation [18-19] that decrease significantly the devices' efficiency.

2-Phenylbenzo[d]thiazole (bt) is a general ligand framework for designing Ir(III) complexes; while $(bt)_2Ir(acac)$ has been the most thoroughly investigated yellow emitter among all Ir(III) complexes. Its popularity is due to the fact that, in combination with a blue emitter, it is particularly promising for fabrication of WOLEDs with a simple design based on emission of two complementary colors [20, 21]. Up to now, many $(bt)_2Ir(acac)$ derivatives with adjustable color, high luminescent efficiency and electroluminescent performances have been synthesized and investigated [10, 21, 22].

All this motivated us to launch a study on the Ir(III) (bt) complexes synthesized earlier [23] that contain different β-diketone ancillary ligands as dopants, and that are intended for producing WOLEDs. The photophysical and electroluminescent properties of these complexes are presented below.

2. Experimental

2.1. Synthesis of iridium complexes

The complexes used in this study, namely, Iridium(III) bis[2-phenylbenzothiazolato-N,C$^{2'}$](acetylacetonate) $(bt)_2Ir(acac)$, Iridium(III) bis[2-phenylbenzothiazolato-N,C$^{2'}$]-(1,3-diphenylpropane-1,3-dionate) $(bt)_2Ir(dbm)$, Iridium(III) bis[2-phenylbenzothiazolato-N,C$^{2'}$]-[1-(4-fluorophenyl)-3-(4-methoxyphenyl)propane-1,3-dionate] $(bt)_2Ir(fmtdbm)$, Iridium(III) bis[2-phenylbenzothiazolato-N,C$^{2'}$]-[4,4,4-trifluoro-1-(thiophen-2-yl)butane-1,3-dionate] $(bt)_2Ir(tta)$, and Iridium(III) bis[2-phenylbenzothiazolato-N,C$^{2'}$]-(1-phenylicosane-1,3-dionate) $(bt)_2Ir(bsm)$ (figure 1)

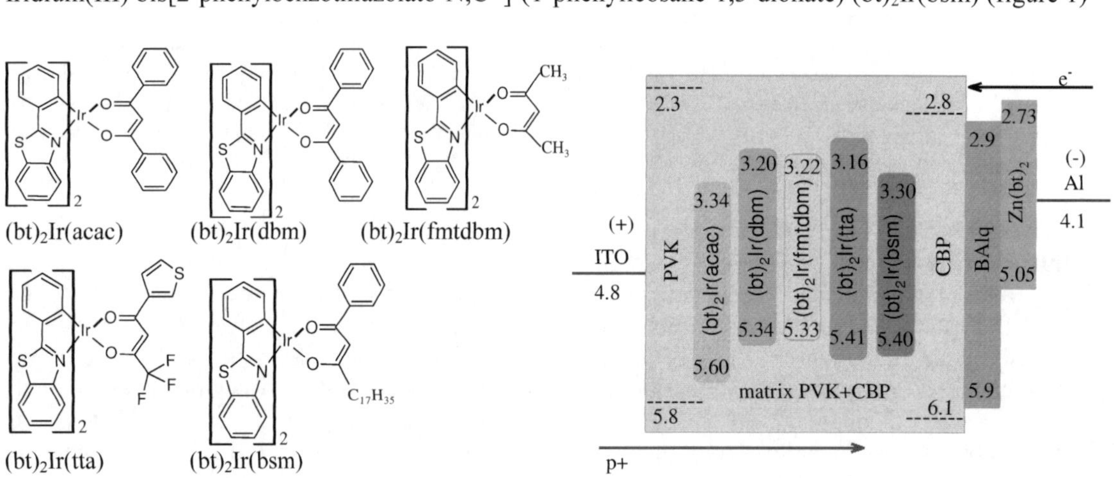

Figure 1. Chemical structure of the Ir(III) complexes used and energy diagram of the OLED investigated. The data for the HOMO/LUMO level are taken from [23].

IOP Conf. Series: Journal of Physics: Conf. Series **992** (2018) 012029 doi:10.1088/1742-6596/992/1/012029

were synthesized according to a procedure suggested by Nonoyama [24]. The synthesis, characterizations and data for HOMO/LUMO levels, obtained by cyclic voltammetry measurements of cyclometalated Ir(III) complexes were published elsewhere [23].

2.2. Device preparation

We investigated Iridium(III) complexes as dopants in an HTL in a multilayered OLED structure: ITO/doped-HTL/ElL/ETL/M. ITO was a transparent anode of In_2O_3:SnO_2, M, a metallic Al cathode; HTL, 4,4'-Bis(9H-carbazol-9-yl)biphenyl (CBP) incorporated in a poly(N-vinylcarbazole) (PVK) matrix; ElL, an electroluminescent layer of aluminum(III)bis(2-methyl-8-quninolinato)-4-phenylphenolate (BAlq); and ETL, an electron-transporting layer of zinc(II) bis(2-(hydroxyphenyl)benzothiazole (Zn(bt)$_2$). Devices were prepared with a size of 1 cm^2 on commercial polyethylene terephthalate (PET) substrate coated with ITO (40 Ω/sq). The layer of a composite film (30 nm) PVK: CBP$_{10\%(\text{relatively to PVK})}$ + Ir(III) complex was deposited by spin-coating at 2000 rpm from 0.75% PVK solution in dichloroethane. All other organic layers: BAlq (40 nm), Zn(bt)$_2$ (35 nm) and the Al cathode (100 nm) were prepared successively by thermal evaporation in vacuum better than 10^{-4} Pa at rates 2-5 Å/s without disturbing the vacuum. The layers' thicknesses were controlled by a quartz crystal microbalance sensor positioned near the PET/ITO substrate.

2.3. Instruments and measurements

The UV and the fluorescence emission spectra of the Ir complexes investigated were taken in a solution by Thermo Spectronic Unicam 500 and Varian Cary Eclipse spectrophotometers, respectively. The photoluminescence quantum yield (QY) measurements were performed by a Jobin–Yvon–Horiba Fluorolog III apparatus on solution samples using an F-3029 integrating sphere. The luminescent decay times (lifetimes τ) of the complexes in solutions were measured by excitation by a ZWGIGL 300/2 nitrogen laser (wavelength 337.1 nm, pulse duration 2.5 ns, pulse energy 150 μJ). The laser beam was focused on the center of a quartz cuvette with 1 cm optical pathway containing solutions of the complexes (the size of the focused UV laser beam spot was about 200 nm). The timing measurements were performed by a Tektronix TDS 1012B digital oscilloscope and a fast PIN photodiode. The lifetimes were determined by the exponential dependence $I(t) = I(0)\exp(-t/\tau)$, where I is the light intensity at moment (t). The response time of the measuring system was about 10 ns, which determined the accuracy of the lifetime values obtained. The electroluminescence (*El*) emission spectra and CIE coordinates of the emitted light were taken by an Ocean Optics HR2000+ spectrometer. The current-voltage (*I-V*) characteristics were measured using a programmable Labview Keithley 2220-30-1 power supply. The electroluminescence was determined in *dc* mode and the light output was detected using an S2281-01 Hamamatsu calibrated silicon photodiode. The current efficiency (η_{El}) used for quantifying the properties of the OLEDs were calculated as

$$\eta_{El} = El/I, \text{ cd/A}, \tag{1}$$

where *El* is the electroluminescence (cd/m^2) and *I* is the current density (A/m^2).

3. Result and discussion

3.1. UV-VIS absorption and photoluminescence

The UV-VIS absorption spectra of the Ir(III) complexes and the photoluminescent (Pl) spectra of the Ir(III) complexes, CBP and PVK in CH_2Cl_2, measured under ambient conditions, are shown in figures 2 and 3; their photophysical data are summarized in Table 1. The transitions observed are assigned in accordance with [10]. The absorption spectra shapes of all complexes are very similar, regardless of the varied ancillary ligand. The spectra show a complicated structure due to the overlapping of various bands and are difficult to interpret. The basic absorption bands are observed at about 270 nm and 326 nm. The bands below 300 nm can be attributed to $\pi \rightarrow \pi^*$ transitions in the benzene ring. The absorption at about 326 nm can be referred to LC (ligand-centered) electronic transitions, which lead

IOP Conf. Series: Journal of Physics: Conf. Series **992** (2018) 012029 doi:10.1088/1742-6596/992/1/012029

to excitations in the bt ligand. The weak bands above 360 nm are assigned to singlet and triplet metal-to-ligand charge transfer bands (^{1}MLCT, ^{3}MLCT), respectively. The low-intensity bands detectable in the absorption spectrum at 450 nm and 480 nm may be interpreted as resulting from effective spin-orbit coupling (SOC) caused by mixing of the singlet and triplet states. Furthermore, as can be seen in figure 2, the matrices of PVK and CBP emitted blue light with peaks at 367 nm and 385 nm, respectively, providing a good spectral overlapping with the MLCT absorption bands of the Ir(III) complexes ranging from 360 nm to 480 nm, i.e. there is an opportunity for effective Förster or Dexter energy transfer from the host (PVK: CBP) to iridium complex guests. The Pl emission spectra of the Ir(III) complexes in non-degassed CH_2Cl_2 without blowing at room temperature obtained under excitation at 330 nm exhibited strong yellow phosphorescence with a maximum emission peak at 558-559 nm and shoulders in 593-601 nm (figure 3).

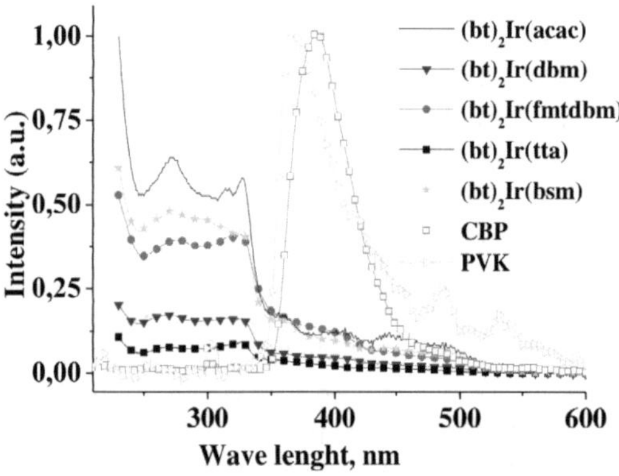

Figure 2. UV-VIS absorption spectra of Ir(III) complexes and Pl emission spectra of PVK and CBP in CH_2Cl_2 under ambient conditions.

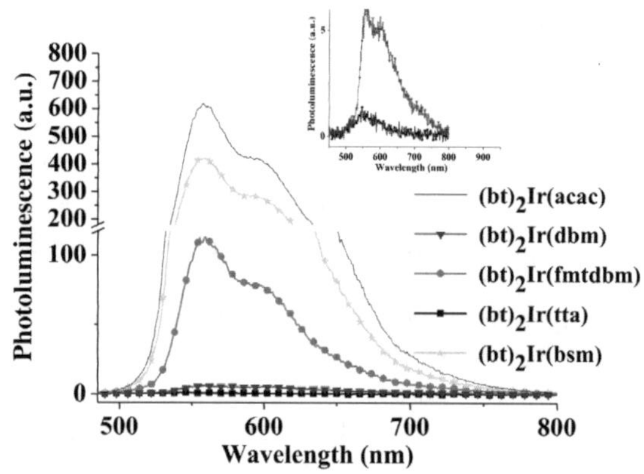

Figure 3. Pl emission spectra of the Ir(III) complexes in non-degassed CH_2Cl_2 under ambient conditions.

Only the peak of $(bt)_2Ir(tta)$ spectrum was blue-shifted by 22 nm in comparison with the other, which was attributed to the presence of electron-withdrawing fluorine atoms on the ancillary ligands. The Pl intensities of the complexes were low, especially these of $(bt)_2Ir(tta)$ and $(bt)_2Ir(dbm)$ (inset in figure 3), because of the presence of oxygen dissolved in the solution, which is known as a quencher for many fluorophores [25, 26]. It was established that the Pl peaks' wavelengths did not shift significantly upon a change of the ancillary ligand in the complexes. The Pl data are consistent with the absorption spectral results (figure 2) and indicate that the ancillary ligands: acac, bsm, dbm, tta and fmtdbm, do not seem to change significantly the bandgaps of their complexes and the main bt ligands are the important factor in determining the emission color of the complexes.

As one can see in table 1, the highest QY (0.50) and the longest fluorescence lifetime ($\tau = 200$ ns) were exhibited by $(bt)_2Ir(acac)$, followed by $(bt)_2Ir(bsm)$ (QY = 0.20 and $\tau = 210$ ns). Most probably, the presence of aliphatic moieties in the ancillary ligands of $(bt)_2Ir(acac)$ and $(bt)_2Ir(bsm)$ imparts some flexibility to the molecules, which reduces the intramolecular quenching and results in the rise observed in the QY and the lifetime. The QYs of $(bt)_2Ir(fmtdbm)$, $(bt)_2Ir(dbm)$ and $(bt)_2Ir(tta)$ were significantly lower and considerably shorter than those of $(bt)_2Ir(acac)$ and $(bt)_2Ir(bsm)$. The substitution of acetylacetonate ancillary ligand with different β-diketone lowers the phosphorescent QY, probably because of the appearance of a number of low-energy vibrations enhancing the nonradiative deactivation pathways. That vibronic coupling is a common quenching mechanism for phosphorescent emission.

3.2. Electroluminescence

The El spectra of the devices undoped and doped with the complexes studied and their electroluminescent characteristics are shown in figures 4a; 5a; 6a; 7a, and figures 4b, 5b; 6b; 7b,

IOP Conf. Series: Journal of Physics: Conf. Series **992** (2018) 012029 doi:10.1088/1742-6596/992/1/012029

respectively. The El spectra and their CIE coordinates were taken at 16 V *dc*. For comparison, the El spectra of devices with HTL and ElL based on only of BAlq or $Zn(bt)_2$ are also presented in figure 5a. As seen, the El spectrum of the device with BAlq has a peak at 496 nm, while that with $Zn(bt)_2$, at 530 nm. Surprisingly, the El spectrum shape of the device with evaporated in-series layers of BAlq/$Zn(bt)_2$ was just the same as that of the OLED with ElL of BAlq. We believe that, since the $HOMO_{Zn(bt)2}$ level (5,05 eV) is lower than the $HOMO_{BAlq}$ level (5,90 eV) (figure 1), the layer of $Zn(bt)_2$ cannot retain the holes, which flow unimpeded to the cathode. Therefore, $Zn(bt)_2$ acts only as an ETL in the BAlq/$Zn(bt)_2$ structure.

Table 1. Photophysical data of the studied complexes in CH_2Cl_2 solution.

Complex	Absorption λ [nm] (lgε)*	Emission λ_{max}/*shoulder*[nm]	Quantum yield (QY)	Lifetime τ [ns]
$(bt)_2Ir(acac)$	272(3.93),315(3.86),327(3.88), 407(3.21), 442(3.19), 489(3.03)	558/593	0.50	210
$(bt)_2Ir(dbm)$	269(4.23),315(4.21),327(4.21), 362(3.75),407(3.64),445(3.45), 492(3.25)	558/600	<0.01	18
$(bt)_2Ir(fmtdbm)$	276(4.60),316(4.61),326(4.61), 400(4.08), 492(3.60)	559/600	0.02	80
$(bt)_2Ir(tta)$	269(3.89),315(3.92),327(3.96), 434(3.23)	546	<0.01	110
$(bt)_2Ir(bsm)$	273(4.68),298(4.66),314(4.64), 327(4.63), 359(4.19), 400(3.98), 444(3.88), 485(3.75)	559/601	0.20	200

* ε is the molar extinction coefficient (molar absorptivity); The absolute photoluminescence quantum yields (QY) were determined by using rhodamine B as a reference.

The doped devices' El spectra were basically the sum of the BAlq emission (at 496 nm) and those of the Ir complexes at 556 nm for $(bt)_2Ir(dbm)$; 557 nm for $(bt)_2Ir(acac)$ and $(bt)_2Ir(fmtdbm)$; and 558 nm for $(bt)_2Ir(bsm)$. The fact that the El peaks' positions of these four dopants in OLEDs were almost identical with their Pl peaks in solution (figure 3) indicated that El and Pl originate from the same excited state and strongly suggested the absence of any appreciable aggregates from complexes in the solid state. The absence of any emission peaks of PVK and CBP pointed to an effective charge trapping on the Ir(III) complexes and/or an effective energy transfer to the Ir(III) complexes. The only exception were devices doped with $(bt)_2Ir(tta)$, in whose El spectra the peak of $(bt)_2Ir(tta)$ at 546 nm was missing, but that at 524 nm was present (figure 8a).

As the dopant concentration was raised, the relative intensity of the greenish-blue emission at 496 nm decreased, while that of the yellow one at 556-558 nm grew (figures 4a-8a). Furthermore, the CIE (*x/y*) coordinates of OLEDs were red-shifted from light blue 0,2015/0,3453 for the undoped device to white for different concentrations of different dopants (6 wt % for $(bt)_2Ir(dbm)$ – 0,3368/0,4152; 6 wt % for $(bt)_2Ir(tta)$ – 0,2851/0,4380; 4 wt % for $(bt)_2Ir(fmtdbm)$ – 0,3338/0,4048; 2 wt % for $(bt)_2Ir(bsm)$ – 0,3378/0,4237; 1,25 wt % for $(bt)_2Ir(acac)$ – 0,2981/0,4390); and to different nuances of yellow for devices doped with 10 wt % of: $(bt)_2Ir(fmtdbm)$ – 0,3562/0,4224; $(bt)_2Ir(dbm)$ – 0,3452/0,4407; $(bt)_2Ir(bsm)$ – 0,3810/0,4512; $(bt)_2Ir(acac)$ – 0,4311/0,4522 and $(bt)_2Ir(tta)$ – 0,2581/0,4015. At the same time, the threshold voltage of all devices initially increased at lower dopant concentration and subsequently decreased, without reaching the value of the respective undoped device within the range of tested dopant concentrations. The other characteristics: electroluminescence (*El*) at

16 V *dc* and current efficiency (η_{El}) at electroluminescence 250 cd/m^2 had just the opposite behavior – initially decreased and then increased with increasing the dopant concentrations (figures 4b-8b).

The best performance was demonstrated by the OLED doped with $(bt)_2Ir(bsm)$ (figure 4b), whose *El* maximum reached 1600 cd/m^2 and η_{El}, 6.5 cd/A at 8 wt % of dopant, followed by the devices doped with $(bt)_2Ir(dbm)$ with $El = 1220$ cd/m^2 and $\eta_{El} = 4.72$ cd/A at 6 wt %; $(bt)_2Ir(acac)$ with $El = 720$ cd/m^2 and $\eta_{El} = 4.67$ cd/A at 10 wt %; $(bt)_2Ir(fmtdbm)$ with $El = 617$ cd/m^2 and $\eta_{El} = 4.28$ cd/A at 8 wt %; and finally $(bt)_2Ir(tta)$ with $El = 667$ cd/m^2 and $\eta_{El} = 4.27$ cd/A at 8 wt %.

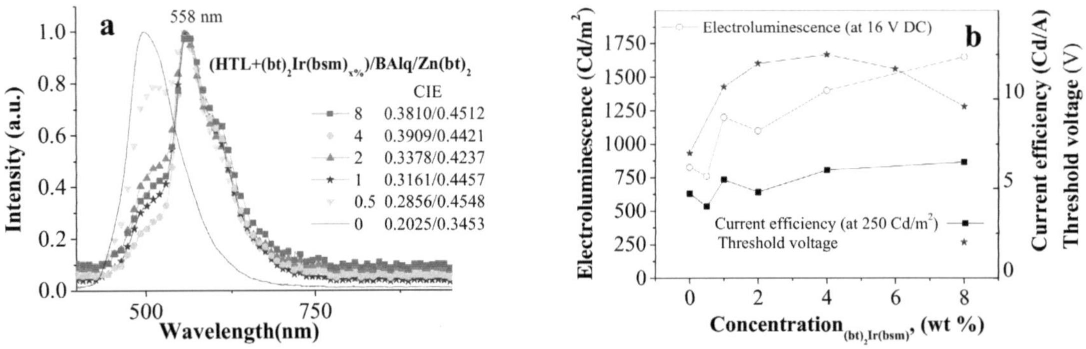

Figure 4. Undoped and doped with $(bt)_2Ir(bsm)$ devices: **a)** El spectra and CIE coordinates at 16 V DC and **b)** electroluminescence, current efficiency and threshold voltage.

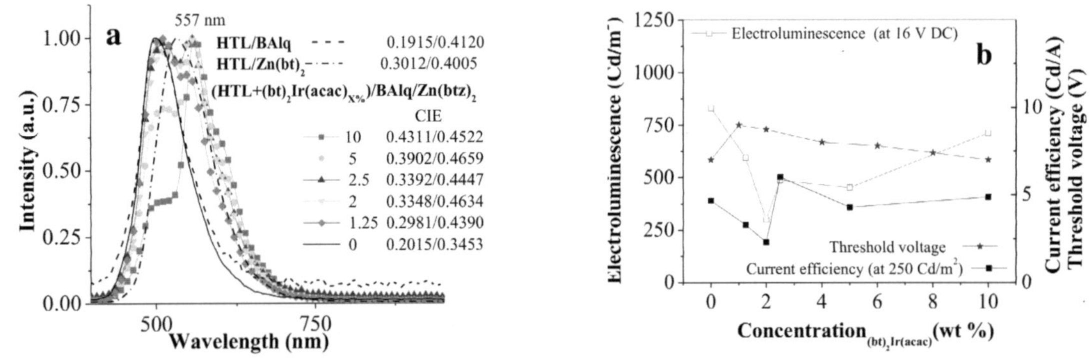

Figure 5. a) El spectra and CIE coordinates at 16 V DC for undoped and doped with $(bt)_2Ir(acac)$devices and OLED structures: HTL/BAlq and HTL/$Zn(bt)_2$; **b)** electroluminescence, current efficiency and threshold voltage.

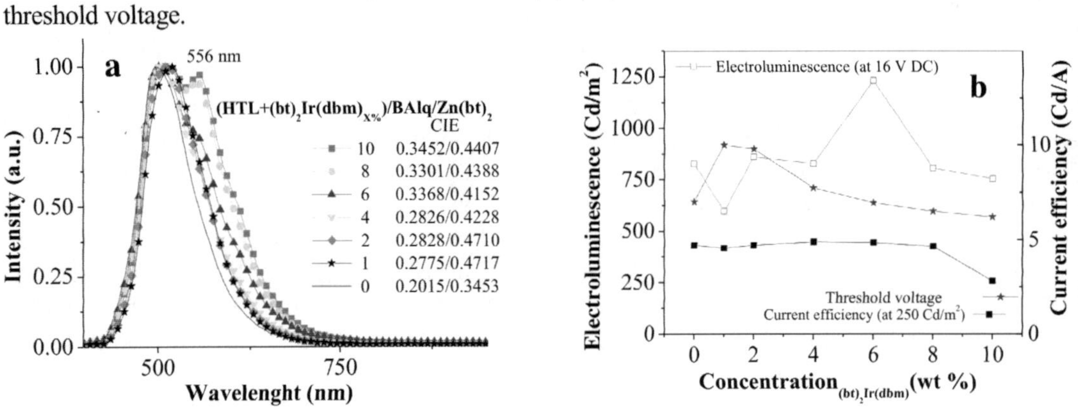

Figure 6. Undoped and doped with $(bt)_2Ir(dbm)$ devices: **a)** El spectra and CIE coordinates at 16 V DC and **b)** current efficiency and threshold voltage.

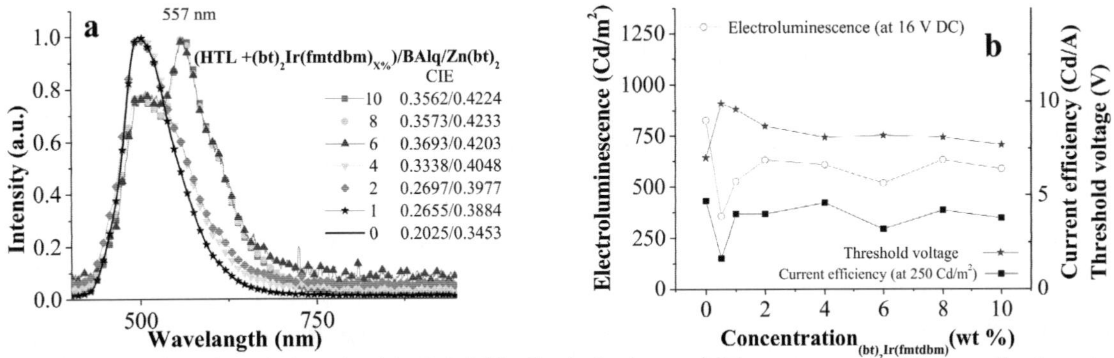

Figure 7. Undoped and doped with $(bt)_2Ir(fmdbm)$ devices: **a)** El spectra and CIE coordinates at 16 V DC and **b)** electroluminescence, current efficiency and threshold voltage.

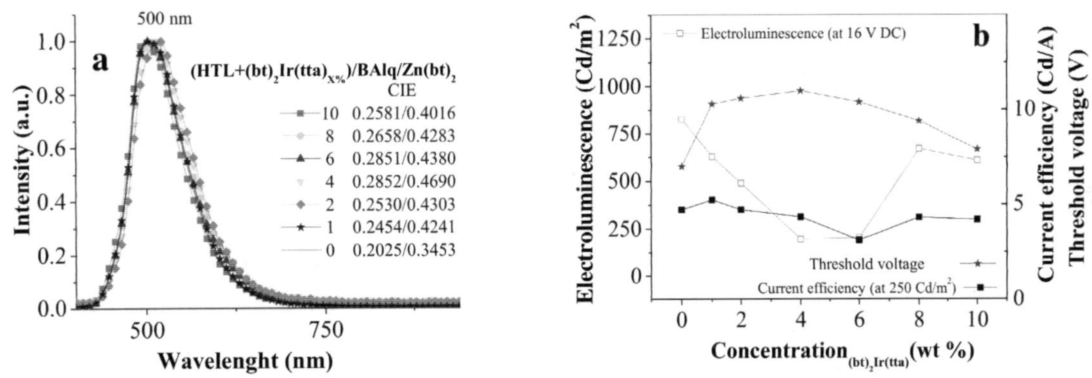

Figure 8. Undoped and doped with $(bt)_2Ir(tta)$ devices: **a)** El spectra and CIE coordinates at 16 V DC and **b)** electroluminescence, current efficiency and threshold voltage.

The decrease of the OLEDs efficiencies as the dopant concentration was increased, on the one hand, and the absence of any PVK and CBP emission in the El spectra of doped devices, even at low dopant concentrations, on the other, are indicative of charge trapping in the Ir complex, rather than energy transfer, as the dominant mechanism in the OLEDs studied. Since the low dopant concentrations cause mobility limitations associated with the motion of a charge from one dopant site to another, a rise in the devices' threshold voltage was originally observed [6]. Increasing the dopant concentration improves the charge mobility; correspondingly, the current density and the light intensity start increasing and the threshold voltage decreases.

4. Conclusions

The effect was studied of the four β-diketone ancillary ligands (bsm, dbm, tta and fmtdbm) on the photophysical and electroluminescent properties of the iridium phenyl-benzothiazole complexes. All complexes exhibited Pl peaks around 560 nm. The change of the ancillary ligand in them did not affect their Pl characteristics significantly, thus evidencing that the base bt ligands play a key role in determining the complexes' emission color.

We investigated the possibilities of applying these complexes in OLEDs and found that they are very promising candidates as dopants for producing white OLEDs. The optimal concentration (x) for various different dopants in the OLED structure investigated varied in the range between 1 wt % and 6 wt %. The best performance was demonstrated by the device doped with 2 wt % of $(bt)_2Ir(bsm)$, which had twice as high electroluminescence and one-and-a-half times as high current efficiency in comparison with the device doped with the already known $(bt)_2Ir(acac)$, at approximately the same CIE (*x/y*) coordinates of the warm white light emitted by the two devices.

References

[1] Baldo M, O'Brien D, You Y, Shoustikov A, Sibley S, Thompson M and Forrest S 1998 *Nature* **395** 151-4
[2] D'Andrade B W and Forrest S R 2004 *Adv. Mater.* **16** 1585- 95
[3] Wang L, Jiang Y, Luo J, Zhou Y, Zhou J, Wang J, Pei J and Cao Y 2009 *Adv. Mater.* **21** 4854–8
[4] Choi J, Jung C, Kwon J, Cho H, Lee J, Lee J-Ik, Chu H and Hwang D 2009 *Synth. Met.* **159** 1517–21
[5] Liu Z, Bian Z, Ming L, Ding F, Shen H, Nie D and Huang C 2008 *Org. Electron.* **9** 171–82
[6] Hu J, Zhang J, Shih H, Jiang X, Sun P and Cheng C 2008 *J. Organomet. Chem.* **693** 2798–802
[7] Bauer R, Finkenzeller W, Bogner U, Thompson M and Yersin H 2008 *Org. Electron.* **9** 641–8
[8] Wang R, Deng L, Zhang T and Li J 2012 *Dalton Trans.* **41** 6833-41
[9] Seo J, Lee S, Kim Y K and Kim Y S 2008 *Thin Solid Films* **517** 1346–8
[10] Lamansky S, Djurovich P, Murphy D, Razzaq F, Lee H, Adachi C, Burrows P, Forrest S and Thompson M 2001 *J. Am. Chem. Soc.* **12** 4304-12
[11] Zhang L, Li L, Shi L and Li W 2009 *Opt. Mater.* **31** 905–11
[12] Tavasli M, Moore T, Zheng Y, Bryce M, Fox M, Griffiths G, Jankus V, Al-Attar H and Monkman A 2012 *J. Mater. Chem.* **22** 6419- 28
[13] Kawamura Y, Goushi K, Brooks J, Brown J, Sasabe H and Adachi C 2005 *Appl. Phys. Lett.* **86** 071104
[14] Adachi C, Baldo M, Forrest S, Lemanski S, Thompson M and Kwong R 2001 *Appl. Phys. Lett.* **78** 1622-4
[15] Ha S and Noh Y 2013 *Jpn. J. Appl. Phys.* **52** 10MB11
[16] Miniak H, Teicher M, Scherf U, Trost S, Riedl Th, Lehnhardt M, Rabe T, Kowalsky W and Lochbrunner S 2012 *Phys. Rev.* B **85** 214204
[17] Baldo M, Lamansky S, Burrows P, Thompson M and Forrest S 1999 *Appl. Phys. Lett.* **75** 4-7
[18] Baldo M and Forrest S 2000 *Phys. Rev.* B **62** 10958-67
[19] Baldo M, Adachi C and Forrest S 2000 *Phys. Rev.* B **62** 10967-78
[20] Liu W and Huang W 2013 *ECS J. Solid State Sci. and Technol.* **2**/1 R16-20
[21] Liu D, Ren H, Deng L and Zhang T 2013 *ACS Appl. Mater. Inter.* **5** 4937–44
[22] Li M, Zeng H, Meng Y, Sun H, Liu S, Lu Z, Huang Y and Pu X 2011 *Dalton Trans.* **40** 7153–64
[23] Petrova P, Ivanov P, Marcheva Y and Tomova R 2013 *Bulgarian Chem. Commun.* **45 B** 159-64
[24] Nonoyama M 1974 *Bull. Chem. Soc. Jpn.* **47** 767-8
[25] Arık M and Yavuz N 2005 *J. Photochem. Photobiol.* A *Chem.* **170** 105–11
[26] Weber G and Lakowicz R 1973 *Chem. Phys. Lett.* **22** 419–23

Ti- and Cr-based hard coatings obtained at low temperatures by unbalanced magnetron sputtering

T M Cholakova[1,5], V Chitanov[1], L P Kolaklieva[1], R D Kakanakov[1], B Ranguelov[2], S Atanasova[2], P K Stefanov[3], G B Atanasova[3] and K Balashev[4]

[1]Central Laboratory of Applied Physics, Bulgarian Academy of Sciences, 61 St. Petersburg Blvd., 4000 Plovdiv, Bulgaria
[2]Acad. R. Kaishev Institute of Physical Chemistry, Bulgarian Academy of Sciences, Acad. G. Bonchev Str., Bl. 11, 1113 Sofia, Bulgaria
[3]Institute of General and Inorganic Chemistry, Bulgarian Academy of Sciences, Acad. G. Bonchev , Str., Bl. 11, 1113 Sofia, Bulgaria
[4]Laboratory of Biophysical Chemistry, Department of Physical Chemistry, Faculty of Chemistry and Pharmacy, St. Kl. Ohridski University of Sofia, 5 J. Bourchier Blvd., 1164 Sofia, Bulgaria

E-mail: ipfban-dve@mbox.digsys.bg

Abstract. We prepared a set of Ti- and Cr-based ternary and quaternary hard coatings at low deposition temperatures (< 200 °C) by closed-field unbalanced magnetron sputtering in a gas mixture of Ar + N_2. The morphological, structural and mechanical properties of the coatings were characterized by means of atomic force microscopy, scanning electron microscopy, X-ray photoelectron spectrometry, a micro-scratch tester and a nanoindentation tester. The results of the analyses performed revealed that the surface of the deposited ternary Ti-Al-N coatings was rougher than that of the CrTiAlN coatings. The metallic elements in the coatings react with nitrogen to form CrN, TiN and AlN nitride phases, as no Cr_2N phase was detected in the Cr-based coatings. The optimized quaternary coating demonstrated mechanical properties superior to those of the ternary ones in terms of hardness; adherence strength of 32 GPa and elastic modulus of 418 GPa were measured for the multicomponent coating of composition $Cr_{0.68}Ti_{0.19}Al_{0.13}N$.

1. Introduction

Hard coatings based on Ti and Cr nitrides are widely used as protective coatings. A large number of studies have been reported on the deposition and properties of CrTiN, TiAlN and CrAlN coatings produced by various PVD techniques. The TiAlN coatings have been developed as an alternative to TiN, because of their higher oxidation resistance, enhanced hardness and higher corrosion resistance [1-5]. The CrAlN coatings have exhibited superior mechanical properties and oxidation resistance in comparison to the traditional binary TiN and CrN coatings [6, 7] and have been successfully used in the industry. Films based on the Ti-Cr-N system have also been deposited using a variety of deposition methods in order to improve their wear and oxidation resistance [8-10]. Further improvements in

[5] To whom any correspondence should be addresed.

IOP Conf. Series: Journal of Physics: Conf. Series **992** (2018) 012030 doi:10.1088/1742-6596/992/1/012030

nitride coatings have been found to result in better performance compared to that of ternary films. The Ti–Al–Cr–N system offers a high variability of composition, ranging from Cr-rich to Al-rich [11-13] coatings. These films have demonstrated improved hardness and thermal stability, oxidation resistance and exhibited lower friction and wear. Among the PVD coating technologies, the closed-field unbalanced magnetron sputtering is an exceptionally versatile technique for the deposition of high-quality, well-adhered films. The main advantages of this technique are as follows: possibility for deposition at low temperatures, including room temperature, use of non-toxic working gases, a high degree of smoothness, uniformity and density of the deposited coatings.

The present study focuses on the investigation of the Ti- and Cr-based coatings obtained by a closed-field unbalanced magnetron sputtering technique. The goal of this work is to establish optimal technological regimes which provide stable mechanical properties of the system "substrate - coating".

2. Experimental details

2.1. Deposition of Ti and Cr-based coatings

The UDP850 (CFUBMS) closed-field unbalanced magnetron sputtering system was equipped with four rectangular cathodes of high-purity (99.99%) Cr, Ti and two Al targets operated in an unbalanced magnetron mode. A schematic illustration of the sputtering system is shown in figure 1 (a). The closed magnetic field coupling results in a high degree of ionization and biased current density. The coatings were deposited onto substrates of high-speed steel (EN: 1.3343), stainless steel (EN: 1.4436) and carbon tool steel with low thermal resistance (EN: CT105). Prior to deposition, the substrates were cleaned in an ultrasonic bath for five minutes at a temperature of 65 °C using commercial cleaning solution to remove oils, residues and protection against corrosion. The substrates were then rinsed in distilled water for two minutes, followed by ultrasonic cleaning in hot isopropanol for five minutes and drying. Generally, the ultrasonically cleaned metal substrates have a thin top oxide layer, which affects the adhesion. Thus, after the wet chemical cleaning, the substrates were loaded in the deposition chamber and the system was pumped down to a base pressure of less than 2.5×10^{-3} Pa. In order to remove the top oxide layer, the substrates were plasma-cleaned in argon ambient (1.7×10^{-1} Pa) for 30 minutes at a negative pulsed substrate bias voltage of 500 V and a frequency of 250 kHz.

In all deposition processes, the substrates were rotated axially at a speed of 5 rpm in order to obtain a homogenous film composition. The distance between the substrates and the targets was kept at 150 mm. All the experiments were conducted without additional substrate heating at a bias voltage of −70 V and a frequency of 150 kHz in a DC work mode for of the Ti and Cr cathodes, and a pulsed mode for the Al cathodes. A mixture of argon and nitrogen gases was introduced into the chamber during the deposition process at a fixed Ar flow-rate of 25 sccm. A nitrogen level of 95–50% was provided with the help of a plasma optical emission monitor with a feedback control system (figure1 (b)).

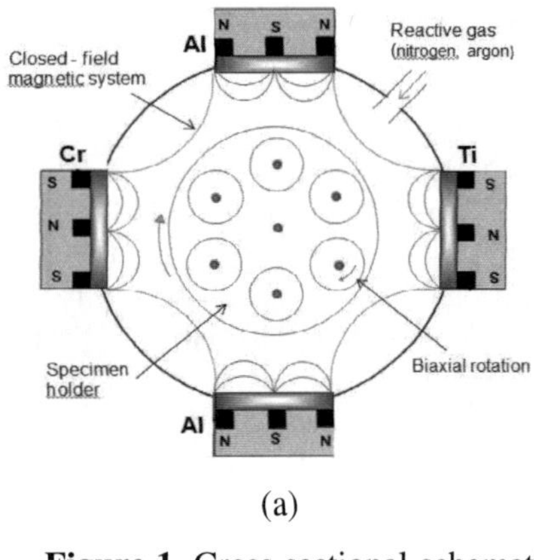

(a)

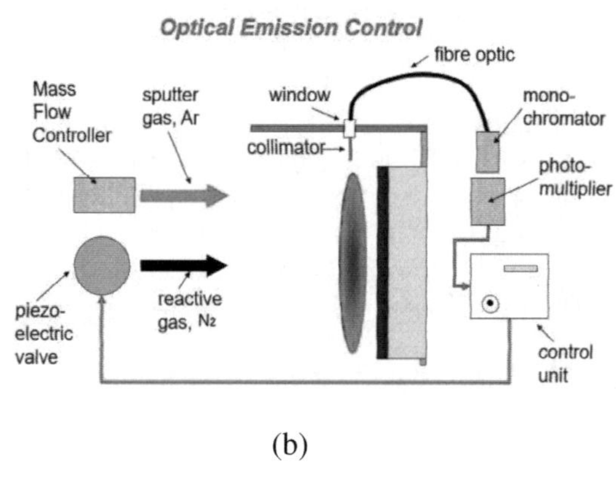

(b)

Figure 1. Cross-sectional schematic drawing of the CFUBMS sputtering system - (a) and the optical emission control scheme - (b).

Three series of experiments were performed for deposition of Ti-based and Cr-based ternary and quaternary coatings using Cr, Ti and Al as the sputtering targets. Pure argon (99,9999%) and nitrogen (99,9999%) were used as the sputtering and reactive gases, respectively. The relative concentration of metals (Cr, Ti and Al) was adjusted by the sputtering power applied to each metallic target. The additional details of deposition conditions used in this study are presented in table 1. In the first series, TiAlN coatings were deposited using Ti as the adhesive layer and TiN as an interlayer. In the second series, CrTiN coatings were deposited using Cr and CrN as the adhesive layer and interlayer respectively. In the third series, Cr-Ti-Al-N coatings were deposited using Ti (or Cr) as the adhesive layer and TiN (or CrN) as an interlayer. Following these two layers, a compositionally-graded Cr-Ti-Al-N top layer was deposited.

2.2. Study of coating properties

The mechanical properties were investigated using Compact Platform CPX (MHT/NHT) CSM Instruments equipment, which includes a nanoindentation module (NHT), a micro-scratch module (MST) and an optical video microscope with a CCD camera, installed together on the same platform. Nanoindentation was performed by a triangular diamond Berkovich pyramid in the loading interval of 10 – 200 mN. The nanohardness (H) and elastic modulus (E) were calculated from nanoindentation load-displacement data at a periodic loading and unloading of the samples at a successively increasing load using an Oliver & Pharr method implemented by software. The adhesion strength of the coatings to the substrate material and the friction coefficient were evaluated using an MST equipped with a spherical Rockwell indenter with a radius of 200 μm under the following test conditions: progressively increasing pressing force in the range of 1 – 30 N; loading rate 0.5 N/min; indenter movement speed 0.02 mm/min; and acoustic emission detector sensitivity of 9. The nature of the resulting damages was assessed by observations with the optical video microscope equipped with a CCD camera.

The morphology and roughness of the deposited coatings were studied using a NanoScope VAFM (Bruker Inc.) atomic force microscope in tapping mode in air. Silicon cantilevers were used with reflective aluminium coating of 30-nm thickness (Tap 300Al-G Budget Sensors, Innovative Solutions Ltd, Bulgaria). The characteristics of the cantilevers were as follows: spring constant from 1.5 to 15 N/m, resonance frequency 150 ± 75 kHz and tip curvature radius of less than 10 nm. The scanning rate was set at 1 Hz. The roughness analysis yields the value R_a, which is an arithmetic average of the absolute values Z_j of the surface height deviations measured from the mean plane, while the image R_q is the root-mean-square average of the height deviations taken from the mean image data plane.

Surface observation and composition analysis were performed on a JEOL JSM 6390 electron microscope equipped with an INCA Oxford EDS energy dispersive detector. Surface images were also obtained in secondary electrons (morphology contrast) and back-scattered electrons (density contrast).

The coatings' chemical bonding was studied by X-ray photoelectron spectrometry (XPS) using a Kratos AXIS Supra photoelectron spectrometer with a monochromatic Al Kα source with energy 1486.6 eV. The base pressure in the analysis chamber was 5×10^{-8} Pa. The binding energies were corrected relative to the C1s peak at 285.0 eV. The elemental concentration was derived on the basis of the core level peak areas, corrected by the corresponding values of the relative sensitivity factor.

3. Results and discussion

3.1. Surface morphology and roughness

Several scans were performed by the AFM in tapping mode over different areas. The AFM data revealed that all the coatings exhibited a densely-packed structure [14] consisting of well-separated grains with apparently spherical form and predominantly equal sizes. The surface morphology of TiAlN and CrTiAlN coatings with different Ti and Al content are shown in figure 2 and figure 3. It is visible that the TiAlN coating surface is rougher than that of the CrTiAlN coating. The average roughness (R_a) and the root-mean-square (R_q) surface roughness obtained on a scanned area of

10 µm×10 µm were as follows: for the ternary coating $R_a = 29.9$ nm and $R_q = 38.8$ nm; for the quaternary coating $R_a = 16.4$ nm and $R_q = 22.0$ nm.

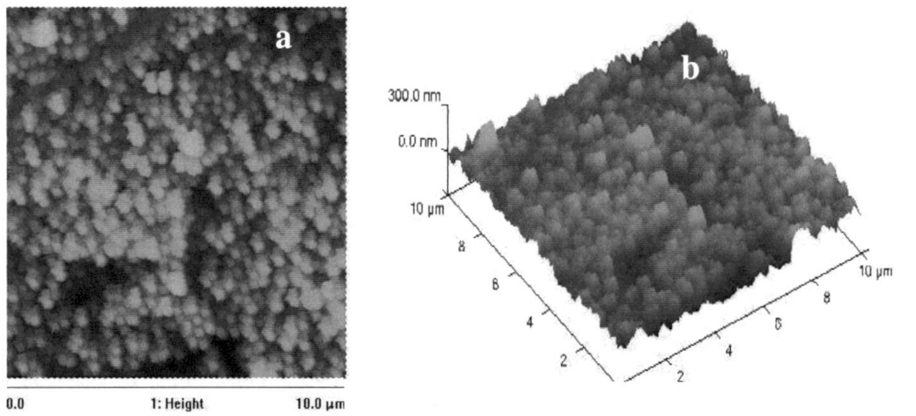

Figure 2. Surface topography of the TiAlN coating (SS substrate) obtained by AFM: a) 2D image; b) 3D image.

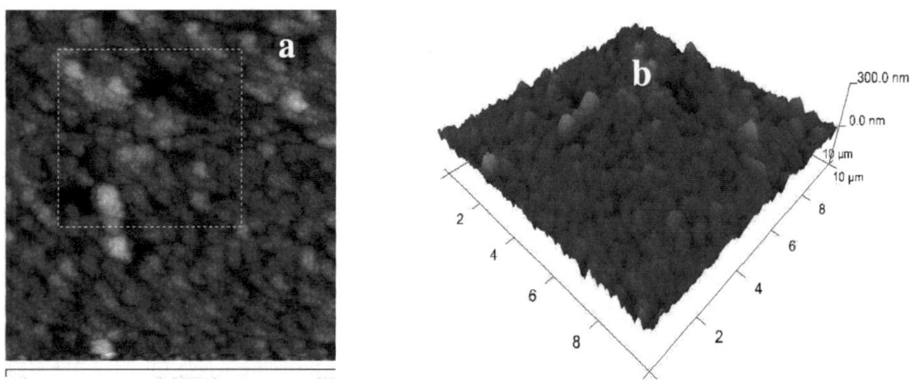

Figure 3. Surface topography of the CrTiAlN coating (SS substrate) obtained by AFM: a) 2D image; b) 3D image.

3.2. Mechanical properties

3.2.1. Nanohardness The mechanical parameters of the developed coatings were investigated in dependence on the nitrogen flow and cathode current. The nanohardness (H) and elastic modulus (E) were calculated from nanoindentation load-displacement data at a periodic loading and unloading of the samples at a successively increasing load using an Oliver & Pharr method. To eliminate the influence of the substrate on the measurement, a load of 10 mN or 15 mN was applied, which corresponded to a maximum indentation depth of less than 10% of the coating thickness. The results showed that, among the technological parameters quoted, the presence of nitrogen affected most strongly the mechanical properties of the coatings. Table 1 lists the mechanical properties of the ternary and quaternary coatings, the latter exhibiting better mechanical properties in terms of hardness and adherence strength.

All the coatings demonstrated hardness higher than that of the TiN and CrN coatings (25 GPA and 18 GPA). We attribute the improved hardness of the CrTiAlN coating to the hardening of the solid solution by the addition of Ti and Al elements in the CrN crystals. The nanohardness obtained is significantly higher than those reported in [15, 16] and agrees with that obtained in [17].

Table 1. Deposition conditions and mechanical properties of deposited coatings.

Coating	OEM [%]	Cathode current $I_{Ti}/I_{Cr}/I_{Al}$ [A]	Substrate temperature T [°C]	Thickness [μm]	Hardness H [GPa]	Elastic modulus E [GPa]	Coefficient of friction [μ]	Critical load Lc1 [N]
TiAlN	45	5/0/3	130	2.0	29	365	0.1	16.5
CrTiN	60	8/5/0	134	1.5	26	331	0.2	24.5
TiCrAlN	50	8/5/3	155	2.5	28	358	0.2	21.0
CrTiAlN	60	8/5/3	147	2.0	32	418	0.1	> 30

3.2.2. Adhesion Three scratches on the coated HSS substrates for each sample were carried out in order to determine the adhesion and coating toughness. During the tests, the load was increased progressively in linear mode from 1 N to 30 N at scratch lengths of 1 mm and 3 mm and constant scratch speeds of 0.02 mm/min and 0.05 mm/min. The critical load (*Lc*) values were determined after the test by optical microscopy observation of the damages formed in the scratch tracks and from the recorded acoustic emission (AE) and friction force (Ft) signals. The critical loads *Lc1* and *Lc2* indicate the appearance of the first cohesive and adhesive cracks, respectively [16]. Figure 4 shows scratch graphs of the TiCrAlN/HSS coating and the optical micrographs of the main parts of the scratch track with cracks at different locations. We found that increasing the scratch length and the normal load results in changes in the values of the friction force and acoustic emission. Very slight semi-circular cracks were observed at a normal load of about 5 N (figure 4e). The first critical load was registered at F_n = 12,5 N (figure 4 f), and the second one, at F_n = 23 N (figure 4g). A distortion of the line of the friction force and the acoustic emission was observed at a higher normal load F_n = 23 N.

As a result of the tests carried out, we established that the critical loads *Lc1* and *Lc2* are observed for the Ti-based coatings, while the Cr-based coatings exhibit very good adhesion properties in the loading range 1 – 30 N. The coefficient of friction of the coatings against a diamond indenter was measured to be μ = 0.1 – 0.2. The best adhesion result without any visible damages within the loading range of 1 – 30 N was exhibited by the CrTiAlN coating (table1).

3.3. Composition and chemical bonding analysis

The chemical bonding structure of the coating samples was investigated by XPS. The TiAlN, CrTiN and CrTiAlN coatings presented similar binding-energy spectra. Typical spectra of TiAlN coating in Ti 2p, Al 2p and N 1s energy ranges are shown in figure 5.

It is seen that the peak associated with Ti (figure 5a) consists of two peaks centered at 457.0 eV and 462.1 eV. These peaks originate from Ti 2p3/2 and Ti 2p1/2 electrons in titanium oxynitride. Figure 5(b) shows the XPS spectra for the corresponding Al2p levels. The contribution in figure 5(b) with a maximum binding energy at 73.7 eV is attributed to the Al-N chemical bonding state within the coating. The fact that the N1s spectrum after etching shows only one peak with a binding energy of approximately 396.3 eV (figure 5c) is attributed to the presence of nitride films (TiN and AlN) [18, 19]. A similar bonding status of Ti2p, Al2p and N 1s and typical XPS spectra were obtained for CrTiN and CrTiAlN samples. The weak Ti2p and Al2p XPS spectra and the stronger spectra of N1s and Cr2p are plotted in figure 6. The spectrum of N1s at 396.5 eV can be related to CrN. In the Cr2p spectrum,

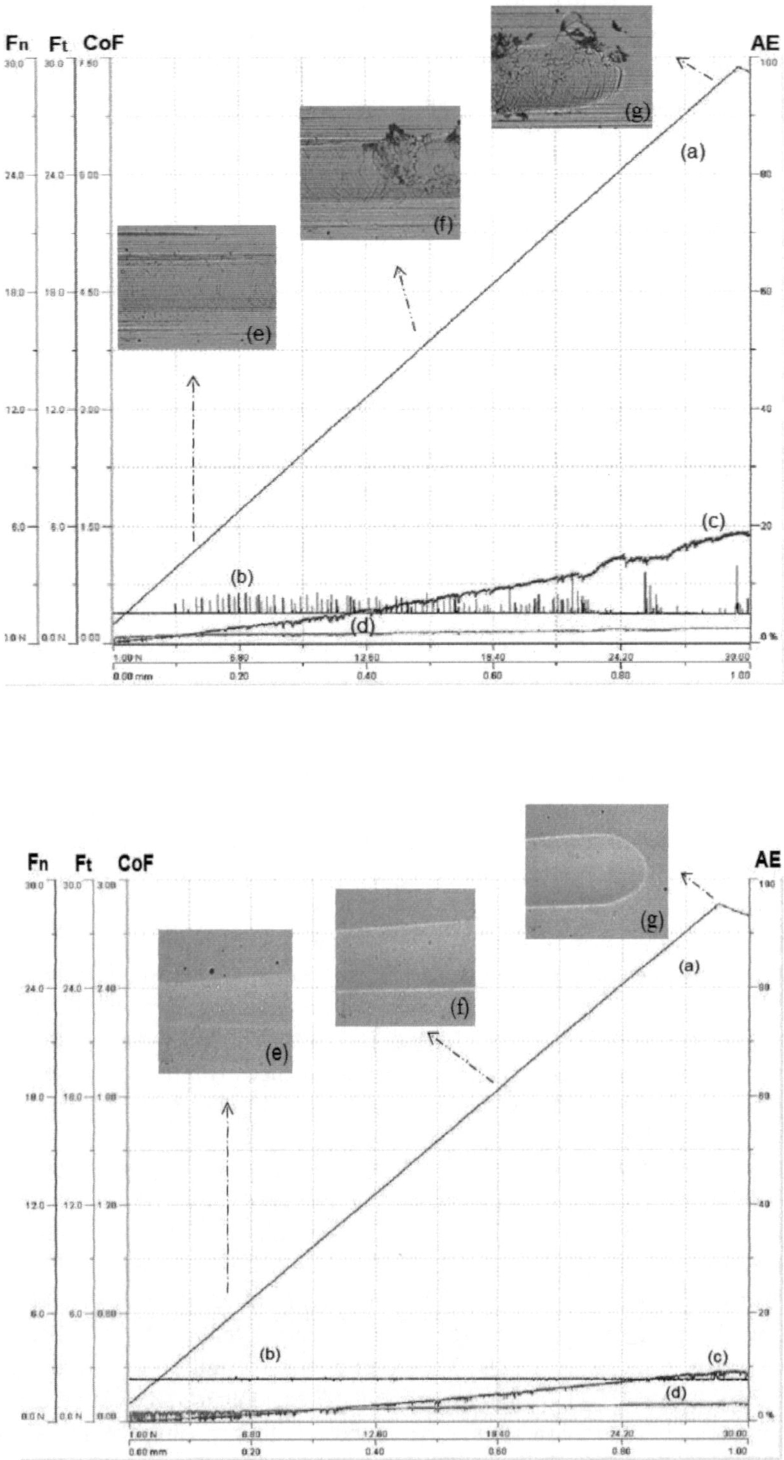

Figure 4. Typical scratch graphs on the Ti- (upper figure) and Cr-based multicomponent coatings and optical micrographs (e, f and g) of the main parts of scratch track with cracks at different locations: a) normal load; b) acoustic emission; c) friction force; d) friction coefficient.

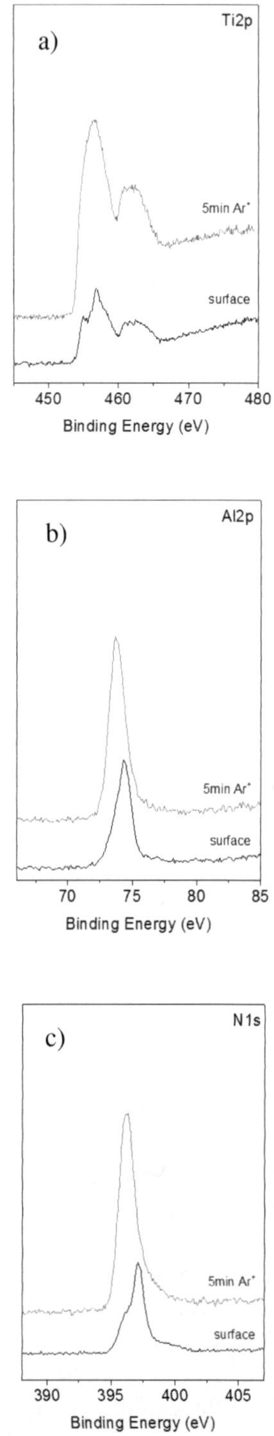

Figure 5. XPS spectra of (a)- Ti 2p, (b)- Al 2p and (c)- N 1s of a TiAlN sample before and after sputter etching in Ar.

IOP Conf. Series: Journal of Physics: Conf. Series **992** (2018) 012030 doi:10.1088/1742-6596/992/1/012030

the peaks centered at 575.1 eV and 585 eV can be attributed to Cr-N bonds. Peaks pertaining to free Cr metal and Cr_2N cannot be observed [21], indicating that the bonding state of chromium is in the form of CrN.

The coatings' composition was determined by EDS and XPS analyses. The elemental concentrations of the coatings determined by EDS (table 2) were also compared with the results obtained by XPS. The XPS analysis was in good agreement with the EDS results (within ± 2%).

Table 2. Composition and chemical formulae of the coatings investigated.

Formulae	EDS analysis results [in at %]			
$Cr_xTi_yAl_zN$	Cr	Ti	Al	N
$Ti_{0.36}Al_{0.64}N$	0	9,54	16,59	45,43
$Cr_{0.7}Ti_{0.3}N$	44,85	19,09	0	30,13
$Cr_{0.68}Ti_{0.19}Al_{0.13}N$	30,33	8,68	5,69	50.72

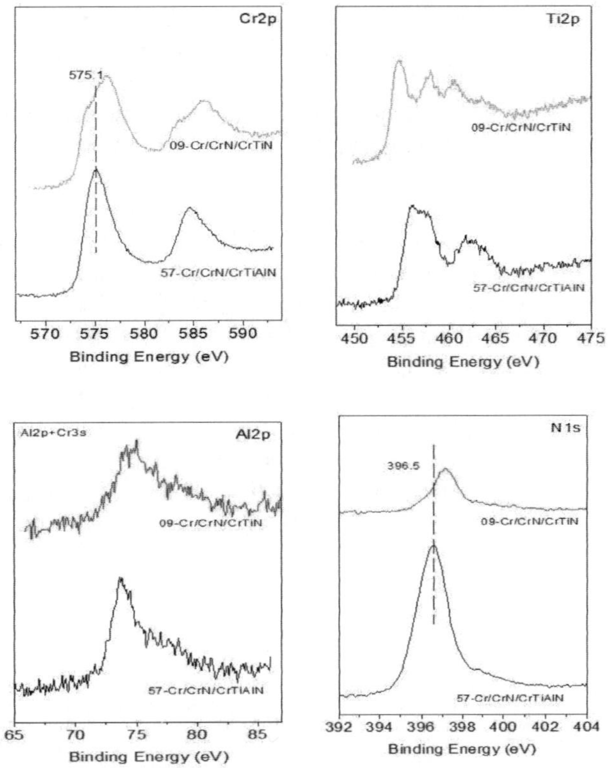

Figure 6. XPS spectra of Cr2p, Ti 2p, Al 2p and N 1s of the CrTiN and CrTiAlN samples.

4. Conclusions

The experimental results demonstrated that proper addition of either Ti and Al into the CrN structure leads to an improvement in the mechanical properties of the coatings. No significant differences were found between the mechanical properties of the coatings obtained on HSS and carbon tool steel substrates. However, the Cr-based coatings demonstrated a lower average surface roughness than the Ti-based coatings. The optimized quaternary Cr-Ti-Al-N coating exhibited enhanced mechanical properties compared to the ternary Cr-Ti-N and Ti-Al-N ones in terms of hardness and adherence strength. The maximum nanohardness of 32 GPa and elastic modulus of 418 GPa for the multicomponent coating were measured for the composition $Cr_{0.68}Ti_{0.19}Al_{0.13}N$. The results showed that among the technological parameters mentioned, the presence of nitrogen was the factor affecting the most strongly the mechanical properties of the coatings.

4. References

[1] Oliveira J C, Manaia A and Cavaleiro A 2008 *Thin Solid Films* **516** 5032
[2] Chokwatvikul C, Larpkiattaworn S, Surinphong S, Busabok C and Termsuksawad P 2011 *J. Metals Mater. and Minerals* **21** 115
[3] Park D G, Cha T H, Lee S H and Yeo I S 2001 *J. Vacuum Sci. Technol.* **19**/6 2289
[4] Barshilia H C, Yogesh K and Rajam K S 2009 *Vacuum* **83** 427-34
[5] Singh A K, Kumari N, Mukherjee S K and Barhai P K 2013 *Int. J. of Recent Res. and Appl. Studies* **14**/3 603
[6] Lin J, Mishra B, Moore J and Sproul W D 2008 *Surf. Coat. Technol.* **202** 3272-83
[7] Longn Y, Zeng J, Yu D and Wunn S 2014 *Ceramics Int.* **40** 9889–94

[8] Aouadia S M, Wong K C, Mitchell K A R, Namavar F, Tobin E, Mihut D M and Rohde S L 2004 *Appl. Surf. Sci.* **229** 387–94
[9] Hsieh J H, Zhang W H , Lia C and Sunc C Q 2001 *Surf. Coat. Technol.* **146-147** 331–7
[10] Kabir M Sh, Munroe P, Zhou Zh and Xie Z 2017 *Surf. Coat.Technol.* **309** 779-89
[11] Sullivan J F, Huang F, Barnard J A and Weaver M L 2005 *J. Vacuum Sci. Technol.* **23** 78-84
[12] Tam P L, Zhou Z F, Shum W and Li K Y 2008 *Thin Solid Films* **516** 5725-31
[13] Fox-Rabinovich G S, Kovalev A I, Aguirre M H, Beake B D, Yamamoto K, Veldhuis S C, Endrino J L, Wainstein D L and Rashkovskiy A Y 2009 *Surf. Coat. Technol.* **204** 489-96
[14] Zhou H, Zheng J, Gui B, Geng D and Wang Q 2017 *Vacuum* **136** 129-36
[15] Kabir M S, Munroe P, Zhou Z and Xie Z 2017 *Surf. Coat. Technol.* **309** 779-89
[16] Shi Y, Long S, Yang S, Pan F 2008 *Appl. Surf. Sci.* **254** 7342–50
[17] Lu L, Wang Q, Chen B, Ao Y, Yu D, Wangn C, Wum S and Kim K H 2014 *Trans. Nonferrous Met. Soc. China* **24** 1800–6
[18] Barshilia H C, Yogesh K and Rajam K S 2009 *Vacuum* **83** 427–34
[19] Shum P W, Zhou Z F, Li K Y and Shen Y G 2003 *Mater. Sci. Eng.* **B100** 204–13
[20] Li X, Wu W and Dong H 2011 *Surf. Coat. Technol.* **205** 3251–9
[21] Emery C, Chourasia P and Yashar J 1999 *J. Electron Spectrosc. Relat. Phenom.* **104** 91

IOP Conf. Series: Journal of Physics: Conf. Series **992** (2018) 012031 doi:10.1088/1742-6596/992/1/012031

Reactive magnetron sputtering of N-doped carbon thin films on quartz glass for transmission photocathode applications

N I Balalykin[1], J Huran[1,2], M A Nozdrin[1,4], A A Feshchenko[1], A P Kobzev[1], V Sasinková[3], P Boháček[2] and J Arbet[2]

[1] Joint Institute for Nuclear Research,
6 Joliot-Curie Str., 141980 Dubna, Moscow Region, Russian Federation
[2] Institute of Electrical Engineering, Slovak Academy of Sciences,
9 Dúbravská cesta , 84104 Bratislava, Slovakia
[3] Institute of Chemistry, Slovak Academy of Sciences,
9 Dúbravská cesta , 84538 Bratislava, Slovakia

E-mail: nozdrin@jinr.ru

Abstract. N-doped carbon thin films were deposited on a silicon substrate and quartz glass by RF reactive magnetron sputtering using a carbon target and an $Ar+N_2$ gas mixture. During the magnetron sputtering, the substrate holder temperatures was kept at 800 °C. The carbon film thickness on the silicon substrate was about 70 nm, while on the quartz glass it was in the range 15 nm – 60 nm. The elemental concentration in the films was determined by RBS and ERD. Raman spectroscopy was used to evaluate the intensity ratios I_D/I_G of the D and G peaks of the carbon films. The transmission photocathodes prepared were placed in the hollow-cathode assembly of a Pierce-structure DC gun to produce photoelectrons. The quantum efficiency (QE) was calculated from the laser energy and cathode charge measured. The properties of the transmission photocathodes based on semitransparent N-doped carbon thin films on quartz glass and their potential for application in DC gun technology are discussed.

1. Introduction

Amorphous and nanocrystalline carbon films possess special chemical and physical properties, such as high chemical inertness, diamond-like properties, and favorable tribological properties. These materials usually consist of graphite and diamond microstructures, and thus possess properties that lie between the two. Amorphous and nanocrystalline carbon films can exist in different kinds of matrices and are usually doped with a large amount of hydrogen. Thus, carbon films can be classified as polymer-like, diamond-like, or graphite-like based on the main binding framework [1]. Electron field emission from diamond, diamond-like carbon, carbon nanotubes and nanostructured carbon has been the object of studies and comparison [2]. The electron emission from carbon materials is based on two effects: field enhancement from conducting nanostructures and barrier lowering due to the negative electron affinity of diamond surfaces [3]. Photocathodes working in a reflection-mode have been made of rich-diamond and rich-graphite nanodiamond (ND) layers, deposited on different conductive substrates by pulsed spraying at low deposition temperatures (120 °C and 150 °C) and starting from

[4] To whom any correspondence should be addressed.

two types of ND particles. The two powders with an average grain size of 250 nm had variable sp^2 (graphite phase) and sp^3 (diamond phase) hybridized carbon content, as assessed by Raman spectroscopy and transmission electron microscopy [4]. The electron emission properties have been investigated of a transmission photocathode DC gun based on a stainless-steel mesh coated with diamond-like carbon films prepared under various technological conditions. Diamond-like carbon films have been deposited on a stainless-steel mesh and a silicon substrate by plasma-enhanced chemical vapor deposition from gas mixtures (CH_4+D_2+Ar, CH_4+H_2+Ar), and by reactive magnetron sputtering using a carbon target and Ar+D_2, Ar+H_2 gas mixtures [5]. All disordered carbons showed common features in their Raman spectra in the 800–2000 cm^{-1} range, the so-called G and D peaks, which lie at around 1560 and 1360 cm^{-1}, respectively. The G peak is due to the bond stretching of all pairs of sp^2 atoms in both rings and chains. The D peak is due to the breathing modes of sp^2 atoms in rings and is activated by disorder [6]. According to the three-stage model of Ferrari and Robertson [7], the visible Raman parameters can be used to derive the sp^3 fraction. There exist various design solutions of photo-injectors for different application requirements, but the common challenge is to obtain a high quantum efficiency (QE), long lifetime and a cold photocathode [8]. A photo-injector prototype for future electron-positron colliders and free-electron lasers (FEL) has been developed at the Joint Institute for Nuclear Research (JINR). A 30-keV photo-gun stand, a transmission (backside irradiated) photocathode concept was described, together with the laboratory studies of such cathodes; a progress report was issued on creating the photo-injector at electron energy of up to 400 keV with a unique 10-ps laser driver [9].

In the study presented here, we focused our attention on preparing N-doped carbon films, making use of the relationship between the carbon film properties and the quantum efficiency of transmission photocathodes based on quartz coated with a semitransparent carbon film. The carbon films were prepared by standard reactive magnetron sputtering with nitrogen as a reactive gas. The films' properties were investigated by Rutherford backscattering spectrometry (RBS), elastic recoil detection (ERD) and Raman spectroscopy. The properties of the photocathodes were examined by measuring the bunch charge and calculating their quantum efficiency.

2. Experiment

The carbon thin films were deposited on a silicon (Si) substrate and quartz glass by radio-frequency (RF) reactive magnetron sputtering (RMS) using a carbon target and Ar+N_2 gas mixtures in a vacuum system equipped with a rotary and a turbomolecular pump with a pumping speed of 500 l/s. The magnetron carbon target was a high-purity graphite disk with a diameter of three inches. Prior to deposition, standard cleaning was used to remove impurities from the silicon surface or the quartz glass. A 5 % hydrofluoric acid was used to remove the native oxide from the Si surface. The Si substrate was then rinsed in deionized water and dried in nitrogen ambient. The quartz glass was cleaned for 30 min in an ultrasonic bath containing acetone and then rinsed in deionized water and dried in nitrogen ambient. The carbon films were deposited at a working pressure of 0.7 Pa and magnetron input RF power of 150 W at 13.56 MHz. The high-purity (6.0) argon and nitrogen flows were kept at 30 sccm and 6 sccm, respectively, while the substrate holder temperature was 800 °C. The carbon film thickness on a silicon substrate was about 70 nm; the carbon films thicknesses on quartz glass were about 60 nm (sample BP1), 50 nm (BP2), 35 nm (BP3), 25 nm (BP4) and 15 nm (BP5).

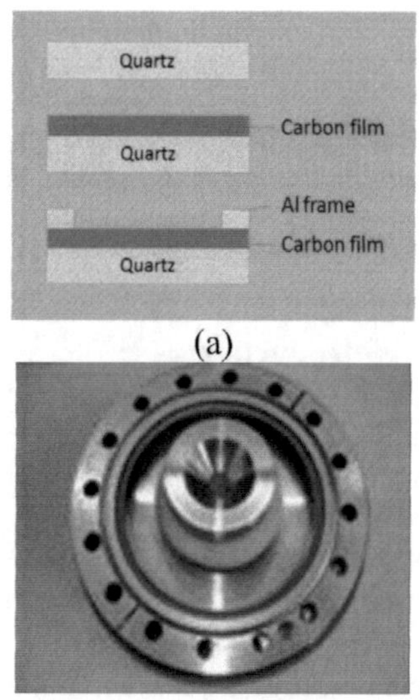

(a)

(b)

Figure 1. Transmission photocathode technology a) and cathode assembly b).

The elemental concentration in the films was determined by RBS and ERD simultaneously [10]. A Thermo Fisher Scientific DXR Raman microscope with a 532-nm laser was used to determine the intensity ratios I_D/I_G of the carbon films' D and G peaks. The peaks were fitted with Gaussians and the intensity was calculated as the peak's area. Al contact frame was formed by thermal evaporation on the carbon/quartz structures, followed by annealing in nitrogen ambient at 400 °C. The photocathodes were placed in the hollow-cathode assembly of a Pierce-structure DC gun to produce photoelectrons. Figure 1 shows a scheme of the main technological steps of a) transmission photocathode preparation and transmission photocathode assembly and b) cathode charge measurements. The bunch charge was measured by a Faraday cup [11]. The quantum efficiency was calculated from the laser energy and cathode charge measured.

3. Results and discussion

The measured and simulated RBS and ERD spectra for the N-doped carbon film with a thickness of about 70 nm are shown in figure 2. In figure 2a, the RBS spectrum shows a leading edge at around 650 ch, corresponding to the Si substrate. The carbon leading edge is observed on the bulk Si profile at 360 ch. The nitrogen and oxygen leading edges are observed on the bulk Si profile at 420 ch and 470 ch, respectively. The hydrogen leading edge, figure 2b, occurs at approximately 610 ch. We assume that the elemental concentrations in the films with different thicknesses are the same, namely, for carbon, ~83 at %; for nitrogen, ~11 at %; for hydrogen, ~4 at %; and for oxygen, ~2 at %. Hydrogen and oxygen can be incorporated into the films during deposition from the vacuum chamber atmosphere, or absorbed from the air atmosphere after deposition.

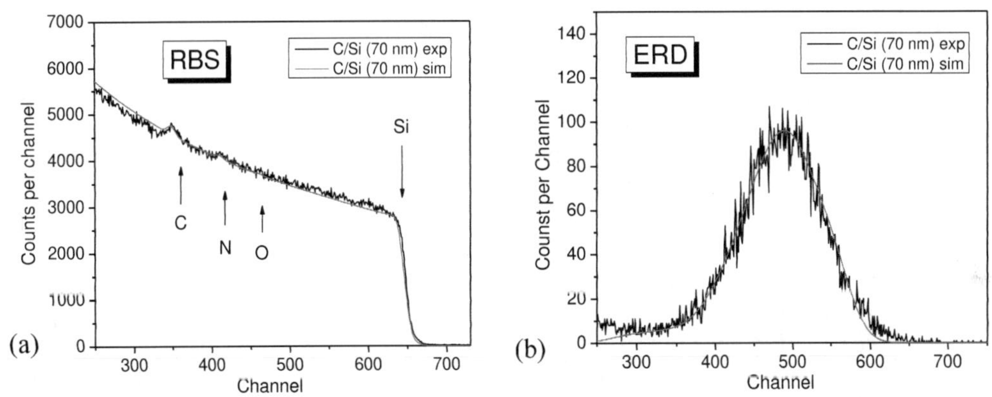

Figure 2. Experimental and simulated RBS a) and ERD b) spectra of BP2 type film on Si.

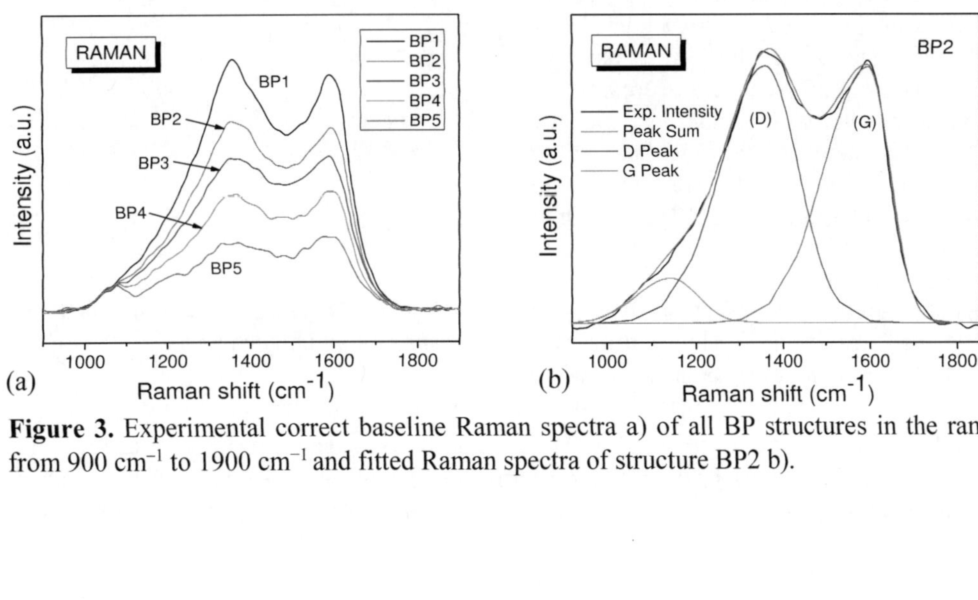

Figure 3. Experimental correct baseline Raman spectra a) of all BP structures in the range from 900 cm^{-1} to 1900 cm^{-1} and fitted Raman spectra of structure BP2 b).

Figure 3 shows Raman spectroscopy results characterizing the atomic bond structure of carbon in the carbon films. Since the π states have a lower energy than the σ states and thus are more polarizable, the sp^2 sites consisting of two π orbits and two σ orbits have a Raman cross-section larger by a factor of 50–230 than the sp^3 sites, which only contain four σ orbits. Figure 3a shows the experimental correct baseline Raman spectra of all carbon-on-quartz structures, BP1-BP5. As seen, the intensity of the broad asymmetric Raman scattering bands in the range from 1000 cm^{-1} to 1700 cm^{-1}, typical of carbon-based films, rises together with the films' thickness. The shapes of the spectra are practically the same and are similar to the spectral shape of graphite-like nanocrystalline diamond (GNCD) [14] and activated carbon [1]. Generally, the asymmetric Raman peaks can be fitted with two peaks: the G peak around 1580 cm^{-1} and the D peak around 1360 cm^{-1}, which correspond, respectively, to the C-C stretching vibration and the symmetric breathing vibration of the aromatic carbon rings [12]. According to the fitted G-peak position, the D-peak-to-G-peak intensity ratio (I_D/I_G) and the G peak's full-width at half maximum (FWHM), one can derive qualitatively the carbon bonding structure in the films, such as bond disorder, sp^3/sp^2 ratio, sp^2 site clustering [13]. For better fitting results, we used three-peak fitting. Figure 3b shows the fitted BP2 Raman spectra, representative of the fitted spectra of all BP structures. The calculated I_D/I_G ratio was in the range 1.2-1.3 for the structures BP1-BP4. For structure BP5, this ratio was about 2.5, but this result is disputable due to the very low Raman spectra intensities. In all Raman spectra, the D and G peaks' positions are practically the same – D, 1338 cm^{-1} and G, 1554 cm^{-1}.

Table 1 presents the measured bunch charge, calculated QE and carbon film' thicknesses. We assume that the electrons are emitted mainly from the carbon surface. Figure 4 shows the QE dependence on the type of transmission photocathodes prepared. As seen, the QE depends on the N-doped carbon film thickness, namely, it increases as the carbon film thickness is reduced from 60 nm to 25 nm and drops at the film thickness of about 15 nm. The theory explaining the photoemission phenomena is based on the quantum photoelectric effect and Spicer's three-step theory [15]. The photoemission process can be divided into three steps: 1) photoexcitation, 2) transport to the surface and 3) escape to vacuum. The free electrons lose part of their energy by scattering on phonons during the transport; photoelectrons are the part of electrons that have energy sufficient to overcome the electron affinity to vacuum. The optimal thickness of our N-doped carbon films was about 25 nm (sample BP4) for laser wavelength 266 nm and energy 0.4 mJ with QE = 1.7×10^{-3} %. For higher carbon film thicknesses, the free electrons' energy loss during their motion to the surface is such that they cannot overcome the electron affinity to vacuum, which results in a lower QE. The laser radiation, being an electromagnetic wave interacting with free electrons, can only penetrate into the conductive film to a depth of the skin layer $d = 1/\sqrt{\pi\mu_0\mu\sigma f}$, where $\mu_0 = 12.566\times10^{-7}$ H/m is the magnetic constant, μ is the magnetic permeability of the film material, which for non-ferromagnetic materials is unity, σ is the conductivity of the film material, f is the wave frequency. The estimated depth of a graphite skin layer with

Table 1. Transmission photo-cathodes' basic data.

Photo cathode	Bunch charge [pC]	QE [%] $\times10^{-4}$	Thickness [nm]
BP1	915	3.1	60
BP2	1005	5.3	50
BP3	1455	9.4	35
BP4	1680	17.3	25
BP5	540	4.4	15

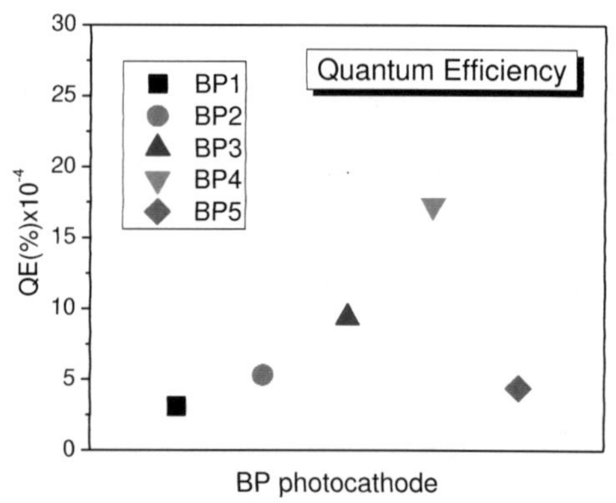

Figure 4. Dependence of QE on the type of transmission photocathode.

conductivity $\sigma = (70–140)\times10^3$ S/m for radiation of wavelength $\lambda = 266$ nm is about 50 nm. The electrons' mean-free-path in the material of the layer must be less than the value of the optical skin layer; thus, only those electrons escape, which have absorbed a photon close to the surface. Or, sufficient charge and high quantum yield can only be obtained when using thin-film photocathodes. The decrease in the recoverable charge (and QE) for BP5 was likely due to the structural properties of very thin carbon films. We plan experiments on achieving a detailed control of the structural properties of very thin carbon films (< 20 nm) by improving the technology of deposition.

4. Conclusions

We explored the performance of transmission photocathodes based on N-doped carbon films with different thickness on quartz glass. The films contained C, N, H and a small amount of oxygen. Raman spectroscopy measurements showed an interesting band between 900 cm^{-1} and 1900 cm^{-1} assigned to carbon-based films. The films' QE depended on the carbon films' thickness. The highest QE, namely 17.3×10^{-4} %, was observed with the transmission photocathode having an N-doped carbon film of 25-nm thickness. The N-doped carbon UV sensitivity is lower than the sensitivity of other widely-used photocathodes (e.g. alkali halides); therefore, this material could be preferable where the QE can be sacrificed in favor of chemical and mechanical stability and robustness.

Acknowledgment

This work was performed in the framework of the Topical Plan for JINR Research and International Cooperation (Project 02-0-1127-2016/2018) and supported by the Slovak Research and Development Agency under contract APVV-0443-12.

References

[1] Chu P K and Li L 2006 *Mater. Chem. and Phys.* **96** 253
[2] Robertson J 2003 *IEICE Trans. on Electr.* **E86-C**/5 787
[3] Nemanich R J, Bilbro G L, Bryan E N, Koeck F A, Smith J R and Tang Y 2010 *Proc. 3rd Int. Conf. Nanoelectr. (IEEE2010)* ed Paul K Chu (2010 Hong Kong) p 56
[4] Velardi L, Valentiny A and Cicala G 2017 *Diamond and Related Mater.* **76** 1
[5] Huran J, Balalykin N I, Feshchenko A A, Kobzev A P, Kleinová A, Sasinková V and Hrubčín L 2014 *Nucl. Instrum. Methods* A **753** 14
[6] Casiraghi C 2011 *Diamond & Related Mater.* **20** 120
[7] Ferrari A C and Robertson J 200HC0 *Phys. Rev.* B **61** 14095
[8] Xiang R and Teichert J 2015 *Phys. Procedia* **77** 58
[9] Balalykin N I, V. F. Minashkin V F, Nozdrin M A, Trubnikova G V, Shirkov G D, Gacheva E I, Katin E V, Khazanov E A, Luchinin G A, Poteomkin A K, Zelenogorskii V V and Huran J 2016 *Phys. of Particles and Nuclei Letters* **13** 897
[10] Kobzev A P, Huran J, Maczka D and Turek M 2009 *Vacuum* **83** S124
[11] Balalykin N I, Huran J, Nozdrin M A, Feshchenko A A, Kobzev A P and Arbet J 2016 *J. Phys.: Conf. Series* **700** 012050
[12] Robertson J 2002 *Mater. Sci. Eng.* **R 37** 129
[13] Dai W, Gao X, Wang Q and Xu A 2017 *Thin Solid Films* **625** 163
[14] Nitti M A, Colasuonno M, Nappi E, Valentini A, Fanizza E, Bénédic F, Cicala G, Milani E and Prestopino G 2008 *Nucl. Instrum. Methods* A **595** 131
[15] Spicer W E 1958 *Phys. Rev.* **112**/1 114

IOP Conf. Series: Journal of Physics: Conf. Series **992** (2018) 012032 doi:10.1088/1742-6596/992/1/012032

Application of TiN/TiO_2 coatings on stainless steel: composition and mechanical reliability

M P Nikolova[1,4], A Genov[1], S Valkov[2], E Yankov[1], D Dechev[2], N Ivanov[2], R Bezdushnyi[3] and P Petrov[2]

[1]Department of Material Science and Technology, A. Kanchev University of Ruse, 8 Studentska Str., 7000 Ruse, Bulgaria
[2]E. Djakov Institute of Electronics, Bulgarian Academy of Sciences, 72 Tzarigradsko Chaussee, 1784 Sofia, Bulgaria
[3]Faculty of Physics, St. Kliment Ohridski University of Sofia, 5 James Bourchier Blvd., 1164 Sofia, Bulgaria

E-mail: mpnikolova@uni-ruse.bg

Abstract. The paper reports on the effect of the substrate temperature (350 °C, 380 °C and 420 °C) during reactive magnetron sputtering of a TiN film on the phase composition, texture and mechanical properties of TiN/TiO_2 coatings on 304L stainless steel substrates. Pure Ti was used as a cathode source of Ti. The texture and unit cell parameters of both TiN and TiO_2 phases of the coating are discussed in relation with the tribological properties and adhesion of the coating. The scratch tests performed showed that the nitride deposited at 380 °C, having the highest unit cell parameter and a predominant (111) texture, possessed the lowest friction coefficient (μ), tangential force and brittleness. The anatase-type TiO_2 with predominant (101) pole density and increased *c* unit cell parameter showed the highest stability on the nitride deposited at 420 °C. The results indicated that the friction coefficient, tangential force and critical forces of fracture could be varied by controlling the coating deposition temperature.

1. Introduction

The demand for better quality of life entails the need of high-quality implant devices making fast recovery possible. The medical grades stainless steel has long been a quite common material used for manufacturing a variety of hard implantable devices, such as orthopedic prostheses, valves, or endo-prostheses, as thin-wall stents, and dental brackets [1]. The AISI grade 304L steel has high corrosion resistance, which is an important aspect for the material biocompatibility; still, concerns persist because of steel's poor tribological properties and the metal ions released from it after prolonged exposure [2] that disturb the cell proliferation and differentiation. An approach to minimizing the corrosion phenomenon and the release of metallic ions is the deposition of coatings of more biocompatible materials, like Ti, TiN, and TiO_2. Moreover, the hardness and wear resistance that are beneficial for the medical device could be thus enhanced [3, 4]. The different material properties will affect the coating adhesion, resistance to plastic deformation, fracture toughness, etc. Hence, the aim of this study was to determine the influence of different deposition temperatures of a magnetron

[4] To whom any correspondence should be addressed.

sputtered TiN/TiO_2 coating on the composition and tribological properties of the implant system. X-ray diffraction (XRD) was employed to characterize the phase composition, texture and crystalline particles of both coating layers. The adhesion, which is the most important element determining the final coating quality, was assessed by means of a macroscratch test analysis in a progressive-load scratching mode.

2. Experimental procedures

Austenitic stainless steel 304L (X2CrNi18-9, 1.4307) with weight composition of C 0.03%, Cr 18%, Ni 8%, Mg 0.2% P 0.045%, Si 0.75% and Fe balance was used for the experiments. The specimens had a dimension of 22.22×19.15×2.9 mm. All the samples were ground by sandpaper down to P800 and mechanically polished.

The TiN/TiO_2 coating was applied by reactive magnetron sputtering. The diameter of the Ti target of 99.8% purity was 100 mm. The TiN layer deposition took a place in Ar-N_2 atmosphere for 90 min at a working pressure of 1.2×10^{-1} Pa. During the deposition process, the substrate was heated to 350 °C, 380 °C and 420 °C. The TiO_2 film was formed in pure O_2 environment for 120 min at the working pressure of 7×10^{-2} Pa, as the substrate temperature was decreased down to 180 °C. The thickness of each coating (TiN and TiO_2) was about 1 µm, as assessed by the resonant frequency of a quartz plate (Intermetalix IL800) coated under the same process conditions. In order to minimize the residual stresses and the oxidation, the samples with deposited coating were retrieved from the vacuum chamber after room temperature was reached.

The coatings were characterized by X-ray diffraction using a URD6 Seiferd&Co diffractometer with CuK_α radiation. The patterns were registered within the 20° – 80° interval at 2θ scale, a 0.1° step and counting time of 10 s per step. The experiments were conducted in symmetrical Bragg-Brentano (B-B) mode.

The scratch tests were performed on the top surface of the samples with a CSM REVETEST Scratch Macrotester equipped with a Rockwell diamond indenter with a 200-µm tip radius. In scratching mode, a progressive load was applied with a normal force ranging from 0 to 50 N at a speed of 10 N/mm. The scratch track was evaluated visually by a Nikon optical microscope, with a 14-megapixel digital camera adapted for image acquisition within the 16-FMME-02 project of University of Ruse [5]. Also, digital-signals were recorded of the friction coefficient (µ), the tangential force (F_t) and the acoustic emission (AE) fluctuations The scratch hardness (H_S) and fracture toughness (K_{IC}) were calculated by the equations quoted in [6] and [7], respectively.

3. Results and discussions

Figure 1 shows diffractograms of the TiN/TiO_2 coatings on 304L substrates, with the nitride sub layers deposited at different temperatures (350 °C, 380 °C and 420 °C). The results indicate the presence of polycrystalline TiN, TiO_2 phases and austenite reflected from the substrate. The TiO_2 phase is anatase-type.

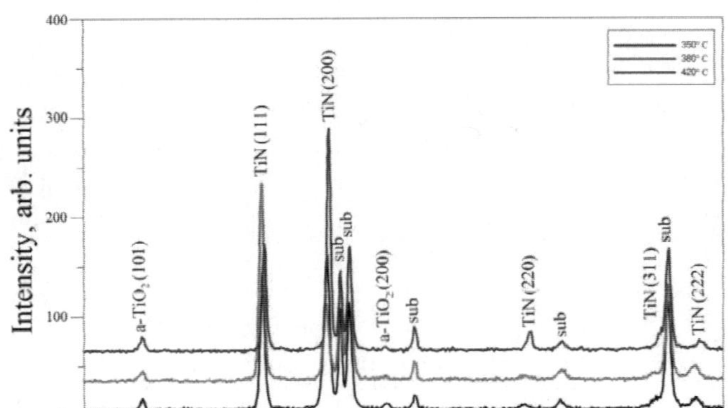

Figure 1. XRD patterns of TiN/TiO_2 coatings deposited on 304L steel substrates under different temperature conditions.

Table 1 illustrates the influence of micro-volume fractions belonging to different texture components, represented by the diffraction pole density P_{hkl} as a function of the substrate temperature during the deposition of TiN. It is clear that at lower temperatures (350 °C and 380 °C) the predominant crystallographic orientation of the TiN coating is (111), while with the substrate temperature increase up to 420 °C the texture changes mainly to the (200) direction. Concerning the TiO_2 film, the texture shifts from the (200) direction at lower temperatures (350 °C and 380 °C) towards the (101) direction at 420 °C. In contrast with the TiN deposited at 350 °C, where the grains size is about 24 nm, the TiN deposited at 380 °C and 420 °C shows larger and similar grain sizes – 27.5 nm and 26.9 nm, respectively. The grain size of TiO_2 rises substantially from 18.86 nm and 18.83 nm for the coatings deposited at 350 °C and 380 °C to 23.3 nm for the one deposited at 420 °C.

Table 1. Texture components presented by diffraction pole density and grain size of the films phases.

T, °C	TiN				TiO_2		Grain size, nm	
P_{hkl},%	$P_{(111)}$	$P_{(200)}$	$P_{(220)}$	$P_{(311)}$	$P_{(101)}$	$P_{(200)}$	TiN	TiO_2
350	41.58	32.96	4.75	20.72	40.86	59.14	24.12	18.86
380	64.54	16.74	4.76	14.02	44.87	55.13	27.52	18.33
420	30.20	46.66	8.25	14.89	58.50	41.50	26.9	23.3

The calculated unit cells parameters of the TiN/TiO_2 coating are summarized in table 2. The unit cell parameters of the TiN films deposited at 350 °C and 380 °C exceed those cited in the ICDD diffraction data. In the case of the substrate temperature of 420 °C, the TiN unit cell is smaller than those quoted in ICDD. For the TiO_2 phase, the *a* unit cell parameters are similar for both temperatures of 350 °C and 380 °C and increase with the increase of TiN deposition temperature. The *c* parameter follows a linear increase in line with the increase of the substrate temperature during the TiN deposition.

Table 2. Calculated unit cells parameters of the deposited films.

	350° C		380° C		420° C		ICDD database	
	TiN	TiO_2	TiN	TiO_2	TiN	TiO_2	TiN	TiO_2
a, Å	4.256	3.776	4.265	3.774	4.237	3.782	4.241	3.785
c, Å	-	9.345	-	9.453	-	9.524	-	9.513

The digital signal recordings from the tribological test of all three coated samples shown in figure 2 a and b indicate that a difference in the friction coefficient of (μ) is seen after a 5-N normal load (figure 2 a). For the film deposited at 350 °C, the friction coefficient's rising trend is accompanied by severe fluctuations of up to 0.4. Once the loading reaches 25 N, μ diminishes because of the joint influence of the nitride and the substrate. The coating deposited at 420 °C shows a smoothly increasing μ behavior with a maximum value of 0.35. After loading to 20 N, the μ values are also dependent on the complex substrate-coating influence. The coating with nitride deposited at 380 °C has the lowest coefficient of friction. The surface film is largely stable up to 50 N despite the coating sinking in the substrate at 22.6 N, which does not cause delamination and μ keeps its low values until the end of the test.

The behavior of F_t (figure 2 b) for all three coated samples is of a similar growing nature, with the highest fluctuations shown by the specimen deposited at 350 °C, and the lowest, by the one deposited at 380 °C. Additionally, the acoustic emission (AE) signals from the scratch test allow one to assess the coatings' hardness and fracture toughness. All three coated samples indicate a similar scratch resistance up to 5 N. As the load progressively increases, the brittleness of the coating deposited at 350 °C causes strong fluctuations in the AE. A similar effect occurs for the coating deposited at 420 °C, but the AE manifests a more uniform variation, while its higher values suggest a higher coating hardness. The AE of lower level and intensity recorded for the coating deposited at 380 °C is indicative of lower internal stresses in the TiN/TiO_2 film.

The scratch track micrographs (figure 2 c) of all three coated samples show distinct first cracks (Lc_1) occurring during the progressive loading; the Lc_1 value increases with the increase of the nitride deposition temperature. The shape of the lateral cracks of all samples indicates ductile tensile cracking arising from the plastic deformation around the indenter. The piling up of material in front and sideways of the residual groove is indicative of low substrate hardness. Although total delamination is not observed up to 50 N for all coated samples, the coatings deposited at 350 °C and 420 °C show a partial detachment at maximum load. The coating deposited at 380 °C is stable under the indenter up to 50 N.

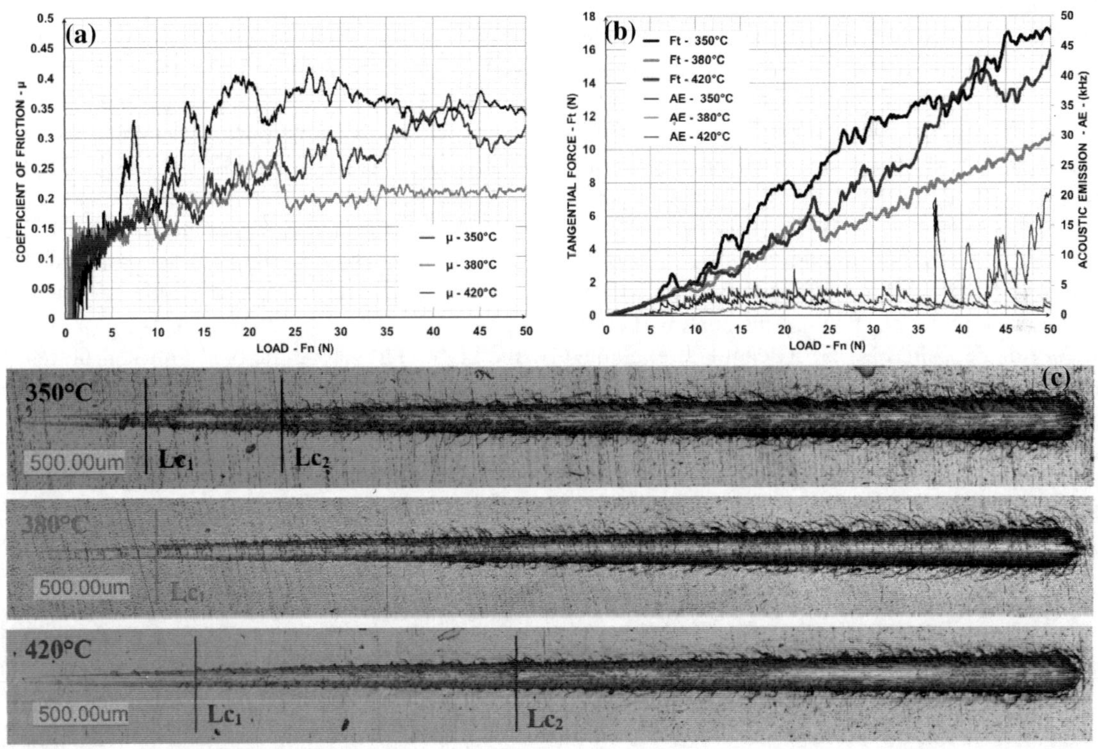

Figure 2. Comparison of the variation in the: **a**) tangential force (F_t); **b**) coefficient of friction (µ) and acoustic emission (AE) and **c**) micrographs of the scratch tracks during progressive 0 N to 50 N scratch tests of the TiN/TiO_2 coatings deposited on 304L steel substrates under different temperature conditions.

As shown in table 3, the scratch hardness values decrease with the increase of the deposition temperature of the nitride. The higher F_t value at Lc_1 for the nitride deposited at 350 ° C proves its fragile behavior, but increases its fracture toughness. The lowest K_{IC} value at L_{c1} calculated for the nitride deposited at 380 °C suggests a higher plasticity of the coating that demonstrates prevailing (111) oriented micro-volume fractions. At 420 °C, the unfavorable slipping planes prevail, which reduces the fracture toughness at L_{c1}. The Lc_2 values reach 14.08 N and 23.36 N for the coatings deposited at 350 °C and 420 °C, respectively, which suggests a better adhesion of the coating deposited at the higher temperature. Nevertheless, the smaller penetration of the indenter because of the higher strength of the surface of the nitride deposited at 350 °C gives rise to its fracture toughness value at L_{c2}. The coating deposited at 380 °C does not show detachment from the substrate, but only deformation and embedding in the stainless steel material.

IOP Conf. Series: Journal of Physics: Conf. Series **992** (2018) 012032 doi:10.1088/1742-6596/992/1/012032

Table 3. Calculated scratch test parameters of the deposited coatings: Lc_1 – the load inducing the first crack of the coating, Lc_2 – the load inducing partial detachment of the coating, F_t – tangential force at the critical loading, H_S – scratch hardness, K_{IC} – fracture toughness.

T, °C	Lc_1, N	F_t at Lc_1, N	H_S, GPa	K_{IC}, MPa.m$^{1/2}$	Lc_2, N	F_t at Lc_2, N	H_S, GPa	K_{IC}, MPa.m$^{1/2}$
350	7.60	2.40	3.85	0.014	14.08	5.02	3.58	0.019
380	7.80	1.53	3.82	0.0062	>50	-	-	-
420	9.52	1.95	3.40	0.0067	23.36	6.18	3.20	0.012

4. Conclusions

The TiN/TiO_2 coatings deposited on 304L stainless steel at different temperatures have the same phase composition, but different texture and tribological behavior. The TiN deposited at 380 °C, which has predominant texturing in the close-packed (111) direction and the highest unit cell parameters and grain size, exhibits the best tribological test results, as its μ, AE, and Ft values are the lowest. The low (200) and (311) pole density of the nitride sub-layer enhances the tribological performance – adhesion of the coating, but decreases its scratch hardness. As the deposition temperature increases, a reorientation to the (200) direction occurs, which is a prerequisite for a higher brittleness of the coating. At the deposition temperature of 350 °C, the allocation of the crystallographic planes of the nitride is more balanced, as the unfavorable directions exceed, in aggregate, the favorable one, which explains the lower plasticity of the coating. On the other hand, the low stability of the anatase oxide on the nitride deposited at 350 °C could be explained by the higher stresses at the oxide-nitride interface and the (200) texture of the surface layer that causes higher scratch hardness and fracture toughness. The oxide obtained on the nitride deposited at 420 °C has the highest stability due to the change of its texture to the (101) direction and the increased *c* unit cell parameter and grain size.

Acknowledgments

The work was supported in part by the Bulgarian National Science Fund (NSF) under contract D 07/3 (2016) “Gradient functional nanocoatings produced by vacuum technologies for biomedical applications”.

References

[1] Santonen T, Stockmann-Juvala H and Zitting A 2010 *Review on toxicity of stainless steel* (Helsinki: Finnish Institute of Occupational Health) p 87
[2] Chew K-K, Zein S H S and Ahmad A L 2012 *Natural Sci.* **4**/3 184-8
[3] Radeva E I, Martev I N, Dechev D A, Ivanov N, Tsaneva V N and Barber Z H 2006 *Surf. Coat. Technol.* **201**/6 2226-9
[4] Dechev D, Ivanov N and Petrov P 2015 *J. Balkan Tribological Association* **21**/2 351-5
[5] http://nis.uni-ruse.bg/FMT-16/#MT-2
[6] Hasan M, Haseeb A and Masjuki H 2012 *Surf. Eng.* **28**/10 778-85
[7] Akono A-T, Reis P and Ulm F-J 2011 *Phys. Rev. Lett.* **106** 20 204302

IOP Conf. Series: Journal of Physics: Conf. Series **992** (2018) 012033 doi:10.1088/1742-6596/992/1/012033

Optical and electrical properties of $TiO_2/Pt/TiO_2$ nanolaminate structures

Ch Dikov[1,4], P Vitanov[1], T Ivanova[1], V Stavrov[2], E Tomerov[2], G Stavreva[2] and I Bineva[3]

[1]Central Laboratory of Solar Energy and New Energy Sources, Bulgarian Academy of Sciences, 72 Tzarigradsko Chaussee, 1784 Sofia, Bulgaria
[2]AMG Technology Ltd., Microelectronica Industrial Zone, 2140 Botevgrad, Bulgaria
[3]G. Nadjakov Institute of Solid State Physics, Bulgarian Academy of Sciences, 72 Tzarigradsko Chaussee, 1784 Sofia, Bulgaria

E-mail: dikov@phys.bas.bg

Abstract. We report the results of a study on the optical and electrical properties of nanolaminate $TiO_2/Pt/TiO_2$ structures fabricated using RF magnetron sputtering. The effect was investigated of the discontinuous Pt layer on the optical transmittance, electrical conductivity and morphology of the $TiO_2/Pt/TiO_2$ structure. On the basis of the theory of electronic tunneling transport between metal granulates in a dielectric matrix, a new technological approach is proposed of preparing high-sensitivity strain sensors based on resistors with a nanolaminate structure containing Pt granulates.

1. Introduction

Recently, we reported the development of an optically transparent and conductive nanolaminated dielectric/metal/dielectric (DMD) structure [1]. The concept is based on the usage of electronic conductivity in materials of granular (discontinuous) type [2]. The granulates are metallic particles of sizes ranging usually from a few to hundreds of nanometers embedded into an insulating matrix. The nanolaminate DMD structure is formed using one or two different dielectrics. The electrical conductivity is modified by plane-located metal granulates. The advantages of this approach include high reproducibility of the technological parameters, easy process control, and selective formation of conductive areas in the dielectric structure. It should be noted also that the conductive nanolaminate structures are transparent in the visible spectral range. The optical transmittance and the sheet resistance are functions of the metal nanoparticles density in the intermediate layer.

Usually, the transparent and conductive oxide (TCO) films exhibit semiconductor behavior, namely, a negative temperature coefficient of resistivity (TCR). In contrast, the nanolaminate structure with metal granulates demonstrates a metallic conductivity, i.e., a positive TCR. Therefore, the charge carriers are delocalized due to thermal activation and the conductivity is dominated by phonon scattering. The strain effects in such nanolaminate structures with Ag granulates have already been reported [3]. The disadvantage of such a structure is that its properties degrade during most of the

[4] To whom any correspondence should be addressed.

common wet chemical and plasma treatments used in the MEMS technologies. Thus, the necessity arises of finding chemically-stable materials for metallic nano-granulates.

In this work, $TiO_2/Pt/TiO_2$ sandwich films were fabricated using RF magnetron sputtering. This technique has many advantages, including large area deposition, high deposition rate, simple equipment and easy operation. The electrical and optical properties of the nanolaminate structures with Pt nanoparticles were studied as depending on the sputtering deposition time of the Pt target. It was found that their electrical and optical properties do not change after thermal annealing at 400 °C for 60 min, as well as during the usual wet chemical and plasma treatments. A new technological approach was developed for preparing high-sensitivity strain sensors based on resistors with a nanolaminate structure containing Pt granulates.

2. Experimental – deposition procedure

The layers were deposited on an RF magnetron sputtering system with three targets. This system was developed to be used for low-temperature high-speed sputtering deposition and is suitable for thin-film deposition of metals and insulating materials. The substrate holder has a size of 200 mm, with three targets (75 mm) located at eccentric positions. The water-cooled (3 l/min) targets and the substrate holder are vertically arranged at a distance of 80 mm. The holder is rotating at a rate of 80 rpm. This configuration allows one to prepare thin films with homogeneous thickness. The three targets provide options to deposit successively three different layers without opening the chamber. In this work, the TiO_2 films were deposited at an RF power of 300 W in Ar_2 ambient at a deposition rate of 10 nm/min. The film thickness was determined by an LEF 3M laser ellipsometer equipped with a He-Ne laser operating at the wavelength of 638.2 nm.

The surface morphology was investigated by AFM measurements carried out using a Multimode V device (Bruker, ex. Veeco, Santa Barbara, CA). The imaging was performed in tapping mode, with a scan rate of 2 Hz and image resolution of 512 lines using silicon cantilevers with a length of 125 μm and Al reflective coating on the backside (MPP-11120-10, Bruker). These probes have a nominal resonance frequency of 300 kHz and a typical force constant of 40 N/m, with the nominal tip radius being less than 10 nm. At least three different areas on the sample surface were studied. The image analysis was performed by means of the SPIP^{tm} 6.1.0 software.

Figure 1 presents AFM images of a nanogranulate coating prepared by sputtering Pt on polished Si wafers with thermally-grown SiO_2 (400 nm) under the following deposition conditions: Ar_2 reactive gas, RF power of 200 W and deposition time of 10 s. The metal nanoparticles with a maximum height approaching 0.9 nm form a discontinuous layer and are randomly distributed.

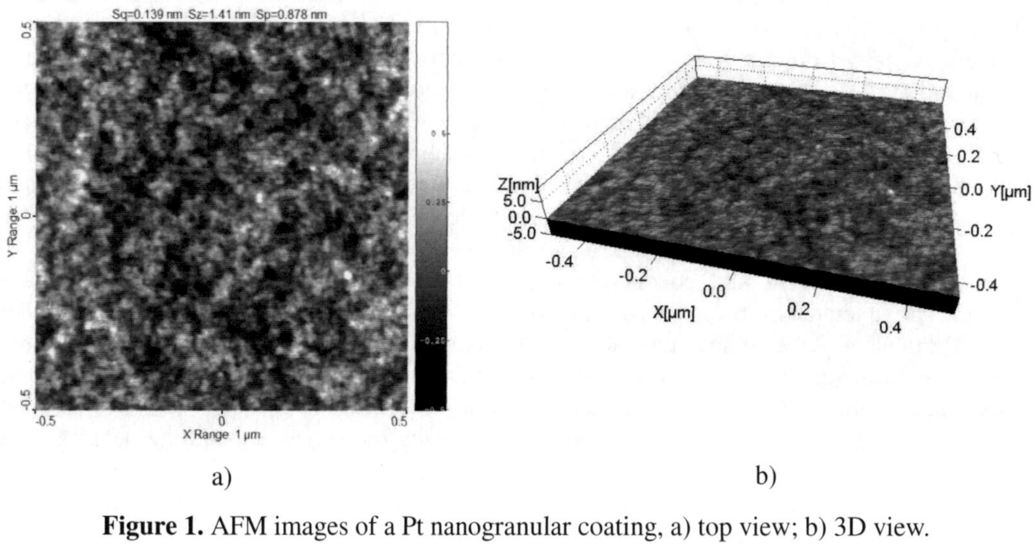

a) b)

Figure 1. AFM images of a Pt nanogranular coating, a) top view; b) 3D view.

IOP Conf. Series: Journal of Physics: Conf. Series **992** (2018) 012033 doi:10.1088/1742-6596/992/1/012033

Figure 2 presents AFM images of the surface of the $TiO_2/Pt/TiO_2$ nanolaminate structure. Its morphology (rough surface) is formed by platinum granulates with a maximal height of 2.63 nm and a grain diameter of about 14 nm.

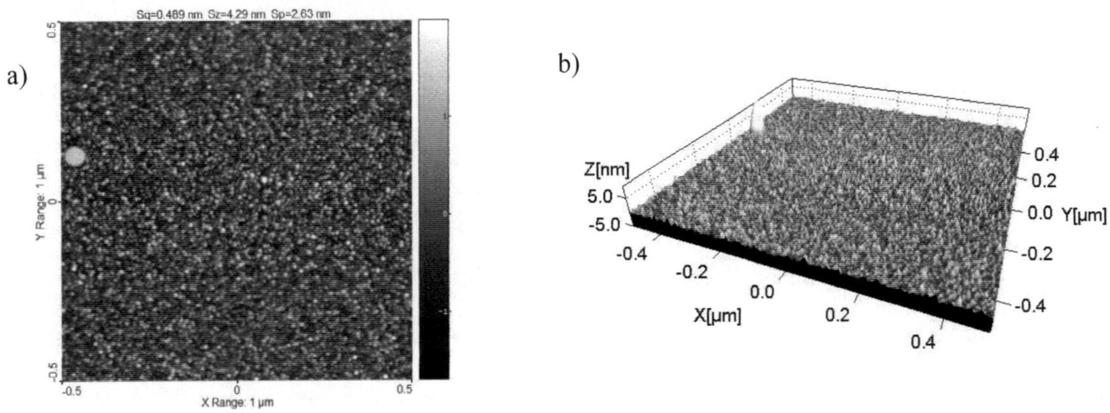

Figure 2. AFM image of a sputtered nanolaminate $TiO_2/Pt/TiO_2$ structure: a) top view; b) 3D view.

3. Results and discussions

3.1. Optical and electrical properties

The two TiO_2 layers in the nanolaminate $TiO_2/Pt/TiO_2$ structures examined had an equal thickness of 20 nm, but we varied the sputtering time Pt (7 s, 10 s and 15 s). The latter determines the size and density of the metal granulates. The two parameters, namely, the optical transmittance and the sheet resistance, depend on the metal nanoparticles density in the intermediate layer. One expects that the optical transmittance will decrease as the sputtering time, i.e., the thickness of the Pt layer, increases.

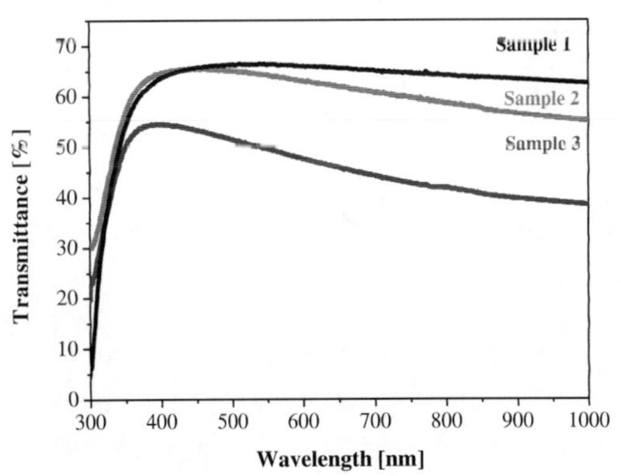

Figure 3. Transmittance spectra of nano-laminate structures, sample 1 refers to $TiO_2/Pt(7s)/TiO_2$, sample 2 to $TiO_2/Pt(10s)/TiO_2$ and sample 3, to $TiO_2/Pt(15s)/TiO_2$. The spectra are measured using a bare glass substrate as a reference.

As demonstrated in figure 3, the optical transmittance does decreases as the Pt layer thickness increases.

The values of sheet resistance for nanolaminated structures with different Pt layers are presented in table 1.

Table 1. Sheet resistance of nanolaminate structures, varying Pt deposition duration.

Sample	Deposition time of discontinuous Pt layer	Sheet-resistance Ω/□
$TiO_2/Pt/TiO_2$	7 seconds	200
$TiO_2/Pt/TiO_2$	10 seconds	140
$TiO_2/Pt/TiO_2$	15 seconds	80

The electrical conductivity values can be explained by the average tunneling conductance G between metal granulates. It is convenient to introduce the dimensionless conductance g, measured in units of the quantum conductance e^2/h, then $g = G/(2e^2/h)$ [2].The samples with $g > 1$ exhibit metallic transport properties, while those with $g < 1$ show insulating behavior. The dependence of the tunneling

IOP Conf. Series: Journal of Physics: Conf. Series **992** (2018) 012033 doi:10.1088/1742-6596/992/1/012033

probability on the inter-granular distance suggests the suitability of the material as a high-sensitivity strain sensor [4, 5]. The resistivity of a dielectric(TiO_2)/metal(Pt)/dielectric(TiO_2) (DMD) resistor increases if the material is tension-strained, as the average distance between the particles increases and the tunneling probability decreases. If the electron transport in the nanogranular metal is of tunneling nature, then the nanolaminate structures can be used as strain resistors.

3.2. Four-cantilever self-sensing test device

Figure 4 displays a top-view of the four-cantilever test device. A single DMD resistor 1 is positioned on each cantilever C_i and all four resistors are connected with patterned metal tracks 2 in a full-bridge configuration [3]. Since all resistors have the same value, the output bridge signal is compensated when the cantilevers do not move, or move synchronously. Because asynchronous oscillations occur at resonance, a bridge misbalance might be observed at the particular resonance frequencies of the cantilevers, if the resistors are strain-sensitive. The frequency response of the device was detected by means of the set-up presented schematically in figure 5. A sweeping signal in the frequency range of 10 kHz to100 kHz from a signal generator was used to drive a piezo actuator that caused the cantilevers to vibrate.

Figure 4. Top view of a four-cantilever test device with a single DMD resistor per cantilever.

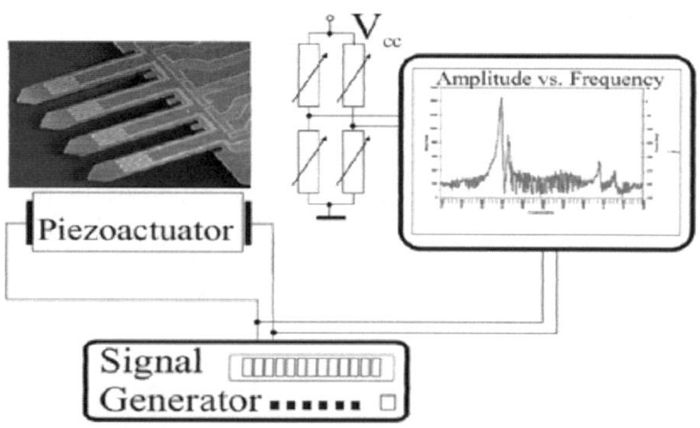

Figure 5. Experimental set-up for measuring the resonance frequency of the four-cantilever device.

Figure 6 shows a plot of the output characteristic of the four-cantilever device. The figure displays also the correlation between the detected frequencies of the bridge misbalance and resonance frequencies of the cantilevers $C_1 - C_4$. The coincidence of the said frequencies proves the sensitivity to strain of the TiO_2/Pt/TiO_2 DMD resistors.

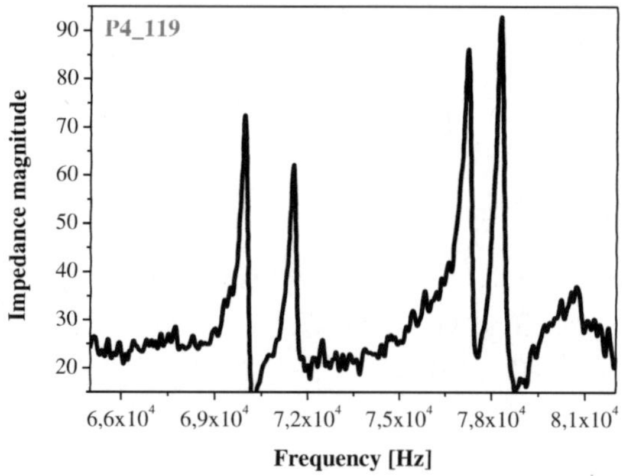

Figure 6. Output characteristic of the four-cantilever test device.

4. Conclusions

We explored the electrical conductivity and optical transparency of TiO_2/Pt/TiO_2 nanolaminate structures. Since the structures possessed high chemical robustness and good compatibility with surface micromachining technologies, we successfully integrated them in a four cantilever test devices. Based on the theory of tunneling electronic transport in nano-granular metals in a dielectric matrix, we were able to verify experimentally that the nanolaminate TiO_2/Pt/TiO_2 resistors are sensitive to strain.

IOP Conf. Series: Journal of Physics: Conf. Series **992** (2018) 012033 doi:10.1088/1742-6596/992/1/012033

References

[1] Dikov Hr, Vitanov P, Ivanova T and Stavrov V 2016 *J. Phys.: Conf. Ser.* **700** 012054
[2] Beloborodov I, Lopatin A, Vinokur V and Efetov K 2007 *Rev. Mod. Phys.* **79** 469
[3] Stavrov V, Stavreva G, Tomerov E, Dikov Hr and Vitanov P 2017 Self-sensing cantilevers with nano-laminated dielectric-metal-dielectric resistors, oral presentation *Proc. ISSE* (May 2017 Sofia Bulgaria)
[4] Huth M 2010 *J. Appl. Phys.* **107** 113709
[5] Dukic M , Winhold M, Schwalb Ch, Adams J, Stavrov V, Huth M and Fantner G 2016 *Nature Comm.* 71 2487

IOP Conf. Series: Journal of Physics: Conf. Series **992** (2018) 012034 doi:10.1088/1742-6596/992/1/012034

Mechanical properties of tantalum-based ceramic coatings for biomedical applications

N Donkov[1,5], J Walkowicz[2], V Zavaleyev[2], A Zykova[3], V Safonov[3], S Dudin[4] and S Yakovin[4]

[1]E. Djakov Institute of Electronics, Bulgarian Academy of Sciences,
72 Tsarigradsko Chaussee, 1784 Sofia, Bulgaria
[2]Faculty of Technology and Education, Koszalin University of Technology,
2 Sniadeckich Str., 75-453 Koszalin, Poland
[3]National Science Center „Kharkov Institute of Physics and Technology",
1 Academicheskaya Str., 61108 Kharkov, Ukraine
[4]Department of Physical Technologies, Kharkov National University,
4 Svobody Sq., 61077 Kharkov, Ukraine

E-mail: nikolaidd@abv.bg

Abstract. The properties were studied of Ta, Ta_2O_5 and Ta/Ta_2O_5 coatings deposited by reactive magnetron sputtering on stainless steel (AISI 316) substrates. The compositional, structural and morphological parameters of the coatings were investigated by means of X-ray photoemission spectroscopy (XPS), energy dispersive X-ray (EDX) spectroscopy, and scanning electron microscopy (SEM). The roughness parameters, adhesion strength, hardness, elastic modulus, and *H*/*E* ratio were evaluated by standard techniques. The hardness parameters of the Ta_2O_5 and Ta/Ta_2O_5 coatings increased in comparison with pure Ta films, while the relatively low Young's modulus was related to high elastic recovery and high resistance to cracking. The tantalum-based coatings possessed good biomechanical parameters for advanced implant and stent applications.

1. Introduction

Tantalum is one of the most attractive materials for biomedical applications [1,2]. However, the clinically-used Ta-based metallic implants and stents have limitations due to their poor biomechanical properties. To solve the problem, different Ta-based coatings have been deposited on the surface of metals and alloys to improve the quality of modern medical products. Tantalum (Ta) [3, 4], tantalum nitride (TaN) [5-8], and tantalum oxide (Ta_2O_5) coatings [9-11] have demonstrated high electrochemical and dielectric properties, and excellent biocompatibility. Coating with Ta is among the most practical ways of improving the long-term stability of biomedical implants made of stainless steel (AISI 316L) [12], Co–Cr [5], Ti-6Al-4V [13] and NiTi [14] by enhancing their *in vitro* bioactivity and biocompatibility, which would widen their applications as orthopedic and dental implants. The results of tests on implanting Ta in both soft and hard rat tissues showed the good osteogenesis of this metal [15]. TaC and TaN possess a relatively high hardness due to the covalent

[5] To whom any correspondence should be addressed.

IOP Conf. Series: Journal of Physics: Conf. Series **992** (2018) 012034 doi:10.1088/1742-6596/992/1/012034

nature of their bond [16], and a superior corrosion resistance [17]. Moreover, the biomedical properties of TaN coatings have also been investigated. The blood compatibility of TaN films was shown to be better than that of Ta [18] for applications in artificial heart valves.

The novel stent types have demonstrated low rates of platelet aggregation and better endothealization, especially in the case of patients with high-risk of in-stent restenosis. The positive effect results in lower rates of myocardial infarction and stent thrombosis [19, 20]. The clinically used metallic stents have limitations because of their low mechanical and bio-inertness properties. As proved in a clinical trial [21], surface treatment of artificial vessels and deposition of coatings reduce the likelihood of thrombosis, control the blood coagulability, and diminish platelet's adhesion compared with metal stents.

Deposition of Ta, TaN, and the Ta/TaN bilayer system as protective coatings has been studied in view of biomedical applications [22]. It was found that the Ta phase improves the corrosion and biomechanical properties of the bilayer system in comparison with the monolayer TaN film. The mechanical properties of functional coatings are strongly correlated with the coatings' properties, such as microstructure, phase and chemical composition, and surface parameters. The hardness (*H*), the Young's modulus (*E*), and the *H/E* ratio of the samples provide an elastic-strain-to-failure value correlated to the wear resistance. The aim of the present work was exploring the effect of the compositional, structural and surface properties of the tantalum-based films on the biomechanical characteristics of Ta_2O_5 and Ta/Ta_2O_5 coatings.

2. Materials and methods

The substrates were made of stainless steel SS (AISI 316) discs with a diameter of 32 mm and a thickness of 3 mm. Before deposition, the samples were cleaned by standard methods in an ultrasonic bath. The tantalum target had a diameter of 170 mm. The deposition of Ta, Ta_2O_5 and Ta/Ta_2O_5 coatings was performed by ion-source-assisted magnetron sputtering in a high-vacuum system with a base pressure of about 10^{-2} Pa. Oxygen for the reactive deposition was delivered through the ICP plasma source at $q = 30 - 60$ sqcm; the magnetron voltage was $U_m = 700$ V, with a magnetron current of about $I_m = 5.7$ A. The ion source was also used to clean the samples' surface before deposition.

The coatings' thickness was measured by a JSM-5500 LV (JEOL) SEM (figure1).The composition of the coatings was analyzed by EDX spectroscopy (Oxford Link ISIS 300). X-ray photoelectron spectroscopy was carried out using ESCALAB MkII (VG Scientific) electron spectrometer with an AlKα X ray source (excitation energy $h\nu = 1486.6$ eV). The surface roughness parameters were measured by a Hommel-Etamic T8000 profilometer, while the hardness and Young's modulus were measured using the Fischerscope HM 2000 nanohardness testing instrument (with a diamond Berkovich tip)/ The adhesive properties were measured using a REVETEST® (CSM) scratch tester and a Rockwell indenter.

3. Results and discussion

Fundamentally, the deposited films' functions should be strongly affected by their microstructure, surface roughness, and chemical composition. We evaluated the thickness and roughness values (R_a, R_z) of the Ta, Ta_2O_5, and Ta/Ta_2O_5 coatings (table 1). In the cases of Ta_2O_5 and Ta/Ta_2O_5 coatings, an increase in the films' thickness and the formation of a bilayer resulted in a decrease in the roughness

Table 1. Thickness and surface parameters of Ta, Ta_2O_5 and Ta/Ta_2O_5 coatings deposited on stainless steel (AISI 316) substrates.

Material/Coating	Thickness and roughness parameters		
	Thickness (μm)	Roughness R_a (μm)	Roughness R_z (μm)
SS / Ta	0.72	0.018	0.280
SS / Ta_2O_5	0.98	0.011	0.083
SS / Ta/Ta_2O_5	1.0	0.008	0.055

parameters. The cross-section and surface structure of the Ta, Ta_2O_5 and Ta/Ta_2O_5 coatings, as studied by SEM, are shown in figure 1. The coatings' surfaces have smooth relief with uniform cross-section structure. The EDX spectra and the elemental compositions of the coatings are presented in figure 2.

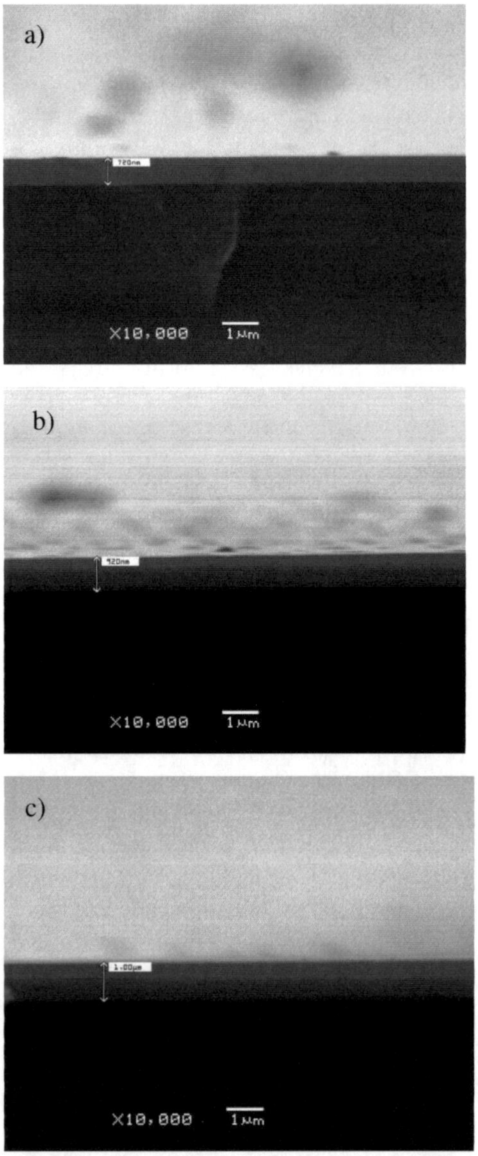

Figure 1. SEM micrographs of coatings: a) Ta, b) Ta_2O_5 and c) Ta/Ta_2O_5.

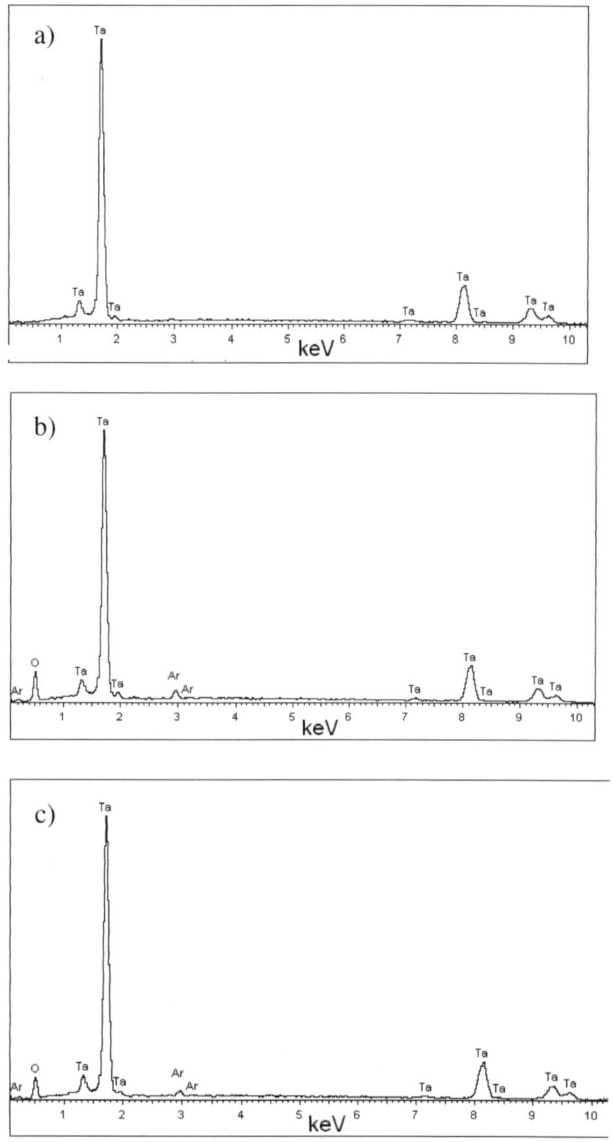

Figure 2. EDX spectra of a) Ta, b) Ta_2O_5 and c) Ta/Ta_2O_5 coatings.

The XPS data on the coatings' surface composition (figure 3) showed various Ta oxidation states. The spectrum includes the photoelectron lines for Ta (4f7/2) and Ta (4f5/2). The Ta^{+5} signals attributed to the stoichiometric oxide were detected at binding energies of 26.8 eV and 28.7 eV (figure 3a). Also, Ta^0 (Ta in its metallic form) was observed for the Ta/Ta_2O_5 bilayer coatings. The doublet peaks (figure 3b) located at binding energies of 22 eV and 26.3 eV correspond to Ta^0 (metallic Ta); and Ta^{5+} (Ta_2O_5), respectively. The O(1s) spectrum is presented in figure 3c. The peak around 530.9 eV can be assigned to the formation of Ta oxide [23, 24].

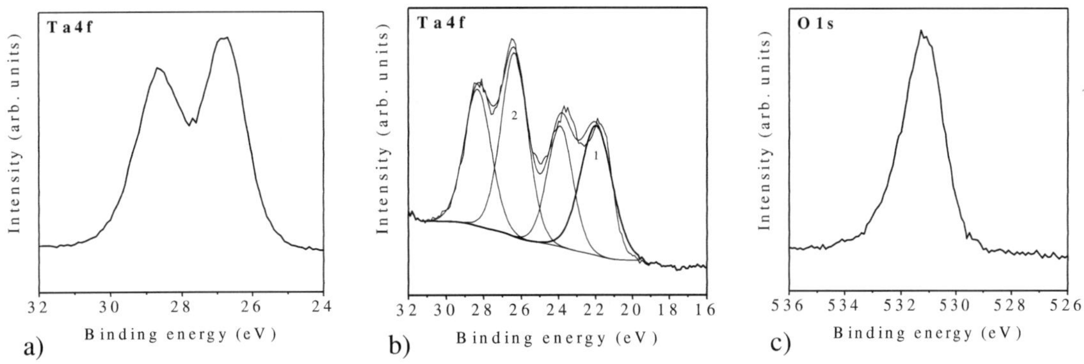

Figure 3. XPS spectra of Ta-based coatings deposited on SS substrates: a) Ta^{+5} for Ta_2O_5, b) Ta^{+5} and Ta^0 for Ta/Ta_2O_5 and c) O1s for Ta_2O_5.

An effect arising from the formation of a very thin Ta_2O_5 film on the surface was discussed in [25]. Ta is a transition metal with a strong affinity to oxygen. Oxygen could be incorporated during the deposition process, thus forming an oxide layer covering the metallic Ta. The presence of oxygen and the possible contamination in the transition layers of the samples with Ta coating may lead to a lower adhesive strength and a higher brittleness of the coatings. In some studies, a poor adherence of the Ta thin film deposited on the AISI 316 steel was reported and attributed to the large mismatch between the steel and Ta lattices. The adhesive strength to the substrate was very poor and delamination was observed as a result of the very high residual stresses. The bilayer deposition has changed the Ta films' adhesion parameters [22, 25].

The coatings' mechanical behavior is strongly correlated with their properties, such as microstructures, phase and chemical composition, surface parameters. The mechanical behavior is characterized by the ratio *H/E* [26, 27], which is proportional to the fracture toughness of the film and to the resistance of the material to plastic deformation. It means that the films with enhanced resistance to cracking and plastic deformation should have lower values of the effective Young's modulus. Materials with a high plastic resistance ratio H^3/E^2 have exhibited a higher yield strength [26]. Moreover, the formation of coatings improves the mechanical properties of the coated materials. The elastic modulus of bilayer coatings is reduced in comparison with monolayer coatings. The hardness parameters of Ta_2O_5 and of bilayer Ta/Ta_2O_5 coatings increase in comparison with those of Ta films. The mechanical advantages of multilayer formations in comparison with monolayer Ta-based coatings for long lifetime's component applications have been reported earlier [22].

The values of the adhesion strength, hardness (*H*), Young's modulus (*E*), *H/E* and H^3/E^2 of the coatings deposited on the stainless steel samples (AISI 316) substrates are summarized in table 2. The Ta-based coatings demonstrated a relatively high hardness and a low Young's modulus, corresponding to a high elastic recovery and a high resistance to cracking. We further calculated the *H/E* ratio, which is related to the wear resistance. The coatings exhibited an elastic-strain-to-failure ratio from 0.05 to 0,08

Table 2. Mechanical characteristics of Ta, Ta_2O_5 and Ta/Ta_2O_5 coatings deposited on the stainless steel (AISI 316) substrates.

Material/Coating	Mechanical parameters (averaged results of 10 tests)				
	Hardness, H_V	Young Modulus (GPa)	*H/E*	H^3/E^2 (GPa)	Adhesion (N)
SS / Ta	681.2	127.3	0.05	0.02	14.6
SS / Ta_2O_5	846.4	108.1	0.07	0.05	24.8
SS / Ta/Ta_2O_5	861.5	109.0	0.08	0.06	15.9

and a plastic resistance ratio from 0.02 to 0.05, higher than the respective values for metals (0.01) and other modern implant materials [28]. The best hardness parameters were found for the bilayer Ta/Ta_2O_5 coatings, while the highest adhesive strength was exhibited by the Ta_2O_5 monolayer.

The above results demonstrate that Ta_2O_5 and Ta/Ta_2O_5 coatings formed by reactive magnetron sputtering deposition on stainless steel (AISI 316) substrates possess structural and mechanical properties differing from those of pure Ta films. Thus, the hardness parameters of Ta_2O_5 and Ta/Ta_2O_5 coatings increase in comparison with pure Ta films, while the highest adhesive strength was shown by the Ta_2O_5 coatings. The insufficient adhesion parameters of Ta films due to possible contamination in the transition layer were improved by forming oxide and bilayer coatings.

4. Conclusions

The microstructure, surface roughness, and chemical composition were determined of Ta, Ta_2O_5 and bilayer Ta/Ta_2O_5 coatings deposited by magnetron sputtering. Increasing the films thickness and forming of a bilayer resulted in a decrease in the roughness of Ta_2O_5 and Ta/Ta_2O_5. The coated surfaces had smooth relief with uniform cross-sectional structure. The elemental compositions of Ta, Ta_2O_5 and bilayer Ta/Ta_2O_5 coatings were also investigated. The XPS spectra obtained revealed the presence of Ta^{+5} (stoichiometric oxide) and Ta^0 (metallic form) in the Ta/Ta_2O_5 bilayer coatings. The hardness parameters of the Ta_2O_5 and bilayer Ta/Ta_2O_5 coatings were higher than those of Ta films. The bilayer Ta/Ta_2O_5 coatings demonstrated a relatively high hardness and a low Young's modulus corresponding to a high elastic recovery and a high resistance to cracking. The coatings exhibited an H/E ratio in the 0.05 – 0.08 range and a plastic resistance ratio ranging from 0.02 to 0.05, higher than for metals and other modern implant materials. The best hardness parameters were found for bilayer Ta/Ta_2O_5 coatings, while the highest adhesive strength was observed in the case of monolayer Ta_2O_5. The tantalum-based coatings are promising candidates as materials for advanced implant and stent applications.

Acknowledgments

The study was supported by the Towards Intelligent Micro-Bearing Tribological Aspects Program (ImBeing-FP7-PEOPLE-2013-IRSES-612593) project and BAS-NASU Cooperation Agreement.

References

[1] Mantripragada V P, Lecka-Czernik B, Ebraheim N A and Jayasuriya A C 2013 *J. Biomed. Mater. Res.* A **101** 3349
[2] Levine B R, Sporer S, Poggie R A, Della Valle C J and Jacobs J J 2006 *Biomater.* **27** 4671
[3] Grosser M and Schmid U 2009 *Thin Solid Films* **517** 4493
[4] Hallmann L and Ulmer P 2013 *Appl. Surf. Sci.* **282** 1
[5] Pham V H, Lee S H, Li Yu, Kim H E, Shin K H and Koh Y H 2013 *Thin Solid Films* **536** 269
[6] Bernoulli D, Müller U, Schwarzenberger M, Hauert R and Spolenak R 2013 *Thin Solid Films* **548** 157
[7] Gladczuk L, Patel A, Demaree J D and Sosnowski M 2005 *Thin Solid Films* **476** 295
[8] Todorova Z, Donkov N, Ristic Z, Bundaleski N, Petrovic and Petkov P 2006 *Plasma Proc. Polym.* **3** 174
[9] Macionczyk F, Gerold B and Thull R 2001 *Surf. Coat. Technol.* **142/144** 1084
[10] Xu G, Shen X, Hu Y, Ma P and Cai K 2015 *Surf. Coat. Technol.* **272** 58
[11] Zykova A, Safonov V, Goltsev A, Dubrava T, Rossokha I, Donkov N, Yakovin S, Kolesnikov D, Goncharov I and Georgieva V 2016 *J. Phys.: Conf. Series* **700** 012027
[12] Díaz B, Światowska J, Maurice V, Seyeux A, Normand B, Härkönen E, Ritala M and Marcus P 2011 *Electrochimica Acta* **56** 10516
[13] Rahmati B, Sarhan A A D, Basirun W J and Abas W A B W 2016 *J. Alloys and Compounds* **676** 369
[14] Li Y, Zhao T, Wei S, Xiang Y and Chen H 2010 *Mater. Sci. Eng.* C **30** 1227

[15] Matsuno H, Yokoyama A, Watari F, Uo M and Kawasaki T 2001 *Biomater.* **22** 1253
[16] Miyazaki T, Kim H M, Kokubo T, Ohtsuki C and Nakamura T 2002 *Biomater.* **23** 827
[17] Georgiev G, Fescheschiev N, Popov D and Uzunov Z 1989 *Vacuum* **36** 595
[18] Leng X Y, Sun H, Yang P, Chen J Y, Wang J, Wan G J, Huang N, Tian X B, Wang L P and Chu P K 2001 *Thin Solid Films* **398/399** 471
[19] Yeh H I et al. 2006 *J. Biomed. Mater. Res.* A **76** 835
[20] Mosseri et al. 2005 *Cardiovasc Revasc. Med.* **6** 2
[21] Windecker et al. 2005 *Circulatio* **111** 2617
[22] Ma G et al. 2003 *J. Phys. Chem.* B **107** 1798
[23] Chun W et al. 2003 *J. Phys. Chem.* B **107** 1798
[24] Zier M, Oswald S, Reiche and Wetzig K 2005 *Appl. Surf. Sci.* **252** 234
[25] Li T C, Lwo B J, Pu N W, Yu S P and Kao C H 2006 *Surf. Coat. Technol.* **201** 1031
[26] Leyland A and Mathews A 2004 *Surf. Coat. Technol.* **177/178** 317
[27] Musil J, Sklenka J and Cerstvy R 2012 *Surf. Coat. Technol.* **206** 4230
[28] Jara A, Fraisse B, Flaud V, Fréty N and Gonzalez 2017 *Surf. Coat. Technol.* **309** 887

IOP Conf. Series: Journal of Physics: Conf. Series **992** (2018) 012035 doi:10.1088/1742-6596/992/1/012035

Structural and mechanical properties of hydroxyapatite coatings formed by ion-beam assisted deposition

A Zykova[1,2,6], V Safonov[1,2], S Dudin[2], S Yakovin[2], N Donkov[3], M H Ghaemi[4], M Szkodo[4], M Antoszkiewicz[4], M Szyfelbain[5] and A Czaban[5]

[1]National Science Center "Kharkov Institute of Physics and Technology", 1 Academicheskaya Str., 61108 Kharkov, Ukraine
[2]Department of Physical Technologies, Kharkov National University, 4 Svobody Sq., 61077 Kharkov, Ukraine
[3]Emil Djakov Institute of Electronics, Bulgarian Academy of Sciences, 72 Tzarigradsko Chaussee, 1784 Sofia, Bulgaria
[4]Gdansk University of Technology, 11/12 G. Narutowicza, 80233 Gdansk, Poland
[5]Gdynia Maritime University, 81/87 Morska, 81225 Gdynia, Poland

E-mail: zykova.anya@gmail.com

Abstract. The ion-beam assisted deposition (IBAD) is an advanced method capable of producing crystalline coatings at low temperatures. We determined the characteristics of hydroxyapatite $Ca_{10}(PO_4)_6(OH)_2$ target and coatings formed by IBAD using X-ray photoemission spectroscopy (XPS), X-ray diffraction (XRD), scanning electron microscopy (SEM), atomic force microscopy (AFM) and energy dispersive X-ray (EDX). The composition of the coatings' cross-section and surface was close to those of the target. The XPS spectra showed that the binding energy values of Ca (2p1/2, 2p3/2), P (2p3/2), and O 1s levels are related to the hydroxyapatite phase. The coatings demonstrate an optimal H/E ratio, and a good resistance to scratch tests.

1. Introduction

The bioactive properties of hydroxyapatite $Ca_{10}(PO_4)_6(OH)_2$ (HAP) materials (powders, ceramics, coatings) are widely applied in the clinical practice. The osteoconductive properties of HAP-coated scaffolds have a great potential for further tissue-engineering use due to the possibilities of bone tissue reconstruction [1-3]. A pressing challenge in biomedical technology is to produce medical implants with highly bioactive and antibacterial surfaces. Modern implants made of hydroxyapatite materials, namely, HAP-coated metal prostheses, bond readily to bone without rejection or inflammatory reactions [4]. Crystalline HAP coatings with homogeneous composition are preferable because of their reduced *in vivo* dissolution [5, 6]. However, the insufficient purity and uniformity of the HAP coatings, the low crystallinity of the films and their poor adhesion to the substrate materials limit the lifetime of implants with HAP coatings. Among the techniques used for HAP coatings deposition, the ion-beam assisted deposition (IBAD) stands out as an advanced method capable of producing crystalline coatings at low-temperature, while avoiding the need of high-temperature annealing for

[6] To whom any correspondence should be addressed.

further crystallization [7, 8]. The relation between HAP coatings growth mechanisms and their properties related to biomedical applications is of great interest. Moreover, the deposition conditions affect significantly the structure, composition, biomechanical properties, and HAP coatings' capacity to promote bone regeneration.

2. Materials and methods

The aim of the present work was to study the deposition conditions effects on the structure and properties of hydroxyapatite $Ca_{10}(PO_4)_6(OH)_2$ ceramic coatings. Glass and silicon samples were used as substrates. The coatings were deposited using ion-beam sputtering of a hydroxyapatite target. The ion beam was generated by a Hall-type ion source with anode voltage of 3 kV. An oxygen ion-beam with a current density of 3 mA/cm^2 bombarded the hydroxyapatite target in the vacuum chamber. The main details of the deposition technique and conditions were presented in [7].

The deposited coatings' thickness was estimated by Calotest measurements. The surface roughness was evaluated using a Hommel Werke T8000 Profilograph. The coatings' surface structure and morphology were observed by a confocal laser microscope (LEXT OLS 4100), a scanning electron microscope (Zeiss EVO MA15 SEM) and an atomic force microscope (NT-206 AFM).

X-ray photoemission spectroscopy (XPS) was carried out using an ESCALAB MkII (VG Scientific) electron spectrometer at a base pressure in the analysis chamber of 5×10^{-8} Pa (1×10^{-6} Pa during the measurements), using an AlKα X-ray source (excitation energy $h\nu = 1486.6$ eV). The pass energy of the hemispherical analyzer is 20 eV. The instrumental resolution measured as the full width at a half maximum (FWHM) of the $Ag3d_{5/2}$ photoelectron peak is 1 eV. The energy scale is corrected to the C1s-peak maximum at 285.0 eV for electrostatic charging. The processing of the measured spectra includes a subtraction of X-ray satellites and Shirley-type background [9]. The peak positions and areas are evaluated by symmetrical Gaussian-Lorentzian curve fitting. The relative concentrations of the different chemical species are determined based on normalization of the peak areas to their photoionization cross-sections, calculated by Scofield [10].

The chemical composition of the coatings was analyzed by energy dispersive X-ray (EDX) spectroscopy (X Flash 6/30, Bruker). The adhesion properties were evaluated using standard methods by a Revetest device (CSM Instrumets). The nano-indentation tests were performed using a nano-indenter fitted with a three-sided pyramid Berkovich indenter. The loading and unloading sequences were standard [11]. Repeated indentations were performed, and the values were calculated as the average of 10 indentations.

3. Results and discussion

The target phase composition was analyzed by XRD in the range θ-2θ by a DRON-3 device in FeKα radiation (table 1). The diffraction pattern is presented in figure 1.

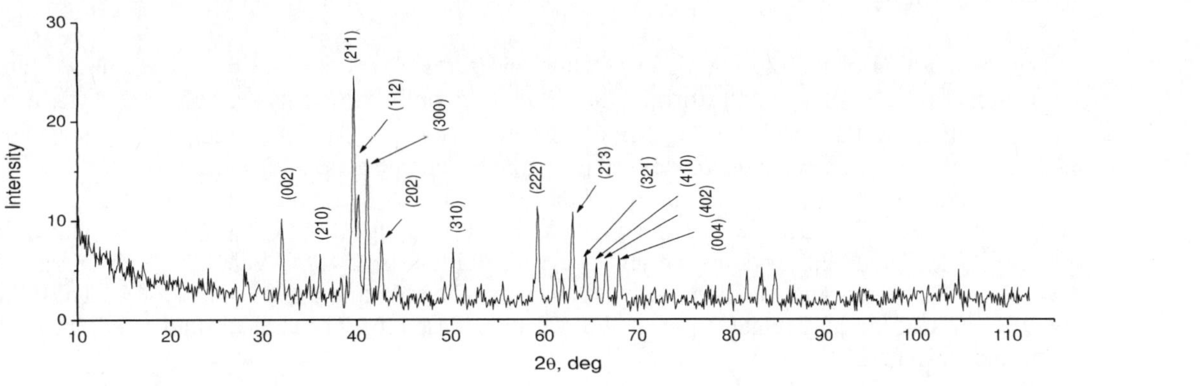

Figure 1. Diffraction peaks of the $Ca_{10}(PO_4)_6OH_2$ target composition.

IOP Conf. Series: Journal of Physics: Conf. Series **992** (2018) 012035 doi:10.1088/1742-6596/992/1/012035

The diffraction lines correspond to hydroxyapatite $Ca_{10}(PO_4)_6OH_2$ of hexagonal syngony. The main peaks are marked. The lattice parameters of the HAP target were evaluated from the position of (300) and (004) lines and were about $a = 0.955 \pm 0.001$ µm and $c = 0.693 \pm 0.001$ µm. In the case of evaluation from the (410) line, the lattice parameter was about $a = 0.924 \pm 0.001$ µm (table 1). No peaks of other compounds were detected.

Table 1. Crystal lattice parameters of the $Ca_{10}(PO_4)_6OH_2$ target.

(hkl)	2θ.deg	d, µm	a, µm	c, µm
(300)	41.10	0.27595	9.55	
(410)	65.50	0.17905	9.24	
(004)	67.97	0.17326		6.93

XRD diffractograms of coatings deposited on glass substrate were presented earlier [12]. The main peaks attributed to the HAP crystalline structure (JPCDS 9-0432) were observed, indicating that a crystalline HAP phase was obtained. This is an important result, because it has been demonstrated that amorphous or less crystalline phases are more soluble in contact with body fluids, leading to a high dissolution rate, which affects the long-term stability of an implant [13-15].

In view of characterizing in further detail the HAP coatings' elemental and stoichiometric composition, we conducted EDX and XPS analyses.

The EDX results revealed the same peaks of Ca, P in the target and in the coating's cross-section and surface (figure 2). The target composition was stoichiometric, with a Ca/P ratio of about 1.67.

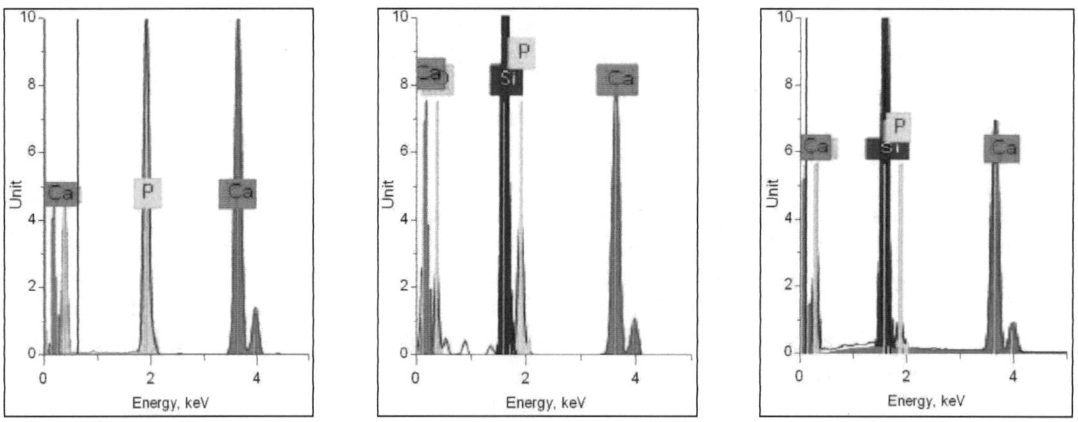

a) b) c)

Figure 2. EDX spectra of $Ca_5(PO_4)_3(OH)$: a) target, b) cross-section, c) surface of coatings deposited on Si substrate.

The composition of the coating's cross-section and surface was close to that of the target. The main elements (Ca, P) were uniformly distributed in both the cross-section and surface of coating, indicating the formation of a homogenous structure. The XPS spectra of O 1s, Ca 2p, and P 2p core levels were recorded in view of analyzing the chemical bonding (figure 3).

The binding energy of O 1s attributed to the PO_4 bond of hydroxyapatite was 531.6 eV. The binding energy of Ca 2p3/2 was 347.8 eV and that of Ca 2p1/2 was 350.7 eV, all related to hydroxyapatite. The P 2p spectrum presented a peak for the p3/2 level with a binding energy of 133.6 eV, which corresponds to P in the hydroxyapatite. The binding energy values of Ca 2p, P 2p, and O1s levels are strongly related to the hydroxyapatite phase [16]. The surface atomic concentration is presented in table 2.

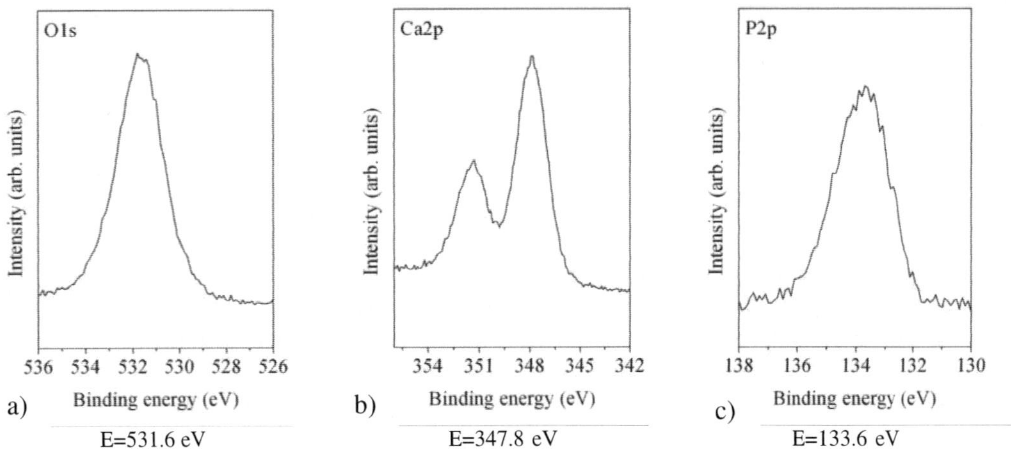

Figure 3. XPS spectra: a) O 1s, b) Ca 2p, c) P 2p core levels.

The surface topography and morphology of the coatings of $Ca_{10}(PO_4)_6(OH)_2$ were observed by SEM and AFM (figure 4). The main parameters of the coatings – thickness, adhesion strength, and surface roughness are presented in table 3.

The nano-indentation tests were conducted at different loads using a constant loading rate. The loading–unloading curves were used to determine the hardness H and the elastic modulus E (table 4).

Table 2. Surface atomic concentration.

Sample\Elements	O 1s, at. %	Ca 2p, at. %	P 2p, at. %
$Ca_{10}(PO_4)_6OH_2$	53	29	18

Table 3. Coating thickness, adhesion strength and surface roughness.

Coating	Thickness, μm	Adhesion strength, N	Roughness Ra, μm	Roughness Rz, μm
$Ca_{10}(PO_4)_6OH_2$	1.8	9.31	0.077	0.602

Table 4. Coating hardness H, elastic modulus E, and plastic resistance ratio parameters.

Coating	H, GPa	E, GPa	H/E	H^3/E^2, GPa
$Ca_{10}(PO_4)_6OH_2$	6.21±0.97	108.6±4.1	0.057	0.021

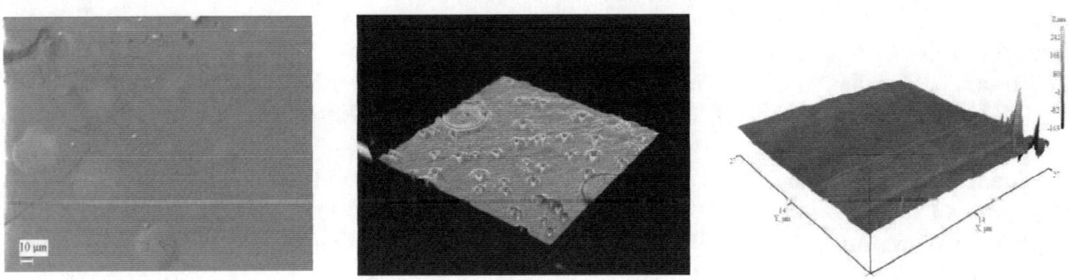

a) b) c)

Figure 4. Surface roughness and morphology of $Ca_{10}(PO_4)_6(OH)_2$ coatings: a) SEM image of a coating surface, b) 3D confocal laser microscope image of the coating's surface, c) 3D AFM image of the coating's surface.

HAP has a higher elastic modulus than natural bone. The difference in this parameter between bone and coating results in greater stress concentration and fractures at the bone-implant interfaces [17].

The H/E ratio can be attributed to a good wear resistance of the material and demonstrates the elastic strain to failure [18]. Materials with a high plastic resistance ratio H^3/E^2 have exhibited a higher yield strength [19]. An optimal H/E ratio results in increased durability of the surface of biomaterials.

The adhesion of the CaP coating to the surface of implants and scaffolds is a very important parameter, which affects the performance and longevity of the medical devices. Adhesion failures result in delamination of the coating and loosening of the implant, thus leading to mechanical failure and bone resorption. Various treatment procedures have been realized to improve the adhesion strength of hydroxyapatite coatings on substrates. Scratch test results have shown that the adhesion strength of the HAP coating on substrates was in the range 2.60 N – 3.14 N depending on the treatment conditions [20]. The effect of an oxide interlayer on the adhesion strength of HAP coating on oxidized Co-Cr-Mo substrate was investigated in [21]. The results demonstrated that the formation of interlayers improves the adhesion strength between coatings and metals in the range 1.40 N – 1.04 N. In the present study, we found an adhesion strength of hydroxyapatite coatings to the substrates of about 9.31 N, a value higher than those reported earlier.

The results presented in the study confirmed that IBAD is a highly reliable, reproducible technique for creating thin, defect-free HAP coatings. The method allows accurate control of the HAP growth parameters at low deposition temperatures with capability of producing highly crystalline coatings [22]. The sputter cleaning process removes contaminants and gas molecules adsorbed on the surface of the substrate to produce a clean, highly-active surface. Furthermore, the process allows superior control over the coatings' microstructure and chemical composition [23].

4. Conclusions

The results demonstrate the strong effect of the sputtering process conditions on the coating structural and surface properties. IBAD using allows one to produce high-quality coatings due to the high-energy ion bombardment process. The general conclusions about the quality of hydroxyapatite $Ca_{10}(PO_4)_6(OH)_2$ coatings were drawn on the basis of the results from EDX, XPS, AFM, SEM measurements. According to the energy-dispersive X-ray analysis, the Ca/P ratio of the deposited HAP coatings reaches values similar to those of the HAP target. The adhesion strength of the hydroxyapatite coatings has values higher than those reported previously. The coatings possess an optimal H/E ratio, a good resistance to scratch tests, and improved biomechanical properties for further advanced biomedical applications.

Acknowledgments

The study was supported under the Towards Intelligent Micro-Bearing Tribological Aspects Project (ImBeing-FP7-PEOPLE-2013-IRSES-612593) and a BAS-NASU Cooperation Agreement.

References

[1] Russo A, Bianchi M, Sartori M, Boi M, Giavaresi G, Salter DM, Jelic M, Maltarello M C, Ortolani A, Sprio, Fini M, Tampieri A and Marcacci M 2017 *Biomed. Mater. Res.* B doi: 10.1002/jbm.b.33836
[2] Huang Y, Zhang X, Zhang H, Qiao H, Jia T, Han S, Gao Y, Xiao H and Yang H 2017 *Ceramics Int.* **43** 992
[3] Zykova A, Safonov V, Smolik J, Rogowska R, Donkov N, Goltsev A and Dubrava T 2012 *J. Phys.: Conf. Series* **356** 012040
[4] Liu H and Webster T J 2007 *Biomater.* **28** 354
[5] Baltag I, et al. 2000 *Biomed. Mater. Res.* **53** 76
[6] Yang Y, Kim K H and Ong J 2005 *Biomater.* **26** 327
[7] Donkov N, Zykova A, Safonov V, Kolesnikov D, Goncharov I and Yakovin S 2014 *J. Phys.: Conf. Series* **514** 012017
[8] Ozeki K, Goto T, Aoki H and Masuzawa T 2015 *Biomed. Mater. Eng.* **26** 139
[9] Shirley D 1972 *Phys. Rev.* B **5** 4709

[10] Scofield J H 1976 *Electron Spectrosc. Relat. Phenom.* **8** 129
[11] Fischer-Cripps A C 2011 *Nanoindentation* p 280 (New York, Springer)
[12] Dudin S, Cotrut C M, Dinu M, Zykova A, Parau A C, Yakovin S and Vladescu A 2017 *Ceramics Int.* doi: 10.1016/j.ceramint.2017.08.016
[13] Zykova A, Safonov V, Yanovska A, Sukhodub L, Rogovskaya R, Smolik J and Yakovin S 2015 *Open Biomed. Eng.* **9** 75
[14] Zhang Q, Chen J, Feng J, Cao Y, Deng C and Zhang X 2003 *Biomater.* **24** 4741
[15] Xue W, Tao S, Liu X, Zheng X and Ding C 2004 *Biomater.* **25** 415
[16] Rulis P, Ouyang L and Ching WY 2004 *Phys. Rev.* B **70** 155104
[17] Roop R, Kumar R and Wang M 2002 *Mater. Sci. Eng.* A **338** 230
[18] Leyland A and Matthews A 2000 *Wear* **246** 1
[19] Musil J 2000 *Surface and Coating Technol.* **125** 322
[20] Grubova I, Priamushko T, Surmeneva M, Korneva O, Epple M, Prymak O and Surmenev R 2017 *J. Phys.: Conf. Ser.* **830** 012109
[21] MasAyu H, Izman S, Daud R, Krishnamurithy G, Shah A, Tomadi H and Salwani MS 2017 *Procedia Eng.* **184** 399
[22] Cui F and Luo Z. 1999 *Surf. Coat. Technol.* **112** 278
[23] Ozeki K, Yuhta T, Fukui Y and Aoki H. 2002 *Surf. Coat. Technol.* **160** 54

Elemental composition and structural characteristics of as-received TriTanium^{TM} orthodontic archwire

I Ilievska[1,5], V Petrov[2], V Mihailov[1], S Karatodorov[1], L Andreeva[2], A Zaleski[3], V Mikli[4], M Gueorgieva[2], V Petrova[1] and A Stoyanova-Ivanova[1]

[1]G. Nadjakov Institute of Solid State Physics, Bulgarian Academy of Sciences, 72 Tzarigradsko Chaussee, 1784 Sofia, Bulgaria
[2]Faculty of Dental Medicine, Medical University of Sofia, St. G. Sofiiski Blvd., 1431 Sofia, Bulgaria
[3]Institute of Low Temperatures and Structure Research, Polish Academy of Sciences, AS, 50-422 Wroclaw, Poland
[4]Institute of Materials and Environmental Technology, Tallinn University of Technology, 19086 Tallinn, Estonia

E-mail: ivannyilievska@gmail.com

Abstract. Orthodontic archwires are among the most important devices of fixed orthodontic therapy. Many types of archwires are made available on the market by various manufacturers with different elemental composition and structural characteristics. Knowing this information is important when choosing a suitable archwire for a particular stage of orthodontic treatment. The aim of our study is to characterize a new type orthodontic archwires (TriTanium^{TM}, American Orthodontics) before their placement in the oral cavity. To achieve the aim, we used modern methods for determining their elemental composition and structural characteristics: laser-induced plasma spectroscopy (LIBS), X-ray diffraction analysis (XRD), scanning electronic microscopy (SEM), energy dispersive X-ray spectroscopy (EDX) and differential scanning calorimetry (DSC). The results obtained from the qualitative elemental analysis by LIBS and the quantitative elemental analysis by EDX showed that Ni and Ti are the main elements in the archwire studied. The room-temperature XRD patterns showed peaks typical for a Ni-Ti alloy with an austenite-type structure. Monitoring the phase transitions by means of DSC measurements in the temperature range from −50 °C to +50 °C, we showed that in TriTanium^{TM} archwires, besides the austenite to martensite transition, there exists a rhombohedral intermediate phase (R phase). This study will be useful in assisting orthodontists in applying appropriate nickel-titanium orthodontic archwires in the clinical practice.

1. Introduction

Fixed orthodontic therapy is based on inducing tooth movement by generating light and continuous forces. Orthodontic archwires are key elements in producing such forces; when selecting them, clinicians should take into consideration both their surface/morphology characteristics and their properties. In the intraoral cavity there exist complex factors that may influence the biomechanical behavior of the archwires during treatment [1]. Nickel-titanium alloys are among the most widely used; their clinical advantages in orthodontics arise from the fact that they can exist in two different

[5] To whom any correspondence should be addressed.

crystal structures: martensite (low-temperature, stress-free phase, with a low-symmetry crystal structure, such as tetragonal, rhombohedral, orthorhombic, monoclinic, or triclinic), and austenite (high-temperature, stress-free phase, with a high-symmetry crystal structure usually based on a cubic lattice) [2]. The two different phases of nickel-titanium are responsible for two clinically significant properties of nickel-titanium: shape memory and super-elasticity [3]. Most of the researchers claim that the manufacturing process allows for more consistent values of the transition temperatures, thus providing controlled force delivery individualized for each patient [4]. The shape memory is resulting from the thermal behavior of the archwires and is closely related to the transition temperature range (TTR) of each alloy. The shape-memory (thermally-activated) Ni-Ti archwires present a great variability in the TTR; therefore, the elastic parameters of each Ni-Ti archwire should be provided by the manufacturers to allow the achievement of the best clinical performance possible [5].

A new Ni-Ti based archwire was recently marketed, namely TriTanium™ (American Orthodontics). It was first used in orthopedics as an outer layer for joint implants, because of its good osteo-compatibility due to its high porosity [6]. The valuable properties of this alloy in orthodontics are the three elasticity zones in the anterior, mid-region and posterior segments. The teeth in those segments have different root systems; thus, an archwire applying different forces to those segments is more effective, especially in the initial phase of the fixed appliance treatment. It is believed that multi-force orthodontic archwires minimize the incidence of root resorption and improve patient comfort [7].

The insufficient literature and manufacturing data on TriTanium™ archwires motivated us to investigate the structure, chemical composition, morphology and thermal transformation temperatures of as-received commercially-available TriTanium™ archwires in the three elasticity zones. This will enable us to compare our results to the results of studies of used TriTanium™ archwires and reveal any changes in their structure and chemical composition.

2. Materials and methods

An as-received orthodontic TriTanium™ archwire (dimension 0.016×0.022 in) was cut in three pieces corresponding to the anterior (A-B), mid-region (B-C) and posterior (C-D) segments (figure 1). The use of the methods such as LIBS, EDX, SEM, DSC and XRD for investigation of materials was a key point of the work performed that did not only deal with theoretical aspects, but also had a strong experimental contribution.

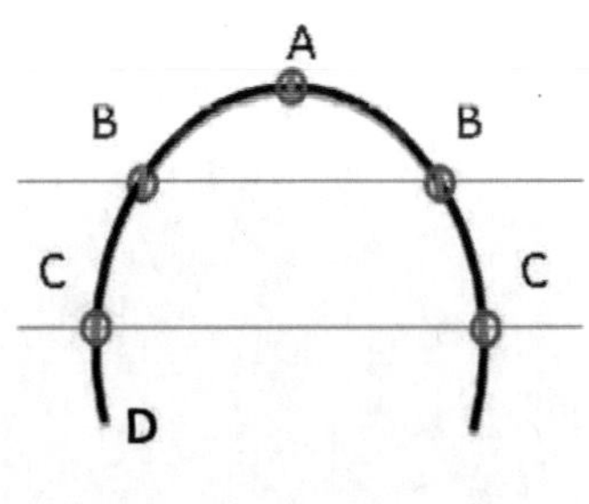

Figure 1. TriTanium™ archwire cuts in the anterior, mid-region and posterior segments.

Laser-induced plasma spectroscopy was used for qualitative determination of the archwires' elemental composition. LIBS allows express elemental analysis of a wide range of samples without preparation. The method is implemented by focusing a powerful laser beam on the sample (dental braces) by a lens of 25 cm focal length. The laser source was a pulsed nanosecond Nd:YAG laser (Quanta Ray GCR3) generating a wavelength of 1064 nm at a repetition rate of 1 Hz and pulse energy 30 mJ. The focused laser beam ablates the sample, yielding laser-induced plasma. The light emitted by the plasma is transmitted through an optical fiber to the registration system – a Mechelle Spectrograph 5000 equipped with an Andor iStar ICCD camera controlled by the Andor SOLIS 4.13 specialized software. Analyzing the plasma spectrum and identifying spectral lines specific to each element make it possible to obtain information about the elements in the target. To quantify the archwires' elemental composition, we used EDS on a Bruker Esprit 1.82 system (accelerating voltage of 20 kV, relative error 0.5 – 1 wt %). The surface morphology and elemental composition were studied using a ZEISS FESEM Ultra 55 SEM (acceleration voltage 4.0 kV). The crystalline structure was assessed by a Bruker D8 Advance powder diffractometer with a Cu-Kα target, within the range from 5 – 80° 2θ at a constant step 0.02° 2θ. The phase transformation characteristics of the archwires segments were studied using a Perkin-Elmer-

8000 DSC device (temperature range –170°C – +600 °C) with a model PE-TGA4000 TGA attachment. Before introducing the sample for each individual test, the DSC apparatus was calibrated with indium. The samples were scanned from –50 °C to +50 °C at a rate of 20 °C /min for heating and cooling.

3. Results and discussions

The LIBS spectra in the range of wavelengths 360 – 370 nm of the as-received TriTanium^{TM} orthodontic archwire are presented in figure 2.

To determine the presence of each element in the sample, the registered spectra are compared to the spectra of pure elements. As seen in figure 2, the sample is made of a Ni-Ti alloy.

The results are qualitative and indicate the presence or absence of an element in the segments examined. The data of the EDS quantitative elemental analysis of the as-received TriTanium^{TM} archwire in all segments are presented in table 1. As seen, the as-received archwire contains pure nickel and titanium with no significant difference in the concentration in the three regions of the archwire. The SEM images (figure 3) also show no significant differences on the surface of the segments. The morphology is quite similar for the three segments; one can also notice the presence of porosity, demonstrating an irregular surface that can influence the archwires properties during treatment.

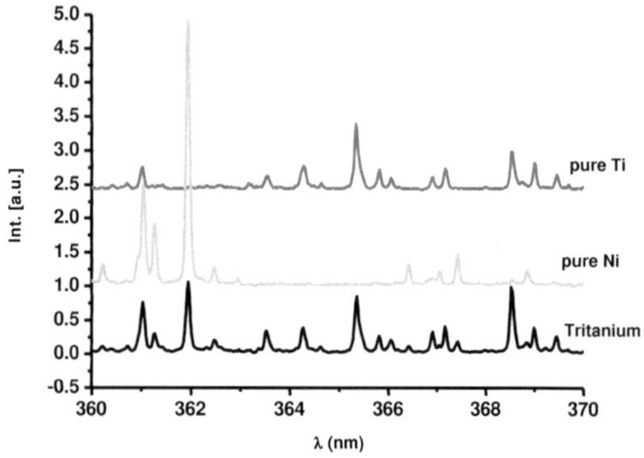

Figure 2. LIBS spectrogram of the TriTanium^{TM} archwire investigated.

Table 1. Elemental content of the TriTanium^{TM} archwire investigated.

TriTanium^{TM}	Region			
	Anterior (A-B)	mid-region B-C	Posterior C-D	Error
Ti	45.50 wt %	45.60 wt %	46.30 wt %	±0.8 %
Ni	54.50 wt %	54.40 wt %	53.70 wt %	±0.8 %

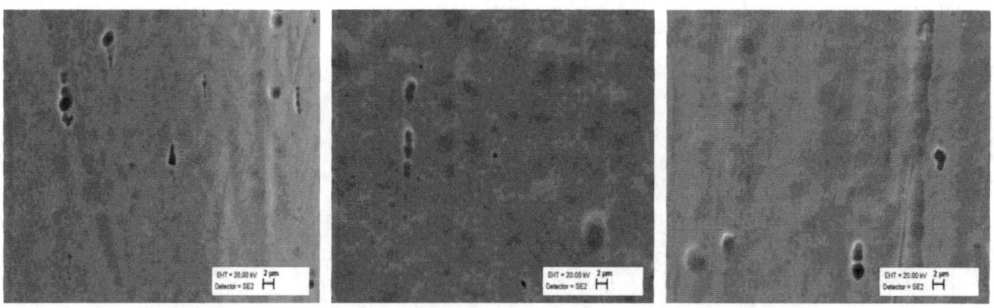

Figure 3. SEM images of investigated TriTaniumTM archwire.

The crystalline structure of the TriTanium^{TM} archwire assessed by XRD made at room temperature (figure 4) showed a cubic structure with parameters $a = b = c = 3.1323$, $ß = 90^{o}$, for all segments. Besides the austenite crystal structure, a small degree of amorphization of the material was seen, which can be due to contamination.

We further performed DSC analyses to determine the thermal behavior of the archwire's three segments in the temperature range of–50 °C to +50 °C, to detect shifts in the transformation temperatures of finished orthodontic archwires that are strained to different extent. This may serve as a marker of the changes and the applicability period of the archwires used by patients. Table 2 shows the measured Austenite, R-phase, Martensite temperatures and the enthalpy changes for phase transformation during cooling and heating within the three segments of the TriTanium^{TM} archwire. The anterior and mid-region segments exhibited higher austenitic temperatures as compared to the posterior segments with final austenitic temperature (Af) ranging from 25.55 °C to 20.82 °C, which is beneficial in the early stages of orthodontic treatment.

Table 2. DSC results within the three segments of the TriTaniumTM archwire.

Wire Brand	Wire Segment	Heating process			
		As temp (°C)	**Peak temp (°C)**	**Af temp (°C)**	**Enthalpy (J/g)**
TriTanium^{TM}	anterior (A-B)	16.16	21.83	25.55	8.5680
	mid-region (B-C)	11.92	18.18	21.23	8.5100
	posterior (C-D)	8.26	14.71	20.82	6.4765

Wire Brand	Wire Segment	Cooling process							
TriTanium^{TM}		**Rs (°C)**	**Peak (°C)**	**Rf (°C)**	**Enthalpy (J/g)**	**Ms (°C)**	**Peak (°C)**	**Mf (°C)**	**Enthalpy (J/g)**
	anterior(A-B)	12.99	9.29	6.93	-1.5011	-21.07	-28.97	-36.80	-3.9354
	mid-region (B-C)	12.76	9.78	6.57	12.89	-30.93	-36.88	-46.17	-2.8194
	posterior (C-D)	12.28	8.05	3.17	-0.8698	-38.95	-43.61	-35.11	-0.3484

4. Conclusions

The results obtained by LIBS provided qualitative information on the chemical composition and showed that the sample investigated is made of a Ni-Ti alloy. The EDS quantitative elemental analysis showed that the TriTanium^{TM} orthodontic archwire contain pure nickel and titanium with no significant difference in the concentration in the three regions of the archwire. The room temperature XRD patterns showed peaks typical for a Ni-Ti alloy with austenite-type structure. The phase transitions were studied by means of DSC in the temperature range of –50 °C –+50 °C. The different force values obtained depend on the archwire segment (anterior, mid-region, posterior). The anterior and mid-region segments show higher austenitic temperatures as compared to the posterior segments; with final austenitic temperature (Af) ranging from 25.55 °C to 20.82 °C, thus presenting benefits for the early stages of orthodontic treatment. We plan comparative analyses of the above results and future results obtained on used samples, in order to provide guidelines for a more precise application of TriTanium^{TM} orthodontic archwires in the clinical practice.

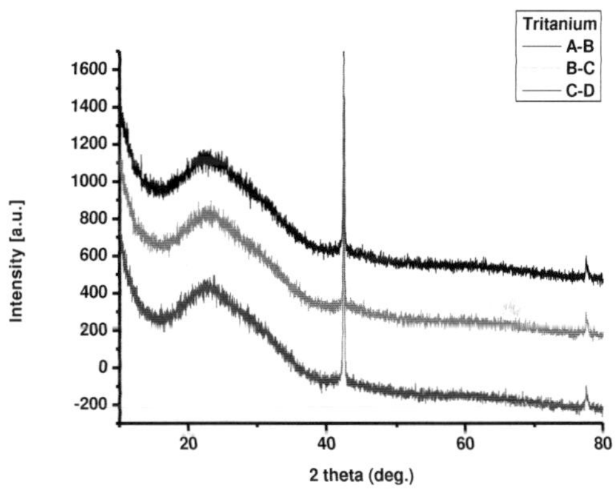

Figure 4. XRD patterns of TriTanium^{TM} archwire in all segments.

Acknowledgments

This work was part of a bilateral project between the Bulgarian Academy of Sciences and Estonian Academy of Science, Tallinn University of Technology (Estonian projects TAR16016 and IUT-T4).

References

[1] Brantley W, Berzins D, Iijima M, Tufekçi E and Cai Z 2017 *A Clinical Guide* **1** 3–38
[2] Marcelo F, Marco L and Borges P 2012 *Dental Press J. Orthod.* **17** 3
[3] Miura F, Mogi M, Ohura Y and Hamanaka H 1988 *Am. J. Orthod. Dentofacial Orthop.* **2** 89-96

[4] Shaw J A, Churchill C B and Iadicola M A 2008 *Mat. Sci. and Eng.* 6th edition
[5] Sobottka T, Pinelli F, Hermont R, Salvatore K and Jardim D 2014 *J. Appl. Oral Sci*. **22**/2
[6] Levine B 2008 *Adv. Eng. Mat.* **10** 788–92
[7] Sudershan A and Mehta K 2015 *Master Thesis* Marquette University

Triblock copolymer micelles as templates for preparation of mesoporous niobia thin films

R Georgiev[1], D Christova[2], L Todorova[2], B Georgieva[1], M Vasileva[1], Ch Novakov[2] and T Babeva[1,3]

[1]Acad. J. Malinowski Institute of Optical Materials and Technologies, Bulgarian Academy of Sciences, Acad. G. Bonchev Str., Bl. 109, 1113 Sofia, Bulgaria
[2]Institute of Polymers, Bulgarian Academy of Sciences, Acad. G. Bonchev Str., Bl. 103-A, 1113 Sofia, Bulgaria

E-mail: babeva@iomt.bas.bg

Abstract. The micelle formation of commercially available Pluronic PE 6800 (PEO-PPO-PEO triblock copolymer) was studied by means of a dye solubilization method, and the solubility of 1,6-diphenyl-1,3,5-hexatriene in aqueous solutions of the copolymer with concentrations ranging from 0.05 to 20 % w/v was studied by recording the transmission and emission spectra of the solutions using UV-VIS and photoluminescence spectroscopy, respectively. Further, mesoporous Nb_2O_5 films using PE 6800 micelles templates were obtained through sol-gel using spin-coating; their morphology, structure and optical properties were studied by TEM, SAED and nonlinear fitting of reflectance measurements. The possible application of the films in optical sensing of vapors was demonstrated and discussed.

1. Introduction

The critical micelle concentration (CMC) is defined as the concentration at which thermodynamically stable polymolecular aggregates, the so-called micelles, start forming. The copolymer studied in this work was commercially available Pluronic PE 6800 (BASF), which consists of a central poly(propylene oxide) (PPO) block flanked by two poly(ethylene oxide) (PEO) blocks with formula $PEO_{72}PPO_{28}PEO_{72}$. Pluronics are regarded as amphiphilic copolymers, where PEO is the hydrophilic part of the macromolecule and PPO, the hydrophobic segment. The micellization of amphiphilic block copolymers is more complex than that of low-molecular weight surfactants and is usually characterized by a lack of a sharp CMC or a critical micelle temperature (CMT) [1]. There are several techniques for CMC determination based on plotting a suitable physical property as a function of the polymer concentration, usually including measurements of absorption, emission, conductivity, size etc. [2, 3]. Among the most commonly used ones is the so-called dye solubilization method, whereby a hydrophobic dye is solubilized in the hydrophobic micelle interior, giving characteristic peaks, whose intensities are plotted versus the polymer concentration, the CMC being determined from the abrupt change in the curve [4].

In the work reported, we used a different approach. Instead of the intensity, we monitored the wavelength shift of the characteristic absorption and emission peaks of the hydrophobic dye 1,6-

[3] To whom any correspondence should be addressed.

diphenyl-1,3,5-hexatriene (DPH) dissolved in aqueous solutions of the studied copolymer with different concentrations, thus determining the CMC from the abrupt change in the curve. The micelle solution was further used as a template for fabrication of mesoporous Nb_2O_5 films following the evaporation-induced self-assembly method [5]. The morphology, optical properties and sensing response to vapors of these films were compared with properties of porous Nb_2O_5 films templated with polymers whose concentration was below the CMC. Accordingly, we demonstrate and discuss the role of the micelles in the formation of porosity.

2. Experimental part

In order to study the micelles formation, Pluronic PE 6800 was dissolved in deionized water in concentrations from 0.05 to 20 % w/v and the hydrophobic fluorescence probe 1,6-diphenyl-1,3,5-hexatriene (DPH) dissolved in methanol in a concentration of 0.4 mM was added into copolymer solutions (25 µl DPH in 2.5 ml solution) and left for at least three hours in dark to equilibrate [3]. Transmission and emission spectra of the copolymer solutions containing DPH were measured using a Cary 5 UV-VIS-NIR spectrophotometer (Varian) and a Fluorolog 3-22 spectrofluorimeter (Horiba JobinYvon), respectively. The hydrodynamic diameter of micelles was studied by dynamic light scattering (DLS) using a Nano ZS Zetasizer (Malvern).

Porous Nb_2O_5 thin films were deposited by spin coating of niobium sol mixed with Pluronic PE 6800 solutions (5 and 20 % w/v) in a volume ratio of 5 to 1 and annealed ata temperature of 320 °C for 120 and 180 min in the cases of 5 and 20 % w/v, respectively. The Nb sol was synthesized using $NbCl_5$ and ethanol according to the recipe in [6]. The effective optical constants (refractive index, n, and extinction coefficient, k, along with the film thickness, d, were determined through non-linear curve fitting of reflectance spectra measured at normal light incidence by the UV-VIS-NIR spectrophotometer [6]. The porosity level in the films was evaluated using both the Bruggemann effective medium theory and the calculated effective refractive indices [7], and additionally visualized by TEM images of the films. The films for TEM analyses were prepared on soluble substrates of NaCl. After dissolution of the substrates in water, the films were transferred on Cu grids for direct observation by TEM (JEOL JEM 2100). The analyte adsorption was studied by measuring the reflectance spectra of the films prior to and after exposure to probe molecules (acetone in our case) in a homemade cell and saturator system [8]. The reflectance spectra measured were further used to determine the refractive index change due to acetone vapor adsorption [7].

3. Results and discussion

To study the micelles formation in a PE 6800 aqueous solution, we used three methods: UV-VIS spectroscopy, photoluminescence and DLS measurements. The idea was to find the concentration at which some property of the copolymer solution (absorbance, emission or particle size) undergoes a sharp change.

In the first two methods, a hydrophobic fluorescence probe (DPH) is used, which has different absorbance and fluorescence in hydrophilic and hydrophobic environments, thus providing a sensitive indicator of micelle formation. The fact that micelles can solubilize relatively large amounts of compounds of poor solubility in water has been used to determine the onset of micelle formation [9]. Figure 1(a) shows typical absorbance of DPH in water and methanol. The influence of the environment is obvious – a distinctive change in the spectra is seen – from one weak peak in water centered at 310 nm to three distinctive peaks in methanol. As Pluronic concentration is increased, solubilization of the dye takes place: the peak at 310 nm shifts toward the longer wavelengths and the characteristics peaks of DPH in methanol become more pronounced (figure 1 (b)). Instead of monitoring the absorption values at 357 nm, as generally accepted, we plotted the wavelength position of the characteristic absorption peak of DPH in water as a function of the copolymer concentration (figure 1(c)). We believe that we thus improved the sensitivity of micelles formation detection, because no substantial amount of DPH needs to be solubilized in order for the characteristics peaks to appear. As seen, the dependence exhibits an abrupt change at PE 6800 concentration of 13.1 %, which we defined as the CMC value.

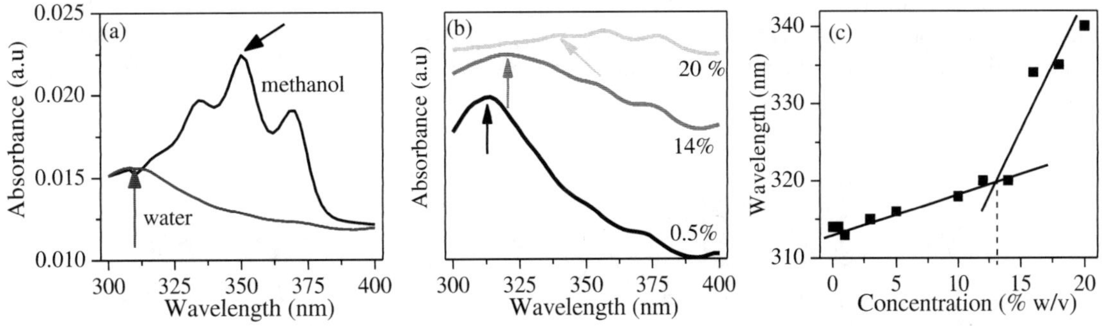

Figure 1. (a) DPH absorbance in water and methanol (25 μl of 0.4 mM stock solution of DPH in methanol is added in 2.5 ml water or methanol); (b) PE 6800 aqueous solutions absorbance with concentrations as denoted mixed with DPH (25 μl DPH stock solution in 2.5 ml aqueous copolymer solution); (c) wavelength position of the DPH characteristic absorption peak in a PE 6800 aqueous solution as a function of the copolymer concentration.

To confirm the CMC values obtained from UV-VIS spectroscopy, we used photoluminescence measurements of DPH in aqueous PE 6800 solutions with different concentrations (figure 2(a)). Instead of tracking the emission intensity with the copolymer concentration, we monitored the shift of the emission peak at 441 nm that is characteristic for DPH in water.

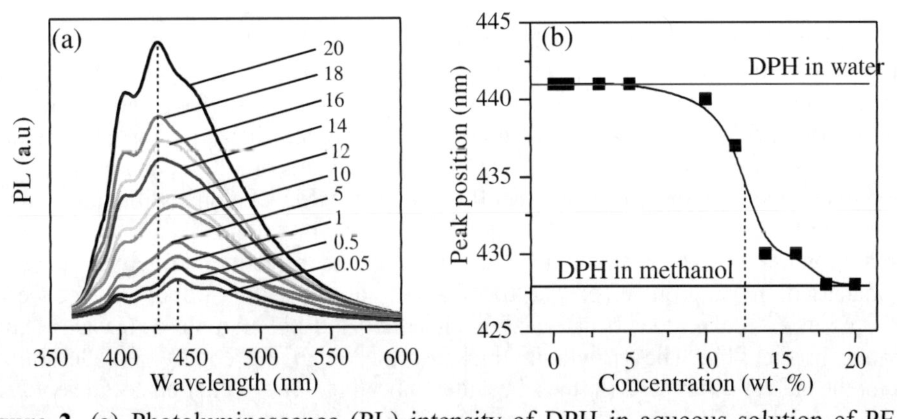

Figure 2. (a) Photoluminescence (PL) intensity of DPH in aqueous solution of PE 6800 with concentrations as denoted in % w/v; (b) peak position of the maximum PL intensity as a function of the PE 6800 concentration.

As the copolymer concentration is raised, the peak shifts toward the shorter wavelengths; at 20 % w/v it matches the characteristics peak of DPH in methanol (428 nm). The plot of peak position as a function of the PE 6800 concentration is presented in figure 2(b). The CMC can be determined from the inflection point of the curve. The value obtained is 12.6 %, which is very close to the value of CMC as determined by UV-VIS spectroscopy (13.1 %).

The DLS measurements conducted (figure 3) confirmed the presence of particles with a hydrodynamic diameter from 4 to 25 nm. The micelles size in PE 6800 (20 % w/v) gradually increased with the temperature, starting from 3 nm and reaching a size of 12 nm. For PE 6800 (5 % w/v) a sharp increase of the size was observed at temperatures above 40 °C, reaching the highest value of 27 nm at 50 °C, where the size began decreasing. At all studied temperatures, a mono-modal distribution of micelle size was observed, except for the temperature range around 40 °C, where micelles with a bi-modal distribution existed.

The next step in our investigation included preparation of mesoporous Nb_2O_5 films using a PE 6800 micelle solution as a template. To check the role of the micelles in the pore formation, we prepared films templated with copolymer aqueous solutions with concentrations 5 and 20 % w/v – one below and another one above the CMC, respectively.

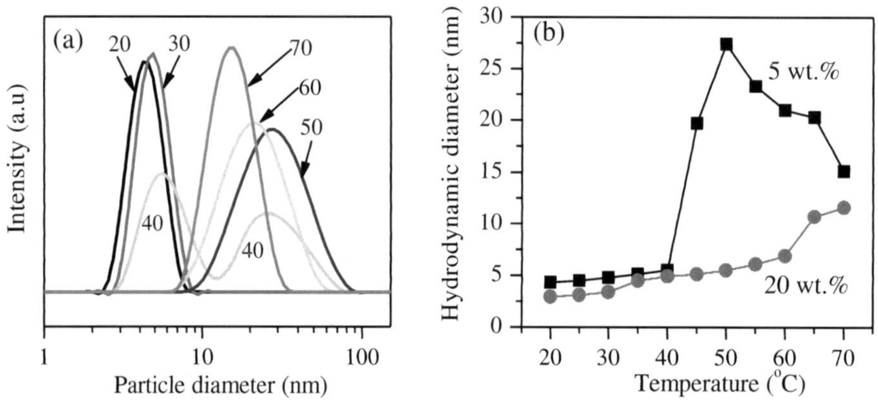

Figure 3. (a) Dynamic light scattering curves (DLS) of 5 % w/v aqueous solution of PE 6800 measured at different temperatures denoted by the numbers on the graph; (b) calculated hydrodynamic diameter versus the temperature for 5 and 20 % w/v aqueous solutions of PE 6800.

As seen in figure 4, adding PE 6800 generated porosity in the thin films, which more pronounced at a copolymer concentration of 20 % w/v. The approximate size of the pores was 3-4 nm in the case of 5 % w/v and 7-8 nm in the case of 20 % w/v. In both cases, the porosity was non-periodic. Selected area electron diffraction (SAED) (inset in figure 4(a)) confirmed that all samples were amorphous. This is in line with our previous results showing amorphous status at 320 °C and crystalline at 450 °C [10]. The refractive index and extinction coefficient of the films, along with their thickness, are presented in table 1.

Adding Pluronic PE 6800, resulted in a decrease in the effective refractive index, while the thickness increased, most pronounced for the films obtained with a copolymer concentration of 20 % w/v. This can be expected, because after annealing the polymer degrades and leaves empty spaces (pores) in the film. The growth in thickness is due to the copolymer addition and is an indication for the lack of contraction in the films after annealing. The extinction coefficient values were

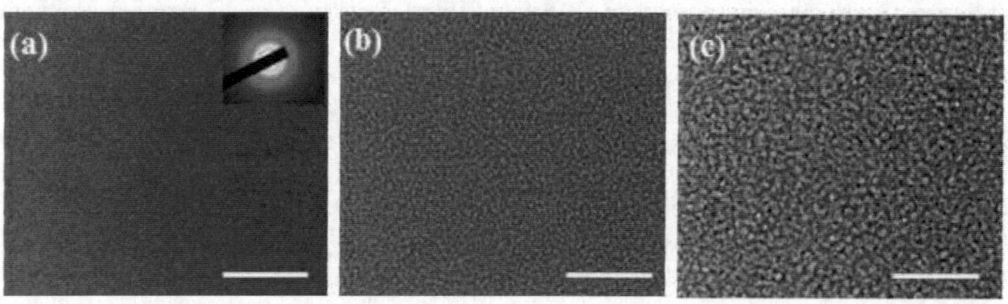

Figure 4. TEM and SAED (inset) images of Nb_2O_5 thin films prepared using PE 6800 aqueous solution as template with concentration of (a) 0, (b) 5 % w/v and (c) 20 % w/v(the bar is 100 nm).

Table 1. Refracive index (n) and extinction coefficient (k) at wavelength of 600 nm, thickness (d), volume fraction of pores (f_{air}), change in n after exposure to acetone (Δn_{ac}) and volume fraction of acetone (f_{ac}) in the Nb_2O_5 films deposited using different concentration of Pluronic PE 6800 as a template.

PE conc.(% w/v)	n_{600}	k_{600}	d (nm)	f_{air} (%)	Δn_{ac}	f_{ac} (%)
0	2.18	0.020	64	0	0	0
5	1.87	0.018	80	24	0.0015	0.4
20	1.49	0.019	146	54	0.028	6.7

the same for all samples. The refractive index dispersion curves calculated were further used to determine the volume fraction of air in the films [7]. In the calculations we assumed the film without template as being the dense phase. The data in Table 1 shows that f_{air} increased from 24 % to 54 % when a copolymer with concentration of 5 or 20 % w/v was used for template, respectively. The refractive index change due to adsorption of acetone vapor in the pores was 0.028 in the case of the film templated with micelles, and only 0.0015 for the one consisting of 5 % w/v polymer. The difference in the acetone volume fraction adsorbed (more details on the calculation can be found in [7]) was also significant: 6.7 % when using micelles templates, and 0.4 % when using a polymer with concentration below the CMC. A possible reason may be the different organization of the pores system in the two films, with the more opened porosity in the first case favoring a higher adsorption of analyte vapor. Another possibility is the different degree of hydrophilicity/hydrophobicity of the samples responsible for different states of the pores: empty or partially filled with water from the ambient. Whatever the reason, more experiments are needed to confirm these assumptions. However, the change obtained in the refractive index of about 0.03 after exposure to acetone vapor is significant and guarantees a successful implementation of the samples studied as sensitive media for chemo-optical sensors.

4. Conclusions

A dye solubilization method, along with UV-VIS transmittance and photoluminescence (PL) measurements, were used to determine the critical micelle concentration (CMC) of the triblock copolymer PEO-PPO-PEO (Pluronic PE 6800). A new approach was adopted for CMC determination, consisting in monitoring the wavelength shift of characteristic absorption and emission peaks of the DPH (1,6-diphenyl-1,3,5-hexatriene) in hydrophilic and hydrophobic environments as a function of the polymer concentration, instead of commonly-accepted observation of the peaks intensities. A very good agreement was achieved between the CMC values obtained by the two methods: 13.1 % determined by UV-VIS spectroscopy, and 12.6 % by PL measurements. The positive role of polymer micelles for pore formation in Nb_2O_5 films was demonstrated by using PE 6800 with concentration bellow and above CMC as templates. The micelle solution led to the formation of 54 % porosity, while in other case the film porosity was 24 %. Moreover, the change in the refractive indices of the porous films obtained using micelles after acetone vapor exposure was about 0.03, while the films deposited with polymer templates below CMC exhibited a very low response, namely, 0.0015.

Acknowledgements

This work was supported by the Bulgarian National Science Fund under Project DN 08/15 (14.12.2016).

References

[1] Alexandridis P and Hatton T 1995 *Colloids and Surfaces* A: *Physicochem. and Eng. Aspects* **96** 1-46

[2] Domínguez A, Fernández A, González N, Iglesias E and Montenegro L 1997 *J. Chem. Educ.* **74** 1227-31

[3] Chattopadhyay A and London E 1984 *Analytical Biochem.* **139** 408-12
[4] Alexandridis P, Holzwarthf J and Hatton A 1994 *Macromolecules* **27** 2414-25
[5] Ren Y, Ma Z and Bruce P 2012 *Chem. Soc. Rev.* **41** 4909–27
[6] Lazarova K, Vasileva M, Marinov G and Babeva T 2014 *Optics & Laser Technol.* **58** 114
[7] Georgiev R, Georgieva B, Vasileva M, Ivanov P and Babeva T 2015 *Adv. Cond. Matter Phys.* 2015 403196
[8] Lazarova K, Awala H, Thomas S, Vasileva M, Mintova S and Babeva T 2014 *Sensors* **14** 12207
[9] Yekta A, Aikawa M and Turro N 1979 *J. Chem. Phys. Lett.* **63** 543
[10] Lazarova K, Vasileva M, Marinov G and Babeva T 2013 *Bulgarian Chem. Commmic.* **45B** 23-27

Optical and sensing properties of sol-gel derived vanadium pentoxide thin films with porous and dense structures

T Babeva[1,3] , H Awala[2], J Grand[2], K Lazarova[1], M Vasileva[1] and S Mintova[2]

[1]Acad. J. Malinowski Institute of Optical Materials and Technologies, Bulgarian Academy of Sciences, Acad. G. Bonchev str., bl. 109, 1113 Sofia, Bulgaria
[2]Laboratoire Catalyse et Spectrochimie (LCS), Normandie Université, ENSICAEN, CNRS, 6 boulevard Maréchal Juin, 14050 Caen, France

E-mail: babeva@iomt.bas.bg

Abstract. The sol-gel and spin-coating methods were used for deposition of thin transparent V_2O_5 films on optical glass substrates and silicon wafers. Different synthesis and deposition conditions, including synthesis temperatures and post-deposition annealing, were used aiming at obtaining transparent films with high refractive index and good optical quality. The surface morphology and structure of the films were studied by SEM and XRD. The optical properties (refractive index, extinction coefficient and optical band gap) and thickness of the V_2O_5 films were determined from their transmittance and reflectance spectra. The potential application of the films as building blocks of optical sensors was demonstrated by preparation of multilayered structures comprising both V_2O_5 and BEA-type zeolite films and testing their response towards acetone vapors.

1. Introduction

High refractive index materials are needed in many areas, such as optics, photonics, photovoltaic etc., for improving the performance of optical devices. Among vanadium oxides, V_2O_5, being the most stable of them, has gained increasing interest due to its fascinating thermal, electric and photochromic properties. V_2O_5 thin films exhibit multicolor electrochromism and have high potential for use in electrochromic display devices, color filters, and smart windows [1-3]. Moreover, V_2O_5 has a wide optical energy gap, a layered structure, and good chemical and thermal stability that make it promising for use in microelectronic, electrochemical and optoelectronic devices [4, 5]. V_2O_5 has already been used as a resistance sensor for NO_2 [6] and ethanol [7], as a catalyst [8], as an infrared detector [9], and a high-performance cathode for lithium ion batteries [10]. Recently, we demonstrated the successful application of V_2O_5 thin films as building blocks of vapor-responsive Bragg stacks for selective optical detection of chloroform [11].

One of the simplest and flexible methods for deposition of V_2O_5 thin films is the sol-gel method, where the structural, morphological and optical properties of the films can be tailored by changing the deposition parameters, such as the precursor type and concentration, the choice of solvent, the duration and temperature of synthesis, the postdeposition annealing etc. In order to improve further the performance of Bragg stacks used for optical sensing [12], it is crucial to optimize the synthesis and

[3] To whom any correspondence should be addressed.

IOP Conf. Series: Journal of Physics: Conf. Series **992** (2018) 012038 doi:10.1088/1742-6596/992/1/012038

deposition conditions so as to obtain materials, which, besides having a high refractive index and good optical quality, are permeable for molecules of the vapors to be detected.

In the present study, we applied the sol-gel and spin-coating methods for deposition of thin transparent V_2O_5 films using various synthesis temperatures and solvents, as well as duration and temperature of the post-deposition annealing. The relationship between the synthesis parameters and the deposition conditions on the one hand, and the structural, morphological and optical properties on the other was thus revealed. The potential application was demonstrated of the films as building blocks for optical sensors.

2. Experimental part

The V_2O_5 films were deposited using both the sol-gel and spin-coating methods. Vanadium sols were synthesized by dissolving 1.5 ml vanadium(V) oxytriisopropoxide (Sigma Aldrich) in 35 ml of 0.2% v/v Pluronic PE 6100 (BASF) solution in isopropanol and stirring for 60 min at temperatures of 25 °C or 55 °C. Alternatively, in some experiments half of the above solvent was replaced with acetone. Thin V_2O_5 films with thickness of 55 nm were deposited on optical glass BK-7 and silicon wafers by spin-coating of 0.3 ml of vanadium sol for 30 s at a rotation speed of 3000 rpm. A post-deposition annealing for 30 min at 320 °C and 450 °C was applied at an acceleration rate of 10 °C/min. The films' surface morphology and structure were studied by scanning electron microscopy (SEM) using a JEOL JSM6700F SEM (accelerating voltage of 30.0 kV, JEOL, Tokyo, Japan) and a PANalytical X'Pert Pro diffractometer with Cu Kα monochromatic radiation ($\lambda = 1.5418$ Å), respectively. The transmittance (T) and reflectance spectra (R) of the films were measured at normal light incidence by a Cary 05E UV-VIS-NIR spectrophotometer (Varian, Australia). The optical properties (refractive index (n) and extinction coefficient (k)), along with the thickness (d) of the V_2O_5 films were determined simultaneously from the T and R measurements of films deposited on both types of substrates with an accuracy of 0.01, 0.005 and 1 nm in n, k and d, respectively, using a previously developed calculating procedure [13, 14]. The optical band gap was determined using already calculated values of the absorption coefficient α ($= 4\pi k/\lambda$, where λ is the wavelength) and Tauc's plots [15].

A stable coating suspension of discrete nanosized BEA-type zeolite synthesized from a colloidal precursor suspension under hydrothermal condition was prepared according to the procedure described elsewhere [16]. The sensing properties of multilayered structures comprising thin films of BEA zeolites and V_2O_5 were studied under their exposure to acetone vapors. An infrared (IR) reactor-cell was used connected to a saturator gas-system with different vapors (concentrations from 5 to 35000 ppm). The IR spectra under adsorption and desorption of analytes were collected in a continuous mode by a Tensor 27 spectrometer (Bruker) equipped with a DTGS detector.

3. Results and discussion

The influence of the temperature of the synthesis of vanadium sol on the optical properties of the deposited thin V_2O_5 films is demonstrated in figure 1(a), where the optical constants (refractive index, n, and extinction coefficient, k) are presented as functions of the wavelength. The plot of $(\alpha E)^2$ versus the energy, E, (the so-called Tauc's plot), where α ($= 4\pi k/\lambda$, λ being the wavelength) is the absorption coefficient, is presented in figure 1(b). The optical band gap values, E_g, for direct transitions are determined by extrapolating the linear part of Tauc's plot to $\alpha = 0$.

As seen in figure 1, the increase of the synthesis temperature from 25 °C to 55 °C results in a decrease of k for wavelengths higher than 500 nm, thus improving the transparency of the films in the visible and near infrared regions. For example, at the wavelength of 1000 nm, k has values of 0.275 and 0.014 at 25 °C and 55 °C, respectively. To the contrary, for smaller wavelengths ($\lambda < 500$ nm) the higher synthesis temperature leads to a stronger absorption associated with the slightly smaller values of E_g in this case (2.66 eV), as compared to the values at 25 °C (2.69 eV). Both films have similar refractive indices values (at 600 nm, n is 2.00 and 1.93 for the films synthesized at 25 °C and 55 °C, respectively); except for the region of shorter wavelengths, where the films synthesized at 25 °C exhibit higher *n*-values (2.34 at 400 nm) compared to $n = 2.04$ in the case of the film prepared at the higher temperature. Our additional studies (not presented here) have shown that the surface morphologies of the films and their structures are very similar, regardless of the synthesis temperature. The slightly different compositions due to the presence of non-stoichiometric oxides, evidenced by the gradually increasing absorption in the near IR region in the case of 25 °C, could be the possible reason for the different optical behavior of films deposited from vanadium sols synthesized at different temperatures [17].

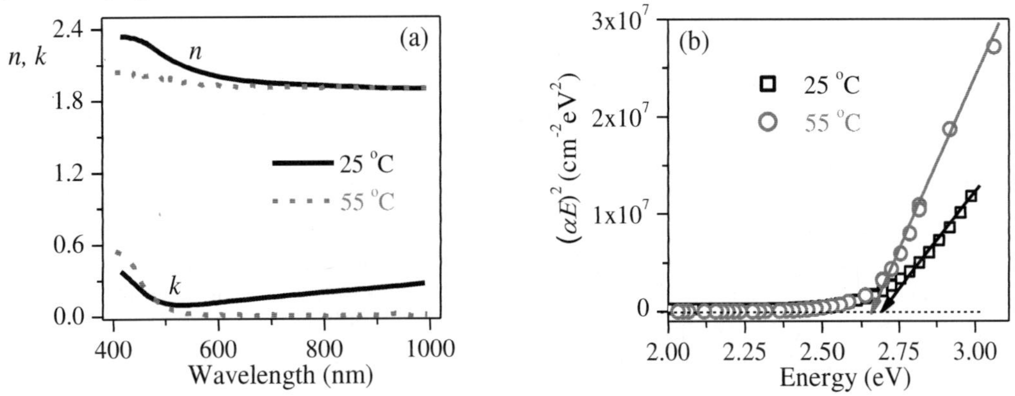

Figure 1. Dispersion curves of the refractive index (*n*) and absorption coefficient (*k*) (a) and Tauc's plots (b) for V_2O_5 thin films synthesized at 25 °C and 55 °C.

The SEM images representing the surface morphology and the XRD patterns of V_2O_5 films annealed at 320 °C and 450 °C for 30 min are shown in figure 2. It is seen that increasing the annealing temperature leads to the formation of grains with a bigger size and improves the crystallinity of the films, as indicated by the sharper XRD peaks in the case the film obtained at 450 °C (figure 2c); these results are consistent with the data reported previously [18]. Unfortunately, the optical quality of the films deteriorates at higher temperatures, and the scattering losses rise due to the rough surface. It is important to mention that for optical sensing applications, where the temporal variation of the specular transmittance or reflectance signals are monitored, the optical quality of thin films is essential for proper operation of the particular device. Thus, the post-deposition annealing of V_2O_5 thin films at temperatures higher than 320 °C were not considered further in our study. Our additional experiments have shown that an annealing duration increase from 30 min to 60 min does not influence the thickness and optical properties of the films. This means that isothermal annealing for 30 min at 320 °C is sufficient for stabilization of the films and there is no need for further annealing.

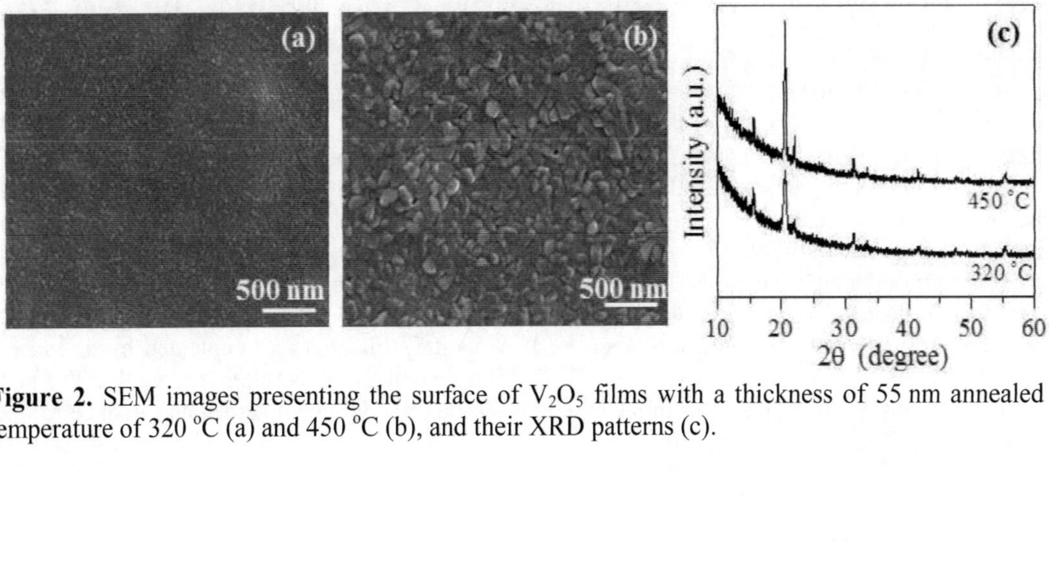

Figure 2. SEM images presenting the surface of V_2O_5 films with a thickness of 55 nm annealed at temperature of 320 °C (a) and 450 °C (b), and their XRD patterns (c).

Figure 3 presents the dispersion curves of *n* and *k* for V_2O_5 films (thickness of 55 nm) obtained by vanadium sol prepared with addition of acetone. This particular synthesis included 1.5 ml of Vanadium(V) oxytriisopropoxide dissolved in a mixture of 17.5 ml acetone and 17.5 ml isopropanol.

The films from this synthesis are referred to here as "dense" to distinguish them from the "porous" films deposited from vanadium sol synthesized using 17.5 ml acetone in addition to 17.5 ml of 0.6 % v/v Pluronic PE 6100 (BASF) in isopropanol. After post-deposition annealing, the added polymer evaporates leaving pores in the film, which increases the volume fraction of empty space inside the film. It is expected that the effective refractive index of the film (porous) will decrease in comparison to the film without Pluronic (dense). At the same time, it is very likely that the permeability of the porous films to analyte molecules will be improved due to generation of additional porosity. As seen in figure 3, the addition of acetone results in a higher refractive index at the wavelength of 600 nm; the refractive index of the dense film is 2.24, while *n* for the porous film is 2.01; i.e., almost equal to that of the film without acetone (figure 1a). However, considering that the concentration of Pluronic is 0.6 % v/v in the former case, we can assume that the porosity of this film is significantly higher as compared to the film without acetone, which will be beneficial if sensing application is contemplated. Besides, the films deposited from sol with acetone are more transparent as compared to those without acetone and have an optical band gap of 2.75 eV.

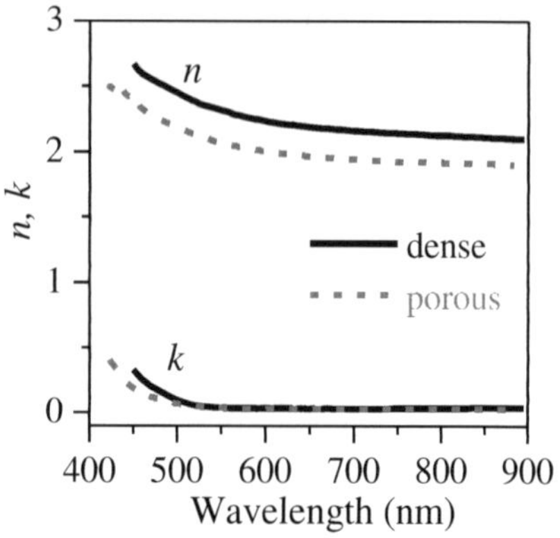

Figure 3. Dispersion curves of the refractive index (*n*) and absorption coefficient (*k*) for V_2O_5 thin films synthesized with addition of acetone, without Pluronic (dense) and with Pluronic (porous).

In addition to high refractive index and transparency, the films should be permeable to the analyte vapors to be suitable for sensing applications. To check the V_2O_5 films' permeability, we deposited two types of double-layered systems on silicon consisting of a V_2O_5 film on top of a porous BEA zeolite film and vice versa, and followed their sensing response to acetone vapors using IR spectroscopy. As the concentration of acetone increases, the intensity of the IR band at 2825-2938 cm^{-1} increases (figure 4) reaching saturation at a concentration of 20 000 ppm. It is seen that in the case of the sample with a top layer of V_2O_5, the adsorption is slightly weaker compared to the structure having a top layer of BEA zeolite; this could be explained by a reduced meso-porosity due to the penetration of the V_2O_5 film into the gap between the zeolite crystals. Nevertheless, the significant response in the case of V_2O_5 on top could be regarded as evidence for the good permeability of the V_2O_5 thin films.

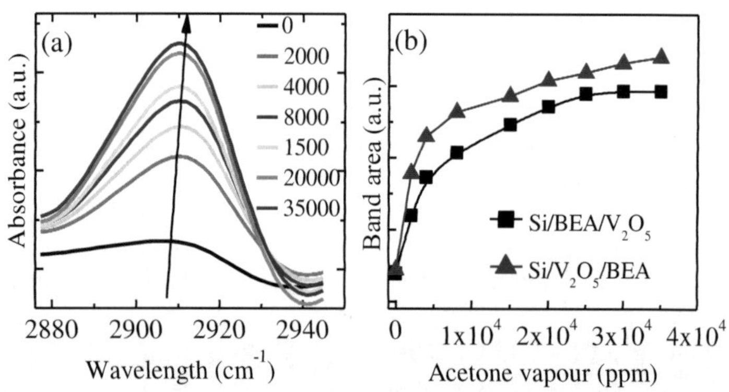

Figure 4. IR absorption spectra of BEA/V_2O_5 double-layered systems at different concentrations of acetone vapors (a); the arrow indicates the concentration increase from 0 ppm to 35 000 ppm; and adsorption isotherms of acetone of the Si/BEA/V_2O_5 (black squares) and Si/V_2O_5/BEA (blue triangles) double-layered systems (b).

4. Conclusions

Vanadium pentoxide films (V_2O_5) with high refractive index and good optical quality were successfully deposited onto optical glass and silicon substrates using both sol-gel and spin-coating methods. The highest refractive index (2.24 at a wavelength of 600 nm) and optical band gap (2.7 eV) were obtained for films synthesized with addition of acetone into the vanadium sol, while the change of synthesis temperature from 25 °C to 55 °C did not lead to a substantial modification of the films' properties. However, the increase of the post-deposition annealing temperature from 320 °C to 450 °C led to deterioration of films' optical quality, i.e., the films became milky due to increased scattering. The suitability of the films studied to optical sensing applications was demonstrated by in-situ IR adsorption measurements of a probe molecule (acetone). The similarity of the adsorption isotherms obtained for two double-layer systems comprising a V_2O_5 film on top of a BEA zeolite film or a BEA zeolite film deposited on top of a V_2O_5 film confirmed the V_2O_5 films' good permeability and their possible further applications in optical sensing.

Acknowledgements

The financial support of Bulgarian National Science Fund (BNSF) under the project DN08/15 (14.12.2016) and DRILA 01/13 (17.04.2015) is highly appreciated.

References

[1] Losurdo M, Bruno G, Barreca D and Tondello E 2000 *Appl. Phys. Lett.* **77** 1129
[2] Ramana C, Smith R, Hussain O, Chusuei C and Julien C 2005 *Chem. Mater.* **17** 1213
[3] Batista C, Ribeiro R and Teixeira V 2011 *Nanoscale Res. Lett.* **6** 1–7
[4] Vernardou D, Paterakis P, Drosos H, Spanakis E, Povey I M, Pemble M E, Koudoumas E and Katsarakis N 2011 *Sol. Energy Mater. Sol. Cells* **95** 2842–7
[5] Luo Z, Wu Z, Xu X, Du M, Wang T and Jiang Y 2011 *Mater. Sci. Eng.:* B **176** 762–6
[6] Mane A and Moholkara A 2017 *Appl. Surf. Sci.* **416** 511–20
[7] Abbasi M, Rozati S, Irani R and Beke S 2015 *Mater. Sci. in Semicond. Processing* **29** 132–8
[8] Justin P and Ranga Rao G 2009 *Catal. Today* **141** 138-43
[9] Rajendra K, Karunagaran B, Mangalaraj D, Narayandass S, Manoravi P and Joseph M 2003 *Mater. Res. Bull.* **38** 1235–40
[10] Yu D, Qiao Y, Zhou X, Wang J, Li C, Chen C and Huo Q 2014 *J. Power Sources* **266** 1–6
[11] Lazarova K, Georgiev R, Vasileva M, Georgieva B, Spassova M Malinowski N and Babeva T 2016 *Opt. Quant. Electron.* **48:310** 1-8
[12] Lazarova K, Awala H, Thomas S, Vasileva M, Mintova S and Babeva T 2014 *Sensors* **14** 12207-18
[13] Panayotov V and Konstantinov I 1994 *Proc. SPIE* **2253**1070–9
[14] Babeva T, Kitova S and Konstantinov I 2001 *Appl. Opt.* **40** 2682–6
[15] Tauc J 1972 *Optical Properties of Solids.* (North-Holland, Amsterdam) 277
[16] Yordanov I, Knoerr R, De Waele V, Mostafavi M, Bazin Ph, Thomas S, Rivallan M, Lakiss L, Metzger T and Mintova S 2010 *J. Phys. Chem.* C **114** 20974
[17] Kana Kana J, Ndjaka J, Vignaud G, Gibaud A and Maaza M 2011 *Opt. Comm.* **284** 807–12
[18] Liu Y, Du X and Liu X 2014 *Surf. Rev. and Lett.* **21** 1450027

IOP Conf. Series: Journal of Physics: Conf. Series **992** (2018) 012039 doi:10.1088/1742-6596/992/1/012039

Optical characterization of antirelaxation coatings

S Tsvetkov[1], S Gateva[1,4], S Cartaleva[1], E Mariotti[2] and K Nasyrov[3]

[1]E. Djakov Institute of Electronics, Bulgarian Academy of Sciences, 72 Tzarigradsko Chaussee, 1784 Sofia, Bulgaria
[2]Department of Physical, Geological and Environmental Sciences, University of Siena and CNISM, via Roma 56, 53100 Siena, Italy
[3]Institute of Automation and Electrometry, SB RAS, Novosibirsk, Russia

E-mail: sgateva@ie.bas.bg

Abstract. Antirelaxation coatings (ARC) are used in optical cells containing alkali metal vapor to reduce the depolarization of alkali atoms after collisions with the cell walls. The long-lived ground state polarization is a basis for development of atomic clocks, magnetometers, quantum memory, slow light experiments, precision measurements of fundamental symmetries etc. In this work, a simple method for measuring the number of collisions of the alkali atoms with the cell walls without atomic spin randomization (Nasyrov et.al., Proc. SPIE (2015)) was applied to characterize the AR properties of two PDMS coatings prepared from different solutions in ether (PDMS 2% and PDMS 5%). We observed influence of the light-induced atomic desorption (LIAD) on the AR properties of coatings.

1. Introduction

Antirelaxation coatings (ARC) are used in optical cells containing alkali metal vapor to reduce the alkali atoms' depolarization after collisions with the cell walls [1]. Characterization of the coating quality is important for the interpretation of the experimental results and development of optical sensors. The long-lived ground-state polarization is the basis for development of atomic clocks, magnetometers, quantum memory, slow light experiments, precision measurements of fundamental symmetries etc. [2,3].

Light-induced atomic desorption (LIAD) is a non-thermal process in which atoms adsorbed at a surface (coated or uncoated) are released upon illumination [4]. It is applied mostly for realization of optical atomic dispensers in the cases when high atomic density at low temperature and/or fast density control is needed – for example, for loading atomic devices as atomic magnetometers, atomic clocks, magneto-optical traps and their miniaturization.

The purpose of this work was to study experimentally the influence of LIAD on the AR properties of coated cells on the example of two PDMS coatings prepared with different solutions in ether (PDMS 2% and PDMS 5%). We showed that the dependence of the number of LIAD collisions is different in the two cells and that this dependence can be used to characterize the process of cell ripening.

[4] To whom any correspondence should be addressed.

IOP Conf. Series: Journal of Physics: Conf. Series **992** (2018) 012039 doi:10.1088/1742-6596/992/1/012039

2. Experimental setup

The experimental setup is given in figure 1. It is described in detail in [5-7]. A 780 nm free-running diode laser (DL) was used. To reduce the optical pumping and saturation, all measurements were conducted on the ^{85}Rb set of absorption lines from the F_g=3 set of transitions. PD1 registered the transmitted 780 nm diode laser light, while PD2, the fluorescence at 780 nm [8].

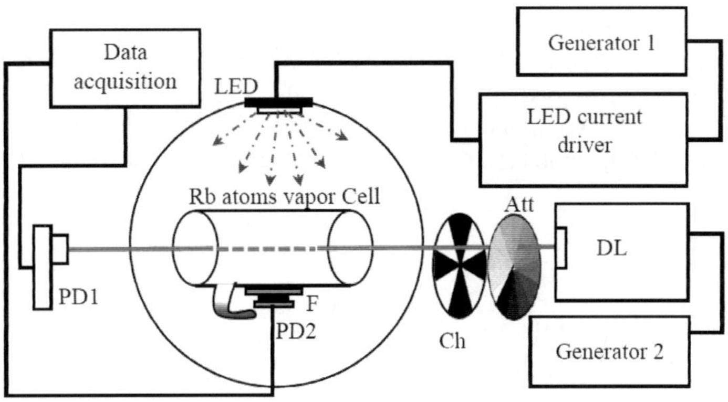

Figure 1. Experimental setup: Att, attenuator; Ch, chopper; DL, 780-nm diode laser; LED, 460-nm light emitting diode; F, filter; PD1 and PD2, photodetectors.

A special sphere with a diffusely reflecting inner surface was used to improve the efficiency of the blue-light illumination by the 460-nm LED [5].

For an optically-thin medium, the change of the Rb density due to the LIAD effect is measured by the absorption coefficient κ_ω of Rb vapor. According to Beer's law, the transmission $T(\omega) = T_\omega$ is:

$$T_\omega = I/I_0 = \exp(-\kappa_\omega L) = \exp(-\sigma_\omega NL), \qquad (1)$$

where I_0 is the 780-nm laser power entering the vapor sample; I, the transmitted power; κ_ω, the frequency-dependent absorption coefficient; σ_ω, the non-saturated, frequency-dependent absorption cross-section; N, the atomic density and L, the cell length.

We applied a simple method for measuring the number of collisions without atom spin randomization $N_c = 1/\varepsilon$ (ε being the probability for atom spin relaxation after one collision with the coating) to characterize the ARCs quality [6]. The method is based on registering the temporal dependence of the fluorescence intensity resulting from alkali atoms irradiated by pulsed light at resonance. The theoretical considerations assuming a cylindrical optical cell with a length larger than the cell radius show that in order to find the number of atom-wall collisions without spin randomization N_c, it is sufficient to know the value of the fluorescence decay rate β and the ratio between the intensities of the fluorescence at the beginning I_i and at the end of the decay I_f, as well as the time between atom collisions with the cell's walls $\tau_c = R_c/2v_T$, where v_T is the atom's thermal velocity.

$$1/N_c = \beta\tau_c I_f/I_i \qquad (2)$$

Table 1 summarizes some parameters of the PDMS-coated cells prepared with two different solutions in ether (PDMS 2% and PDMS 5%).

Table 1. Parameters of the Rb cells: length, diameter, volume, surface, mean distance between two collisions with the cell walls.

Cell coating	Length [cm]	Diameter [cm]	Volume [cm^3]	Surface [cm^2]	L [cm]
PDMS 5	4.5	2.6	24	47	2.0
PDMS 2	6.0	2.6	32	60	2.1

IOP Conf. Series: Journal of Physics: Conf. Series **992** (2018) 012039 doi:10.1088/1742-6596/992/1/012039

3. Results and discussion

The dependence of the number of collisions without spin relaxation N_c on the LIAD measured in the PDMS5 cell (figure 2) shows that the number of collisions decreases with LED illumination. There are two different parts in this dependence. As the LED power is increased, the number of collisions decreases to a certain value; at higher powers it does not change significantly. This dependence correlates with the measured Rb atoms density N on LIAD (figure 3). The Rb density rises very fast to its value in an uncoated cell at this temperature; it then rises linearly with a steepness lower by more than an order of magnitude than at low LED currents. This dependence can be explained by the results of previous studies of the processes of atom-coating interactions [9-12], which have shown two different components: a slow component, related to interaction of the atoms with the glass or chemically active impurities, and a fast one, associated with the atoms-coating interaction.

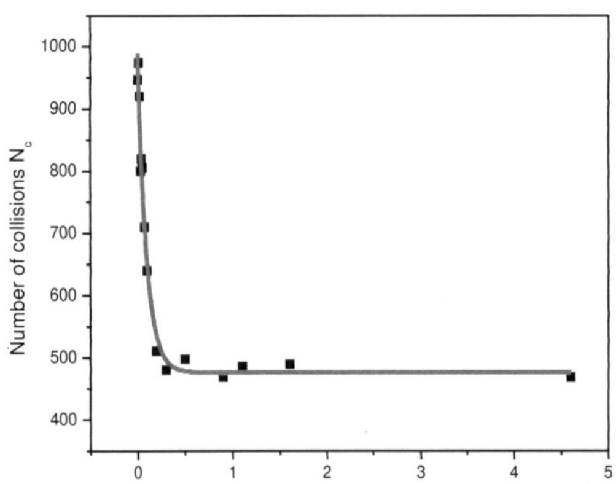

Figure 2. Number of collisions N_c dependence on LIAD in the PDMS5 coated cell.

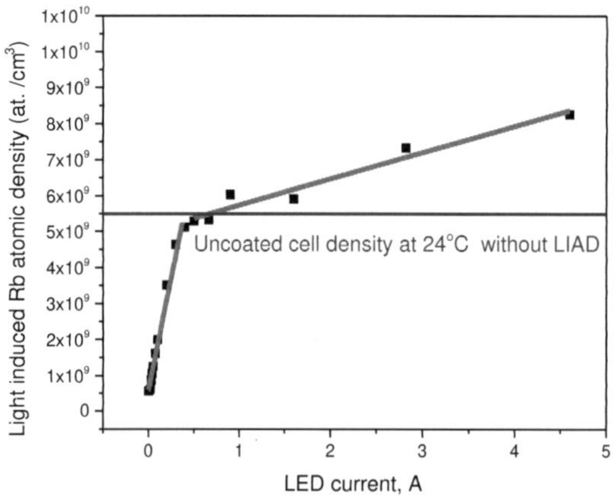

Figure 3. Rb atoms density dependence on LIAD in the PDMS5 coated cell.

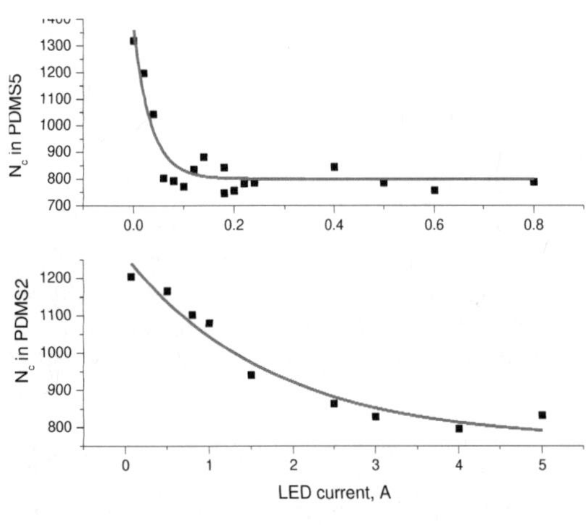

Figure 4. Number of collisions w/o relaxation N_c dependence on LIAD in PDMS5 and PDMS2 coated cells.

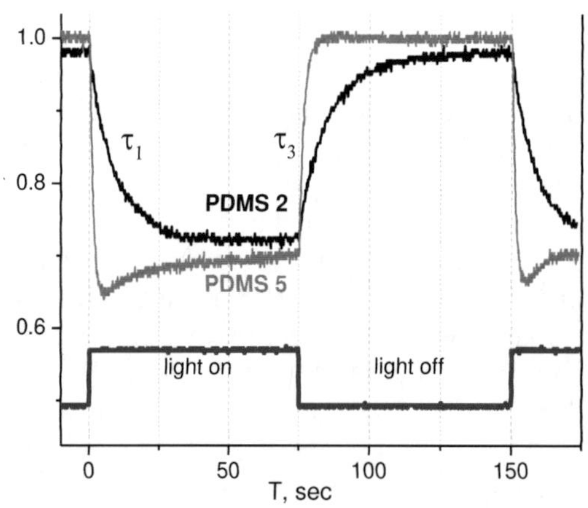

Figure 5. LIAD dynamics in PDMS2 and PDMS5 cells: the normalized transmission measured.

Figure 4 compares the AR properties of two PDMS coatings prepared with different solutions in ether (PDMS 2% and PDMS 5%). As seen, no fast component is present in the AR characteristic of the PDMS2 coated cell. Our previous investigations of LIAD dynamics (figure 5) have shown that the desorption and adsorption rates in a PDMS2-coated cell are about an order of magnitude lower than in

IOP Conf. Series: Journal of Physics: Conf. Series **992** (2018) 012039 doi:10.1088/1742-6596/992/1/012039

a PDMS5-coated cell, which we explained by the different morphology of the surface, thickness of the coating, and volume density. At equal Rb atoms density, the number of collisions N_c is the same [7].

The study of the relaxation rates of atomic polarization during LIAD in a paraffin-coated Cs cell [9] has shown no significant change in the relaxation rates beyond that expected from the faster rate of spin-exchange collisions due to the increase in the Cs density. The authors use relaxation in the dark registered via optical rotation to characterize the AR properties of the coating at Cs densities higher than 12×10^9 atoms/cm^3, which were higher by an order of magnitude than in our experiment.

4. Conclusions

A system for optical characterization of coated cells was developed, allowing one to measure the atomic density in a cell Rb containing alkali atoms via the absorption and the AR properties of coated cells by means of the simple method, proposed in [6], for measuring the number of collisions without spin randomization.

The system was applied to experimental investigation of LIAD influence on the AR properties of two PDMS coatings prepared with different solutions in ether (PDMS 2% and PDMS 5%). It was shown that in both cells LIAD reduces the number of collisions without spin relaxation, and the dependence of the number of collisions on LIAD is different in the two cells. These results illustrate the influence of the alkali atoms' penetrating from the vapor phase into the AR coating, and can be applied for characterization of the coating and the process of ripening.

The studies performed and its results are of value in gaining further understanding of the light-atom-surface interactions, in view of developing new optical elements for photonics applications, LIAD loaded atomic devices and their miniaturization, and new methods for surface and coating diagnostics and monitoring.

Acknowledgements

This work was supported by the Bulgarian National Scientific Fund under Grant No DO08-19/2016. ST and SG acknowledge the funding by the Program for Young Scientists Support of the BAS within Grant DFNP-17-76/2017.

References

[1] Bochiat M and Brossel J 1966 *Phys. Rev.* **147** 41
[2] Karaulanov T, Graf M T, English D, Rochester S M, Rosen Y J, Tsigutkin K, Budker D, Alexandrov E B, Balabas M V, Kimball D F J, Narducci F A, Pustelny S and Yashchuk V V 2009 *Phys. Rev.* A **79**/1 012902
[3] Nasyrov K, Gozzini S, Lucchesini A, Marinelli C, Gateva S, Cartaleva S and Marmugi L 2015 *Phys. Rev.* A **92**/4 043803
[4] Lucchesini A, Gozzini S, Marinelli C and Marmugi L 2016 *Coatings* **6**/4 47 60400476
[5] Tsvetkov S, Taslakov M and Gateva S 2017 *Appl. Phys.* B **123**/3 92
[6] Nasyrov K, Entin V, Nikolov N, Petrov N and Cartaleva S 2015 *Proc. SPIE* **9447** 944704
[7] Tsvetkov S, Gateva S, Nasyrov K, Gozzini S, Mariotti E and Cartaleva S 2016 *Bulg. Chem. Commun.* **48** 173-7
[8] Gateva S, Taslakov M, Sarova V, Mariotti E and Cartaleva S 2013 *Proc. SPIE* **8770** 877000
[9] Graf M T, Kimbal D F, Rochester S M, Kerner K, Wong C and Budker D 2005 *Phys. Rev.* A **72** 023401
[10] Balabas M V and Tret'yak O Yu 2012 *Technical Physics* **57**/9 1257-65
[11] Sekigishi N and Hatakeyama A 2016 *Appl. Phys.* B **122**/4 1
[12] Kumagai R and Hatakeyama A 2016 *Appl. Phys.* B **122**/7 1

IOP Conf. Series: Journal of Physics: Conf. Series **992** (2018) 012040 doi:10.1088/1742-6596/992/1/012040

$Bi_{12}TiO_{20}$ crystallization in a Bi_2O_3-TiO_2-SiO_2-Nd_2O_3 system

S Slavov[1,3] and Z Jiao[2]

[1]Department of Physics, University of Chemical Technology and Metallurgy, 8 Kl. Ohridski Blvd., 1756 Sofia, Bulgaria
[2]School of Environmental and Chemical Engineering, Shanghai University, Huanhua Building, 333 Nanchen Road, 200444 Shanghai, P.R. China

E-mail: stanislavslavov@hotmail.com

Abstract. Polycrystalline mono-phase bismuth titanate was produced by free cooling from melts heated to 1170 °C. The control over the initial amounts in the starting compositions in the system $Bi_2O_3/TiO_2/SiO_2/Nd_2O_3$ and over the thermal gradient of the heat process resulted in the formation of specific structures and microstructures of monophase sillenite ceramics. The main phase $Bi_{12}TiO_{20}$ belongs to the amorphous network groups based on oxides of silicon, bismuth and titanium. In this work, we demonstrated a way to control the crystalline and amorphous phase formation in bulk poly-crystalline materials in the selected system.

1. Introduction

Bismuth-titanate-based materials became very popular in recent years with a number of important applications, such as capacitors, sensors [1], piezoelectric, electro-optical and pyroelectric materials, relaxers, FERAM and DRAM storage devices and semiconductor devices [2].

The most popular methods for preparation are solid-state reactions, co-preparation, molten salt synthesis, mechanochemical synthesis. Addition of suitable additives is one way to modify their properties by controlling the phase formation. Jose Pineda-Flores et al. [3] have studied the ways to modify the properties of bismuth-titanate-based materials by controlling the phase formation and adding suitable additives, as Pr, Nd, Gd. Murugan et al. [4] discussed the glass-phase content influence on the dielectric properties of novel low-permittivity, fine-grain, pore-free and nano-structured materials. In general, the properties of bulk ceramic materials are affected by the phases involved, the grains size, the amount of glass phase and the grain boundary effects [5]. In the present work, we explored the possibility to obtain monophase polycrystalline ceramics using a bismuth-silicate amorphous network as a boundary-matrix on order to control the properties of these new bismuth-titanate materials.

2. Experimental

The synthesis of the bulk materials of the Bi_2O_3-TiO_2-SiO_2-Nd_2O_3 system began by homogenizing for 15 min the starting oxides (Bi_2O_3, TiO_2, SiO_2 and Nd_2O_3, Alfa Aesar 99.99%). The melting was done in aluminum crucibles in a KTM-GSL1700X SiC tube furnace at a temperature of 1170°C.

[3] To whom any correspondence should be addressed.

This study was part of a larger experiment aimed at obtaining monophase polycrystalline samples by melting and controlled cooling to room temperature. In the selected system, it is possible to obtain the following phases: $Bi_4Ti_3O_{12}$, $Bi_2Ti_2O_7$, $Bi_2Si_2O_7$, $Bi_{12}TiO_{20}$, $Bi_{12}SiO_{20}$, depending on the starting composition and the melting temperature. The remaining starting oxides (table 1) were selected with the idea of obtaining only one crystalline phase. $Bi_{12}TiO_{20}$, in combination with a Bi/Si/O amorphous network. In order to control the temperature gradient of the free cooling, it was carried out in graphite crucibles.

Table 1. Starting phase composition and melting conditions of selected samples in the system Bi_2O_3-TiO_2-SiO_2-Nd_2O_3.

Sample index	Initial oxide composition (Raw materials in mol %)				t, °C / time, min	Phases obtained after melting, according to X-Ray Diffraction
	Bi_2O_3	TiO_2	SiO_2	Nd_2O_3		
A	77	14	0	9	1170/15	$Bi_{12}TiO_{20}$
B	74	14	3	9	1170/15	$Bi_{12}TiO_{20}$
C	74	3	14	9	1170/15	$Bi_{12}TiO_{20}$

The phase composition was determined by X-ray diffraction using a Ridacu D/MAX2500V + PC apparatus, CuKa radiation (1.5406 Å). The microstructure was observed by scanning electron microscopy (SEM Hitachi SU1510). The structure of the samples was examined by Thermo Nikolet – Avatar 370 FT-IR equipment.

3. Results and discussion

As seen in the X-ray patterns (figure 1) only one phase, $Bi_{12}TiO_{20}$. was present in the three samples chosen. The SEM images (figure 2) show that the samples examined had a similar dense microstructure,

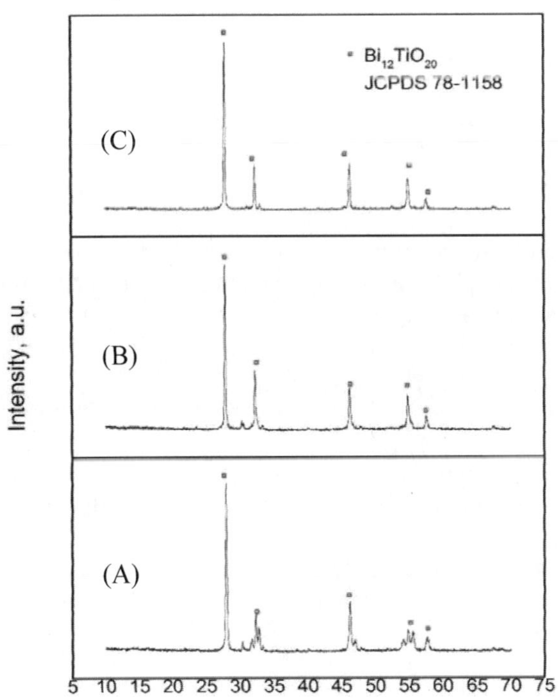

Figure 1. XRD patterns of samples A, B and C.

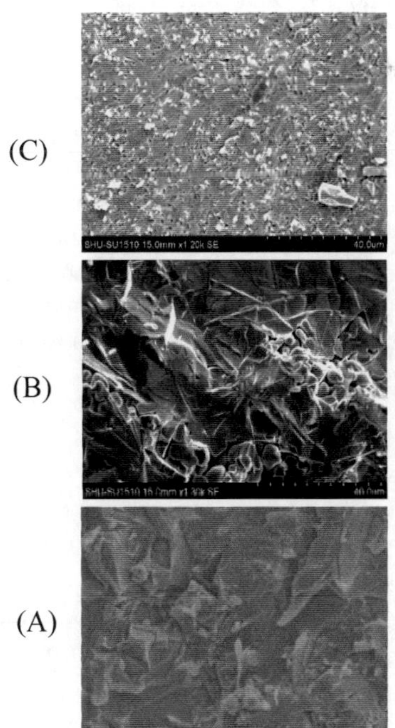

Figure 2. SEM images of samples A, B and C.

without visible cracks or cavities. The formation of separate crystals of different size is also seen.

According to the IR spectra (figure 3), the structures of samples A and B are similar; the band at about 830 cm^{-1} is associated with the symmetric vibrations of the Ti-O linkages; and that at 600 cm^{-1}, with asymmetric and deformation vibrations of the TiO_6 octahedron [6]. Betch and White [7] reported that the band of 850 cm^{-1} is related to symmetric vibrations of Ti-O bonds in addition to Nd together with the TiO_6 octahedral. The bands around 560 cm^{-1} and 820 cm^{-1} characterize a complex combination of a MO_4 tetrahedral and a BiOn polyhedral [8], typical for the sillenite phase $Bi_{12}TiO_{20}$. On the other hand, the bands near 900-800 cm^{-1} and 900-1000 cm^{-1}, and the band near 1050-1100 cm^{-1} are typical for structures with four terminal oxygens SiO^{4-} [13]. Jagannath Roy et al. [14] reported that the appearance of antisymmetric stretching vibrations of Si-O-Al and Si-O-Si networks can be connected with the peaks around 832 cm^{-1} and 1112 cm^{-1}.

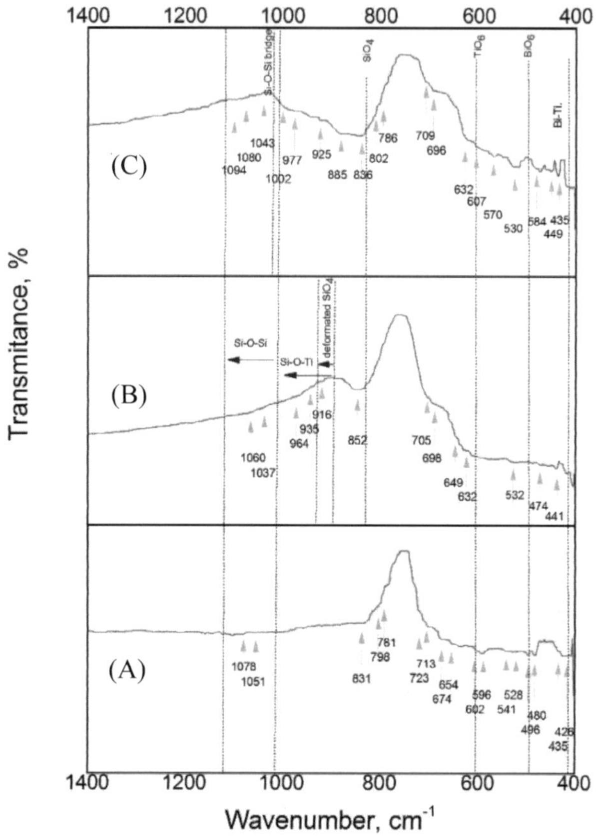

Figure 3. IR spectra of samples A, B and C.

The formation of a multi-component amorphous matrix with Si-O-Si-linkages (1034 cm^{-1}, 1098 cm^{-1}), BiO_6 (480 cm^{-1}), Si-O-Ti linkages (900 cm^{-1}, 1034 cm^{-1}) and of depolymerized SiO_4 groups (890 cm^{-1}, 920 cm^{-1}) is due to the increased amount of SiO_2 in the starting composition of sample C. Thus, the different crystal sizes observed in he SEM images could be due to different SiO_2 content. In this sense, the results obtained by us directly correspond to the literature data, namely, that besides the formation of the sillenite bismuth-titanate phase $Bi_{12}TiO_{20}$, bismuth-silicate and titanium-silicate multi-component amorphous structures are formed.

4. Conclusions

In this work we have investigated

1. The possibility to obtain monophase $Bi_{12}TiO_{20}$ polycrystalline ceramics,
2. The type and composition of the bismuth-silicate amorphous network as boundary-matrix between the crystalline phases.

This investigation indicates a way to create new monophase polycrystalline materials of the system Bi_2O_3-TiO_2-SiO_2-Nd_2O_3 using melting and controlled cooling to room temperature.

Acknowledgement

This work was supported by the Swap and Transfer, Erasmus Mundus Action 2 Mobility Lot 12, Grant ID number SAT_2542

IOP Conf. Series: Journal of Physics: Conf. Series **992** (2018) 012040 doi:10.1088/1742-6596/992/1/012040

References

[1] Sedlar M and Sayer M 1996 Structural and electrical properties of ferroelectric bismuth titanate thin films prepared by the sol-gel method *Ceram. Inter.* **22** 241

[2] Buhay H, Sinharoy S, Kasner W, Francombe M, Lampe D and Stepke E 1991 Pulsed laser deposition and ferroelectric characterization of bismuth titanate films *Appl. Phys. Lett.* **58** 1470

[3] Pineda-Flores J, Chavira E, Reyes-Gasga J, Gonzalezc A and Huanosta-Tera A 2003 Synthesis and dielectric characteristics of the layered structure $Bi_{4-x}R_xTi_3O_{12}$ (R_x=Pr, Nd, Gd, Dy) *J. Eu. Ceram. Soc.* **23** 839–50

[4] Murugan G, Subbanna G and Varma K 1999 Nanocrystallization of ferroelectrics bismuth unstated in lithium borate glass matrix *Mater. Sci. Lett.* **18** 1687-90

[5] Huanosta A, Alvarez-Fregoso O, Amano Mexico E, Tabares-Muñoz C, Mendoza-Alvarez M and Mendoza-Alvarez J 1991 AC impedance analysis on crystalline layered and polycrystalline bismuth titanate *J. Appl. Phys.* **69** 404

[6] Du Y, Zhang M, Chen Q and Yin Z 2003 Investigation of size-driven phase transition in bismuth titanate nanocrystals by Raman spectroscopy *Appl. Phys.* A **76** 1099

[7] Betch D and Write W 1997 Vibrational spectra of bismuth oxide and the sillenite-structure bismuth oxide derivative *Spectrosc. Acta* A **34** 505

[8] Radaev S and Simonov V 1992 Structure of sillenite and atomic mechanisms of isomorphic substitution in them *Crystalograpy* **37** 4

[9] Valant M and Suvorov D 2004 A stoichiometric model for sillenites *Chem. Mater.* **14** 3471-6

[10] Valant M, Medeu A and Suvorov D 2004 Isomorphic A-site substitution on sillenite-type compounds *J. Am. Ceram. Soc.* **84**/4 677

[11] Efendiev Sh, Kulieva T, Zomonov V, Chikgov M, Grandolfe M and Vecchia P 1981 Crystal structure of bismuth titanium oxide $Bi_{12}TiO_{20}$ *Phys. Stat. Soc.* A **74**

[12] Abrahams S, Jamiesom P and Bernstain J 1967 Crystal structure of piezoelectric bismuth germanium oxide $Bi_{12}GeO_{20}$ *J. Chem. Phys.* **47**/10 4034

[13] Lazraev A 1972 *Vibrational spectra and structure of silicates* (Consultant Bureau, New York) pp 123-216

[14] Roy J, Bandyopadhyay N, Das S and Maitra S 2011 Studies on the formation of mullite from diphasic Al_2O_3-SiO_2 gel by Fourier transform infrared spectroscopy *Iran J. Chem. Eng.* **30** 1

Electrical characterization of thin nanoscale SiO_x layers grown on plasma hydrogenated silicon

E Halova[1], N Kojuharova[1,3], S Alexandrova[1] and A Szekeres[2]

[1]Department of Applied Physics, Technical University of Sofia, 8 Kl. Ohridski Blvd., 1797 Sofia, Bulgaria
[2]G. Nadjakov Institute of Solid State Physics, Bulgarian Academy of Sciences, 72 Tzarigradsko Chaussee, 1784 Sofia, Bulgaria

E-mail: nkojuharova@tu-sofia.bg

Abstract. We analyzed the electrical characteristics of MOS structures with a SiO_x layer grown on Si treated in plasma without heating. The hysteresis effect observed indicates the presence of traps spatially distributed into the oxide near the interface. The shift and the shape of the curves reveal a small oxide charge and low leakage currents, i.e. a high-quality dielectric layer. The generalized *C-V* curve was generated by applying the two-frequency methods on the *C-V* and *G-V* characteristics at frequencies in the range from 1 kHz to 300 kHz and by accounting for the series resistance and the leakage through the oxide layer. The energy spectra of the interface traps were calculated by comparing the experimental and the ideal theoretical *C-V* curves. The spectra showed the presence of interface traps with localized energy levels in the Si bandgap. These conclusions correlate well with the results on this oxide's mechanical stress level, composition and Si-O ring structure, as well as on the interfacial region composition, obtained by our previous detailed multi-angle spectral ellipsometric studies. The ellipsometric data and the capacitance in strong accumulation of the *C-V* curves were used to calculate the thickness and the dielectric constants of the oxide layers.

1. Introduction

SiO_2 nanostructures exhibit excellent characteristics, such as electrical insulation and visible light transmission, making them promising for applications in optics (photoluminescence, optical isolation, optical waveguides), photochemistry, biomedicine [1], [2]. In general, low-dimensional nanostructured materials have recently attracted considerable attention due to their unique physical, chemical, optical, electrical, magnetic and photocatalytic properties, to name but a few [3].

The modern semiconductor nano- and microelectronics is still largely dominated by the silicon technology. Structures such as Si/SiO_2 have found applications in solar cells, graphite films and carbon nanotubes, new 2D geometries, double gate devices, ballistic nanotube transistors, etc. A detailed discussion of the future development of nano devices can be found in the recent reviews [4-6]. One can conclude that at least in the next decade Si/SiO_2 will remain an important building block in many applications.

[3] To whom any correspondence should be addressed.

Thermal oxidation at lower temperatures reduces the film's growth rate and quality. The Si interface is of lower quality, higher gate leakage and interface trap density. One method of improvement is the incorporation of hydrogen, which has been widely used in the past decade due to different beneficial effects, such as stabilization of the device characteristics through gettering of impurities and saturation of defects. Hydrogenation is expected to attract further interest in future nanoelectronic applications, as exfoliation of silicon in smart cut process, increasing the carriers' lifetime in solar cells, etc. Applied to silicon oxide layers on a nanometric scale, hydrogenation can have a beneficial impact by improving the defect and leakage characteristics. Technologically, Si processing, especially pre-oxidation cleaning of the silicon surface, can play a decisive role in growing high-quality thin SiO_2 layers at lower oxidation temperatures.

The aim of the present paper is electrical characterization of the interface (111) n-Si/SiO_x formed on Si wafers by exposure to RF hydrogen plasma prior to oxidation. The interface traps density spectra are expected to depend on the amount of incorporated hydrogen. For this purpose, Si/SiO_x structures were incorporated in a MOS structure. During the plasma treatment, the Si wafers were heated up to 300 °C. SiO_x layers with a thickness of about 9 nm were formed by thermal oxidation at 850 °C. As a tool in the studies, multiple frequency C-V and G-V measurement techniques were used.

2. Experimental details

2.1. Sample preparation

The structures investigated were MOS capacitors formed on 5-10 Ω cm n-type (111)-oriented single-crystal Si wafers. The Si substrates were cleaned using a standard RCA procedure (H_2SO_4/H_2O_2 solution followed by a dip in diluted HF and a rinse in deionized H_2O). Some of the substrates were subsequently hydrogenated by exposure to RF hydrogen plasma in a planar plasma unit. The RF generator (13,56 MHz) was capacitively coupled to the reactive chamber and delivered input power of 15 W. The hydrogen gas pressure was 133 Pa. The substrates were kept on the lower electrode. Part of the substrates was hydrogenated without heating; other substrates were heated to a 300 °C. The plasma exposure duration was 15 min. Some of the wafers were left with RCA cleaning only, to serve for comparison revealing the effect of hydrogenation and possible advantages of the plasma technology proposed; these samples are referred to below as RCA oxides. Oxides with a thickness of about 9 nm were grown on RF plasma hydrogenated Si by thermal oxidation at 850 °C in dry O_2 ambient. The thickness of the SiO_x layers, determined by ellipsometric measurements, varied depending on the conditions of the pre-oxidation treatment, the hydrogenation level at different temperatures, or the wet RCA, but was in the nanoscale range. For electrical characterization, Al dots with a density of 1.96×10^{-3} cm^2 on the oxide surface and a continuous Al film on the Si back-side were deposited by vacuum evaporation to form MOS structures.

2.2. Measurements

We obtained information on the concentrations of the electrically-active defects and their location in the Si/SiO_x interface region by analyzing the multiple frequency dispersion of the capacitance-voltage *C-V* and conductance-voltage *G-V* characteristics taken at room temperature in the frequency range from 1 kHz to 300 kHz and a 30-mV test signal. The measurement device used was a Wayne-Kerr 6425 Precision Component Analyzer. The interface trap density $D_{it}(E)$ was evaluated by the standard high-frequency method and by comparing the experimental and the ideal theoretical *C-V* characteristics curves of the MOS structures [7]. The two-frequency method was developed to explore thin and/or leaky oxides [8]. The energy spectra of the interface traps $D_{it}(E)$ were obtained by comparing the experimental and the generalized *C-V* curves. The MOS capacitor was modelled as a three-element circuit. Thus, the generalized frequency-independent *C-V* curve was evaluated.

Using the values measured on the *C-V* and *G-V* curves at two frequencies in the range 1 kHz – 300 kHz, the dissipation D was calculated from

$$\mathbf{D} = \frac{\mathbf{G}}{\boldsymbol{\omega}\mathbf{C}}, \tag{1}$$

where G is the parallel conductance and C is the capacitance measured at the different frequencies. From the method presented by Yang and Hu [8] the generalized capacitance C′ is

$$\mathbf{C}' = \frac{\mathbf{f}_1^2\mathbf{C}_1(1+\mathbf{D}_1^2)-\mathbf{f}_2^2\mathbf{C}_2(1+\mathbf{D}_2^2)}{\mathbf{f}_1^2-\mathbf{f}_2^2}, \qquad (2)$$

where C_1 and D_1 refer to the values measured at the frequency f_1, and C_2 and D_2, to the values measured at the frequency f_2.

Additional information about the oxide and the interface region with the Si was obtained using a Rudolf Research spectroscopic ellipsometer in the wavelength range 280 – 820 nm at an incidence angle of 70°. The thickness was determined with an accuracy of ± 0.2 nm. The capacitance evaluated from the *C-V* curves and the values measured by the ellipsometer were used to calculate the thickness and dielectric constants of the oxide layers. The refractive index was calculated assuming the oxide-Si as being a single-oxide layer system.

3. Results and discussion

The analysis of the electrical characteristics showed that the MOS structures with a SiO_x layer grown on Si treated in plasma without heating exhibit a hysteresis effect indicating the presence of traps spatially distributed into the oxide near the interface. The shift and the shape of the curves revealed small oxide charge and low leakage currents, i.e., a high-quality dielectric layer.

We applied the two-frequency method of obtaining the *C-V* and *G-V* characteristics [8] at frequencies ranging from 1 kHz to 300 kHz and generated the generalized *C-V* curve for every sample. In this curve, the series resistance and the leakage through the oxide layer are taken into account. The generalized *C-V* curves for all samples examined are presented in figure 1.

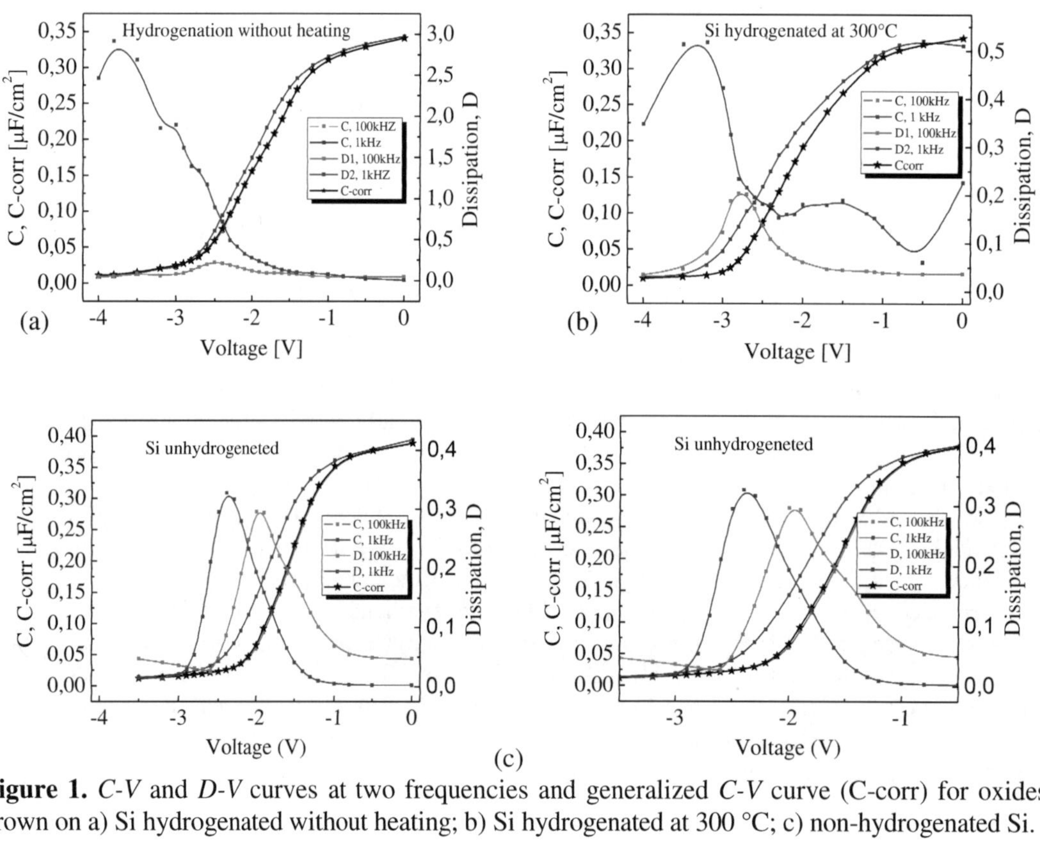

(c)

Figure 1. *C-V* and *D-V* curves at two frequencies and generalized *C-V* curve (C-corr) for oxides grown on a) Si hydrogenated without heating; b) Si hydrogenated at 300 °C; c) non-hydrogenated Si.

The role of the plasma pre-oxidation treatment on the electrically-active interface defects is illustrated in figures 1a and 1b for MOS structures with oxides grown on plasma-exposed substrates without heating, or heated at 300 °C, respectively. The shape of the curves of plasma-treated Si shows variations typical of a relatively high density of interface traps. In the shape of the curves of Si oxides hydrogenated without heating (figure 1a) and at 300 C (figure 1b), deviations are observed from the ideal shape, typical for a high density of interface traps, even higher than for the oxide of non-hydrogenated Si (figure 1c).

Hydrogenation of the Si wafers at 300 °C results in oxide with *C-V* and generalized *C-V* curves with a regular shape, as seen in figure 1b. The *C-V* curves of this sample are also steeper in comparison to the *C-V* curves of the oxides on Si not heated during the plasma hydrogenation, which indicates an overall lower density of traps, which can be attributed to the higher degree of an annealing or passivation effect.

The energy spectra of the interface traps D_{it} over the Si bandgap were calculated for the idealized case without the presence of interface traps over Si bandgap by comparing the generalized *C-V* curves with the theoretical *C*-V curves. The results for all oxides grown on hydrogenated without heating, hydrogenated at 300 °C and non-hydrogenated Si substrates are shown in figure 2. As can be seen, the interface trap density spectra show the presence of interface traps with localized energy levels in the Si bandgap. The overall high density of traps in all samples is not unexpected bearing mind the low oxidation temperature and the fact that the oxides were not subjected to any post-oxidation anneals. For all samples, the densities of the interface traps around midgap are less than 3×10^{12} eV^{-1} cm^{-2}. Also, the spectrum of the oxide grown on non-hydrogenated substrate reveals a set of energy levels in the Si bandgap with a relatively high density near the conduction band edge. It should be noted that this oxide has not been subjected to any heat treatment to reduce the trap density.

The localized interface levels for the hydrogenated samples are observed at approximately the same energy positions in the bandgap, with few exceptions, as compared to the non-hydrogenated samples. The oxide grown on hydrogenated Si shows a decrease of the density near the midgap. The position and density of the interface trap levels depend on the hydrogenation temperature of the Si substrate and are assumed to be related to particular structural defects at the interface and in a very thin region within the oxide. In the case of the oxide grown on a Si substrate heated to 300 °C during plasma hydrogenation, the density of the traps decreases even more than in the non-hydrogenated and hydrogenated without heating cases. The explanation could be due to the combined effect of the plasma, the higher temperature of the Si substrate and the incorporation of the hydrogen during the hydrogenation in the plasma, which serves to restore the destroyed chemical bonds in the interface region. For the oxide grown on a Si substrate hydrogenated without heating, although showing a positive effect as seen in figure 2, the inclusion of hydrogen only is not sufficient to reduce the density of the traps compared to the oxide grown on non-hydrogenated substrate.

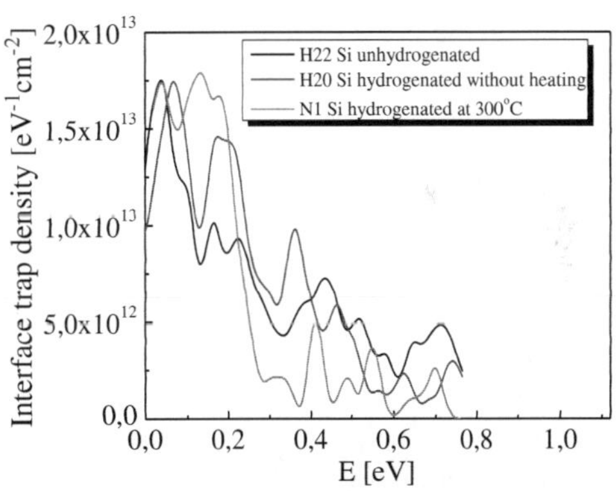

Figure 2. Interface trap density spectra.

The localized trap levels observed in the Si bandgap can be associated with the interface region where a transition from crystalline Si to amorphous silicon oxide occurs. It can be assumed that the oxide layer is not stoichiometric SiO_2, because the low oxidation temperature is insufficient for growing oxide with complete Si oxidation. The presence of hydrogen in the hydrogenated Si layer influences the mechanism of oxide growth and, hence, its structure. This is why in the interface spectra there appears a set of distributed localized states due to deformed bonds, rather than single

IOP Conf. Series: Journal of Physics: Conf. Series **992** (2018) 012041 doi:10.1088/1742-6596/992/1/012041

levels. The latter are usually attributed to a well-defined dangling bond at the interface of Si with high-temperature SiO_2.

For all samples, two peaks in the interval 0,05÷0,2 eV are evident at an energy position in the upper half of the bandgap, the one closer to E_c being with higher intensity, with the exception of the sample hydrogenated at 300 °C. Theoretically [9], these interface defect centers are considered oxygen vacancies and weak Si-Si bonds that appear energetically in this region. The peak at about 0.4 eV, observed for hydrogenated samples, could be linked to dangling Si-bonds, again based on the same theoretical and experimental results [9].

An interesting effect is the peak around the midgap found in the non-hydrogenated sample and in the sample hydrogenated at 300 °C, usually attributed to the dangling Si-bonds [9]. A peak at 0.68 eV is also observed, which is characteristic of the non-hydrogenated sample and the one hydrogenated at 300 °C and can be ascribed to weak Si-O bonds.

The basic oxide parameters characterizing the MOS structures, such as the oxide thickness, refractive index, oxide stress and width of the transitional interface region are given in table 1. These parameters resulting from ellipsometric measurements indicate that grown oxides must be regarded as SiO_x, where $x < 2$. The values of the refractive index at 633 nm for the oxides exceedr 1,46, as istypical of nonstoichiometric SiO_x. For our oxides, x was estimated from simulated data for the refractive index for SiO_x by varying the oxygen content [10]. The results are given in the table. Similar values for x were also found for SiO_x films deposited by reactive sputtering [11]. These refractive index values are higher as compared to stoichiometric SiO_2 and substantially lower than for Si monoxide. This implies that the oxide composition can be characterized rather as sub-stoichiometric SO_2 with $x < 2$.

The thickness of the SiO_2 layers, calculated from the ellipsometric data, varied around ~ 9 nm, depending on the pre-oxidation treatment, either RCA cleaning or plasma exposure. The oxides grown on plasma exposed substrates showed slightly greater thicknesses, as already observed in our previous studies [12]. This can be explained by the fast growth of the first monolayers of SiO_2 due to the presence of a plasma-modified surface layer in the Si substrate. This layer has been suggested to contain hydrogen decorated voids [13]. The refractive indices were higher than 1.46, the value typical for stress-free oxide.

Table 1. Parameters of oxides grown on Si with different pre-oxidation treatment.

Si wafer pre-oxidation treatment	Refractive index n	Oxide thickness d_{ox} (nm)	Dielectric constant ε	$\sigma.10^8$ (N/m^2)	x
Plasma hydrogenation: unheated	1,458	9,54	3,24	3,1	1,35
Plasma hydrogenation: 300°C	1,490	9,19	3,69	2,0	1,33
Without hydrogenation	1,580	8,30	3,70	4,0	1,25
SiO_2	1,460	-	3,90	-	2,00
SiO	1,960	-	5,00	-	1,00

The dielectric constants of the oxides were calculated from the capacitance in strong accumulation of the generalized *C-V* curves. The results for the dielectric constant of the oxide are also summarized in the table. For comparison, the table also presents data for ε and n of stoichiometric SiO_2 and SiO [12, 13, 14]. From the data for the dielectric constant, one can also draw conclusions about the oxides structure. One could assume that the lower dielectric constants are due to the presence of voids (ε = 1) in the upper Si-layer formed during the hydrogenation plasma, which persist even after oxidation [15].

The parameters of oxides on hydrogenated Si can also differ from those of oxides grown on non-hydrogenated Si before oxidation, as evidenced by the data in from the table.

IOP Conf. Series: Journal of Physics: Conf. Series **992** (2018) 012041 doi:10.1088/1742-6596/992/1/012041

4. Conclusions

Conclusions on the quality of the oxide and the interface to the Si substrate were drawn from the results of the electrical characterization performed on silicon oxides grown on RF plasma hydrogenated Si. The densities of the interface traps, as evaluated from *C-V* and *G-V* measurements using a two-frequency technique to generate frequency and leakage independent curves, depend on the degree of hydrogenation. The *C-V* and *G-V* analysis demonstrated the generation of localized interface traps due to dangling Si bonds and to weak and distorted Si-O bonds. A series of localized states acting as interface traps was found that characterize the interface of Si as a sub-stoichiometric SiO_x layer. For the oxide grown on hydrogenated Si at higher wafer temperature, most of the traps were annealed. The dielectric constants were determined from generalized *C-V* curves. We believe that the oxides grown on hydrogenated Si are non-stoichiometric and contain voids that reduce the dielectric constant. We further concluded that the density of interface defects can be optimized by heating the Si to 300°C during plasma hydrogenation.

References

[1] Ran G Z, Chen Y, Yuan F C, Qiao Y P, Fu J S, Ma Z C, Zong W H and Qin G G 2001 *Solid State Commun.* **118**/11 599-602
[2] Swain B S, Swain B P, Lee S S and Hwang N M 2012 *J. Phys. Chem.* C **116**/41 22036-42
[3] Qian K K and Bogner R H 2012 *J. Pharm. Sci.* **101** 444-63
[4] Saraswat K How far can we push Si CMOS? http://www.ohio.edu/people/starzykj/network/class/ee516/slides/Future%20Devices.pdf
[5] Iwai H 2013 Ultimate CMOS scaling *Proc. Korean Int. Summer School on Nanoelectr.* (2–5 July 2013 Daejon Korea) http://www.iwailab.ep.titech.ac.jp/pdf/iwaironbun/201307nanokiss.pdf
[6] Balestra F 2014 *Beyond CMOS Nanodevices* (Wiley & Sons – Interscience, New York)
[7] Nicollian E H and Brews J R 1982 *MOS Phys. and Technol.* (Wiley, New York)
[8] Yang K J and Hu Ch 1999 *IEEE Trans. Electron Dev.* **46** 1500
[9] Flietner H 1985 *Phys, Stat. Solidy* A **91** 153–7
[10] Tomozeiu N 2011 *Silicon oxide (SiOx, 0 < x < 2): a challenging material for optoelectronics, Optoelectronics - materials and techniques* ed Predeep P
[11] Miyazaki H 2010 Phys. Chem. of Glasses - *European J. Glass Sci. Technol.* B **51** 136
[12] Szekeres A, Alexandrova S, Lytvyn P and Kompitsas M 2005 *J. Phys.: Conf. Series* **10** 246
[13] Asoka-Kumar P, Stein H J and Lynn K G 1994 *Appl. Phys. Lett.* **64** 1684
[14] Szekeres A, Paneva A, Alexandrova S, Lisovskyy I, Litovchenko V and Mazunov D 2003 *Vacuum* **69** 355
[15] Kaschieva S, Halova E, Vlaikova E, Alexandrova S, Valcheva E and Dmitriev S 2006 *Plasma Process. Polym.* **3** 237

IOP Conf. Series: Journal of Physics: Conf. Series **992** (2018) 012042 doi:10.1088/1742-6596/992/1/012042

Surface modification of an aluminum alloy by electron beam introducing TiCN nanoparticles

M Kolev[1,3], R Dimitrova[1], St Parshorov[1], St Valkov[2], R Lazarova[1] and P Petrov[2]

[1]Acad. A. Balevski Institute of Metal Science, Equipment and Technology with Hydro- and Aero-dynamic Center, Bulgarian Academy of Sciences, 67 Shipchenski Prohod Blvd., 1574 Sofia, Bulgaria
[2]E. Djakov Institute of Electronics, Bulgarian Academy of Sciences, 72 Tzarigradsko Chaussee, 1784 Sofia, Bulgaria

E-mail: mihail1kolev@gmail.com

Abstract. TiCN nanopowder deposited in an appropriate way on the surface of an $AlSi_{12}Cu_2NiMg$ substrate was incorporated in the matrix using an electron beam technology. The samples were studied by means of light microscopy, SEM, and EDX; their microhardness was also determined. The formation was found of a uniform and dense coating with a thickness of 7 – 10 μm with a good adherence to the substrate. A modified zone appeared under the coating with a thickness of 100 – 150 μm containing dendrites of an α-solid solution and a fine eutectic between them, as well as primary silicon crystals. The microhardness of this modified zone was up to 2.4 times higher than that of the matrix. The results of SEM and EDX studies revealed unambiguously the presence of titanium in the coating and in the zones below it. Obviously, the electron beam treatment resulted in the TiCN nanoparticles penetrating into the coating and the substrate immediately below the coating.

1. Introduction

A large number of works reported the possibility to prepare coatings with improved properties by depositing nanoparticles in an appropriate way on a metal substrate followed by an irradiation treatment [1-6]. It is known that nanoparticles having a crystal lattice with dimensions and orientation similar to those of a metal act as nucleus centers in the melt during cooling [7]. When they are well distributed and in a sufficient concentration in the melt, they hinder the atoms' diffusion, thus limiting the grains' growth. As a result, the metal solidifies in a finer structure. Also, the nanoparticles serve as barriers to the motion of dislocations, thus increasing the strength of the solid material [8–10].

The aim of the present work was to obtain a coating with improved properties by introducing TiCN nanoparticles in an aluminum alloy substrate through electron beam irradiation. The $AlSi_{12}Cu_2NiMg$ alloy is widely used for pistons and cylinder liners, which requires good friction and tribological properties of the surface. The above references, together with our previous experience on applications involving nanoparticles in metal melts [11-13] on the one hand, and on preparing highly wear-resistant nanosized layers by electron beam treatment [14-17] on the other, prompted us us to develop the approach discussed here. In our opinion, the electron beam treatment is a technique suitable for

[3] To whom any correspondence should be addressed.

incorporating nanoparticles deposited on a metal substrate. Combining a circular electron beam motion with a translational motion of the substrate, one can cause re-melting, stirring and mixing of the metal surface, thus incorporating the nanoparticles into the matrix. The result expected is reinforcement of the metal matrix by the hard and refractory nanoparticles.

2. Experimental procedure

2.1. Materials

2.1.1. Coating material. The refractory TiCN powder (Neomat Co., Latvia, mean particle size 40±5 nm) was chosen as a coating material due to its excellent properties, such as high melting point, extreme hardness, and high heat conductivity [18]. Like aluminium, it has an FCC lattice, with a constant close to that of aluminium.

2.1.2. Substrate material. The chemical composition of the aluminum alloy used as a substrate is shown in table 1. This alloy is used for production of pistons operating under heavy duty conditions of friction and wear.

Table 1. Chemical composition of the substrate alloy.

Element	Cu	Mg	Si	Ni	Al
Wt %	0,930	0,331	11,31	0,109	rest

2.2. Procedure

Metallographic samples of the $AlSi_{12}Cu_2NiMg$ alloy were ground by grinding paper 1200. They were then smeared with a mixture of three drops of $CHCl_3$, plastic rasping and TiCN powder to obtain a film with concentration 0.015 mg/mm^2 TiCN. The film was uniform and well adherent to the aluminum substrate. The as-prepared samples were subjected to electron-beam treatment in a type ESW300/60-15 installation (Leybold-Heraeus). The electron beam scanned the sample in a circular rotating motion, while the sample moved in a straight line. We were thus able to maintain a liquid pool for a long time, with the melt being well stirred and the nanoparticles mixing with the molten metal. The electron beam treatment conditions are given in table 2.

Following the electron beam treatment, the cross-sections of the metallographic samples perpendicular to the coating were wet-grinded by silicon carbide grinding paper 1200 and 4000 and etched by 0.5% water solution of HF. The samples were studied by means of light microscopy, SEM, and EDX. The micro-hardness of all samples was also determined.

Table 2. Electron beam treatment parameters.

I_b [mA]	v [cm/s]	f [kHz]	U [kv]	I_f [mA]	P [W]
30	2,5	10	52	472	1560

The equipment used was as follows: a PolyvarMet metallographic microscope with magnification up to 1000×; a MicroDuromat (Reichert-Jung) HV microhardness device with a load of 20 g, time for reaching the load of 10 s and holding time of 10 s; SEM/FIB LYRAI XMU, TESCAN, equipped with an EDX detector (Quantax 200, Brucker).

3. Results and discussion

3.1. Light microscopy

The microstructure of a treated sample is shown in figure 1(a) and (b). A dense coating with a uniform thickness of about 7 – 10 μm is seen on the substrate surface. It is well adherent to the underlying material. The microhardness was not measured because of the relatively small thickness and the relief. There is a modified zone with a thickness of 100 – 150 μm and a dendritic structure under the coating. It consists of α-solid solution dendrites and a fine eutectic between them, as well as of primary Si crystals. The microhardness of this modified zone was up to 2.4 times higher than that of the matrix – figure 1(c).

We should note here that the microhardness measurement unit, given as imprint in figure 1 (c) is kgf/mm^2. The microhardness of the modified zone in MPa is approximately 2893 MPa, while the same parameter for the matrix is 1206 MPa.

3.2. SEM and EDX

The mapping zone and the mapping chart are presented in figure 2 (a) and (c). The titanium is distributed mainly in the coating and slightly underneath. The elements' spectra are given in figure 2 (b), where the titanium and aluminum peaks are well visible. The SEM and EDX results (figure 3) clearly show the presence of titanium in the coating and in the areas below it. Thus, the TiCN nanoparticles were embedded in the substrate material as a result of the electron beam treatment. The higher hardness of the coating is due not only to the residual stresses after the electron beam irradiation, but also to the presence of nanoparticles, which obstruct the motion of dislocations and increase the strength of the metal matrix. We believe that this is the main material strengthening mechanism, although other mechanisms are also possible in this case, such as, for example, the grain refinement mechanism, as the microstructure of the surface is visibly refined.

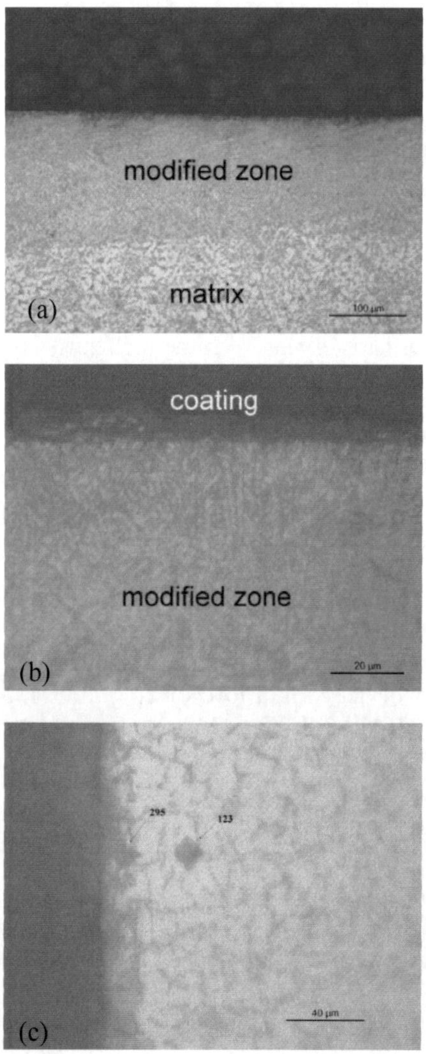

Figure 1. Microstructure of a treated sample (a, b) and microhardness imprints (c).

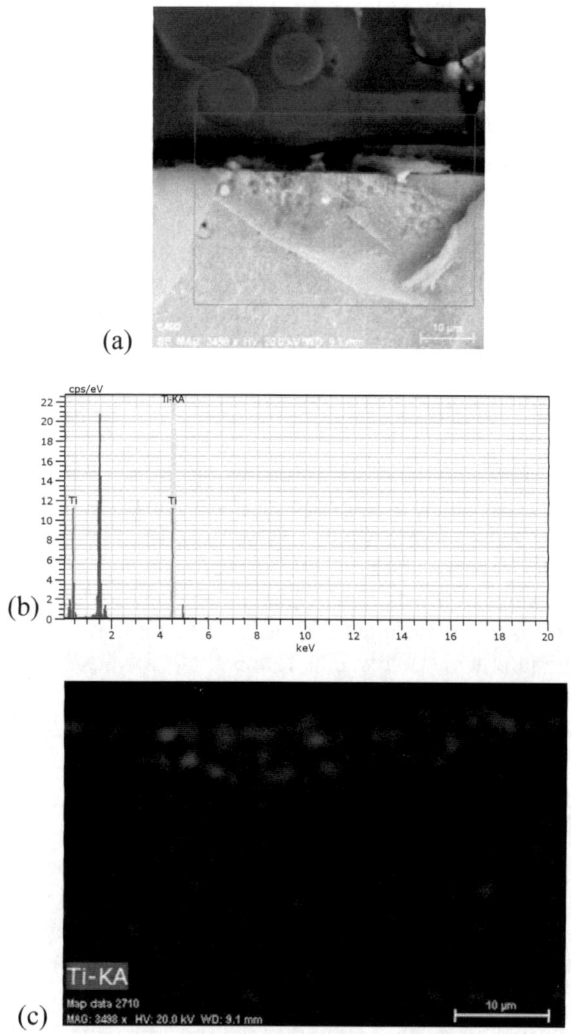

Figure 2. SEM image of TiCN coated sample with mapping zone (a), elements spectra (b) and composition map of Ti (c).

IOP Conf. Series: Journal of Physics: Conf. Series **992** (2018) 012042 doi:10.1088/1742-6596/992/1/012042

The Ti content at the marked point, which is in the substrate immediately beneath the coating, is 0.24 wt % – table 3. The conclusion one should draw is that the nanoparticles have penetrated into the substrate due to the electron beam treatment.

Table 3. Elemental composition in the marked point – figure 3 (a).

Acquisition 6531 Date:6/23/2017 11:25:05 AM HV:20.0 kV Puls th.:3.89 kcps

El	AN	Series	unn. [wt.%]	C norm. [wt.%]	C Atom. [at.%]	C Error [%]
C	6	K-series	23.09	19.21	32.18	3.9
Al	13	K-series	49.43	41.13	30.67	2.4
F	9	K-series	14.21	11.82	12.52	2.5
Si	14	K-series	19.78	16.46	11.79	0.9
O	8	K-series	10.85	9.03	11.35	16.6
Mg	12	K-series	1.45	1.21	1.00	0.1
Cl	17	K-series	0.44	0.37	0.21	0.0
Cu	29	K-series	0.64	0.53	0.17	0.1
Ti	22	K-series	0.29	0.24	0.10	0.0
	Total:		120.18	100.00	100.00	

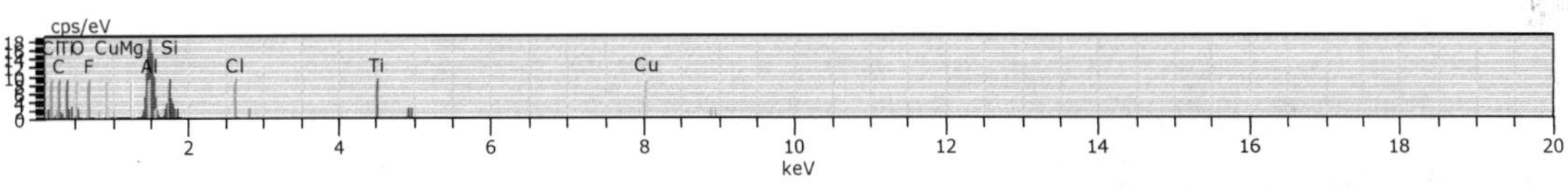

(b)

4. Conclusions

Electron beam treatment was successfully applied to produce a TiCN coating on the surface of an alloy used for fabrication of pistons.

The nanoparticles penetration in the metal matrix was proved by EDX analysis.

The coating showed good adherence to the substrate and uniform thickness.

The underlying zone beneath the coating possessed hardness higher than that of the matrix.

The major possible cause of the coatings' increased strength is the presence nanoparticles which obstruct the motion of dislocations. The grains refinement could also contribute to the improved properties of the coatings.

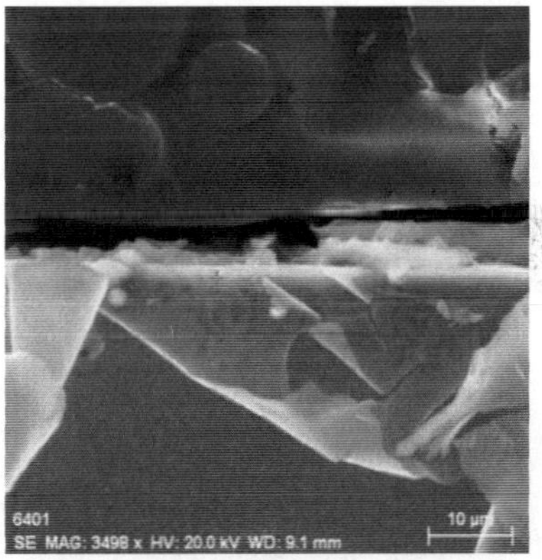

(a)

Figure 3. EDX analysis results in the marked point (a) and elements spectra (b).

Acknowledgements
This work was financially supported in part by the Bulgarian National Science Fund under Contract DH07/16.

References

[1] Murzakov M, Chirikov S and Markushov Y 2016 *J. Phys.: Conf. Series* **747** 012062
[2] Xiaopeng L, Mohedano M, Blawert C, Matykina E, Arrabal R, Kainer K and Zheludkevich M 2016 *Surf. Coat. Technol.* **307** 1165–82
[3] Petrov P 1997 *Vacuum* **48**/1 49-50
[4] Valkov S, Petrov P, Lazarova R, Bezdushnyi R and Dechev D 2016 *Appl. Surf. Sci.* **389** 768–74

[5] Valkov S, Petrov P and Lazarova R 2017 *Proc. SPIE* **10226** 102260I-1
[6] Wei D, Xin W and Wang R 2010 *Proc. Int. Conf. Adv. Technol. Des. Manuf.* 357-60
[7] Cherepanov A and Ovcharenko V 2015 *Phys. Metals and Metallography* **116**/12 1279-84
[8] Kowalczyk B, Bishop K, Lagzi I, Wang D, Wei Y, Han S and Grzybowski B 2012 *Nat. Mater.* **11** 227
[9] Chen L, Xu J, Choi H, Konishi H, Jin S and Li X 2014 *Nat. Commun.* **5** 3879
[10] Wang K, Jiang H, Jia Y, Zhou H, Wang Q, Ye B and Ding W 2016 *Acta Mater.* **103** 252
[11] Kuzmanov P, Dimitrova R, Lazarova R, Cherepanov A and Manolov V 2014 *J. Nanoeng. Nanosystems* **228**/1 11-8
[12] Dimitrova R, Kuzmanov P, Lazarova R and Manolov V 2013 *Adv. Mater. Res.* **629** 284-91
[13] Lazarova R, BojanovaN, Dimitrova R, Panov I and Manolov V 2016 *Int. J. Metalcasting* **10**/4 466-76
[14] Petrov P, Parshorov St, Lazarova R, Dechev D, Ivanov N and Chobanov P 2015 *J. Mater. Sci. Technol.* **23**/3 251–6
[15] Petrov P, Valkov S, Lazarova R, Parshorov S, Bezdushnyi R, Detchev D and Ivanov N 2016 *J. Mater. Sci. Technol.* **24**/2 92-98
[16] Petrov P, Parshorov S, Lazarova R, Dechev D, Ivanov N and Chobanov P 2015 *J. Mater. Sci. Technol.* **23**/3 251–63
[17] Petrov P, Ormanova M, Dechev D, Ivanov N and Chobanov P 2015 *J. Mater. Sci. Technol.* **23**/2 153–9
[18] Caicedo J C, Amaya C, Yate L, Gómez M E, Zambrano G, Alvarado-Rivera J, Muñoz-Saldaña J and Prieto P 2010 *Appl. Surf. Sci.* **256** 5898–904

IOP Conf. Series: Journal of Physics: Conf. Series **992** (2018) 012043 doi:10.1088/1742-6596/992/1/012043

Shelf life and outdoor degradation studies of organic bulk heterojunction solar cells

R Gergova[1,3], M Sendova-Vassileva[1], G Popkirov[1], V Gancheva[2] and G Grancharov[2]

[1]Central Laboratory of Solar Energy and New Energy Sources, Bulgarian Academy of Sciences, 72 Tzarigradsko Chaussee, 1784 Sofia, Bulgaria
[2]Laboratory of Structure and Properties of Polymers, Institute of Polymers, Bulgarian Academy of Sciences, Acad. G. Bonchev Str., Bl. 103-A, 1113 Sofia, Bulgaria

E-mail: rgergova@lakeheadu.ca

Abstract. We studied the degradation of different types of bulk heterojunction devices, in which the materials comprising the active layer and/or the materials used for the back electrode are varied. The devices are deposited on ITO covered glass and have the structure PEDOT:PSS/BHJ/Me, where PEDOT:PSS is the hole transport layer, BHJ (bulk heterojunction) is the active layer comprising a polymer donor (e.g. PTB7, PCDTBT) and a fullerene derivative acceptor (e.g. PC60BM, PC70BM) deposited by spin coating, Me is the metal back contact, which is either Ag or Al deposited by magnetron sputtering or thermal evaporation. The device performance was monitored after storage in the dark at ambient conditions by following the evolution of the J-V curve over time. Results of real conditions outdoor degradation studies are also presented. The stability of the different solar cell structures studied is compared.

1. Introduction

Organic solar cells (OSCs) have been intensively studied as a low-cost alternative to the solar cells based on inorganic semiconductors. The goal of power conversion efficiency of above 10 % has been achieved by successful device optimization of bulk heterojunction solar cells [1]. However, device lifetime continues to be a key limitation for their commercial application. Long-term stability of OSCs is currently one of the major topics for this technology and combinations of various donor and acceptor materials are being studied in order to obtain more stable devices.

A considerable body of work has been published in recent years describing the stability and degradation behavior of the most stable but relatively low-efficient poly(3-hexylthiophene) P3HT:fullerene-based organic solar cells [2]. On the other hand, low band-gap polymers are researched as promising materials that are able to achieve higher efficiencies than the P3HT:fullerene system, but few reports have appeared dealing with their stability upon exposure to real outdoor conditions.

Here we report on the shelf life and outdoor degradation of devices based on thieno[3,4-b]thiophene-*alt* -benzodithiophene (PTB7) and the fullerene derivatives [6,6]-phenyl C71-butyric acid

[3] To whom any correspondence should be addressed.

methyl ester (PC70BM) and [6,6]-phenyl-C_{61}-butyric acid methyl ester (PC60BM). Devices based on PTB7 in combination with PC70BM have achieved efficiency over 7 %, but the blend was found to suffer from poor stability [3]. Pristine PTB7 is very susceptible to photooxidation and further research and modifications of the polymer are needed to prevent this degradation.

A promising material which has been shown to have high degree of photochemical stability is the donor-acceptor copolymer poly[N-9'-heptadecanyl-2,7-carbazole-alt-5,5-(4',7'-di-2-thienyl-2',1',3'-benzothiadiazole)] PCDTBT [6]. The most studied device structure based on this polymer is glass/ITO/PEDOT:PSS/PCDTBT:PC70BM/Al. It is known that thermal annealing is not desirable for this structure, in contrast to the devices based on the crystalline P3HT polymer [5]. We explored the stability under outdoor sun irradiation and in the dark of annealed PCDTBT:PC70BM devices with thermally evaporated Al cathodes and compared their performance over time to identical unannealed devices. We have also compared these results to the behavior of similar devices with magnetron-sputtered Ag back contacts.

2. Experimental

The solar cells were prepared on glass/ITO substrates. Prior to the HTL deposition, the substrates were cleaned with distilled water, acetone and isopropanol followed by UV ozone treatment. PEDOT:PSS was spin-coated from water solution and annealed for 10 minutes at 130 °C in Ar. The PCDTBT:PC_{70}BM BHJ was deposited by spin-coating from dichlorobenzene solution with concentration 10 mg/ml (1:4 weight ratio) in ambient conditions. For the cells with the sputtered Ag contacts, a concentration of 30 mg/ml was used. The PCDTBT polymer and the substrates were purchased from Ossila Ltd. The PTB7:PC60BM and PTB7:PC70BM blends were prepared in a 1:1.5 weight ratio in 25-mg/ml solution of dichlorobenzene and spin-coated in air. The PTB7 polymer used for the experiment was synthesized in the Institute of Polymers of the Bulgarian Academy of Sciences. After spin-coating, the samples were annealed in Ar for 15 minutes at 50 °C. Finally, an aluminium layer was thermally evaporated on top of the active layer for back contact. For some of the PCDTBT: PC70BM solar cells, Ag contacts were deposited by magnetron sputtering. The samples thus prepared represent a matrix of six independent cells of area 6 mm^2 each. The devices were encapsulated using a cover glass on top of the back metal contact glued by epoxy UV-curable resin from Ossila Ltd. After the encapsulation, the devices comprising PTB7:PC70BM (PC60BM) active layers were annealed in air for 10 minutes at 120 °C. Some of the PCDTBT:PC70BM solar cells were annealed for 10 minutes at 110 °C, while others were left as deposited.

The J-V characteristics were acquired in periods of ten to twenty days in order to determine the lifetime in the dark of the devices stored in ambient environment. The J-V measurements were conducted using a fully-automated computerized system including a halogen lamp as an illumination source adjusted to deliver 100 mW/cm^2.

The experiment for outdoor degradation study was performed by exposing the samples under open circuit conditions every day on the rooftop of the laboratory for five hours around midday (between 10 a.m. and 3 p.m.) in July and August 2017. Five cells of each kind were studied for this experiment. Their performance was recorded before and after each exposure. The temperature range during irradiation was from 35 °C to 45 °C.

3. Results and discussion

3.1. Degradation behavior of devices stored in ambient environment

Figure 1 shows the J-V characteristics of two typical glass/ITO/PEDOT:PSS/PCDTBT:PC70BM/Al devices. One of them was subjected to post-fabrication annealing in air at 110 °C; the other was not annealed. Both devices were stored in the dark and characterized again after two weeks. The open-circuit voltage (Voc) of the annealed sample decreased by nearly 100 mV after the annealing, while the short-circuit current (Jsc) remained almost unchanged. The decline in the performance of PCDTBT-based BHJs after annealing has been observed and studied in [5]. In a previous work [6], we studied PCDTBT:PC70BM bulk heterojunction solar cells with magnetron-sputtered Ag contacts and observed a slight improvement in the J-V curves after annealing at 120 °C. In figure 2, the normalized quantum efficiency (QE) spectra measured before and after annealing at 110 °C are compared. The shape of the curve has not changed after thermal annealing; this provides evidence that the thermal degradation in this case was not related to morphological changes or oxidation in the active layer of the device, which would have resulted in a change of the shape of the spectral response.

After 15 days storage in the dark, a sharp drop in the Voc of the unannealed sample was observed, which reached the Voc of the annealed device. We speculate that this behavior might be related to chemical changes at the interface between the active layer blend and the cathode.

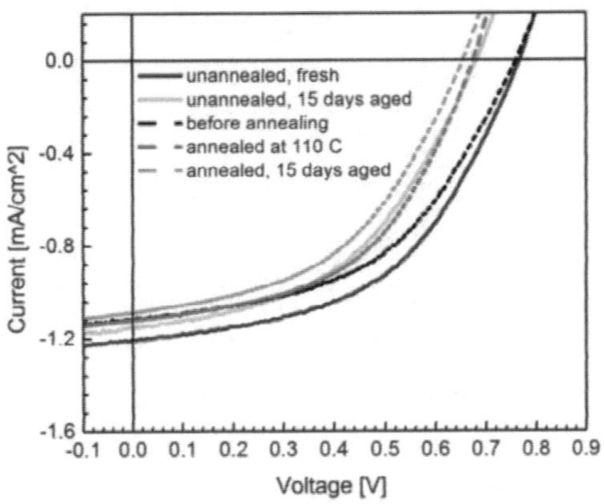

Figure 1. J-V characteristics of glass/PEDOT:PSS/PCDTBT:PC70BM/Al annealed and unannealed devices after 15 days of storage in the dark.

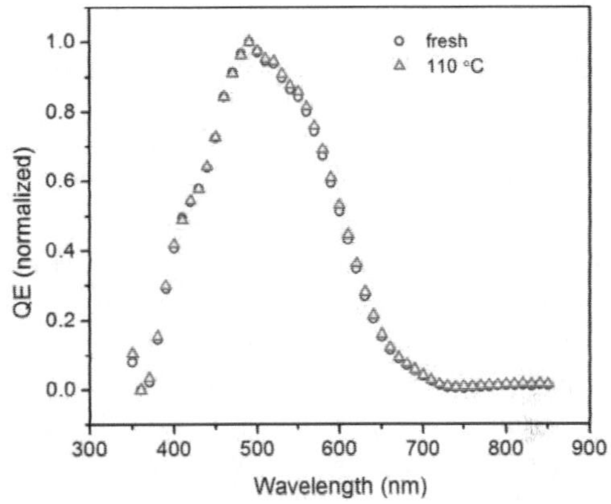

Figure 2. Normalized quantum efficiency of PCDTBT:PC70BM solar cells of freshly pre-pared device and after annealing at 110 °C.

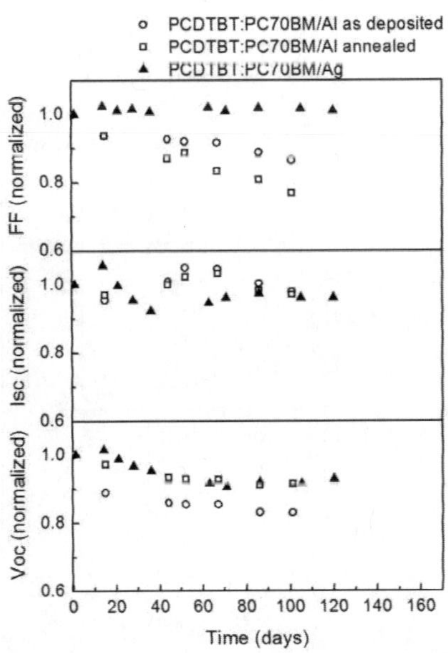

Figure 3. Evolution of the parameters obtained from J-V curves for three solar cells with PCDTBT:PC70BM active layers, stored in the dark.

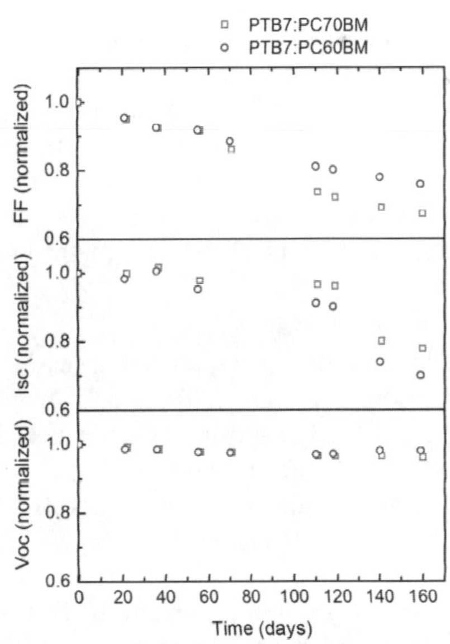

Figure 4. Evolution of the parameters obtained from J-V curves for solar cells with PTB7:PC70BM (PC60BM) active layers, stored in the dark.

IOP Conf. Series: Journal of Physics: Conf. Series **992** (2018) 012043 doi:10.1088/1742-6596/992/1/012043

The performance of the solar cells studied was measured frequently for 120 days in order to observe their behavior when stored in ambient environment. The evolution of the main parameters obtained from the J-V measurements and normalized to their initial value is shown in figure 3. For the unannealed device, after an initial sharp drop during the first fifteen days, the Voc stabilizes and for the next measurements decreases very slowly. The annealed device, in contrast, does not show such drastic drop in Voc during the first fifteen days, but it declines gradually and seems to saturate after about 60 days. For both devices, there is a slight decrease in the initial Jsc; afterwards it remains unchanged. The fill factor (FF) decreases linearly with time which could be an indication of increasing series resistance. The parameters of a third structure with silver sputtered back contact are also plotted for comparison (filled symbols). As the graph illustrates, this device is more stable than the devices with thermally evaporated aluminium electrodes, which is in agreement with our previous work [6], although it should be noted that this device had initially a higher series resistance.

The devices with active layers PTB7:PC70BM and PTB7:PC60BM showed improve-ment in their J-V characteristics after thermal annealing at 120 °C with an increase in their FF. The devices were stored in ambient environment and were monitored frequently for 160 days. The different lengths in the aging periods for the various devices arise from the different time of their fabrication. The evolution of the main parameters is presented in figure 4. It is evident, that the FF of the sample with the PC70BM acceptor degrades faster than FF of the sample having PC60BM in the active layer. The FF of both devices decreases by about 30 % for 160 days. Also a decline is observed in the Jsc, while the Voc remains unchanged.

3.2. *Outdoor degradation*

Five representative samples for the structures discussed in the previous section were selected and their behavior was monitored upon daily exposure to sun irradiation.

Figure 5 and figure 6 show the time evolution of the device parameters for cells with PCDTBT:PC70BM and PTB7:PC70BM(PC60BM) active layers, respectively. Only results from the inner cells of the 6-pixel matrix were selected and presented in this work because during the experiment we found that the edge cells degraded very fast as a result of insufficient edge encapsulation. As seen in the figures, the devices having PTB7:PC70BM(PC60BM) active layers experience more abrupt burn-in degradation in the first five hours of exposure than the PCDTBT:PC70BM-based devices. By the

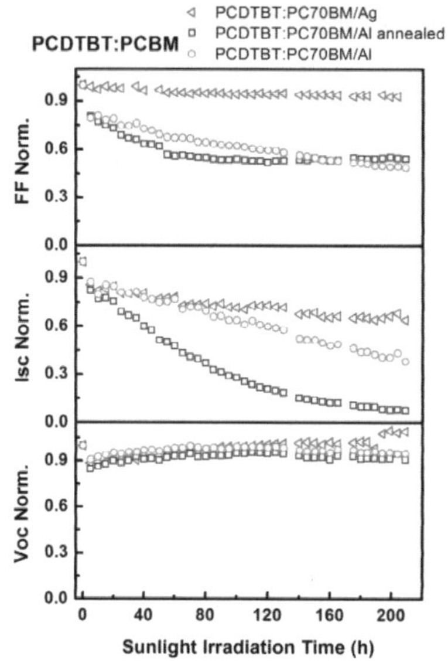

Figure 5. Device characteristics of PCDTBT:PC70BM device with sputtered Ag contact (triangles), annealed PCDTBT:PC70BM with thermally evaporated Al contact (squares) and unannealed PCDTBT:PC70BM with thermally evaporated Al contact over 200 hours under sunlight irradiation.

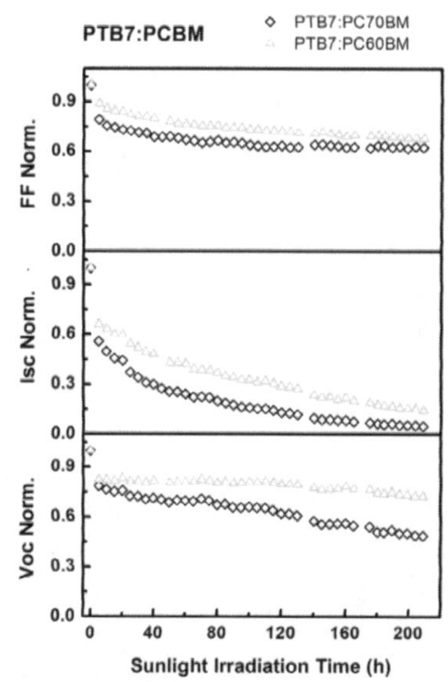

Figure 6. Device characteristics of PTB7:PC70BM-based structures (diamonds) and PTB7:PC60BM-based structures (triangles) over 200 hours under sunlight irradiation.

end of the experiment, after 200 hours of exposure to outdoor conditions, all of the devices studied with thermally evaporated Al back contacts decreased their performance significantly. Both PTB7:PC70BM- and PTB7:PC60BM-based solar cell experienced severe degradation after the first five hours of exposure to sun irradiation. Their short-circuit current dropped by 45% and 35%, respectively, which is almost double the amount of current drop in PCDTBT:PC70BM-based structures.

The PCDTBT:PC70BM-based solar cell with silver sputtered cathode showed the most stable behavior in comparison to the other two similar samples. The annealed and unannealed devices showed similar behavior in the first 20 hours of exposure to outdoor conditions which suggests that the initial stage of the burn-in process does not depend on the post-deposition annealing.

4. Conclusions

We studied and compared the shelf life and outdoor degradation of two types of organic bulk heterojunction solar cells that were spin-coated in air. The PCDTBT:PC70BM samples experience a significant decrease in the open-circuit voltage after the first two weeks of storage in the dark, which we attribute to possible chemical changes at the active layer/electrode interface that seem to be accelerated by thermal annealing. PCDTBT:PC70BM samples with silver sputtered contacts show superior resistance to outdoor weathering compared to the samples with thermally evaporated aluminum back contacts. Samples with aluminium back contacts experience longer lifetime upon sunlight exposure when they are not subjected to post-deposition thermal treatment.

The solar cells based on PTB7:PC60BM degrade faster than the cells using the PC70BM acceptor. Both types of devices degrade very fast upon exposure to sunlight irradiation.

Acknowledgments

The authors would like to acknowledge the support of the Bulgarian National Science Fund (Contract DFNI E02/5)

References

[1] Liu C, Yi C, Wang K, Yang Y, Bhatta R S, Tsige M, Xiao S and Gong X 2015 *Appl. Mater. Interfaces* **7**/8 4928
[2] Jorgensen M, Norrman K, Gevorgyan S A, Tromholt T, Andreasen B and Krebs F C 2012 *Adv. Mater.* **24** 580
[3] Razzell-Hollis J, Wade J, Tsoi W C, Soon Y, Durrant J and Kim J-S 2014 *J. Mater. Chem.* A **2** 20189
[4] Peters C H, Sachs-Quintana I T, Kastrop J P, Beaupre S, Leclerc M and McGehee M D 2011 *Adv. Energy Mater.* **1** 491
[5] Beiley Z M, Hoke E T, Noriega R, Dacuna J, Burkhard G F, Bartlet J A, Salleo A, Toney M F and McGehee M D 2011 *Adv. Energy Mater.* **1** 954
[6] Sendova-Vassileva M, Popkirov G, Vitanov P, Grancharov G, Gancheva V, Dikov H, Lazarova E and Gergova R 2016 *J. Phys.: Conf. Ser.* **700** 012053

IOP Conf. Series: Journal of Physics: Conf. Series **992** (2018) 012044 doi:10.1088/1742-6596/992/1/012044

Structural and morphological properties of sol-gel ZnO:Ni films

T Ivanova[1,5], A Harizanova[1], T Koutzarova[2], B Vertruyen[3] and B Stefanov[4]

[1]Central Laboratory of Solar Energy and New Energy Sources, Bulgarian Academy of Sciences, 72 Tzarigradsko Chaussee, 1784 Sofia, Bulgaria
[2] E. Djakov Institute of Electronics, Bulgarian Academy of Sciences, 72 Tzarigradsko Chaussee, 1784 Sofia, Bulgaria
[3]LCIS/SUPRATECS, Institute of Chemistry B6, University of Liege, Sart-Tilman, B-4000 Liege, Belgium
[4]Department of Engineering Sciences, The Ångström Laboratory, Uppsala University, P.O. Box 534, Uppsala, Sweden

E-mail: tativan@phys.bas.bg

Abstract. Ni doping induces modifications and considerable changes in the optical, electrical and magnetic properties of ZnO films. In this work, the influence is discussed of Ni-doping (two nickel concentrations) and annealing temperature (ranging from 300 °C to 800 °C) on the structural and optical properties of sol-gel derived ZnO:Ni films. Uniform and smooth films were obtained by spin-coating on quartz and Si substrates. The ZnO:Ni films were crystallized in wurtzite phase with no impurity phases found for annealing temperatures up to 600 °C. The size of the crystallites was strongly affected by the Ni content and the heat treatment. Furthermore, the Ni doping improved the optical transparency of the sol-gel films, while the AFM studies showed that the film morphology and the roughness were influenced by the nickel doping.

1. Introduction

ZnO is considered as being one of the most important semiconductors due to its wide bandgap (3.37 eV), large exciton energy (60 meV), thermal and chemical stability and hexagonal crystal structure. The ZnO nanostructured films possess desirable optical, structural and electronic properties for applications in thin-film transistors, solar cells, gas sensors, UV light emitting diodes, field emitters, piezoelectric devices, photo detectors, flat screen displays, sun screens, spintronic devices, etc. [1].

Doping the ZnO films modulates and changes their morphology, density of states, transmittance, bandgap, electrical conductivity, etc. [2]. Nickel is an important dopant, as Ni^{2+} has an ionic radius of 0.069 nm, which is slightly smaller than that of the Zn^{2+} ion (0.074 nm); thus, Ni^{2+} ions should be able to substitute Zn^{2+} in a ZnO matrix. ZnO:Ni films are being studied for potential applications in solar cells and memory devices [3]. Ni doping is a viable method of improving the sensing ability of ZnO by tuning the electronic band structure and modifying its energy levels and surface states. Since gas sensing is a surface phenomenon, where the sensing element‘s energy band bends upwards or

[5] To whom any correspondence should be addressed.

downwards based on its interaction with an oxidizing or reducing gas, Ni can enhance the gas sensing properties [4]. ZnO films doped with Co, Ni, Fe have been widely studied as diluted magnetic semiconductors (DMS) [5]. Doped ZnO films exhibit magnetic behavior due to doping with transition metals and/or manifest magnetic properties through structural defects caused during the doped films' synthesis. The origin of ferromagnetism in nanostructured DMO's is still being explored and discussed [6].

ZnO doped films are deposited by pulse laser deposition, r.f. magnetron sputtering, molecular-beam epitaxy, CVD and ALD [7]. Among these methods, the sol-gel deposition has several advantages, such as inexpensive equipment, precise microstructural and chemical control, reproducibility, possibilities of engineering the films' properties. Preparation of high-quality ZnO-based nanostructured films by the sol-gel technology has been reported earlier [8].

In the present work, the effect is studied of adding Ni (two Ni concentrations) and varying the annealing temperature (300 °C – 800 °C) on the structural, optical and morphological properties of sol-gel Ni-doped ZnO films.

2. Experimental

The preparation of sols for ZnO and nickel-doped ZnO deposition were described previously [9]. Briefly, 0.4 M Zn sol and 0.2 M Ni sol were mixed in two volume ratios: ZnO:Ni 0.4 (19.6 ml Zn sol:0.4 ml Ni sol) and ZnO:Ni 1 (19 ml Zn sol :1 ml Ni sol). The films were obtained by spin coating at 4000 rpm on Si and quartz substrates. The layer deposition procedure was repeated five times with pre-heating of 300 °C/10 min between layers. The ZnO:Ni films were annealed at temperatures ranging from 300 °C to 800 °C.

A Bruker D8 XRD diffractometer (grazing angle 2°, step time 8 s) was used for the structural study. The FTIR spectra were taken by an IRPrestige-21 Shimadzu FTIR Spectrophotometer. The optical measurements were carried out by a Shimadzu 3600 UV-VIS spectrophotometer. The atomic force microscopy (AFM) observation was performed by employing a PSIA XE150 SPM/AFM instrument equipped with an ACTA silicon cantilever (AppNano), a tip radius 6 – 10 nm operating in non-contact mode.

3. Results and discussions

The XRD patterns of ZnO and ZnO:Ni films annealed at 300 °C and 600 °C are presented in figure 1; as seen, the annealing temperature influences significantly the crystallization behavior of ZnO-based films. Thc undoped ZnO films improve their crystallization gradually as the annealing temperature is raised. The Ni-doped ZnO films exhibit a more complicated crystallization evolution. ZnO:Ni 0.4 and ZnO:Ni 1 films manifest an enhanced crystallinity for the annealing temperatures of 300 °C, 400 °C and 500°C; then the crystallization degree deteriorates after the high-temperature treatment at 600 °C.

Our earlier studies [9] revealed that the ZnO:Ni 0.4 films achieved the best crystallization after treatment at 700 °C, while the ZnO:Ni 1 films exhibited the most intense XRD peaks after annealing at 800 °C,; or, doping with Ni affects significantly the crystallization behavior of ZnO:Ni films. For the

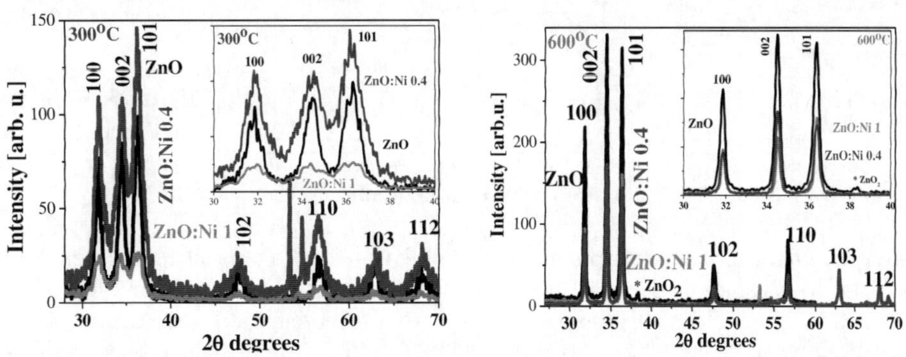

Figure 1. XRD patterns of ZnO, ZnO:In 0.4 and ZnO:Ni 1 films treated at 300 and 600°C.

films annealed at 300 °C (figure 1), the smaller Ni concentration induces a film crystallization comparable to that of the ZnO film, as the ZnO:Ni 1 films manifest weaker and broader XRD peaks, a sign of a less crystallized structure and possible traces of an amorphous phase. After annealing at 600 °C, the ZnO:Ni 0.4 and ZnO:Ni 1 films show lower crystallization compared to undoped ZnO films. The ZnO:Ni 0.4 and ZnO:Ni 1 films are polycrystalline, with all XRD peaks observed in their XRD patterns attributed to a hexagonal wurtzite phase of ZnO (according to JCPDS 01-07-8070) regardless of the annealing temperature, from 300 °C to 600 °C. No diffraction reflections assignable to Ni oxides or mixed Zn-Ni fractions are registered within the XRD detection limits. The ZnO:Ni 1 film treated at 800 °C reveals a broad peak at 2θ = 43.2° related to NiO bunsenite (JCPDS 00-004-0835) [9]. The presence of a NiO fraction indicates that the solubility of Ni in ZnO matrix is limited and dependent on the deposition method [10].

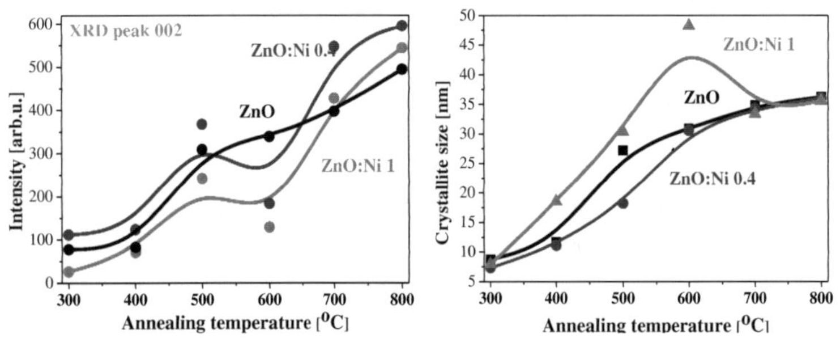

Figure 2. Intensity of the 002 main XRD peak and estimated crystallite size for ZnO and ZnO:Ni films depending on the annealing temperatures.

Figure 2 presents the annealing dependence of the 002 peak's XRD intensity and of the average crystallite size as estimated from the XRD peaks (100, 002 and 101 reflections) by using Scherrer's formula. It is seen that the XRD intensity depends strongly on the annealing temperature. The crystallites grow bigger as the annealing temperature is raised up to 600 °C for both the ZnO and ZnO:Ni films. After annealing at 700 °C and 800°C, the ZnO:Ni crystallites grow smaller, their sizes become comparable to those of the ZnO films treated at these temperatures. The XRD analysis leads to the conclusion that Ni doping results in improved film crystallinity. The higher doping concentration (ZnO:Ni 1 films) leads to narrower XRD lines, i.e., greater crystallites compared to those of the ZnO and ZnO:Ni 0.4 films.

We further conducted FTIR measurements to examine the effect of Ni doping on the vibrational properties of sol-gel ZnO films; the results are shown in figure 3.

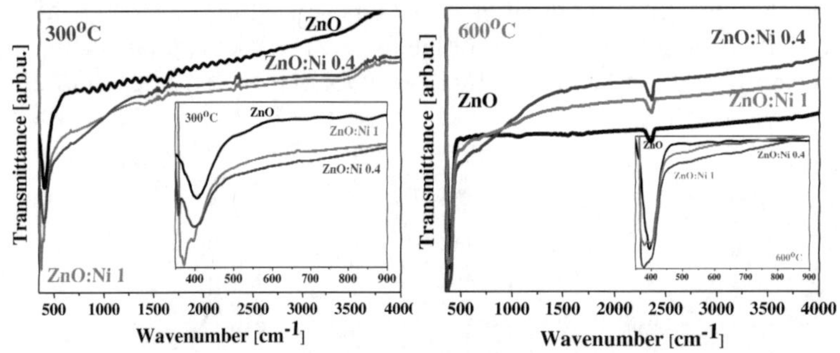

Figure 3. FTIR spectra of ZnO and ZnO:Ni films treated at 300 °C and 600 °C.

Metal-oxygen characteristic absorptions are exhibited in the fingerprint region below 1000 cm^{-1}. Absorptions due to organic residues and water inclusions appear above 1000 cm^{-1}. The ZnO:Ni films annealed at 300 °C show clear absorptions at 3780 cm^{-1} (adsorbed water molecules [11]) and broad bands near 3450 cm^{-1} (O-H stretching vibrations [12]). The ZnO film reveals a weak line at 3450 cm^{-1}. These bands vanish after high-temperature annealing. The weak band at 1640 cm^{-1} is assigned to the bending δ(HOH) vibration (seen in ZnO, ZnO:Ni films). The peak (seen in all spectra) at 2370 cm^{-1} is due to CO_2 absorbed from the atmosphere [13]. The absorption bands below 1000 cm^{-1} of the ZnO films annealed at 300 °C are located at 760 cm^{-1} (due to Zn-O), 520 cm^{-1} (corresponding to wurtzite ZnO), and a very strong and broad line at 407 cm^{-1} with a shoulder at 370 cm^{-1}. The absorption bands at 407 cm^{-1} and 520 cm^{-1} are active infrared modes that have been confirmed theoretically for wurtzite ZnO [14]. The main absorption line of the ZnO:Ni 0.4 film centered at 402 cm^{-1} is very strong and broad with an additional peak at 360 cm^{-1}. Increasing the nickel content (ZnO:Ni 1) results in a modification of the main absorption feature: the band is located at 372 cm^{-1} with two lines at 396 cm^{-1} and 367 cm^{-1}. Annealing at 600 °C induces stronger main absorption bands, shifted to 395 cm^{-1} (ZnO) and 376 cm^{-1} for the ZnO:Ni films. Weak lines at 680 cm^{-1} and 520 cm^{-1} are detected for all films. Adding Ni clearly results in a broadening of the absorption bands. The shifting to the infrared frequencies as the amount of the Ni^{2+} dopant is increased may be due to a difference in the bond length, which confirms the incorporation of Ni^{2+} in the ZnO lattice [15]. The absorption bands broadness can also point to a nanocrystalline nature of the films [12]. The FTIR studies did not indicate a presence of Ni oxide phases. This conclusion is in accordance with the XRD characterization.

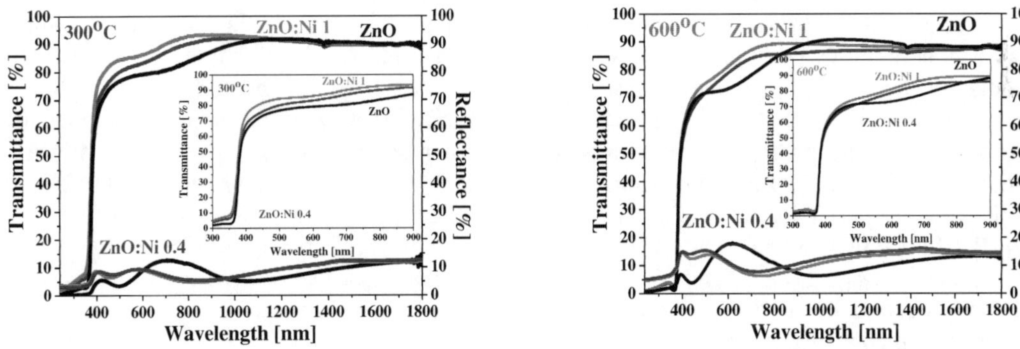

Figure 4. Transmittance and reflectance spectra of ZnO, ZnO:Ni films treated at 300 °C and 600 °C.

Figure 4 presents a comparison of the optical spectra of ZnO and ZnO:Ni films. The estimated values of the optical bandgap (E_g) and the transmittance at 550 nm are given in table 1. Increasing the Ni doping improves the transparency, the reflectance being below 15 % for the ZnO:Ni films. The transmittance reaches up to 80 – 84 % for the sol-gel ZnO:Ni 1 films.

The ZnO films' thickness was found to be 160 nm. The film thicknesses of the ZnO:Ni 0.4 and ZnO:Ni 1 films are 150 nm and 140 nm, respectively. The difference in the thickness of the films studied was small and thus, the optical properties could be compared.

The transparency of ZnO and ZnO:Ni films worsened with the temperatures, which became more pronounced after the treatments at 700 °C and 800 °C. The optical bandgap values diminished as the annealing temperatures was increased, due to the bigger crystallite sizes. The optical bandgaps of

Table 1. Optical bandgap values, transmittance at λ = 550 nm for ZnO and ZnO:Ni films as a function of the annealing temperatures.

$T_{Annealing}$ [°C]	ZnO	ZnO/NiO 0.4	ZnO/NiO 1	ZnO	ZnO/NiO 0.4	ZnO/NiO 1
		E_g [eV]			T [%] at λ=550 nm	
300°C	3,29	3,31	3,32	78,4	81,9	85,2
400°C	3,29	3,29	3,29	80,2	82,0	84,1
500°C	3,28	3,27	3,28	75,3	78,1	82,5
600°C	3,27	3,28	3,28	72,3	75,5	78,1
700°C	3,28	3,28	3,28	67,0	72,1	77,9
800°C	3,27	3,28	3,28	65,6	73,6	74,6

ZnO:Ni films are slightly higher than the corresponding ZnO values. The estimated E_g values are within the range of the data reported previously [4, 16].

Figure 5. AFM micrographs of ZnO:Ni 0.4 and ZnO:Ni 1 films treated at 600 °C. The inset images are 3D micrographs.

Figure 5 presents AFM micrographs of the ZnO:Ni 0.4 and ZnO:Ni 1 films treated at 600 °C (scan area of 5×5 μm). The AFM study of a ZnO film performed at a scan area of 2×2 μm reveals a smooth and uniform morphology with a grained film structure and an uneven size distribution. The grains are well-packed, which results in a small r.m.s. roughness R_q = 5.115 nm for the undoped ZnO film (not shown here). It is known that the roughness parameters are higher when the scan area is increased, because longer-wavelength topography components are then sampled [17]. Thus, direct comparison with the results for the ZnO:Ni films cannot be done. The AFM micrographs of the ZnO:Ni films show different morphological features. Their average and r.m.s. roughness is given in Table 2. The ZnO:Ni 0.4 film shows bigger grains with greater heights compared to the ZnO:Ni 1 films (see 3D images, figure 5). On the other hand, the XRD analysis reveals that the ZnO:Ni 0.4 film annealed at 600 °C possesses smaller crystallites compared to ZnO:Ni 1, which can be a sign of micro-strain broadening of the XRD peaks.

Table 2. AFM parameters of sol-gel ZnO, ZnO:Ni 0.4 and ZnO:Ni 1 films treated at 600°C.

AFM parameter	**ZnO** (2x2 μm)	**ZnO:Ni 0.4** (5x5 μm)	**ZnO:Ni 1** (5x5 μm)
Average roughness	4,215	8,246	7,00
r.m.s. rougnness	5,115	10,368	8,455

ZnO:Ni 1 shows lower roughness parameters values, indicating a smoother surface with a closely-packed granular structure. This may explain its improved transparency. On the other hand, due to its rougher surface and, hence, more intensive light scattering, the ZnO:Ni 0.4 films exhibit a lower transmittance in the visible range. The conclusion we draw from the AFM study is that Ni-doping affect substantially the film surface morphology.

4. Conclusions

The work presented demonstrates that the sol-gel approach is a successful simple technology for depositing Ni-doped ZnO films. The ZnO:Ni films thus prepared are uniform, smooth, and polycrystalline with a wurtzite crystal phase. The FTIR analysis shows that Ni-doping changes the characteristic light absorption features. The transparency is improved by Ni incorporation and the optical bandgap of the ZnO:Ni films is slightly wider than that for ZnO films.

References

[1] Muniyandi I, Mani G K, Shankar P and Rayappan J B B 2014 *Ceramics Inter.* **40** 7993
[2] Chaitra U, Kekuda D and Rao K M 2017 *Ceramics Inter.* **43** 7115
[3] Kayani Z N, Naz F, Riaz S and Naseem Sh 2017 *J. Saudi Chem. Soc.* **21** 425

[4] Mani G K and Rayappan J B B 2014 *Appl. Surf. Sci.* **311** 405
[5] Ramesh J, Pasupathi G, Mariappan R, Kumar V S and Ponnuswamy V 2013 *Optik* **124** 2023
[6] Montes-Valenzuela I, Romero-Paredes G, Vázquez-Agustín M A, Baca-Arroyo R and Peña-Sierra R 2015 *Mater. Sci. Semicond. Processing* **37** 185
[7] Siddheswaran R, Netrvalová M, Savková J, Novák P, Očenášek P, Šutta P, Kováč Jr J and Jayavel R 2015 *J. Alloys Comp.* **636** 85
[8] Srinet G, Kumar R and Sajal V 2013 *J. Appl. Phys.* **114** 033912
[9] Ivanova T, Harizanova A, Koutzarova T and Vertruyen B 2015 *Proc. 15th Int. Conf. Nanotechnol.* (IEEE-NANO 2015) **7388970** 250
[10] Goswami N and Sahai A 2013 *Mat. Res. Bulletin* **48** 346
[11] Tang A, Li X, Zhou Zh, Ouyang J and Yang H 2014 *J. Alloys Comp.* **600** 204
[12] El-Kemary M, Nagy N and El-Mehasse I 2013 *Mater. Sci. Semicond. Processing* **16** 1747
[13] Malika A N, Reddy A R, Babu K S and Reddy K V 2014 *Ceramics Int.* **40** 12171
[14] Iordanescu C, Tenciu D, Feraru I, Kiss A, Bercu M, Savastru D, Notonier R and Grigoresc C 2001 *Digest J. Nanomater. Biostr.* **6** 863
[15] Shinde K P, Pawar R C, Sinha B B, Kim H , Oh S and Chung K C 2014 *Ceramics Int.* **40** 16
[16] Xu K, Liu C, Chen R, Fang X, Wu X and Liu J 2016 *Physica* B*: Cond. Matter* **502** 155
[17] Elam J W, Sechrist Z A and George S M 2012 *Thin Solid Films* **414** 43

IOP Conf. Series: Journal of Physics: Conf. Series **992** (2018) 012045 doi:10.1088/1742-6596/992/1/012045

Study of impedance spectra parameters of PCDTBT:PC70BM solar cells as a function of the deposition solution concentration

M Sendova-Vassileva[1,3], G Popkirov[1], R Gergova[1], G Grancharov[2] and V Gancheva[2]

[1]Central Laboratory of Solar Energy and New Energy Sources, Bulgarian Academy of Sciences, 72 Tzarigradsko Chausse, 1784 Sofia, Bulgaria
[2]Institute of Polymers, Bulgarian Academy of Sciences, Acad. G. Bonchev Str., Bl.103 A, 1113 Sofia, Bulgaria

E-mail: marushka@phys.bas.bg

Abstract. In this contribution, we apply impedance spectroscopy (IS) to study a series of bulk-heterojunction PCDTBT:PC70BM solar cells deposited by spin-coating from dichlorobenzene solution of different concentrations. The samples are deposited on ITO covered glass and have the structure PEDOT:PSS / PCDTBT:PC70BM/ Me where Me is Ag or Al deposited by either magnetron sputtering or thermal evaporation. The impedance spectra are measured upon one sun illumination at short-circuit conditions as a function of the concentration of the solution used for the deposition of the active layer. For one of the samples, measurements are made as a function of the applied bias as well. The measured spectra are fitted using an equivalent circuit. The dependences of the parameters derived from the fits on the concentration of the solution and on the bias voltage are presented and discussed in correlation with the quality of the solar cells.

1. Introduction

Polymer based solar cells are the subject of intense study in recent years as they are perceived as promising for future applications. They can be deposited on flexible substrates and can be made semi-transparent for application in buildings, shades, etc. Among the advantages are low cost, ease of preparation and non-toxicity. The disadvantages include the still comparatively low efficiency and stability.

The polymer poly[N-9'-heptadecanyl-2,7-carbazole-alt-5,5-(4',7'-di-2-thienyl-2',1',3'-benzothiadiazole] (PCDTBT) is one of the widely used compounds in polymer solar cells. In combination with PC70BM as an acceptor, it provides a high open-circuit voltage of the solar cell [1], as well as high internal quantum efficiency [2].

Impedance spectroscopy has been used to characterize solar cells and yield information about the processes of carrier generation and extraction in addition to that obtained by the standard measurements of current-voltage characteristics. In all cases, however, an equivalent circuit describing adequately the device studied has to be devised.

[3] To whom any correspondence should be addressed.

IOP Conf. Series: Journal of Physics: Conf. Series **992** (2018) 012045 doi:10.1088/1742-6596/992/1/012045

In this contribution, we used impedance spectroscopy to study PCDTBT:PC70BM bulk heterojunction (BHJ) polymer solar cells as a function of the solution concentration for the active layer. We used two different equivalent circuits to fit the results and correlate the parameters obtained with the changes in the cells' fill factor and efficiency.

2. Experimental

Bulk heterojunction PCDTBT:PC70BM solar cells were produced using spin-coating on ITO covered glass. All devices had normal configuration with a PEDOT:PSS hole transport layer deposited directly on the ITO. The active layer is PCDTBT:PC70BM deposited from a solution in dichlorobenzene in a weight ratio of 1:4 of the two components. The concentration of the solution was varied between 10 mg/ml and 30 mg/ml, which led to variation in the thickness and, possibly, the structure of the active layer [3]. The back contacts were either sputtered Ag or evaporated Al. Each cell had an area of 6 mm^2.

Our home-made FFT-impedance spectrometer uses a multi-sine perturbation signal with low amplitude applied to the sample under test. Fast 16-bit analog-to-digital converters are used for simultaneous data acquisition of the perturbation voltage and response current, respectively. The two series of data obtained in the time domain are then transformed into the frequency domain using the fast Fourier transform (FFT) algorithm and, consequently, utilized to calculate the impedance spectra. The EIS Spectrum Analyzer Software [4] was applied to fit the spectra obtained to the impedance of a simplified equivalent circuit used to model bulk heterojunction solar cells.

3. Results and discussion

We first investigated BHJ solar cells with sputtered Ag back contacts. A series of impedance spectra measurements were performed on samples produced using different solution concentrations of PCDTBT:PC70BM in dichlorobenzene. The impedance spectra were measured at short circuit conditions, i.e. at controlled d.c. bias of 0 V under halogen lamp illumination equivalent to one sun. As an example, figure 1 shows the impedance data (the dots) obtained for the sample deposited from a 20-mg/ml solution of PCDTBT:PC70BM in dichlorobenzene.

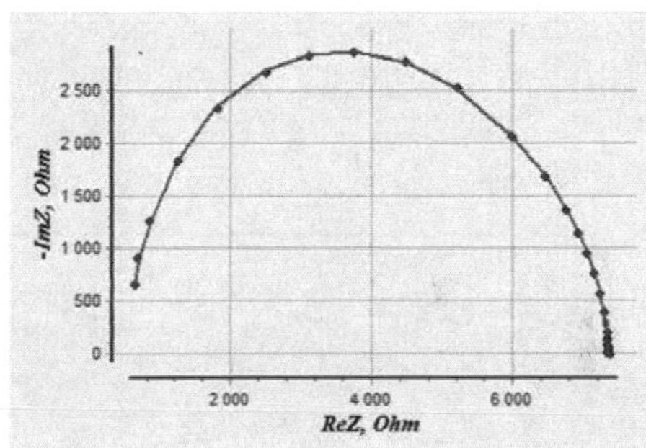

Figure 1. Example of a fit of the experimental spectrum for a 20 mg/ml solution concentration using the equivalent circuit presented in figure 2.

The equivalent circuit shown in figure 2, and described in [5], was successfully used to fit the impedance spectra obtained. In this circuit, R_s is the series resistance, R_1 is the resistance of the bulk, C_1 is the capacitance of the bulk and R_2 is the recombination resistance. A constant phase element (CPE) is used instead of a capacitor to account for a possible distribution of relaxation times. Generally, one could attribute this to the inhomogeneous nature of the BHJ system. However, we will not speculate on the nature of the inhomogeneities in our case: is it surface roughness, varying composition, non-uniform current distribution and/or charge carrier lifetime. Any one of those, single or in combination, could lead to the appearance of a depressed semi-circle in the respective impedance spectrum. The $R_2||CPE$ combination has the following impedance Z_{CPE}:

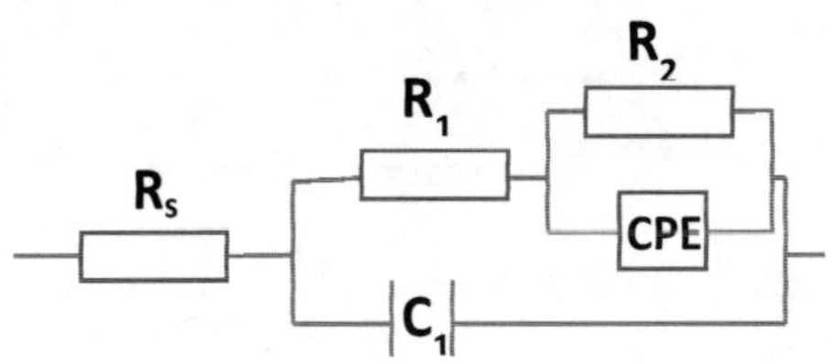

Figure 2. Equivalent circuit used for the fits.

IOP Conf. Series: Journal of Physics: Conf. Series **992** (2018) 012045 doi:10.1088/1742-6596/992/1/012045

$$Z_{CPE} = \frac{R_2}{1 + R_2 Q (i\omega)^n}, \tag{1}$$

where Q is the magnitude of the CPE and n is its ideality factor, smaller than or equal to 1. The equivalent capacitance of the CPE element (C_{eq}) in this equivalent circuit is

$$C_{eq} = \frac{\tau_{avg}}{R_2} = \frac{(R_2 Q)^{1/n}}{R_2}. \tag{2}$$

According to [5], C_{eq} is related to the so-called "chemical capacitance" [6], and the characteristic relaxation time τ_{avg} is linked to the average lifetime of the charge carriers, closely related to bimolecular recombination in the active layer. The time constant associated with the $R_1||C_1$ pair is connected with the transit time of the charge carriers through the device.

Figure 3 and figure 4 show the dependence of R_1, R_2, and τ_{avg} on the concentration of PCDTBT:PC70BM in the dichlorobenzene solution used for the production of BHJ solar cells with sputtered Ag back contact. The values were obtained by fitting the experimental spectrum of each sample to the equivalent circuit in figure 2. As mentioned above, all measurements were made under short-circuit conditions under illumination equivalent to one sun. The different points for one and the same concentration correspond to different equivalent devices produced during one deposition cycle.

The value of R_1 decreases slightly with the concentration, while the recombination resistance R_2 shows no dependence on the concentration, with the exception of the lowest concentration, when it grows considerably and reaches a value similar to and greater than R_1. It is considered [5] that when R_2 is smaller than R_1, carrier recombination in the device dominates over carrier extraction. In our study, this is the case for most of the studied concentrations with the exception of the lowest one. This correlates with the slight improvement of the fill factor of the solar cells at this concentration. The average charge carrier lifetime τ_{avg} was found [5] to correlate with the efficiency of the devices. This is true in our case as well.

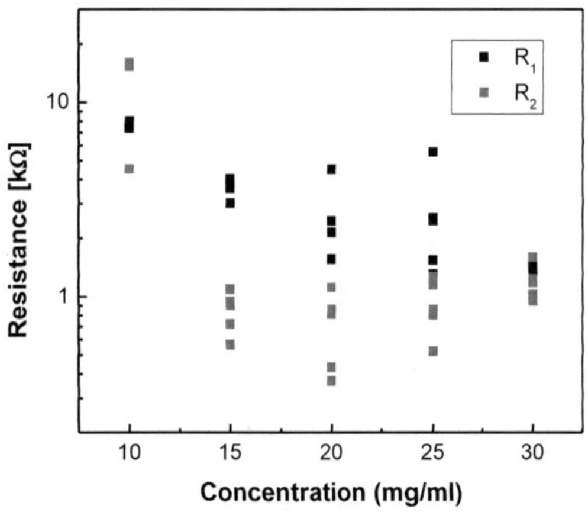

Figure 3. Dependence of R_1 and R_2 on the solution concentration for PCDTBT:PC70BM devices. Measurement under short circuit conditions, one sun illumination.

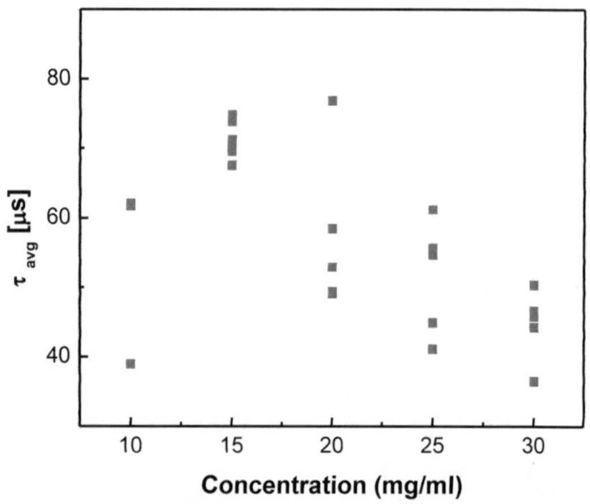

Figure 4. Dependence of τ_{avg} on the solution concentration for PCDTBT:PC70BM devices. Measurement under short circuit conditions, one sun illumination.

The solar cells from the above experimental series produced with the lowest concentration of the PCDTBT:PC70BM – dichlorobenzene solution showed the best fill-factor values. Unfortunately, these samples exhibited lower efficiencies due to the very thin active layer (a result of spin-coating with the most diluted solution) and, thus, relatively low light absorption.

In an earlier study [7], we found that evaporated Al is superior to sputtered Ag when used as back contact of our BHJ cells. Thus, in a second experiment, BHJ solar cells were produced as described above, but with back contacts deposited by evaporation of Al and using the lowest concentration of the PCDTBT:PC70BM – dichlorobenzene solution of 10 mg/ml. The solar cells prepared indeed had considerably higher fill-factor and efficiency. The impedance spectra of these cells, measured at the same conditions as for the case of Ag electrodes, showed a simpler, almost ideal semicircle in the Nyquist plot representation ($-Z_{imag}$ vs Z_{real}), as shown in figure 5. This encouraged us to perform a set of impedance measurements in a broad range of bias voltages applied to the solar cell upon illumination.

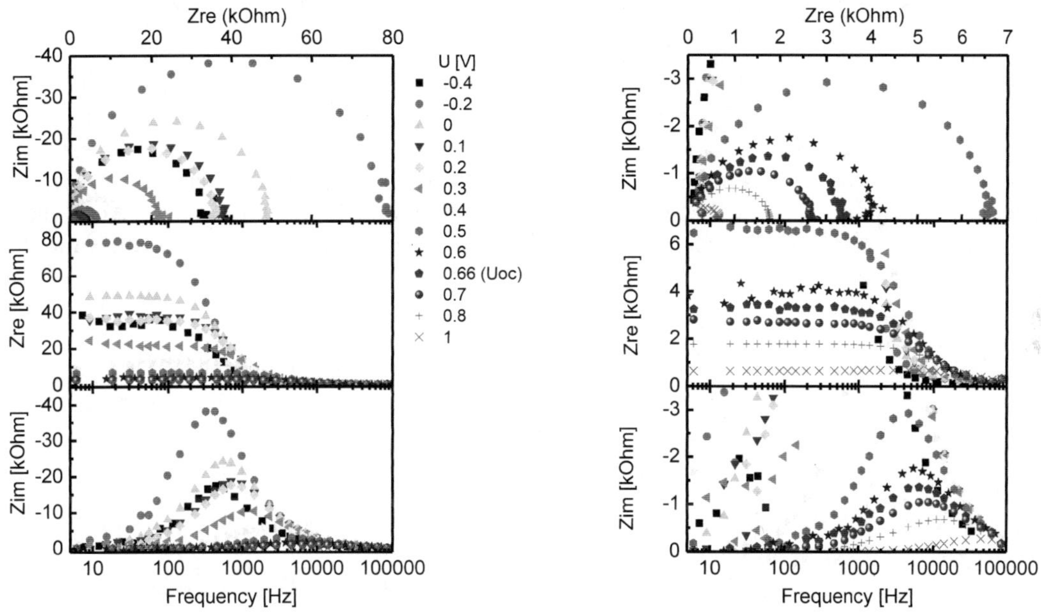

Figure 5. Impedance spectra dependence on the applied voltage upon one sun illumination for a PCDTBT:PC70BM device with Al back contact.

In the spectra obtained, the semicircle was not depressed, leading us to the suggestion to use a simple equivalent circuit, as shown in figure 6, for the fit of the data. We still used a CPE element instead of a capacitor for the fit procedure, but the fit results showed a high degree of ideality of the semicircles in the whole bias-voltage range under investigation. Figure 7 is added as an example of the quality of the fit for short circuit conditions.

The fit results obtained provide data for the series resistance R_1, the parallel (shunt or recombination) resistance R_2 and the CPE element. As could be expected for a solar cell, it can be seen in figure 5 that the parallel resistance R_2 decreases with the increase of the bias voltage starting from a reverse bias of –0.2 V. This resistance is connected with the bulk resistance of the active layer. The R_2 value

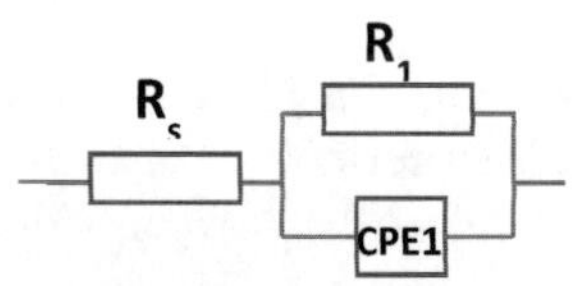

Figure 6. Equivalent circuit used for the solar cell with Al back contact.

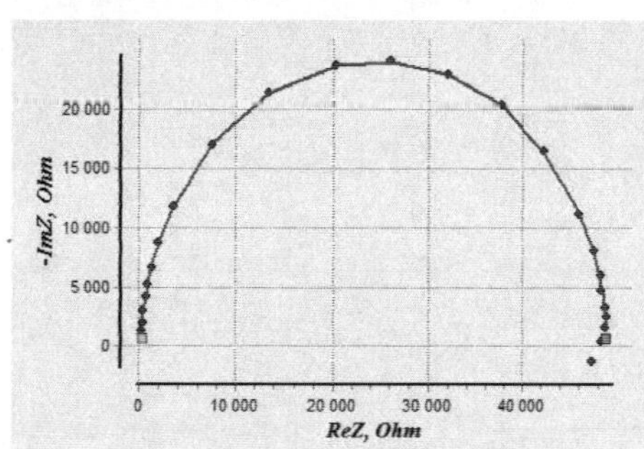

Figure 7. Typical impedance spectrum and its fit using the equivalent circuit in figure 6.

IOP Conf. Series: Journal of Physics: Conf. Series **992** (2018) 012045 doi:10.1088/1742-6596/992/1/012045

obtained from the spectrum measured under short circuit conditions (bias = 0 V) is almost five times higher than the value of R_2 obtained in the previous experiment using Ag-back contact. We are inclined to explain this effect with the much better interface of the active layer with the metal cathode in the case of evaporated Al. The capacitance C_{eq} of the CPE element does not vary much with the bias and has a value between 5 nF and 6 nF, while the ideality factor of the CPE is close to 1. This result supports further the suggestion that evaporated Al leads to a much better interface between the back contact and the active layer compared to sputtered silver.

4. Conclusions

It is shown that impedance spectroscopy can be used as a tool for additional express characterization of polymer bulk-heterojunction solar cells. There is a correlation between the parameter τ_{avg} derived from the fitting of the impedance spectra and the efficiency of the solar cells studied. For the PCDTBT:PC70BM devices with low FF, the recombination resistance is lower than the bulk resistance R_1 except for the lowest solution concentration, where the FF improves, which suggests that in the low FF devices carrier recombination dominates over carrier extraction. The dependence of the impedance spectra on the applied voltage in the case of a PCDTBT:PC70BM cell with Al back contact could be fitted at most voltages by a single RC circuit and a series resistance. This is attributed to the better interface between the back contact and the active layer.

Acknowledgements

The work was supported in part by the Bulgarian National Scientific Fund under Contract DFNI E02/5.

References

[1] Nelson J 2011 *Materials Today* **14** 462

[2] Park S H, Roy A, Beaupre S, Cho S, Coates N, Moon J S, Moses D, Leclerc M, Lee K and Heeger A J 2009 *Nature Photonics* **3** 297

[3] Fang G, Liu J, Fu Y, Meng B, Zhang B, Xie Z and Wang L 2012 *Organic Electronics* **13** 2733

[4] Bondarenko A S and Ragoisha G A 2005 *Progress in Chemometrics Research* ed Pomerantsev A L (New York: Nova Science Publishers) p 89 (the program is available online at http://www.abc.chemistry.bsu.by/vi/analyser/)

[5] Leever B J, Bailey C A, T. Marks J, Hersam M C and Durstock M F 2012 *Adv. Energy Mater.* **2** 120

[6] Bisquert J 2003 *Phys. Chem. Chem. Phys.* **5** 5360

[7] Sendova-Vassileva M, Popkirov G, Vitanov P, Dikov C, Gancheva V, Tsocheva D and Mokreva P 2012 *J. Phys.: Conf. Ser.* **398** 012049

Synthesis of bimetallic nanostructures by nanosecond laser ablation of multicomponent thin films in water

R G Nikov[1,3], N N Nedyalkov[1], P A Atanasov[1] and D B Karashanova[2]

[1]E. Djakov Institute of Electronics, Bulgarian Academy of Sciences, 72 Tzarigradsko Chaussee, 1784 Sofia, Bulgaria
[2]Acad. J. Malinowski Institute of Optical Materials and Technologies, Bulgarian Academy of Sciences, Acad. G. Bonchev str., bl. 109, 1113 Sofia, Bulgaria

E-mail: rosen_nikov@abv.bg

Abstract. The paper presents results on nanosecond laser ablation of thin films immersed in a liquid. The thin films were prepared by consecutive deposition of layers of different metals by thermal evaporation (first layer) and classical on-axis pulsed laser deposition (second layer); Ni/Au, Ag/Au and Ni/Ag thin films were thus deposited on glass substrates. The as-prepared films were then placed at the bottom of a glass vessel filled with double distilled water and irradiated by nanosecond laser pulses delivered by a Nd:YAG laser system at $\lambda = 355$ nm. This resulted in the formation of colloids of the thin films' material. We also compared the processes of ablation of a bulk target and a thin film in the liquid by irradiating a Au target and a Au thin film by the same laser wavelength and fluence ($\lambda = 355$ nm, $F = 5$ J/cm^2). The optical properties of the colloids were evaluated by optical transmittance measurements in the UV–VIS spectral range. Transmission electron microscopy was employed to estimate the particles' size distribution.

1. Introduction

The synthesis of metallic and bimetallic nanoparticles (NPs) with controlled size, morphology, and composition attracts considerable interest from both the scientific community and the industry because of their unique physical and chemical properties [1–4]. A special interest has been devoted to the fabrication of multicomponent NPs consisting of noble and ferromagnetic metals with potential applications in catalysis, biophotonics and magneto-optics [5–7]. Ag/Co, Ag/Ni, and Au/Ni particles show well-expressed plasmon effects combined with magnetic properties [8]. In particular, the bimetallic NPs of Ag and Ni can be used as catalysts, electrical contacts, switches, electromagnetic wave absorbers, and in plating of conducting materials [9, 10]. Functionalized magnetic particles, such as Ni, Fe or magnetite can also be incorporated in Au-based NPs to enhance the efficiency and flexibility of drug treatment by applying an external magnetic field [11]. The position of the surface plasmon resonance (SPR) of NPs of Au/Ag alloys can be precisely tuned in a wide spectral range by varying the Au/Ag ratio. In [12], a linear dependence was reported of the position of the plasmon absorption maximum on the amount of one of the metals included. This effect can help in increasing the efficiency of the surface-enhanced Raman spectroscopy, the photothermal cell therapy, and the bio-imaging based on such nanostructures.

[3] To whom any correspondence should be addressed.

The preparation of bimetallic NPs is based on different techniques; the most popular among them were reported in [13]: wet chemical synthesis, photochemical synthesis, sputter deposition, sonochemical synthesis, electroless plating, and electrochemical synthesis. However, these synthesis techniques usually involve several steps and require toxic reagents.

The method known as pulsed laser ablation of a target in liquid (PLAL) has emerged as one of the most interesting techniques of preparing nanostructures in a wide variety of liquids, since it is a simple, environmentally friendly, inexpensive and clean procedure [14-16]. Its most important advantage is the fact that the resulting colloidal NPs are chemically pure, containing elements of the target and the liquid only. A disadvantage of the PLAL method is the relatively broad size distribution of the NPs obtained compared to the NPs synthesized via chemical methods. However, it has been reported [17] that laser ablation of thin metal films could result in the formation of NPs with a narrow size distribution; i.e., this drawback can be overcome.

In the work presented, we applied laser ablation in a liquid in a configuration that allows easy fabrication of colloidal bimetal NPs. For this purpose, thin films were previously deposited on a glass substrate by consecutive deposition of two layers of different metals (Au, Ag and Ni). The first layer was deposited by thermal evaporation, a method characterized by fast and relatively easy implementation, allowing deposition of large surfaces. The second layer was deposited by the classical pulsed laser deposition (PLD) technique. It allows one to deposit films with different thickness by controlling the laser radiation parameters – number of pulses, laser fluence and wavelength. The multicomponent films produced were then immersed in double distilled water and irradiated by ь nanosecond Nd: YAG laser. In another experiment, we followed the influence of the film thickness on the morphology and size distribution of the NPs formed in the solution. For this purpose, laser ablation of a thin Au film and a Au target in water was performed under identical experimental conditions. The morphology, composition and optical properties of the colloidal NPs prepared were studied and discussed.

2. Experimental

The preparation of the colloidal metal and bimetallic NPs was carried out in two steps. In the first one, we prepared thin Au, Ni/Ag, Ni/Au, and Ag/Au films. The second step consisted in ablation of thin films in double distilled water, which resulted in the production of colloidal NPs. The multicomponent films preparation involved a consecutive deposition of two layers of different metals on glass substrates. The first layer was formed by thermal evaporation, and the second, by the classical on-axis PLD technique, using the third harmonic (355 nm) of a Nd:YAG laser (pulse duration 15 ns, repetition rate 10 Hz, fluence 1.5 J/cm^2. Both films were deposited in vacuum at an ambient pressure of about 10^{-3} Pa for six minutes. The thickness of the multicomponent films thus produced was measured by optical profilometer to be in the range of 160 ÷ 200 nm.

The thin-film samples were placed at the bottom of a glass vessel under 5 mm of double distilled water and ablated by the same Nd:YAG laser, whose third harmonic was focused on the metal film surface by a fused quartz lens and scanned by an XY stage for 5 min. The laser fluence was chosen to be 5 J/cm^2 to avoid optical breakdown in the liquid, which would drastically change the laser-target interaction and reduce the efficiency of the ablation process. The absorbance spectra of the colloids were recorded immediately by a UV–Vis spectrometer (Ocean Optics HR 4000) in the 300 – 900 nm range. The nanostructures in the dried colloids were visualized by a transmission electron microscope (JEOL JEM 2100, accelerating voltage of 200 kV).

3. Results and discussion

3.1. Laser ablation of bulk target and thin film

We first studied the influence of the film thickness on the size distribution of the NPs formed during laser ablation in water. We ablated a Au bulk target (purity of 99.99 %, thickness of ~1 mm) and a Au thin film (deposited by thermal evaporation, thickness of 120 nm) under the conditions described above. The TEM images of the dried colloids are presented in figure 1, (a) and (b), with the particles size distribution histograms shown on the right-hand side. The predominantly spherical or spherical-like shape of the NPs is evident in both TEM images, which is typical of particles formed in a liquid environment [18]. The histogram of the size distribution of the particles obtained by ablation of the bulk target revealed a mean particles size of 7.2 nm, with a relatively high standard deviation of 5.7 nm. In the case of the thin Au film (figure 1-b), the mean size of the formed Au NPs was 4.1 nm and the standard deviation, 1.5 nm. Moreover, in the case of bulk target ablation, large particles with dimensions over 500 nm were also observed (*inset* in figure 1-a). Generally, the ablation of the thin film yielded NPs with a smaller mean size and a more uniform (narrower) size distribution compared to the NPs produced by ablation of a bulk target.

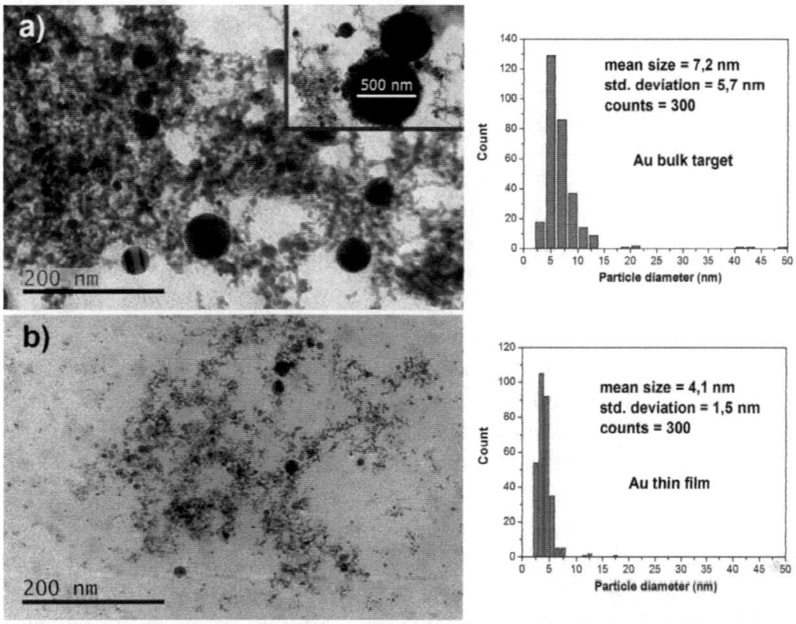

Figure 1. TEM images and corresponding histograms of the particles size distribution of Au nanoparticles produced by laser ablation of Au bulk target (a) and Au thin film (b) in water.

The optical absorption spectra of the colloids produced by ablation of the bulk target and the thin film are presented in figure 2. A single SPR band in the 500 – 600 nm range, characteristic of the presence of nanosized Au particles, is seen in both spectra. The slight red-shift and the higher intensity

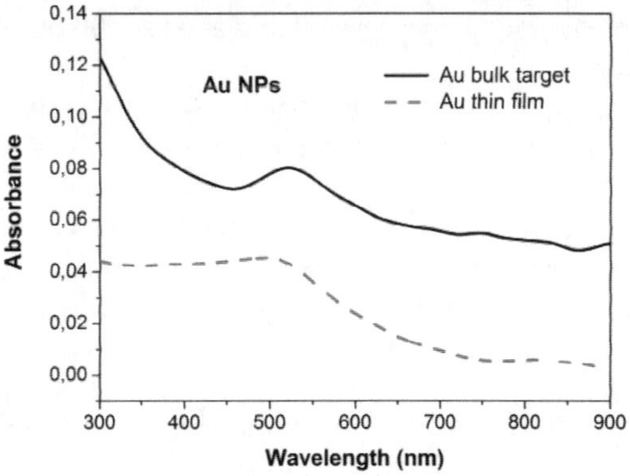

Figure 2. Optical absorption spectra of colloids produced by laser ablation of Au bulk target and Au thin film in water at laser wavelength of 355 nm and laser fluence of 5 J/cm^2.

of the SPR maximum in the spectrum of the colloid fabricated by ablation of the bulk target, compared to the thin-film case, could be attributed to the larger mean size of the NPs in the former case [19].

3.2. Laser ablation of multicomponent thin films

In this section we consider the laser ablation of multicomponent thin films of different metals. Figure 3 shows the TEM images and the corresponding size-distribution histograms of the NPs produced by laser ablation of Ni/Au (a), Ag/Au (b) and Ni/Ag (c) thin films in water. Formation of separated spherical and spheroidal shaped NPs with a relatively narrow size distribution is observed in the case of ablation of Ni/Au (top layer/bottom layer) thin film (figure 3-a). The histogram of the particle size distribution indicates a mean particles diameter of 3.8 nm and a standard deviation of 1.4 nm. In the case of Ag/Au thin-film ablation, besides spherical NPs, nanostructures with a chain-like shape (nanowires) are also observed (figure 3-b). The mean size of the spherical NPs and the particles forming the nanowires is 5.2 nm and the standard deviation is 2.1 nm. These elongated nanostructures are probably formed by joining of particles already created in the colloid. The laser wavelength used (355 nm) falls within the absorption band of the Ag and Au NPs. Thus, the absorption of laser radiation by the NPs may lead to partial melting of the particles' surface, whose melting temperature is significantly lower than that of the bulk material [20]. The shape of the particles in the case of ablation of the Ni/Ag thin film is predominantly spherical or spheroidal (figure 3-c). However, it should be noted that the larger particles have a rounded and asymmetrical shape, which is probably due to aggregation of smaller particles at a later stage, after the condensation of the plasma plume. According to the histograms, the mean particle size and the standard deviation are 19.4 nm and 16.1 nm, respectively.

The specific properties of the colloidal nanostructures prepared were explored by optical transmission measurements. Figure 4 shows the optical absorption spectra of the colloids produced by ablation of Ni/Au, Ag/Au and Ni/Ag thin films in water. It should be mentioned that the concentration of the particles in the colloidal solution is relatively low in all cases considered. A single plasmon band with a maximum located at 450 nm is observed in the spectrum of the NPs produced by ablation of the Ag/Au film. One should also bear in mind that the position of this SPR band falls in between those related to pure Au and Ag NPs [19]. This fact confirms the fabrication of bimetallic NPs [21] and the absence of monometallic Au and Ag NPs. Further, it is well known that the Ag/Au alloy phase is stable and can exist in a solid or liquid state. Therefore, Au-Ag alloy NPs are formed as a result of ablation of the bimetallic Ag/Au films. Two absorbance peaks with positions at 416 nm and 574 nm are observed in the optical spectrum of the colloid obtained by ablation of the Ni/Ag film. It is known that pure Ag NPs exhibits SPR absorption around 400 nm [19], while there is some controversy about the optical absorption of pure Ni NPs. Creighton et al. [22] have calculated the spectrum of Ni NPs and found a single SPR between 300 nm and 400 nm. In what concerns experimental works, Lee et al. have not reported any absorption band for Ni NPs [9]. However, Xiang et al. [23] and Carja et al. [24] have found absorption bands around 370 nm related to oxidized nickel nanoparticles, as well as a broad absorption band at 550–700 nm related to absorptions of nanostructures containing Ni^{2+} ions.

Nevertheless, the main reason for the appearance of an absorption band at 416 nm is excitation of SPR of Ag-containing NPs. The position of this band corresponds to the presence of NPs with sizes up to 60 nm in the solution [19]. The histogram in figure 3-c confirms that most of the NPs formed have such dimensions. The appearance of a second peak at 574 nm could be due to two factors: the presence of NPs with significantly larger sizes (over 80 nm) and particles with a shape deviating from spherical. Both statements are confirmed by the TEM image in figure 3-c. It thus appears that additional studies are necessary to clarify the nature and composition of the nanostructures obtained in the case of ablation of Ni/Ag thin films. In the optical absorption spectrum of the NPs produced by ablation of the Ni/Au thin film, a wide absorbance band with a maximum at 467 nm is seen. The position of this band is blue-shifted with respect to the SPR band of pure Au NPs. Similar shifting of the SPR band of Au-containing NPs has already been seen in the spectrum of the nanostructures obtained by ablation of the Ag/Au film. The reason for this behavior could be the same, namely, formation of alloyed NPs, in this case Ni-Au. However, further research is necessary to confirm this statement.

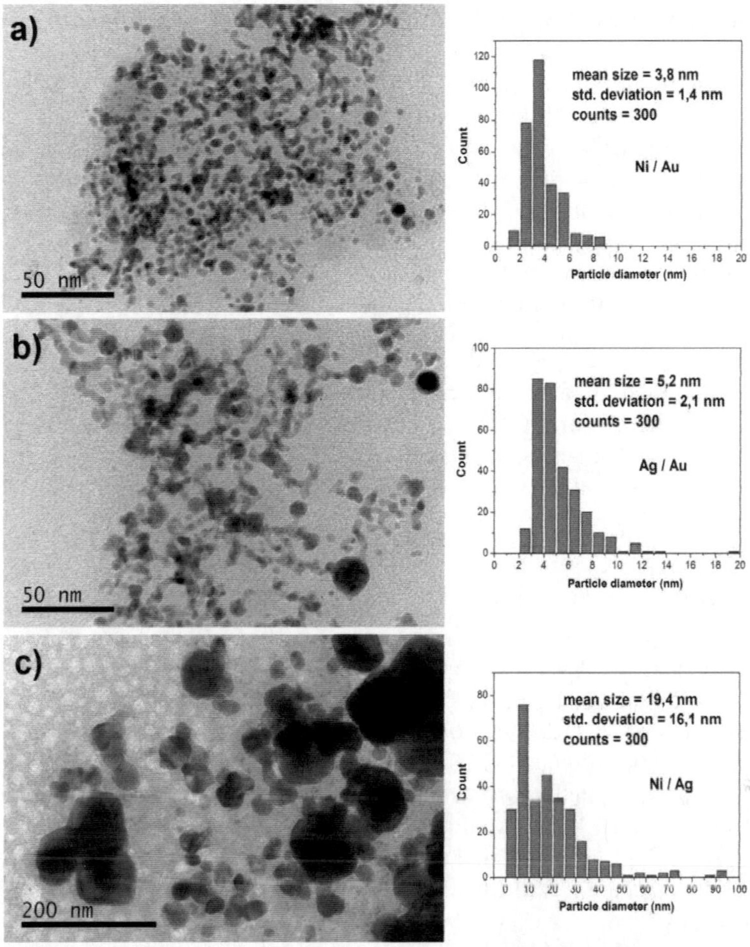

Figure 3. TEM images and corresponding size distribution histograms of nanoparticles produced by laser ablation in water of multicomponent thin films: Ni/Au (a), Ag/Au (b) and Ni/Ag (c). The ablation is performed by using laser wavelength of 355 nm and laser fluence of 5 J/cm^2.

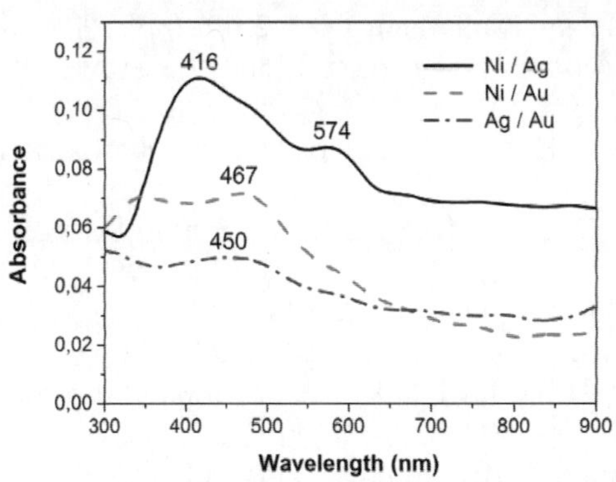

Figure 4. Optical absorption spectra of colloids produced by laser ablation of multicomponent (Ni/Ag, Ni/Au, and Ag/Au) thin films in water at the laser wavelength of 355 nm and laser fluence of 5 J/cm^2.

4. Conclusions

We presented the basic characteristics of a method for synthesizing multicomponent colloidal nanostructures. The method is based on laser ablation in double distilled water of a bimetallic thin film prepared by a consecutive deposition of layers from different metals by thermal evaporation and PLD on a glass substrate. TEM analysis reveals that the ablation of the Ag/Au film leads to the formation of both nanowires and spherical NPs, with their size ranging from 2 nm to 20 nm. The optical absorption spectrum of the colloid produced by ablation of the Ag/Au film reveals that the SPR maximum of the resulting NPs lies between the resonance maxima of pure Ag and Au NPs. This demonstrates that the NPs obtained are composed of both Au and Ag. In the case of Ni/Au thin-film ablation, spherical NPs with a relatively small size (up to 10 nm) are predominantly observed. In contrast, the ablation of the Ni/Ag film results in the formation of larger NPs with a broad size distribution and asymmetrical shape. Also, the Au thin-film ablation results in producing NPs with a more uniform size distribution compared to that of NPs formed by ablation of the corresponding bulk target. The method proposed appears to be an efficient alternative to the known methods of fabrication of bimetal NPs in solutions.

Acknowledgments

The work was supported under project DFNP-17-117 "Preparation of complex colloidal nanostructures by laser ablation of thin films and structures in liquid" under the Assistance of Young Scientists Program of the Bulgarian Academy of Sciences.

References

[1] Addato S D, Grillo V, Altieri S, Tondi R, Valeri S and Frabboni S 2011 *J. Phys.: Condens. Matter* **23** 175003
[2] Xiao Q, Yao Z, Liu J, Hai R, Oderji H Y and Ding H 2011 *Thin Solid Films* **519** 7116
[3] Xia L, Hu X, Kang X, Zhao H, Sun M and Cihen X 2010 *Colloids Surf.,* A **367** 96
[4] Guo H, Chen Y, Chen X, Wen R, Yue G-H and Peng D-L 2011 *Nanotechnol.* **22** 195604
[5] Ortigoza M A and Rahman T S 2008 *Phys. Rev.* B **77** 195404
[6] Wang L L and Johnson D D 2009 *J. Am. Chem. Soc.* **39** 14023
[7] Kilimis D A and Papageorgiou D G 2010 *Eur. Phys. J.* D **58** 189
[8] Gaudry M, Cottancin E, Pellarin M, Lermé J, Arnaud L, Huntzinger J R, Vialle J L, Broyer M, Rousset J L, Treilleux M and Mélinon P 2003 *Phys. Rev.* B **67** 155409
[9] Lee C-C, Cheng Y-Y, Chang H Y and Chen D-H 2009 *J. Alloys Compd.* **480** 674
[10] Ahmed J, Ramanujachary K V, Lofland S E, Ruriato A, Gupta G, Shivaprasad S M and Ganguli A K 2008 *Colloids Surf.* A **331** 206
[11] Gould P 2004 *Mater. Today* **2** 36
[12] Link S, Wang Z L and El-Sayed M A 1999 *J. Phys. Chem.* B **103** 3529
[13] Major K J, De C and Obare S O 2009 *Plasmonics* **4** 61
[14] Amendola V and Meneghetti M 2013 *Phys.Chem. Chem. Phys.* **15** 3027
[15] Zeng H, Du X-W, Singh S C, Kulinich S A, Yang S, He J and Cai W 2012 *Adv. Funct. Mater.* **22** 1333
[16] Nikov R G, Nedyalkov N N, Atanasov P A and Karashanova D B 2017 *Appl. Phys.* A **123** 490
[17] Amoruso S, Nedyalkov N N, Wang X, Ausanio G, Bruzzese R and Atanasov P A 2014 *Thin Solid Films* **550** 190
[18] Zhu X P, Suzuki T, Nakayama T, Suematsu H, Jiang W and Niihara K 2006 *Chem. Phys. Lett.* **427** 127
[19] Kreibig U and Vollmer M 1995 *Optical properties of metal clusters* (Berlin: Springer-Verlag)
[20] Sakai H 1996 *Surf. Sci.* **351** 285
[21] Shin K S, Kim J H, Kim I H and Kim K 2012 *J. Nanopart. Res.* **14** 735
[22] Creighton J A and Eadon D G 1991 *J. Chem. Soc., Faraday Trans. 2* **87** 3881
[23] Xiang X, Zu X T, Zhu S, Zhang C F and Wang L M 2006 *Nucl. Instrum. Methods Phys. Res., Sect.* B **250** 229
[24] Carja G, Nakajima A, Dranka C and Okada K 2010 *J. Nanopart. Res.* **12** 3049

Optical properties modification induced by laser radiation in noble-metal-doped glasses

N Nedyalkov[1,5], N E Stankova[1], M E Koleva[1], R Nikov[1], P Atanasov[1], M Grozeva[2], E Iordanova[2], G Yankov[2], L Aleksandrov[3], R Iordanova[3] and D Karashanova[4]

[1]E. Djakov Institute of Electronics, Bulgarian Academy of Sciences, 72 Tzsarigradsko Chaussee, 1784 Sofia, Bulgaria
[2]G. Nadjakov Institute of Solid State Physics, Bulgarian Academy of Sciences, 72 Tzarigradsko Chaussee, 1784 Sofia, Bulgaria
[3]Institute of General and Inorganic Chemistry, Bulgarian Academy of Sciences, Acad. G. Bonchev Str., Bl. 11, 1113 Sofia, Bulgaria
[4]Acad. J. Malinowski Institute of Optical Materials and Technologies, Bulgarian Academy of Sciences, Acad. G. Bonchev Str., Bl. 109, 1113 Sofia, Bulgaria

E-mail: nned@ie.bas.bg

Abstract. We present results on laser-induced color changes in gold- and silver-doped glass. The doped borosilicate glass was prepared by conventional melt quenching. The study was focused on the change of the optical properties after irradiation of the glass by femtosecond laser pulses. Under certain conditions, the laser radiation induces defects associated with formation of color centers in the material. We studied this process in a broad range of laser radiation wavelengths from UV to IR, and observed changes in the color of the irradiated areas after annealing of the processed glass samples, the color being red for the gold-doped glass red and yellow for the silver-doped glass. The structural and morphological analyses performed indicated that this effect is related to formation of metal nanoparticles inside the material. The results obtained show that femtosecond laser processing of noble-metal-doped glasses can be used for fabrication of 3D-nanoparticles systems in transparent materials with application as novel optical components.

1. Introduction

The research interest in nanocomposites has been continuously growing in recent years. These complex materials composed of nanostructures embedded into a host matrix express novel optical, mechanical and electric properties that open new possibilities for development of advanced devices in the field of sensors, biophotonics and MEMS, and contribute to enhancing the efficiency of light harvesting, battery performance, catalytic reactions [1-3]. The composite materials consisting of noble-metal nanoparticles are of great interest due to the specific optical properties that they express. The interaction of the electromagnetic field with the nanoparticles under the condition of resonance of the incident field frequency with that of the plasmon is characterized by a huge increase of the far-field extinction coefficient, which can reach orders of magnitude [4]. Thus, structures composed by

[5] To whom any correspondence shoul be addressed.

IOP Conf. Series: Journal of Physics: Conf. Series **992** (2018) 012047 doi:10.1088/1742-6596/992/1/012047

nanoparticles incorporated in a host material can present novel, more complex properties. The experimental and theoretical studies conducted in this field have shown that the plasmon frequency can be continuously tuned in a wide range by changing the particle's parameters and the conditions of interaction with the incident radiation. In addition to the optical characteristics of the metal nanoparticles that can be observed in the far-field zone, the properties of the electromagnetic field in the nanoparticle's near field zone also attract significant interest. This interest is based on some basic properties that are not expressed at distances larger than approximately the wavelength of the incident radiation. In this zone, the intensity of the electromagnetic field is enhanced and, under optimal conditions for plasmon excitation, it can be several orders of magnitude higher than the incident one [5]. The specific properties of the electromagnetic field in the near-field zone are the basis of effects observed in composite materials consisting of noble-metal nanoparticles, such as a large increase of the third-order nonlinear optical susceptibility in oxide/Au nanoparticles composites, an enhancement of the photoluminescence signal in ZnO/Ag, ability of modifying the conductivity in some semiconductors by embedding particles and varying their concentration [6-8]. The complex study and development of reliable applications based on such materials are still hampered by the lack of efficient fabrication techniques. The chemical and physical bottom-up methods developed so far are still expensive and slow and do not offer efficient manipulation of the composite material morphology in three dimensions.

In this paper, we discuss a method for real 3D fabrication of noble-metal nanoparticles in glass. It is based on femtosecond-laser reduction of noble metal ions to atoms that form nanoparticles upon subsequent annealing. The method has been demonstrated for different glass types containing silver and gold ions when 800-nm ultrashort laser pulses were used [9-11]. Processing by irradiation at other wavelengths has not been considered so far for this application. Furthermore, the method has rarely been applied to fabricating gold particles in borosilicate glass. In this study, we used irradiation at four wavelengths, namely 266 nm, 355 nm, 500 nm and 800 nm, in order to follow the influence of this parameter to the glass response and to the formation of nanoparticles.

2. Experimental

The samples studied were made of borosilicate glass fabricated by melt quenching with composition 50% SiO_2, 20% Al_2O_3, 20%B_2O_3, 5% CaO, 2% Li_2O, 3% MgO (in wt %). Gold and silver ions were included as $HAuCl_4 \cdot 3H_2O$ or $AgNO_3$ added to the initial mixture. The mixed material was melted in a Pt crucible and kept at a temperature of 1450 ^{0}C for three hours. Using different amounts of the noble-metal donor compounds, glasses with gold concentration of 0.015 and 1 % and silver of 1% were prepared and used in the experiments. The irradiation of the samples was performed by a Ti:Sapphire femtosecond laser system (Spectra Physics) that includes an optical parametric amplifier and emits at a central wavelength of 800 nm with a pulse duration of 35 fs and 1-kHz repetition rate. The optical parametric amplifier covers the wavelength range from 240 nm to 2600 nm. In the present experiments we selected the wavelengths of 266 nm, 355 nm, 500 nm and 800 nm and focused the laser beam by a 20 cm-lens. The transmission spectra of the samples in the range 200 – 1000 nm were taken by an HR 4000 Ocean Optics spectrometer. The thermal annealing of the glass samples was performed in a standard oven with controllable heating parameters. The structural characteristics of the samples were visualized by a JEM 2100 JEOL TEM, while the optical images of the processed areas were obtained by an Optica B-150 optical microscope.

3. Results and discussion

The glass samples fabricated were cut in pieces with a thickness of 2 mm and polished. The material was transparent and colorless. Figure 1 shows the transmission spectra of the samples considered in this study. It is seen that the glasses are opaque at wavelengths shorter than approximately 300 nm; i.e., the difference in their composition concerning the content of gold and silver affected negligibly their optical properties. The laser wavelengths used were chosen to represent different cases of laser radiation interaction with the investigated material. Thus, the wavelength of 266 nm is in the range of

IOP Conf. Series: Journal of Physics: Conf. Series **992** (2018) 012047 doi:10.1088/1742-6596/992/1/012047

high absorption of the material; 355 nm is the wavelength where the glass transmits about 50% of the incident radiation; at the wavelengths of 500 nm and 800 nm the glass samples are transparent. Further, the wavelength of 500 nm falls in the absorption band corresponding to the typical defects (color centers) induced in silicate glasses by irradiation [12], while the wavelength of 800 nm, being typical for ultrashort laser systems, was used for comparison with other works.

The irradiation by femtosecond laser pulses induced clear brown coloring of the samples. Figure 2 shows the optical transmission spectra of glass with gold concentration of 0.015% after irradiation by 40 000 femtosecond laser pulses at the wavelengths of 266, 355, 500 and 800 nm. The laser fluences used were 1.5 J/cm^2, 1.8 J/cm^2, 1.8 J/cm^2, and 1.9 J/cm^2, respectively.

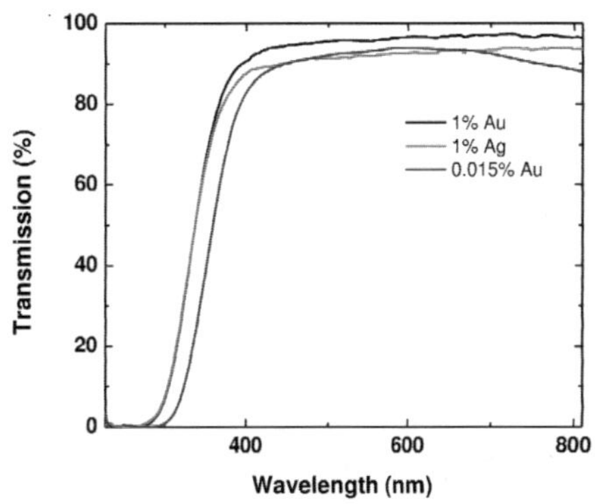

Figure 1. Transmission spectra of the glass samples used in this study.

The transmission curves are plotted with respect to the transmission of a non-irradiated sample. One can distinguish two pronounced zones: the first one is at wavelengths lower than approximately 450 nm and a dip centered at about 500 nm. These regions can be attributed to the absorption of color centers formed in the sample and related to intrinsic oxygen-deficiency defects in a glass matrix [11, 12]. The lower-wavelength absorption can be associated with two types of defects – an unpaired electron in sp3 of a single silicon atom with an oxygen vacancy, and a hole trapped in an oxygen vacancy. The defects that absorb in the visible spectral range are related to non-bridging oxygen hole centers. A bond modification in B_2O_3, whose concentration in the samples is 20%, could also have a small contribution to the spectra. Defects induced in borate oxide have strong absorption in the visible range at about 590 nm [13]. The spectra presented in figure 2 indicate that the application of laser radiation at different wavelengths produces similar types of changes in the transmission spectra, i.e. formation of similar types of defects. In the case of 266 nm, the laser energy is absorbed in the surface layer due to the strong absorption. The color changes are also localized in the surface layer. At the other wavelengths, defects are formed when clear evidence of nonlinear interaction is seen. The color modification observed when the wavelength of 500 nm or 800 nm were used was accompanied by a spectral broadening of the laser pulses. Figure 3 presents the spectrum of the radiation (wavelength of 800 nm, fluence of 1.9 J/cm^2) transmitted through the glass sample with a gold concentration of 0.015%. The spectral broadening is a complex phenomenon and could be the result of different mechanisms, as self-modulation, THz generation, or wave mixing [14]. The spectral broadening in the case of irradiation at 355 nm is barely seen due to the glass absorption in the shorter-wavelengths range. It should be mentioned that, under the conditions described, the brown-colored areas expanded throughout the entire sample thickness for the wavelengths of 355 nm, 500 nm and 800 nm.

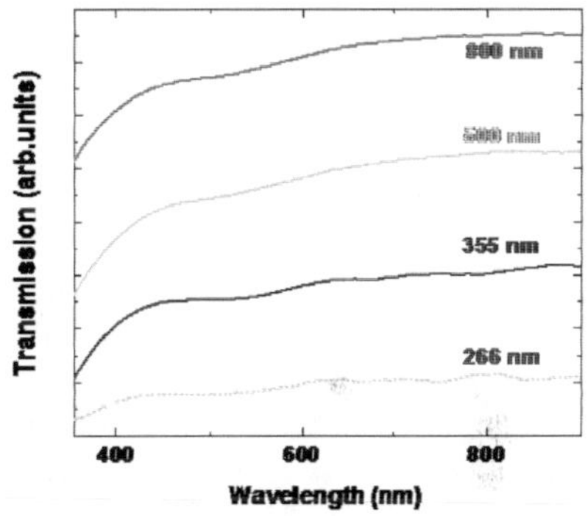

Figure 2. Transmission spectra of glass samples with gold concentration of 0.015% irradiated by 40 000 laser pulses at different wavelengths.

Figure 4 shows the transmission spectra of glass samples with different compositions after irradiation by 40 000 laser pulses at the wavelength of 800 nm and fluence of 1.9 J/cm^2. The spectrum of the glass without inclusion of a noble metal is also shown (BS0). The appearance is seen of a dip at about 700 nm with the increase of the gold concentration from 0.015 to 1 %. The spectrum of the silver-containing glass differs from that of the gold-containing one, namely, a sharp decrease of the transmission is observed at wavelengths below approximately 520 nm. Extracting the transmission spectrum of the glass without noble metal inclusion, on can find that this transmission reduction is due to a dip centered at about 400 nm. Its nature is not clear, but one may speculate that in this case the laser radiation induces the formation of Ag nanoparticles that have a plasmon absorption band at a wavelength of about 400 nm [4]. Further analyses are needed to clarify the validity of this proposition.

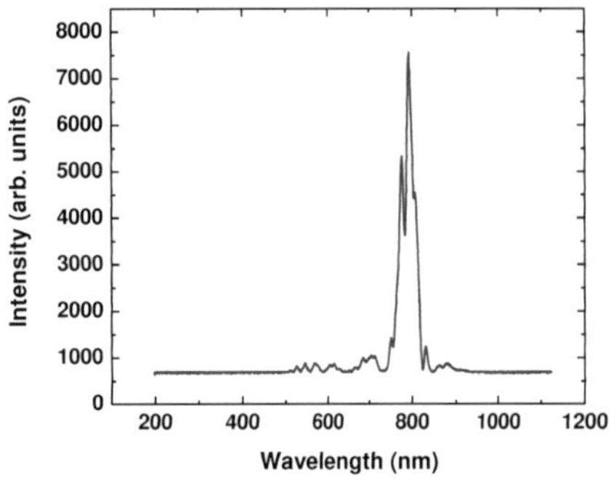

Figure 3. Spectrum of an 800-nm fs laser pulse transmitted through a glass sample with gold concentration of 0.015%. The laser fluence is 1.9 J/cm^2.

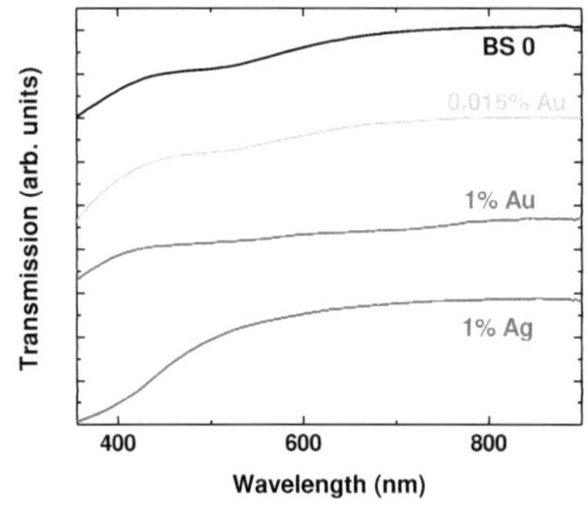

Figure 4. Transmission spectra of glass samples with different noble-metal concentration irradiated by 40 000 laser pulses at wavelength of 800 nm and laser fluence of 1.9 J/cm^2.

After irradiation, the samples were annealed. The optimal conditions were established by experiments performed at different temperatures in the range from 500 °C to 800 °C. After annealing for 30 min above 650 °C, the gold containing samples acquired a red coloration. Annealing of the same duration at temperatures below 550 °C did not induce any change in the sample color. Annealing the irradiated samples at a temperature of 600 °C resulted in the appearance of red color in the irradiated areas only, while the rest of the material remained colorless. Thus, a red coloring was observed only when the laser radiation induced the formation of defects, i.e. in the brown areas only. Figure 5 shows an optical microscope image of colored areas obtained by scanning the sample. The wavelength used was 500 nm, the sample contained gold at a concentration of 0.015%. The scanning conditions ensured the overlap of about 50 000 laser shots.

Figure 6 is a TEM image of the material in the colored area. It shows a part of a gold nanoparticle with a size of about 20 nm: i.e., the red color of the treated areas has to do with the formation of gold nanoparticles in the sample. Their optical properties, as defined by the plasmon resonance, give rise to the specific color of the area. Figure 7 shows the transmission spectra of areas irradiated at different wavelengths and annealed at 600 °C for 30 min. The laser fluences used were 1.5 J/cm^2, 1.8 J/cm^2, 1.8 J/cm^2, and 1.9 J/cm^2, for the wavelengths of 266 nm, 355 nm, 500 nm, and 800 nm, respectively.

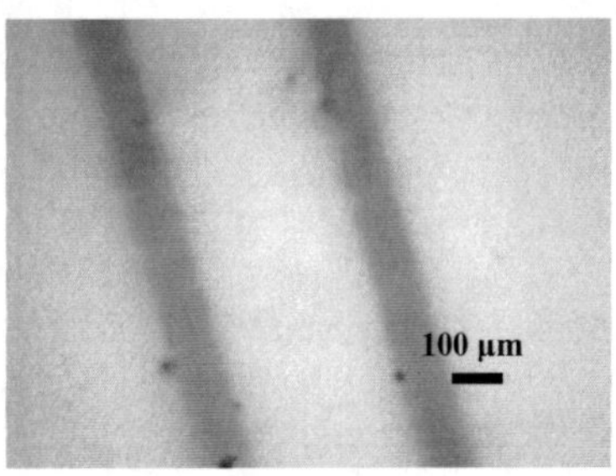

Figure 5. Optical microscope image of area irradiated by laser radiation at λ = 500 nm and annealed at 600 °C for 30 min. The laser fluence is 1.5 J/cm^2.

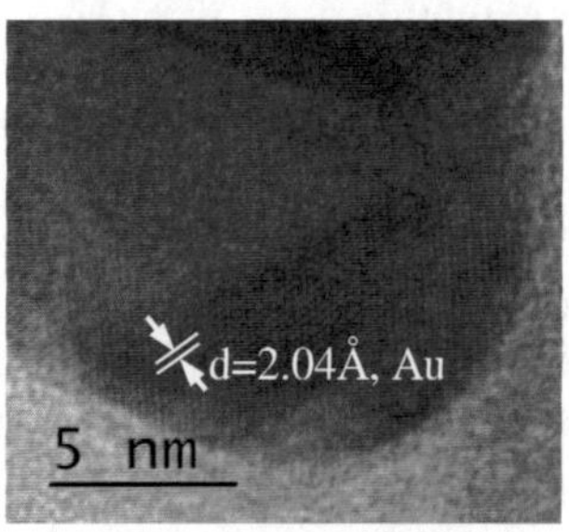

Figure 6. TEM image of a gold nanoparticle in the area irradiated by laser radiation at λ = 500 nm and annealed at 600 °C for 30 min.

The spectrum of that glass sample containing Ag is also shown. In this case the irradiation was at 800 nm with a fluence of 1.9 J/cm^2.

The results indicate that in all cases of different wavelengths used for irradiation, a selective formation of nanoparticles can be achieved. The well-pronounced transmission dip is an evidence of the formation of nanoparticles possessing specific optical properties due to the plasmon resonance. The irradiation at 266 nm results in the appearance of areas containing nanoparticles on the surface of the sample, while for the other wavelengths such areas spread over the whole sample. However, using tight focusing, one can obtain a localized nanoparticles-containing area, whose position can be tuned in 3D. The clear correlation between the position of the color centers induced after laser irradiation and the areas containing nanoparticles after annealing suggest that laser processing is a necessary condition for nanoparticle formation. It has been proposed [10, 11] that the electrons released after laser irradiation may induce reduction of the noble-metal ions. The annealing process then provokes the coalescence of atoms into nanoparticles. We showed that the formation of defects and, subsequently, of nanoparticles does not depend significantly on the irradiation wavelength. In addition, we showed that the wavelength of the transmission dip could be tuned by varying the irradiation time and laser fluence, since the effect of irradiation at the different wavelengths used was similar. The ability of implementing the method at different wavelength opens a possibility of fabricating wavelength-dependent structures in glasses, such as ripples, that can be used as selective optical components.

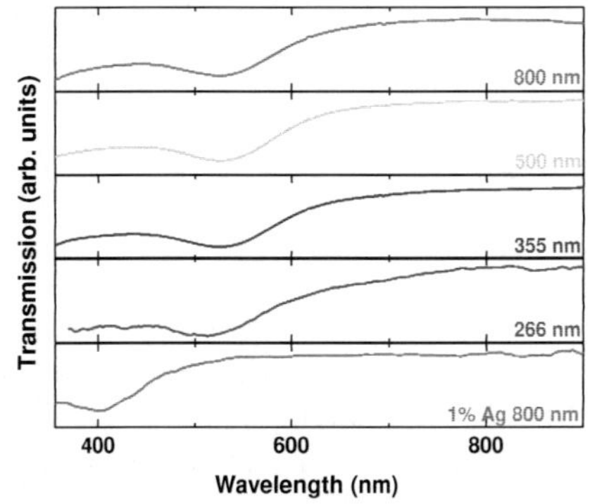

Figure 7. Transmission spectra of areas of glass containing 1% Au irradiated by 40000 laser pulses at different wavelengths and annealed at 600^{0}C for 30 min. The spectrum of glass containing 1% Ag is also shown.

4. Conclusions

A technique is presented for space-selective fabrication of noble metal nanoparticles inside glass samples. The method includes laser irradiation of ion-doped glass samples and subsequent annealing. The laser irradiation is performed by femtosecond laser pulses at different wavelengths – in the UV, VIS, and IR spectral range. Under certain conditions, the laser irradiation induces defect (color) centers in the glass samples. After annealing at 600 °C for 30 min, the color centers change their color from brown to red in the case of Au-doped glass, and to yellow for Ag-doped glass. The analysis performed indicates that this change is related to formation of nanoparticles, as their optical properties define the color observed. The process of nanoparticle formation can be realized at the different wavelengths used. Thus, the mechanism of laser absorption and the photon energy seem to play a minor role in the process when ultrashort pulses are used. It is found that the formation of color centers is the mark which indicates the possibility of formation of nanoparticles at the annealing stage. The method presented can be used for fabrication of 3D structures in glasses that may find applications in data storage, as optical components and sensors.

Acknowledgment

This work was supported by the Bulgarian National Science Fund under project DN 08/16.

References

[1] Petronella F, Truppi A, Ingrosso C, Placido T, Striccoli M, Curri M L, Agostiano A and Comparelli R 2017 *Catalysis Today* **281** 85
[2] Zhan C, Yu G, Lu Y, Wang L, Wujcikb E and Wei S 2017 *J. Mater. Chem.* C **5** 1569
[3] Beecroft L L and Ober C K 1997 *Chem. Mater.* **9** 1302

[4] Noguez C 2007 *J. Phys. Chem.* C **111** 3806
[5] Plech A, Leiderer P and Boneberg J 2008 *Laser Photon. Rev.* **2** 1
[6] Hanemann Th and Szabó D V 2010 *Materials* **3** 3468
[7] Koleva M E, Nedyalkov N N, Dikovska A Og, Atanasov P A, Avdeev G V, Shimizu H, Terakawa M, Obara M, Pallotti D, Orabona E, Maddalena P and Lettieri S 2014 *JOAM* **16** 144
[8] Sasai J and Hirao K J 2001 *Appl. Phys.* **89** 4548
[9] Dai Y, Yu G, He M, Ma H, Yan X and Ma G 2011 *Appl. Phys.* B **103** 663
[10] Almeida J M P, Tribuzi V, Fonseca R D, Otuka A J G, Ferreira P H D, Mastelaro V R, Brajato P, Hernandes A C, Dev A, Voss T, Correa D S and Mendonca C R 2013 *Opt. Mater.* **35** 2643
[11] Hu X, Zhao Q, Jiang X, Zhu C and Qiu J 2006 *Sol. State Commun.* **138** 43
[12] Salh R 2011 Defect related luminescence in silicon dioxide network: A Review *in Crystalline Silicon - Properties and Uses* Basu S ed (Rijeka Shanghai InTech) ISBN: 978-953-307-587-7
[13] Saka S 1970 *Bull. Inst. Chem. Res. Kyoto Univ.* **48** 53
[14] Tan D, Sharafudeen K N, Yue Y and Qiu J 2016 *Progress in Mater. Sci.* **76** 154

Femtosecond laser surface texturing of 3D poly-ε-caprolactone matrices for bone tissue engineering applications

A Daskalova[1,5], I Bliznakova[1], A Zhelyazkova[1], B Ostrowska[2], A Trifonov[3], I Buchvarov[3], L Avramov[1] and W Husinsky[4]

[1]E. Djakov Institute of Electronics, Bulgarian Academy of Sciences
72 Tzarigradsko Chaussee, 1784 Sofia, Bulgaria
[2]Faculty of Materials Science and Engineering, Warsaw University of Technology, 141Woloska Str., 02-507 Warsaw, Poland
[3]Department of Physics, St. Kl.Ohridski University of Sofia,
5 J. Bourchier Blvd., 1164 Sofia, Bulgaria
[4]Institute of General Physics, Vienna University of Technology,
Wiedner Haupt Str. 8-10/134, A-1040 Wien, Austria

E-mail: a_daskalova@code.bg

Abstract. Fibrous 3D matrices were fabricated from poly-ε-caprolactone (PCL) by fused deposition modeling. Femtosecond laser irradiation was then used to demonstrate the possibility to affect the porosity of the 3D PCL fiber meshes. The surface characteristics were analyzed by scanning electron microscopy (SEM) and confocal microscopy. The interrelationship was examined between the laser processing parameters (number of pulses, pulse energy applied) and the response of the biomaterial. The formation was demonstrated of well-defined micropores, while the original fiber structure was retained. The study of cells cultivation on the laser-modified scaffolds showed good adhesion compared to a non-modified scaffold. The results obtained showed that femtosecond laser processing can be used as an alternative non-contact tool in enhancing the porosity of artificial constructs, thus influencing the cell adhesion into fibrous meshes.

1. Introduction

The deterioration of tissues or organs is a serious medical problem in trauma and orthopedic surgery, which is yet to be solved satisfactorily. Bone regeneration is proving invaluable in several applications and has given rise to hope in situations that had thus far been deemed beyond salvation, such as repair or replacement of irreversibly damaged and fully differentiated bone defects resulting from trauma, inflammation or tumor resection [1]. The limitations of the established techniques, such as distraction osteogenesis and implantation of autografts or allografts, involve problems with risk of infection, immune reaction and pain [1,2]. To address these challenges, diverse tissue engineering approaches have been attempted aimed to develop functional three-dimensional (3D) tissues matrices. In this respect, the scientific field of tissue engineering has emerged as an important technique for bone regeneration; it is based on forming a temporary porous matrix to guide the regeneration of the tissue

[5]To whom any correspondence should be addresed.

into the desired three-dimensional shape. The extra-cellular matrix (ECM) is used as a micro-environment onto which cells can adhere and proliferate. In the field of regenerative medicine, a basic task is to design new biomaterials and methods for controlling their interface characteristics for guided tissue regeneration [3].

Synthetic matrices have evolved to become viable alternative materials for bone reconstruction. An ideal bone substitute should possess certain properties, including porosity, interconnectivity of the scaffold, osteoconductivity, biodegradability, as well as adequate mechanical properties; all this limits the particular material's ability to stimulate cell growth and generate new tissue [1,4]. Over the past years, a wide range of innovative synthetic materials have been developed to overcome the problems associated with autologous bone grafts. Newly developed implant materials are being studied as alternatives, including bioceramics, such as calcium phosphates (CaPs) and bioactive glasses, polymers (natural – collagen-I and synthetic – poly-caprolactone) and hybrid materials (a mixture of bioceramics and polymers), partly in combination with growth factors, bone marrow or mesenchymal stem cells [5,6]. A scaffold made of ceramic, such as calcium phosphate and calcium sulfate, is the most commonly used material for bone regeneration due to its bioactive properties. However, its brittleness and fast resorption rate pose clinical concerns. Biodegradable polymers are another promising type of potential bone graft substitutes. One such suitable candidate, namely poly-ε-caprolactone (PCL), is a biodegradable polymer, approved by the U.S. Food and Drug Administration, that has a low degradation rate compared with poly (lactic-co-glycolic acid) (PLGA) and polylactic acid (PLA) [7-9].

A current challenge in this field has to do with creation of 'smart' biomimetic, synthetic or natural materials (polymers, ceramics, and metals) by means of various chemical or physical techniques with the purpose of inducing enhanced surface functionalization by way of tailored topographic surface modification. Applying PCL scaffolds in bone regenerative medicine is mainly limited by problems related to volume vascularization of the artificial constructs. An additional laser-assisted treatment leading to surface modification by inducing surface roughness could provide enhanced cell infiltration [5,10]. Laser tailoring of the surface topography by modifying the respective material's surface offers advantages over both the chemical and physical methods. Moreover, the laser-assisted technique enables precise modification of certain surfaces that are difficult to treat with conventional chemical methods. Among its advantages is also the fact is that the modified surfaces are free from contaminants [11].

This article discusses the laser induced post-modification of poly-ε-caprolactone (PCL) scaffolds surface as a potential bone graft substitute in bone defects. The scaffolds design and fabrication, its surface modification and the influence on the cellular activities were evaluated by scanning electron microscopy (SEM), confocal microscopy and fluorescence microscopy measurements.

2. Material and methods

2.1. Fabrication and processing of 3D – PCL scaffolds

Three-dimensional (3D) scaffold constructs (5×5 mm in width and 3 mm in height) from poly-ε-caprolactone (PCL), (Sigma-Aldrich, St. Louis, MO) with a molecular weight (Mn) of 45 kDa, melt index of 0.95 g/ 10 min at 95 °C (ISO 1133), melting temperature of 57 °C, and a glass-transition temperature of –60 °C were designed using Solid Works 3D CAD Design Software. The geometrical model was fabricated by a Bioscaffolder machine (SYSENG, Germany) as an STL file. The internal architecture of the tested scaffolds was then designed with lay-down patterns 0/90/180.

To create the wood-pile structures, the PCL granules were heated to $T = 70$ °C. Once the material melted, air at a pressure of 5 bar was applied to the reservoir in order to transfer the molten polymer for extrusion. The layer-by-layer plotted fibrous scaffolds were characterized by theoretical parameters, namely, spacing between fibers in the same layer (140 μm) and layer thickness (130 μm).

The femtosecond laser modification experiments were carried out by a Ti:Sapphire laser (Quantronix-Integra-C) delivering 30-fs pulses at a central wavelength $\lambda = 800$ nm and tunable repetition rate. The number of laser pulses applied (*N*) was controlled by a computer-driven fast mechanical shutter synchronized by the controlling software. The experiments were performed in air with the laser beam focused to a focal spot with a diameter of 50 μm using a lens of focal length 20 cm at normal incidence, figure 1.

The focusing lens was placed on a translation stage equipped with a micrometer screw for fine adjustment of the focus position on the specimen's working surface. The sample was positioned perpendicular to the focusing beam on a high-precision XYZ translation stage. The experimental setup was controlled via specially written LabView software. The effect of the laser processing parameters on the created surface microstructures was followed by scanning electron microscopy (Microscope FEI, Quanta 200F) and confocal microscopy- μsurf explorer (Nanofocus). We thus examined the laser-induced modification effect on the adhesion of an MG63 osteoblast-like cell line. The cell culturing was initiated by seeding MG63 osteoblast-like cells at a density of 5×10^5 cells/cm^2. The cell viability was monitored after three days of culturing in Dulbecco's Modified Eagle Medium (DMEM). The imaging of the cells was achieved by adding Rhodamine Phalloidin conjugated staining to visualize the actin cytoskeleton, and DAPI (blue) to visualize the cell nuclei.

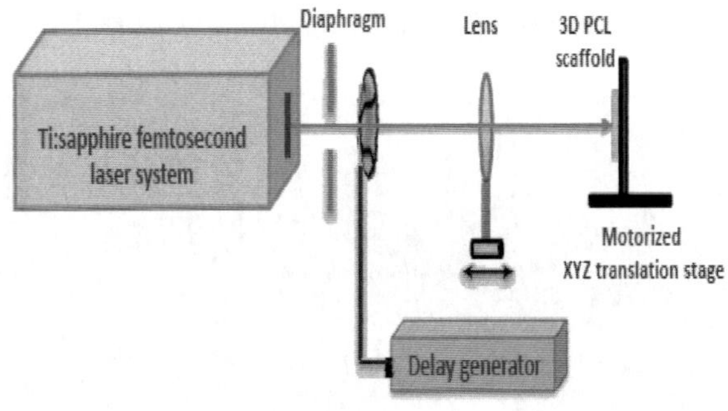

Figure 1. Experimental setup for femtosecond laser irradiation of PCL scaffold.

3. Results and discussion

3.1. Femtosecond laser ablation and characterization of modified fibrous scaffolds

We varied the laser pulse energy and effective pulse number and followed the effect on the size and structure of the modified zone of the PCL scaffolds. To study the effect of laser fluence on the PCL fiber meshes, the struts were irradiated at a fluence of 0.105 J/cm^2 and 0.629 J/cm^2. The confocal microscopy observations showed that irradiation by a single laser pulse induced the formation of bump

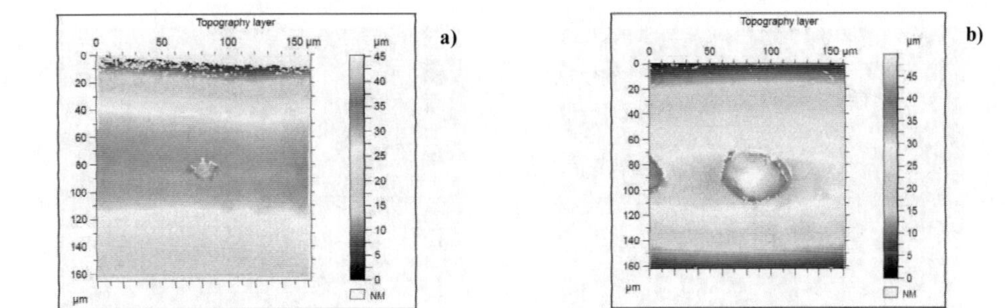

Figure 2. Confocal microscopy image of the laser induced modification; a) topography and cross-section image for $N = 1$, $F = 0.105$ J/cm^2, $\tau = 30$ fs, $\nu = 1$ kHz; b) topography and cross-section image for $N = 1$, $F = 0.629$ J/cm^2, $\tau = 30$ fs, $\nu = 1$ kHz.

like structures with a diameter of 25 μm and 59 μm, and a height over the surface level of 10 μm and 17 μm, for laser fluences of 0.105 J/cm^2 and 0.629 J/cm^2, respectively (figure 2).

We further monitored the changes in the surface of scaffold fibers initiated due to irradiation with increasing number of laser pulses. The porosity of the mesh was altered when the number of pulses (*N*) exceeded five. Figure 3 is an example of PCL fiber meshes irradiated by $N = 5$, 10 and 30.

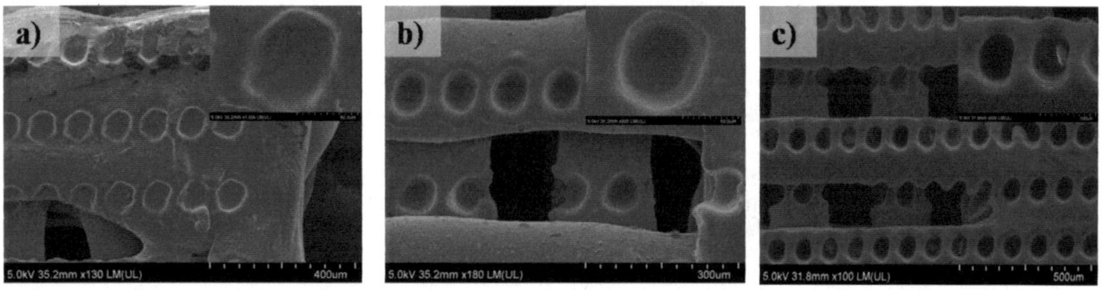

Figure 3. SEM micrographs of PCL fiber meshes (a) $N = 5$, $F = 0.105$ J/cm^2 (b) $N = 10$, $F = 0.105$ J/cm^2; (c) $N = 30$, $F = 0.105$ J/cm^2.

The results showed that pores with an average diameter size of 70 μm can be formed without additional side effects to the scaffold structure by optimizing the laser energy and number of laser pulses applied.

The SEM investigations confirmed that fs laser irradiation of printed scaffolds resulted in a minimal collateral damage and thermal side effects. Regarding the effect of the number of laser pulses applied (*N*), this was varied from 1 to 30 pulses. Differences were observed in the depths of the created craters when the number of applied laser pulses was increased while the laser fluence was kept constant. The most pronounced modification in the volume of the sample was achieved by irradiation by the highest number of pulses, figure 3(c).

3.2. Cellular adhesion on topographically structured PCL matrix

Previous studies [12,13] have demonstrated that topographical post-modification of scaffold surfaces with microscale features influences the cells' behavior. We observed by confocal laser microscopy the morphologies of MG63 osteoblasts adhering to PCL constructs, figure 4.

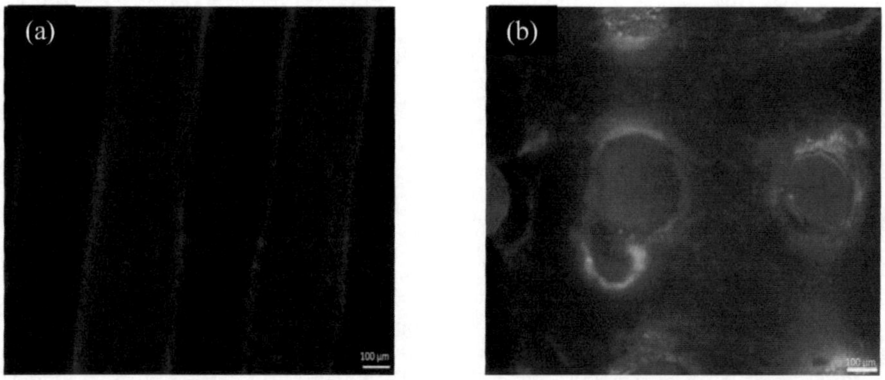

Figure 4. Confocal laser microscopy images of MG63 osteoblast cells cultured for three days on 3D PCL matrices, demonstrating the actin staining and visualizing the fibrillar network of actin cytoskeleton a) non-irradiated surfaces, scale bar 100 μm b) laser-modified fiber surface, scale bar 100 μm.

Fluorescence microscopy images showed that the MG63 osteoblast cells adhere homogeneously to untreated and treated scaffold fibers. These basic cellular studies demonstrated that cells tend to grow completely over the areas where a new structured zone is formed. These preliminary results need a further detailed study, which is out of the scope of this article.

4. Conclusions

The results reported here show that the laser-induced modification can be optimized by controlling the laser processing parameters. The physical integrity of the 3D PCL fiber meshes is thus preserved, as confirmed by SEM analysis. Additional sample modification can affect the mechanical properties of the produced matrices and provide a path to enhancing the cell's infiltration in the volume for improved vascularization. Therefore, femtosecond laser processing could be used as an alternative method for micropatterning of polymeric meshes with a high-level of precision, thus offering a way of enhancing the surface functionalization in tissue applications.

Acknowledgments

This work was supported by the Bulgarian National Science Fund under project DN08/5 „Bioactivity improvement of biomimetic materials by texturing with ultra-short laser pulses" (2016-2018); and under projects DNTS/Austria/01/1/-2013–2018, DFNI-B02/9/2014, COST MP1301 NEWGEN and bilateral project FWO/BAS "Functionalization of biomaterials modified by femtosecond pulses for cell adhesion and guidance improvement" (2016-2019).

References

[1] Rentsch B, Bernhardt R, Scharnweber D, Schneiders W, Rammelt S and Rentsch C 2012 Biomatter**2:3** *Landes Biosci.* 158–65
[2] Amini A, Laurencin C and Nukavarapu S 2012 *Critical Rev. in Biomedical Eng.* **40**/5 363–408
[3] Mano J, Silva G, Azevedo H, Malafaya P, Sousa R, Silva S and Reis R 2007 *J. Royal Soc. Interface* **4**/17 999–1030
[4] Jin Woo Lee 2015 *J. Nanomater.* Article ID 213521, 14 pages
[5] Roohani-Esfahani S, Newman P and Zreiqat H 2016 *Sci. Rep.***6** 19468
[6] He Y and Lu F 2016 *Stem Cells Int.* 5786257
[7] Wong H, Chu P, Leung F, Cheung K, Luk K and Yeung K 2014 *Progress in Natural Sci.: Mater. Int.* **24** 561–7
[8] Applications of Polymers in Drug Delivery 2014 *Handbook* ed S Rapraed A Misra and A Shahiwala
[9] Daskalova A, Bliznakova I, Iordanova E, Yankov G, Grozeva M and Ostrowska B 2016 *J. Phys.: Conf. Ser.* **682** 012006
[10] Morouço P, Biscaia, S, Viana T, Franco M, Malça C, Mateus A and Alves N 2016 *BioMed. Res. Int.*1596157
[11] Tiaw K, Goh S, Hong M, Wang Z, Lan B and Teoh S 2005 *Biomater.* **26/7** 763–9
[12] Matsugaki A, Aramoto G, Ninomiya T, Sawada H, Hata S and Nakano T 2015 *Biomater.* **37** 134-43
[13] Malinauskas M, Rekštyt S, Lukoševicius L, Butkus S, Balciunas E, Peciukaityte M, Baltriukiene D, Bukelskiene V, Butkevicius A, Kucevicius P, Rutkunas V and Juodkazis S, 2014 *Micromachines* **5** 839-58

Optical properties of Ag/ZnO nanocomposite coatings

A Og Dikovska[1,4], G B Atanasova[2], G V Avdeev[3] and N N Nedyalkov[1]

[1]Emil Djakov Institute of Electronics, Bulgarian Academy of Sciences, 72 Tzarigradsko Chaussee, 1784 Sofia, Bulgaria
[2]Institute of General and Inorganic Chemistry, Bulgarian Academy of Sciences, Acad. G. Bonchev Str., Bl. 11, 1113 Sofia, Bulgaria
[3]Rostislaw Kaischew Institute of Physical Chemistry, Bulgarian Academy of Sciences, Acad. G. Bonchev Str., Bl. 11, 1113 Sofia, Bulgaria

E-mail: dikovska@ie.bas.bg

Abstract. In this work, ZnO nanostructures were fabricated on Ag-coated substrates by pulsed laser deposition. Ag was incorporated into ZnO nanostructures during the growth by using a composite, or mosaic, target. The effect was studied of Ag doping on the morphology, structure, composition, and photoluminescence of the ZnO nanostructures. It was found that the increase of the dopant concentration leads to a deterioration of the crystalline structure, as well as to a change of the sample morphology. It was also found that the manner of incorporating Ag into the ZnO matrix affects the photoluminescent performance of the nanocomposite coatings.

1. Introduction

One of the most versatile materials used in nanotechnology is zinc oxide (ZnO), a semiconductor with an exceptional potential for applications in optics, energy conversion, biomedicine, etc. [1]. ZnO exhibits numerous remarkable characteristics, such as functional catalytic and optical properties, electrical, chemical and thermal stability, low dielectric constant, antimicrobial and antibacterial effects, and has formed probably the richest family of nanostructures among all semiconducting materials [2-4]. Modification of the oxide nanostructures through the introduction of appropriate additives is a key factor in optimizing the existing, or developing novel, properties [5,6]. When the additives are in the form of nanoparticles, a new composite structure is obtained, which possesses specific properties defined, on the one hand, by the nanoparticles composition, shape, and size, and, on the other, by the nanosized oxide matrix surface, morphology, and structure. A huge variety of composite materials are being synthesized and investigated for various optical applications ranging from third-order nonlinearity to fabrication of new transparent magnetic materials [7-9]. In this regard, it is essential to clarify the fundamental phenomena and study the optical properties of these nanocomposite materials, as they are expected to form the basis of future technologies.

A wide variety of techniques has been employed to fabricate oxide nanostructures with different morphologies [10]. Among those, pulsed laser deposition (PLD) stands out as a simple, less expensive and more flexible technique for producing well-crystallized structures than many rival gas-surface-

[4] To whom any correspondence should be addressed.

based growth methods [11]. Furthermore, because laser ablation maintains the stoichiometry of the target material, PLD is especially suitable for obtaining a doped material using pre-doped targets.

The aim of this work was to fabricate Ag/ZnO nanocomposite coatings by PLD. Ag was incorporated into ZnO nanostructures during their growth by using a composite, or mosaic, target. We also studied the effect of Ag doping on the morphological, structural, compositional, and optical properties of the nanostructured ZnO coatings.

2. Experimental

The experiments were performed in a standard on-axis PLD configuration. The third harmonic of a Nd:YAG laser ($\lambda = 355$ nm and a repetition rate of 10 Hz) was used for ablation of a ceramic target. The laser fluence applied was 0.4 J cm^{-2}. Two different targets were used: composite or mosaic. The mosaic target was formed by a ZnO target partly covered by a Ag slide (covering ratio 1/3 Ag:2/3 ZnO). The samples produced by laser ablation of the pure ZnO and the mosaic targets are hereinafter referred to as "sample A" and "sample B", respectively. Composite targets with three different Ag contents were sintered by using ZnO and $AgNO_3$ powders as initial materials. The samples prepared by ablation of the composite targets with Ag content of 0.5, 1, and 3 at % are referred to as "sample C", "sample D", and "sample E", respectively. The distance between the target and the substrate was kept at 60 mm. All experiments were performed in oxygen atmosphere at a pressure of 5 Pa, a substrate temperature of 600 °C and for a deposition time of 90 min. The substrate was made of (001)-oriented silica (SiO_2) covered by a Ag layer. The catalyst layers were produced by laser ablation of a metal target in the same PLD set-up, but under different experimental conditions, namely, at room substrate temperature in vacuum and a target-substrate distance of 35 mm. The average thickness of the covering layers was 100 nm.

The surface morphology of the samples was analyzed using scanning electron microscopy (SEM). X-ray diffraction (XRD) was applied in order to investigate the crystalline structure of the ZnO samples. The elemental composition and chemical state of the sample surfaces were studied using X-ray photoelectron spectroscopy (XPS). The optical absorbance spectra of the samples were determined by a UV-VIS spectrophotometer. The photoluminescent (PL) spectra of the ZnO samples were recorded at room temperature by using an excitation source at $\lambda = 325$ nm.

3. Results and discussion

SEM images of the ZnO samples prepared from the pure and the mosaic targets are presented in figure 1. As seen in figure 1(a), ablating the pure ZnO target under the experimental conditions described above results in the fabrication of nanorods with diameters within 70 – 200 nm and lengths in the 200 – 250 nm range.

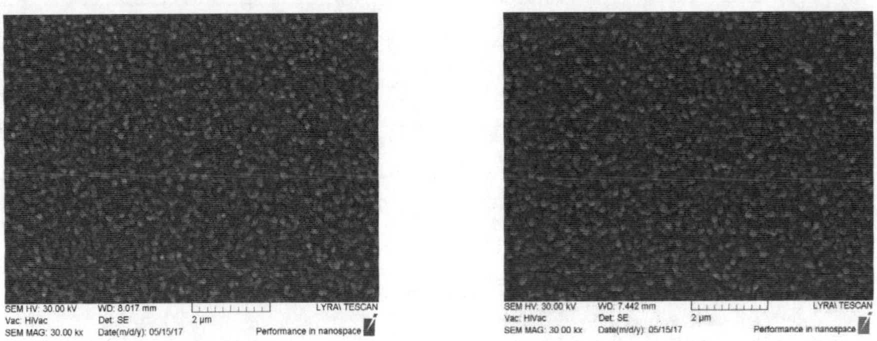

(a) (b)

Figure 1. SEM images of Ag/ZnO samples prepared from: (a) pure – sample A, and (b) mosaic target – sample B.

IOP Conf. Series: Journal of Physics: Conf. Series **992** (2018) 012049 doi:10.1088/1742-6596/992/1/012049

Ablation of the mosaic target (figure 1(b)) did not lead to a significant change in the samples' morphology. In the latter case, adding Ag into the ZnO samples during deposition resulted in nanostructures with a larger mean diameter and a wider range of nanorods' diameters (130 – 250 nm).

SEM images of the ZnO samples produced from composite targets with different Ag concentrations are presented in figure 2. Figure 2(a) shows the SEM image of the sample deposited from the composite target with the smallest Ag content. It is seen that the sample's morphology (and the particles' mean diameter as well) did not differ from that of the sample produced from pure ZnO target (figure 1(a)). The increase of the Ag concentration in the target led to the deposition of a denser structure of shorter nanorods with diameters ranging from 150 to 250 nm (figure 2(b)).

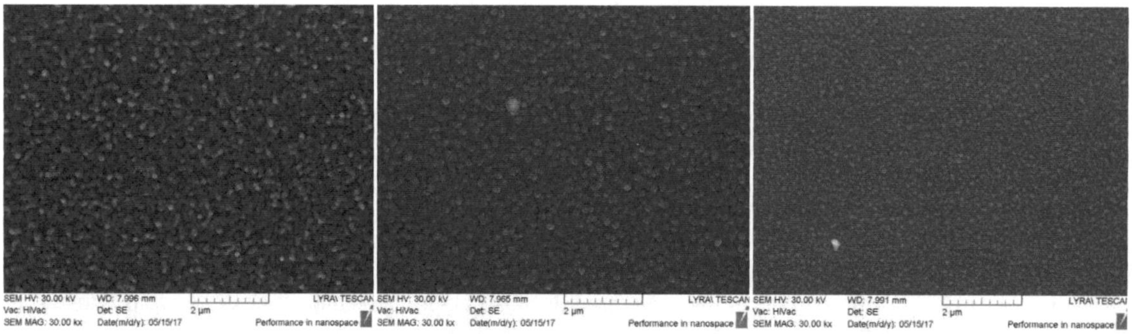

(a) (b) (c)

Figure 2. SEM images of Ag/ZnO coatings prepared from composite targets: (a) sample C, (b) sample D, and (c) sample E.

Increasing further the Ag content in the composite target changed the morphology of the sample (figure 2(c)). Thus, the sample's morphology presented a denser grain structure with smaller nanorods diameters in the range of 60 – 120 nm; i.e., increasing the Ag dopant concentration incorporated into the ZnO structure during deposition (by using a composite target) changes the morphology of the samples and impedes the growth of nanostructures.

XRD spectra of the Ag/ZnO nanostructured coatings produced from the mosaic and the composite target are presented in figure 3. The patterns can be indexed to a hexagonal structure of bulk ZnO (ICSD 98-002-6170). All samples were slightly textured along the [002] direction; we should point out that Ag was present in all spectra as a separate phase (ICSD 98-004-4387).

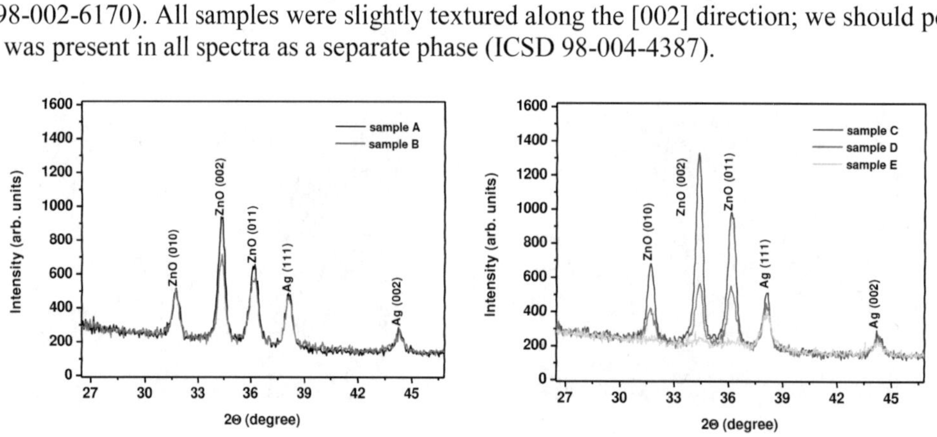

(a) (b)

Figure 3. XRD spectra of the Ag/ZnO nanostructures prepared from: (a) mosaic and (b) composite targets.

The XRD spectra of the sample produced from the pure and the mosaic target are compared in figure 3(a). As it is seen, the diffraction peaks intensity for the predominant [002] direction decreased slightly with the use of a mosaic target, i.e. with adding Ag. The average crystallite size, as evaluated for the dominant diffraction peak with (002) symmetry, was about 20 nm. Figure 3(b) presents XRD spectra of the samples produced from composite targets with different Ag content. As seen, the increase of the dopant concentration in the targets led to a substantial decrease of the samples' crystalline quality. In this case, an intensity decrease was observed for all diffraction peaks. It should be mentioned that both the ZnO and Ag phases were homogeneously distributed in the depth of all samples, independently of the target used for ablation.

The ZnO structures' physico-chemical composition was studied by XPS analysis of the samples' surface (not shown). The $Zn2p_{3/2}$ photoelectron peak located at 1021.4 eV was symmetrical in shape, implying the existence of only one zinc state. On the contrary, the O1s peak outline was asymmetrical and presented an apparent shoulder. The O1s peak was deconvoluted into two components at 530.1 eV and 531.5 eV. The main peak centered at 530.1 eV belongs to the lattice oxygen of the wurtzite structure of ZnO, while the shoulder peak located at 531.5 eV could be associated with O^{2-} ions in the oxygen deficient regions within the ZnO matrix [12,13]. A peak at 367.4 eV was also observed which is due to the binding energy of Ag3d. This result indicated the existence of Ag dopant in the ZnO nanostructures. The XPS analyses showed the presence of oxygen deficient states in the samples; the increase of the Ag content in the targets increased the oxygen deficiency on the samples' surface. The XPS analyses were also used for quantification of the atomic composition of the samples. According to the experimental XPS peak areas, the Ag/Zn atomic ratios in Ag/ZnO nanostructures deposited from composite targets were approximately: 0.06 % of Ag/Zn in sample C, 0.1 % of Ag/Zn in sample D, and 0.3 % of Ag/Zn in sample E. The Ag/ZnO sample produced from the mosaic target (sample B) had a Ag/Zn ratio of 0.1 % (the same as sample D).

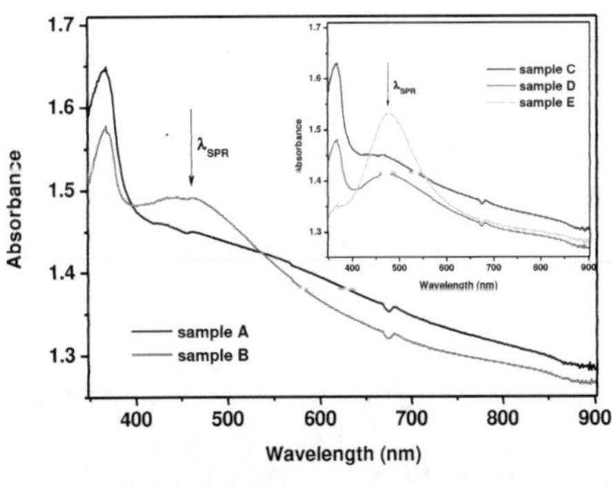

Figure 4. UV-VIS spectra of ZnO samples prepared from pure ZnO and a mosaic target. Inset: optical spectra of samples produced from composite targets.

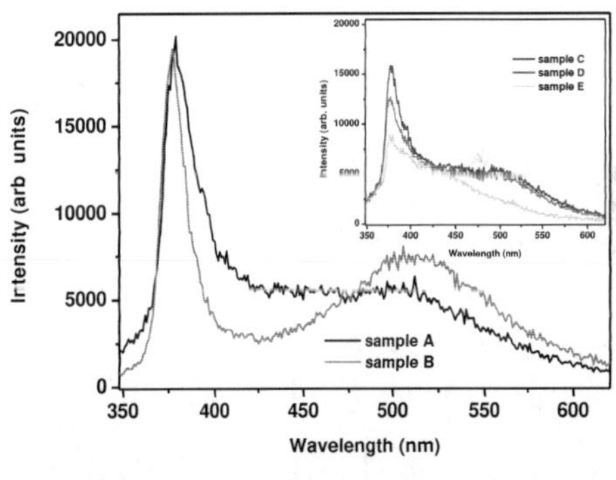

Figure 5. PL spectra of ZnO samples prepared from pure ZnO and a mosaic target. Inset: PL spectra of the samples produced from composite targets.

The UV-VIS absorbance of the pure ZnO and Ag/ZnO nanostructures (figure 4) were obtained to provide evidence of the presence of Ag nanoparticles in the samples. A well-defined surface plasmon resonance (SPR) band was observed in the Ag/ZnO sample produced from the mosaic target. A broad SPR peak was centered at 460 nm (sample B). The inset in figure 4 illustrates the influence of the Ag content in the ZnO samples on their optical absorbance. As seen, all samples possessed an SPR peak, with a monotonic red shift observed in the SPR resonance with the increase of the Ag content, indicating that Ag was present in the ZnO nanostructured coatings in the form of nanoparticles, independently of the manner of the incorporation.

Figure 5 presents PL spectra of the pure and the Ag-doped ZnO nanostructures. They exhibited a strong UV emission and a broadband visible emission with a peak at ~500 nm. It is well known that the UV emission corresponds to the near band-edge emission, while the green emission peak is commonly referred to as a deep-level or a trap-state emission [14,15]. Incorporating Ag nanoparticles in ZnO nanostructures by using a mosaic target led to a change in the emission spectrum as compared with that of the pure ZnO nanostructure (figure 5). The UV emission kept its intensity, but the full width at half maximum (FWHM) of the peak was substantially narrower. At the same time, the intensity of the green emission increased, while the FWHM of its peak decreased. The inset in figure 5 shows the dependence of the Ag/ZnO samples' PL performance on the Ag concentration. As seen, increasing the Ag content reduces the intensity of the PL signal from the samples, which is probably due to a decrease in the crystalline quality of the samples. It should be mentioned that, despite the same crystalline quality and Ag/Zn atomic ratio of samples B and D, their PL spectra are different, or the manner of incorporating Ag in the ZnO matrix affects the photoluminescent performance of the nanocomposite samples. Clarifying the effect of incorporating Ag nanoparticles in a ZnO matrix on the PL performance will require further detailed studies of these nanocomposites' microstructure.

4. Conclusions

Pure ZnO and Ag/ZnO nanocomposite coatings were grown by means of PLD. Ag was incorporated in ZnO nanostructures during the growth by using a composite, or mosaic, target. It was found that the use of a mosaic target for ablation did not change significantly the samples' morphology. Furthermore, the Ag doping did not change the growth direction of the nanostructures on the samples. Ag was present in the ZnO nanostructure coatings in the form of nanoparticles, which was confirmed by detecting an SPR peak in the absorbance spectra of all nanocomposites. The PL spectra exhibited a strong UV emission and a broadband green emission with a peak at ~500 nm. The manner of incorporating Ag in the ZnO matrix affected the photoluminescent performance of the nanocomposite samples.

Acknowledgements

The uthors wish to thank Vladimir Georgiev and Nadya Stoykova for the special equipment provided for registering the optical spectra of the samples. The authors also acknowledge the financial support of the Bilateral Join Research Project "Laser-assisted fabrication of composite nanostructures" between the Bulgarian Academy of Sciences and CNR-SPIN Consiglio Nazionale delle Ricerche.

References

[1] Wang Z 2007 *Appl. Phys.* A **88** 7
[2] Huang M, Mao S, Feick H, Yan H, Wu Y, Kind H, Weber E, Russo R and Yang P 2001 *Scinece* **292** 1897
[3] Law M, Greene L, Johnson J, Saykally R and Yang P 2005 *Nat. Mater.* **4** 455
[4] Aal N, Al-Hazmi F, Al-Ghamdi F, Al-Ghamdi A, El-Tantawy F and Yakuphanoglu F 2015 *Spectrochimica Acta Part* A*: Molecular and Biomolecular Spectroscopy* **135** 871
[5] Zhang X, Chen Y, Liu R and Tsai D 2013 *Rep. Prog. Phys.* **76** 046401
[6] Baffou G and Quidant R 2014 *Chem. Soc. Rev.* **43** 3898
[7] Beecroft L and Ober C 1997 *Chem. Mater.* **9** 1302
[8] Khan R, Uthirakumar P, Bae K, Leem S and Lee I 2016 *Mater. Lett.* **163** 8
[9] Lee M, Kim T, Kim W and Sung Y 2008 *J. Phys. Chem.* C **112** 10079
[10] Wang Z 2004 *J. Phys.: Cond. Matter.* **16** R829
[11] Eason R 2007 *Pulsed Laser Deposition of Thin Films: Applications-Led Growth of Functional Materials* (Hoboken: Wiley)
[12] Zhang L and Yin Y 2013 *Sens. & Act.* B **183** 110
[13] Xu X, Chen D, Yi Z, Jiang M, Wang L, Zhou Z, Fan X, Wang Y and Hui D 2013 *Langmuir* **29** 5573
[14] Li C, Fang G, Fu Q, Su F, Li G, Wu X and Zhao X 2006 *J. Crystal Growth* **292** 19
[15] Yang P, Yan H, Mao S, Russo R, Johnson J, Saykally R, Morris N, Pham J, He R and Choi H 2002 *Adv. Funct. Mater.* **12** 323

IOP Conf. Series: Journal of Physics: Conf. Series **992** (2018) 012050 doi:10.1088/1742-6596/992/1/012050

SERS analysis of Ag nanostructures produced by ion-beam deposition

P A Atanasov[1,4] , N N Nedyalkov[1], Ru G Nikov[1], Ch Grüner[2], B Rauschenbach[2] and N Fukata[3]

[1]Emil Djakov Institute of Electronics, Bulgarian Academy of Sciences, 72 Tzarigradsko Chaussee, 1784 Sofia, Bulgaria
[2]Leibniz Institute of Surface Modification (IOM), Permoserstrasse 15, D-04318 Leipzig, Germany
[3]International Center for Materials for NanoArchitectonics (MANA), National Institute for Materials Science (NIMS), 1-1Namiki, Tsukuba 305-0044, Japan

E-mail: paatanas@hotmail.com

Abstract. This study deals with the development of a novel technique for formation of advanced Ag nanostructures (NSs) to be applied to high-resolution analyses based on surface enhanced Raman scattering (SERS). It has direct bearing on human health and food quality, e.g., monitoring small amount or traces of pollutants or undesirable additives. Three types of nanostructured Ag samples were produced using ion-beam deposition at glancing angle (GLAD) on quartz. All fabricated structures were covered with BI-58 pesticide (dimethoate) or Rhodamine 6G (R6G) for testing their potential for use as substrates for (SERS).

1. Introduction

The properties of noble metallic nanostructures (NSs) have been the subject of considerable fundamental and technological interest. Metal nanoparticles (NPs) play an important role in scientific research and nanotechnology. The interest in nanoscience and nanotechnologies has grown dramatically – the number of the various objects of nano research and applications has increased exponentially during the last decade [1]; the nanosystems find nowadays application in many areas, such as chemistry, optics, biology, medicine, microelectronics, etc.

The excitation spectrum of a given noble metallic sub-wavelength structure is determined by its surface plasmon resonances. The energies of plasmon resonances depend strongly on the shape and composition of the nanostructures. The tunability of the plasmon resonances of noble metallic NPs can be exploited to position the optical resonances at specific wavelength regions of interest and has led to a wide range of applications. The strong local electromagnetic field enhancement arising at a surface plasmon resonance has also been used to manipulate light-matter interactions, and noble metallic sub-wavelength structures have widely been applied to SERS [2,3]. The enhancement of the Raman signal may reach factors of up to 10^8-10^{12}, so that the method has been applied to detection of even single molecules [2]. The enhancement of the Raman signal is a result of the local enhancement of the

[4] To whom any correspondence should be addressed.

IOP Conf. Series: Journal of Physics: Conf. Series **992** (2018) 012050 doi:10.1088/1742-6596/992/1/012050

electromagnetic field in the vicinity of a structured surface due to the excitation of local and surface plasmons. In addition to their fundamental importance, plasmonic nanostructures have received a great deal of attention for potential applications in areas such as sub-wavelength waveguides, optical nanoantennas, photovoltaic technology for efficient light coupling into solar cells, metamaterials, chemical and biological sensing, and biomedicine [4-7].

Three-dimensional (3D) sculptured thin films (STF's) are truly nanostructured inorganic materials with anisotropic properties. They show a potential of being designed in a controllable manner using physical vapor deposition methods (PVD) for potentially multifunctional materials and products. Glancing angle ion-beam deposition (GLAD) is a sophisticated method of creating 3D chiral NSs with a tailored geometry. GLAD is based on controlling the self-shadowing between the growing crystallites/nanostructures. The self-organized nanostructure growth is based on a concurrent growth mechanism due to geometrical shadowing in combination with kinetic limitation for surface adatoms [8-10]. This technique requires a particle flux reaching the substrate under an extremely oblique angle of incidence (typically $\theta \sim 80$ deg). The growth conditions support a columnar growth, and the samples consist of 3D needles, which are slanted into the direction of the particle flux. The instantaneous change of the growth direction due to a simple variation of the incident vapor flux (by substrate rotation) allows for the fabrication of 3D nanostructures with manifold morphologies. Since the growth process is mainly determined by the shadowing length and the surface diffusion length, materials ranging from insulators to metals to semiconductors can be grown with 3D morphologies like spirals, chevrons, screws, pillars, needles etc. [11].

Chiral nanostructures are of particular interest due to their theoretically-predicted optical properties; uses have been proposed for such materials, e.g., as circularly-polarized light filters, in close similarity to liquid crystals [12]; or as sensors, filling liquids or gaseous substances into the open and porous structure of a sculptured thin film.

Zhou et al. [13] reported detection of polychlorinated biphenyls in soil and distilled spirit samples by surface-enhanced Raman scattering using Ag nanorod arrays as substrates produced by a glancing-angle e-beam technique. They detected polychlorinated biphenyls to a concentration of 5 μg/g in dry soil samples within one minute. Based on simulation, trace amounts were identified of an analyte in acetone solutions at a concentration of 10^{-6} mol/L.

Au or Ag NSs produced by laser methods were applied successfully to SERS. Thus, enhanced Raman spectrum has been observed of Rhodamine 6G (R6G) on Au nanocolumns created by off-axis PLD. R6G concentration as low as 1 nM [3] was measured and a maximum enhancement exceeding 10^5 was achieved.

SERS has been used for trace analyses of residues of different organophosphorous pesticides and insecticides [14-19]. Various nanostructures have been tested as substrates for the SERS analyses: solution of 100 nm Ag nanocubes [15], standard Klarite Au-coated SERS-active substrate [16-18]. The performance of SERS analyses has been compared with that of traditional analyses, such as chromatography, fluorescence polarization immunoassay, multi-enzyme inhibition assay, and biosensors. Although these methods can be used to detect trace amounts of pesticides, they are time consuming, labor-intensive, and expensive, making these analytical approaches less attractive and limiting laboratory, real-time, and field detections to some extent [16]. Among the large variety of pesticides, dimethoate is a very widely studied substance. The lowest concentration detected so far has been $5 \div 10$ $\mu g \cdot mL^{-1}$ using confocal Raman micro-spectrometry with Klarite substrates [16]. Generally, low concentration levels of about 10^{-6} have been reached.

This paper describes a study and the results of the development of a novel technology – CLAD ion-beam deposition – for formation of advanced Ag nanostructures (NSs) on quartz, which were applied to high-resolution analysis (SERS) of the BI-58 insecticide (dimethoate) and R6G as probing agents. The technology has direct bearing on human health and food quality, e.g., monitoring small amount or traces of pollutants or undesirable additives.

IOP Conf. Series: Journal of Physics: Conf. Series **992** (2018) 012050 doi:10.1088/1742-6596/992/1/012050

2. Experimental

A set of ion-beam deposited at glancing angle (GLAD) Ag NSs on quartz samples was prepared under different conditions (angle of incidence of the flow with respect to the substrate and rotation speed of the substrate).

The as prepared samples, to be used for SERS analyses, were covered by an aqueous solution of 0.2 % (800 µg/mL, i.e. 3.49 mM) of the insecticide with trade name BI-58 (containing dimethoate - $C_5H_{12}NO_3PS_2$), or an alcohol solution of 200 nM of R6G, and then were dried. The organophosphate insecticide dimethoate (commercially known as BI-58 – BASF SE, Ge) is one of the most widely used such agents. This compound is a basic substance at different concentration in a large number of insecticides used for treating plants. The standard colloid concentration of dimethoate in marketed BI-58 is 400 g/L (400 mg/mL). For plants treatment, it is diluted in water at a concentration of 0.2 %, i.e. 800 µg/ml. It is important to note that BI-58 consists of 37,4 % (W/W) dimethoate, 4,7 % (W/W) xylene and 48,4 % (W/W) cyclohexanone. The latter two compounds are solvents; when drops of BI-58 diluted in water were deposited on the surface of the substrates for analyses, they evaporated and disappeared during drying. Further, 200 nM R6G was used in order to check the sensitivity of the ion-beam deposited samples.

Table 1 presents a summary of the produced and analyzed samples. Samples **A** were covered by BI-58 (dimethoate), and $\mathbf{A_1}$, with R6G; θ is the glancing angle, the rotation speed is given in revolutions per minute (rpm).

Table 1. Summary description of the samples ion-beam-deposited at glancing angle and analyzed.

Samples A	**Ag NSs**	**3 samples:**	**Covered by 0.2% dimethoate – 3.49 mM**
		01a	400 nm; θ = 78°; normal thickness 1.5 µm; rotation speed 10 rpm
		01b	420 nm; θ = 78°; normal thickness 1.5µm; rotation speed 0 rpm
		01c	480 nm; θ = 86°; normal thickness: 2.5µm; rotation speed: 0 rpm
Samples A₁	**Ag NSs**	**3 samples:** 01a'; 01b'; 01c'	**Same as samples A, covered by 200 nM R6G**

The morphology of the samples was observed by scanning electron microscopy using a FE-SEM (Zeiss Ultra55, Gemini) microscope. The Raman spectra were obtained by an RMS-310 µ-Raman spectrometer (Photon Design, Japan). It uses a laser excitation power of 1 mW for detecting dimethoate and 0.0001 mW for detecting R6G, respectively, at a wavelength of λ = 532 nm and with a resolution of 0.2 cm^{-1}. The image dimension of the exciting laser beam on the surface of the checked substance is ~1 μm^2. It is worth noting that such a low excitation power was used to get the SERS spectra since the analytes are very sensitive to heating and are damaged even at very low excitation power.

3. Results and discussions

The morphology of the Ag nanostructures GLAD ion-beam-deposited on quartz substrates is presented in figure 1. As is seen, the structural properties are quite different and depend strongly on the deposition conditions.

All structures prepared were analyzed by µ-Raman spectrometry. SERS peaks of dimethoate appeared in all cases of structures deposited by the GLAD technology. Figure 2 a) is the Raman spectrum of dimethoate at the highest concentration (1.7 M), as purchased from an agriculture store, and is given for reference. Figure 2 b) and c) present two parts of the spectrum detected from the structure shown in figure 1 a). Furthermore, figure 2 d) is a SERS spectrum of dimethoate taken from the Ag NSs presented in figure 1 c). The parameters of the SERS peaks originating from the Ag NSs, and the Raman peaks recorded from the highest dimethoate concentration are summarized in table 2.

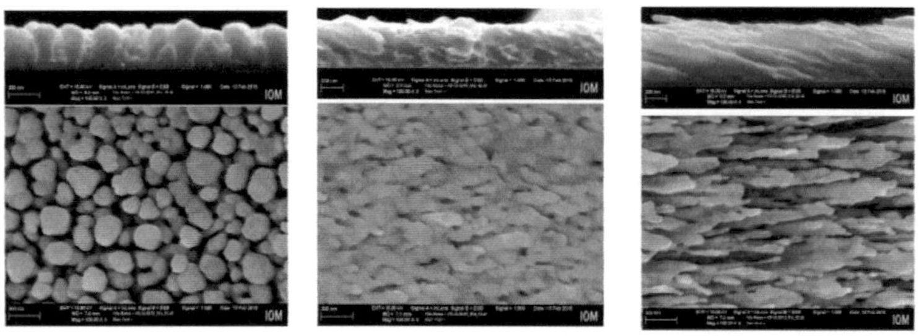

(a) (b) (c)

Figure 1. SEM images of Ag nanostructures deposited on quartz. **a)** film 400 nm (thickness of the structure at normal incidence 1.5 μm), θ = 78°, rotation 10 rpm; **b)** film 420 nm (thickness of the structure at normal incidence 1.5 μm), θ = 78°, rotation 0 rpm; **c)** film 480 nm (thickness of the structure at normal incidence 2.5 μm), θ = 86°, rotation 0 rpm.

Table 2. Summary of some important vibration modes observed (in cm^{-1}) of dimethoate. Relative intensity: vs – very strong; s – strong; m – middle; w – weak; vw – very weak; and br - broad. The results according to Ref. [16] are given for comparison and assignments of the vibrations; ν, δ and r denote stretching, deformation and torsion.

SERS peaks of 01a, 3.49 mM	**SERS peaks of 01c, 3.49 mM**	**Raman peaks, 1.7 M**	**Ref. [16]**			
			Peaks		**Assignments**	
			Raman	**SERS**	**Raman**	**SERS**
409, m	411, m	409, w	407	401	δ(P-O-C)	δ(P-O-C)
502, vs	503, vs	498, m	498	497	r(CH_3)	r(CH_3)$_v$
656, vs	656, w	656, vs	650	650	ν(P=S)	ν(P=S)
749, vs	750, vwbr	749, w				
	771, vs	773, vs				
	853, w	853, m				
1164, vs	1164, vs	1156, vw	1166	1160-1165	ν(C-C)	ν(C-C)
1276, s	1276, s	1266, w				
1417, m	1417, m	1417, wbr				
		1450, wbr	1449	1450	ν(O=C-N)I	ν(O=C-N)II
1520, vs						

The strongest peaks in the SERS of BI-58 (figure 2) are at 502, 656 and 1164 cm^{-1}, related according to Liu et al. [16] to CH_3 torsion, P=S stretching and C-C stretching vibrational modes, respectively. However, two other very strong peaks at 749 and 1520 cm^{-1} are also observed. According to Liu et al [17], the characteristic peaks of dimethoate are located at 494, 651 and 1640 cm^{-1}. The shallow peak (figure 2 c) at 1628 cm^{-1} can be assigned to the symmetric C=O stretching mode.We should note here that the slight differences between the peaks' positions in the μ-Raman and SERS spectra were recorded in our experiments with BI-58, when those reported in Refs. 16 and 17 were obtained from pure dimethoate.

In order to check the sensitivity of the ion-beam-produced structures for SERS analysis; 100 nM R6G was used. Figure 3 a) and b) show parts of the μ-Raman spectra of R6G produced by the samples presented in figure a) and c). The well-expressed peaks at 612 and 774 cm^{-1} are typical for this analyte [20]. Additionally, the intensities of the peaks taken using substrate 1 b (figure 1 c) are the highest, which is probably because of the needle structure (figure 1 c).

In what concerns the SERS sensitivity, the Ag NSs shown in figure 1 a and c exhibited the best properties. This is probably due to the character of these 3D structures – needle-like, which presents “hot spots” on their surfaces.

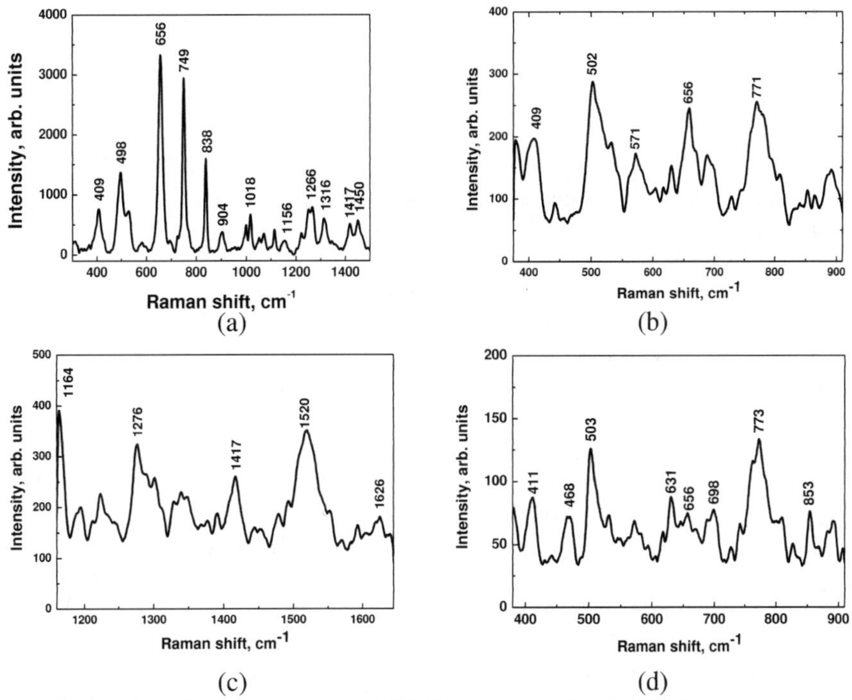

Figure 2. μ-Raman spectra of BI-58 a) at standard colloid concentration (400 mg/ml) for reference; b) and c) parts of SERS spectra of BI-58 deposited on Ag NSs which corresponds to the structure 01a in figure 1 a); d) for the structure 01c in figure 1 c). Excitation power was 1 mW.

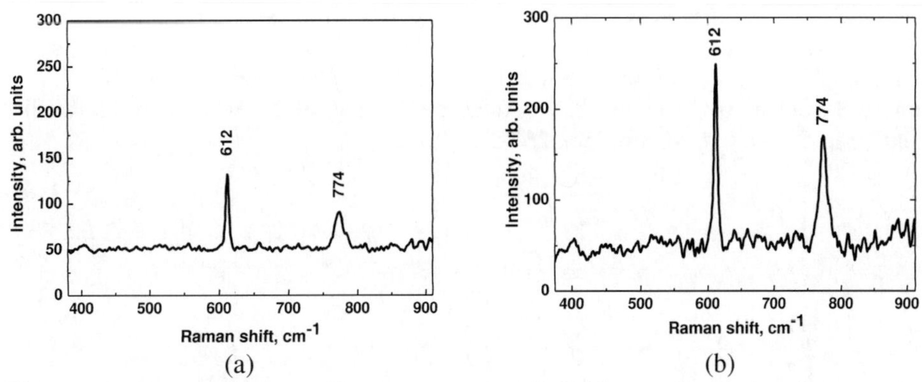

Figure 3. Parts of the μ-Raman spectra of R6G produced by the samples: a) **01a** (0.0001 mW, 20 s) and b) **01b** (0.0001 mW, 20 s).

4. Conclusions

Ion-beam deposition at glancing angle was used to form advanced Ag nanostructures (NSs) on quartz. They were applied to high-resolution analyses (SERS) of insecticide BI-58 (dimethoate) and R6G as probing agents. The lowest concentrations of the analytes studied using Ag NSs were 3.49 mM for BI-58 and 100 nM for R6G. This advanced technique of generating active substrates for SERS was thus proven to be very attractive and useful for high-resolution analyses of different analytes. However, further improvement of the properties of the ion-beam-deposited at glancing angle structures is needed in order to achieve a higher resolution.

Acknowledgements
The work was supported in part under projects DNTS 01/1 and TO 2/24 funded by the Bulgarian National Science Fund. P. Atanasov highly acknowledges the financial support of JSPS under contract JSPS Fellow ID No. S16152.

References

[1] Bajpai S K and Yallapu M M eds 2009 *Recent Advances in Nanoscience and Technology* (Bentham Sci. Pub.)
[2] Nie S and Emory S R 1997 *Science* **275** 1102
[3] Sakano T, Tanaka Y, Nishimura R, Nedyalkov N N, Atanasov P A, Saiki T and Obara M 2008 *J. Phys.* D*: Appl. Phys*. **41**/23 235304
[4] Pissuwan D, Valenzuela S M and Cortie M 2006 *Trends Biotechnol.* **24**/2 62
[5] Huang X, El-Sayed I H, Qian W and El-Sayed M A 2006 *J. Am. Chem. Soc.* **128** 2115
[6] Cho W J, Kim Y and Kim Y K 2012 *ACS Nano* **6**/1 249
[7] Beliatis M J, Henley S J and Silva S R P 2011 *Opt. Lett.* **36** 1362
[8] Robbie K, Brett M J and Lakhtakia A 1995 *J. Vac. Sci. Technol.* A **13** 2991
[9] Smy T, Dew S K and Joshi R V J 2001 *Vac. Sci. Technol.* A **19**/1 251
[10] Robbie K, Beydaghyan G, Brown T, Dean C, Adams J and Buzea C 2004 *Rev. Sci. Instrum.* **74** 1089
[11] Patzig C, Rauschenbach B, Fuhrmann B and Leipner H J 2008 *J. Appl. Phys.* **103**/2 024313
[12] Robbie K, Broer D J and Brett M J 1999 *Nature* **399** 764
[13] Zhou Q, Zhang X, Huang Yu, Li Z and Zhang Z 2011 *Sensors* **11** 10851
[14] Guerrini L, Sanchez-Cortes S, Cruz V L, Martinez S, Ristori S and Feis A 2011 *J. Raman Spectrosc.* **42**/5 980
[15] Costa J, Ando R, Sant'Ana A and Corio P 2012 *Phys. Chem. Chem. Phys.* **14**/45 15645
[16] Liu Y, Ye B, Wan C, Hao Y, Lan Y and Ouyang A 2013 *Am. Soc. of Agricultural and Biological Eng.* **56**/3 1043
[17] Liu Y and Liu T 2011 *Int. Feder. for Inform. Processing AICT* **347** 427
[18] Bin L, Peng Z, Xiaoming L, Xin S, Hao L, Mengshi L 2013 *Food Bioprocess Technol.* **6** 710
[19] Zhang P X, Xiaofang Z, Andrew Y S, Cheng Y F 2006 *J. Phys.: Conf. Series* **28** 7
[20] Imamova S, Nedyalkov N, Dikovska A, Atanasov P, Sawczak M, Jendrzejewski R, Śliwiński G and Obara M 2010 *Appl. Surf. Sci.* **257**/3 1075

Laser-induced surface modification of biopolymers – micro/nanostructuring and functionalization

N E Stankova[1,6], P A Atanasov[1], N N Nedyalkov[1], Dr Tatchev[2], K N Kolev[2], E I Valova[2], St A Armyanov[2], K Grochowska[3], G Śliwiński[3], N Fukata[4], D Hirsch[5] and B Rauschenbach[5]

[1]E. Djakov Institute of Electronics, Bulgarian Academy of Sciences, 72 Tzarigradsko Chaussee, 1784 Sofia, Bulgaria
[2]Acad. R. Kaishev Institute of Physical Chemistry, Bulgarian Academy of Sciences, Acad. G. Bonchev Str., Bl. 11, 1113 Sofia, Bulgaria
[3]Photophysics Department, The Szewalski Institute, Polish Academy of Sciences, 14 Fiszera Str., 80-231 Gdańsk, Poland
[4]International Center for Materials for NanoArchitectonics (MANA), National Institute for Materials Science (NIMS), 1-1Namiki, Tsikuba 305-0044, Japan
[5]Leibniz Institute of Surface Modification (IOM), 15 Permosestrasse, D-04318 Leipzig, Germany

E-mail: nestankova @yahoo.com

Abstract. The medical-grade polydimethylsiloxane (PDMS) elastomer is a widely used biomaterial in medicine for preparation of high-tech devices because of its remarkable properties. In this paper, we present experimental results on surface modification of PDMS elastomer by using ultraviolet, visible, and near-infrared ns-laser system and investigation of the chemical composition and the morphological structure inside the treated area in dependence on the processing parameters – wavelength, laser fluence and number of pulses. Remarkable chemical transformations and changes of the morphological structure were observed, resulting in the formation of a highly catalytically active surface, which was successfully functionalized via electroless Ni and Pt deposition by a sensitizing-activation free process.

The results obtained are very promising in view of applying the methods of laser-induced micro- and nano-structuring and activation of biopolymers' surface and further electroless metal plating to the preparation of, e.g., multielectrode arrays (MEAs) devices in neural and muscular surface interfacing implantable systems.

1. Introduction

Polydimethylsiloxane (PDMS) is medical grade silicone and is one of the most widely-used polymeric materials in biomedical implantable devices, such as substrate insulator carriers and/or for packaging of such devices [1-4]. The popularity of PDMS in such medical application is based on its distinctive characteristics, which include tensile strength, chemical and biological inertness, non-toxicity,

[6] To whom any correspondence should be addressed,

IOP Conf. Series: Journal of Physics: Conf. Series **992** (2018) 012051 doi:10.1088/1742-6596/992/1/012051

permeability to gasses, high dielectric constant and breakdown field, optical transparency in the UV, VIS and NIR spectral ranges, which determine its high biocompatible and biostable nature [4]. Owing to these advantageous properties and the simple and inexpensive fabrication process, PDMS is a suitable and preferred material for fabrication of long-term implantable devices [5]. Overall, PDMS-based elastomers are extremely stable and keep their high flexibility. They are excellent insulators and have clinical approval according to USP class VI for unrestricted use in chronic implants. Photo-detectable PDMS elastomers are also available that increase the absorption in the UV region (below 365 nm), thus improving the processing conditions under irradiation by such light sources (lasers, excimer lamps). However, they are not applicable in the implantable grades. Surface processing of PDMS elastomers is needed to improve their surface properties, like wettability or adhesion features for further functionalization, e.g., metallization. Highly-flexible polymeric multielectrode arrays (MEAs) based on PDMS-elastomer scaffolds are applied as neural interfacing technologies for monitoring and/or stimulation of neural activity by connecting the neurons to electronic circuitry [6-8]. Elastomeric MEAs can be rolled and flexed, thus offering an improved structural interface with the neural tissues. UV, VIS and NIR ns- and fs-laser processing of the PDMS elastomer is a powerful method for surface modification and activation without altering the material's bulk properties [9-18]. Generally, laser micromachining is based on direct laser ablation or surface modification for fabrication of tracks, holes and complex 2D/3D structures with dimensions of several tens of microns or less in a wide range of materials. It is a low-cost technique, unique with the versatile control of the processing parameters. This approach opens up new interesting possibilities for rapid patterning of delicate materials on micro- and nano-scale over large areas, especially of biodegradable polymers like PDMS for tissue engineering application. Selective metallization has been reported of polymers (such as polyamide, polyurethane or polydimethylsiloxane) after laser modification (chemical activation and ablation) with Cu, Ni or Pt by using electroless plating [17-20]. Since the major demands to implants are biocompatibility and biostability, Pt is the most suitable element used in metallization via electroless deposition [17, 19].

The aim of this work was to provide complex data on the modification and activation of the PDMS elastomer surface by using the traditional method of ns-laser processing followed by successful functionalization of the laser-modified surface. The effects of the laser parameters on the chemical structure and the morphology after UV, VIS and NIR treatment were studied and compared. The trenches produced were successfully metalized by Ni or Pt deposition by an improved autocatalytic coating bath – a tin- and palladium-free process. Our work can contribute to the research efforts to achieve successful direct laser writing of micro channels via ns-laser processing of polymer materials and their functionalization for application as MEAs in neural interface technologies.

2. Experimental

For surface processing of medical grade PDMS elastomers (MED 4860 and SSPM823), we used a Q-switched Nd:YAG laser multimode system oscillating at the fundamental wavelengths of 1064 nm and its harmonics: second (532 nm), third (355 nm) and fourth (266 nm), with a pulse duration of 15 ns and a repetition rate of 10 Hz. The thickness of the sheets was between 150 μm and 200 μm. The laser beam was focused normally to the sample by UV fused-silica lenses with different focal lengths (between 10 cm and 25 cm). In order to obtain a larger laser beam spot, lenses with a higher focal length combined with a homogenizer (with relevant parameters) were used. The samples were mounted on a stepper-motor computer-controlled *x-y* stage, which allowed fabrication of single trenches by overlapping 10 to 220 consecutive laser pulses. The number of pulses depended on the laser spot size on the sample surface and the stage speed of motion. The laser processing was performed in air at ambient temperature. The laser fluence was a function of the wavelength and the beam spot size. The data on the laser processing parameters are presented in table 1.

The absorption coefficient (α) and the penetration depth (δ) were calculated following the Beer-Lambert law; as the scattering is ignored, the values represent the linear absorption coefficient. After the laser treatment, the PDMS samples were successfully metallized with Ni or Pt via electroless deposition (autocatalytic bath). The metallization was performed without chemical sensitization by Sn and activation by Pd of the laser-treated surface. The electroless plating process is based on a hydrazine hydrate reducer at different concentrations used to increase the active sites without electrochemical methods. The details of this procedure will be described in a forthcoming communication. It should be emphasized that the electroless Ni deposition is the best and the cheapest analog of the classic Pt electroless plating process.

Table 1. Summary of the laser processing parameters.

Wavelength (nm)	266	355	532	1064
Laser Fluence (J cm^{-2})	0.5 – 1.5	2.5 – 6.0	9.5 – 12.0	13.0 – 16.0
α (cm^{-1})	14.9	7.38	3.58	2.86
δ (μm)	670	1354	2794	3502

The surface topography of the samples was investigated by a VK-9700K color 3D laser microscope (KEYENCE, Japan) and scanning electron microscopy (SEM) (Hitachi SU-70 with field-emission gun) and SEM/FIB (Lyra/Tescan dual-beam system); the structure modifications were analyzed by a μ-Raman spectrometer (Invia, Renishaw) using 514-nm excitation wavelength and beam spot on a sample ~5 μm×5 μm for a 100× objective.

3. Results and discussions

3.1. Raman spectroscopy inside the trenches

The influence of the laser radiation in the UV, VIS and NIR spectral ranges on the chemical composition of the PDMS elastomer samples was investigated by μ-Raman spectroscopy. The spectra were acquired by measuring at different points in each trench. The spectra of areas treated by different laser parameters (see Table 1) differ significantly from the spectrum of the untreated sample (figures 1 and 2). Since the PDMS elastomer belongs to the silicones' polymeric family, its backbone is composed by an alternate succession of Si–O atoms joined by a strong covalent bond of 4.7 eV. Two organic CH_3 radicals are also coupled to the Si atoms by a Si–C bond with a bond energy of 3.3 eV. The Raman peaks of native PDMS characterize the following vibration modes: 488 cm^{-1} (Si-O-Si symmetric stretching); 685 cm^{-1} (Si-CH_3 symmetric rocking); 709 cm^{-1} (Si-C symmetric stretching); 787 cm^{-1} (CH_3 asymmetric rocking + Si C asymmetric stretching); 859 cm^{-1} (CH_3 symmetric rocking); 1262 cm^{-1} (CH_3 symmetric bending); 1411 cm^{-1} (CH_3 asymmetric bending); 2909 cm^{-1} (CH_3 symmetric stretching); and 2970 cm^{-1} (CH_3 asymmetric stretching). Obviously, the laser treatment leads to chemical modification of the PDMS surface. A typical tendency in the spectra is observed at all wavelengths applied. The intensity of the O-Si-O bond at 488 cm^{-1} decreases or almost disappears and a new sharp and strong peak appears between 515 cm^{-1} and 519 cm^{-1}. This peak is attributed to mono and/or polycrystalline or only to monocrystalline silicon (c-Si) [21]. This can be explained by chemical transformations occurring during the laser processing – breaking of the O-Si-O bonds and formation of crystalline Si as a result of the higher pressure and temperature in the laser spot area. It is worth noting that the intensity of the other peaks of the native PDMS elastomer decreases. This indicates that the Si-CH_3, CH_3, and Si-C bonds are also broken and probably contribute to the c-Si formation. The comparison of the spectra of trenches obtained at different processing parameters show complex dependence on the wavelength, laser fluence and number of pulses. At the wavelengths of 266 nm and 355 nm and low fluences, the c-Si peak appears after 110 or more pulses, but its intensity is still weak. As the laser fluence increases, the O-Si-O peak's intensity is reduced, while the intensity of c-Si peak is strongly enhanced. Also, this peak becomes very intensive and sharp even at low number of pulses (5-18) when higher fluences are applied. This tendency is also observed at the wavelengths of 532 nm and 1064 nm (figure 2); however, the fluence values needed to break the Si-O-Si bond significantly increase compared with the UV wavelengths (between 9.5 J cm^{-2} and 12 J cm^{-2} for 532 nm, and between 13 J cm^{-2} and 16 J.cm^{-2} for 1064 nm) at number of pulses between 11 and 33. The chemical transformations observed after UV laser irradiation is in accordance with the known data of the photon energy of 4.7 eV for the wavelength of 266 nm (and probably 3.5 eV for 355 nm), which is needed for the direct breaking of the covalent bonds, like Si-O-Si (4.7 eV), and/or C-H (4.3 eV), and/or S-C (3.3 eV), respectively. Probably, photochemical dissociation of the covalent bonds can take place during laser irradiation at the wavelength of 266 nm. However, at the longer wavelengths (355 nm, 532 nm and 1064 nm), the photon energy (3.5 eV, 2.3 eV and 1.2 eV, respectively), is not sufficient to cause photochemical dissociation, especially for the Si-O-Si bond. Therefore, the chemical decomposition observed can rather be the result of heat accumulated with each subsequent pulse per unit area. In the current experiments on ns-laser irradiation of the PDMS polymer, the ablation mechanism can be due predominantly to the thermal (vibrational) relaxation of the excited states, especially at the wavelength of 355 nm and longer. It can also be assumed that during the ultraviolet irradiation at 266 nm, photothermic reactions contribute to the mechanism of ablation simultaneously with the photochemical reactions. The ratio between them depends both on the laser parameters and the properties of the polymer. In summary, all μ-Raman spectra provide evidence that the PDMS elastomer surface is chemically transformed and activated after ns-laser processing.

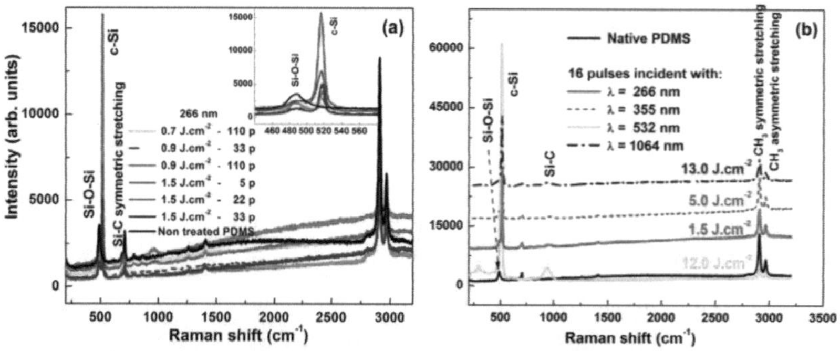

Figure 1. μ-Raman spectra of native (non-treated) and ns-laser treated PDMS elastomer at: a) 266 nm; b) comparison between treatment at 266, 355, 532 and 1064 nm.

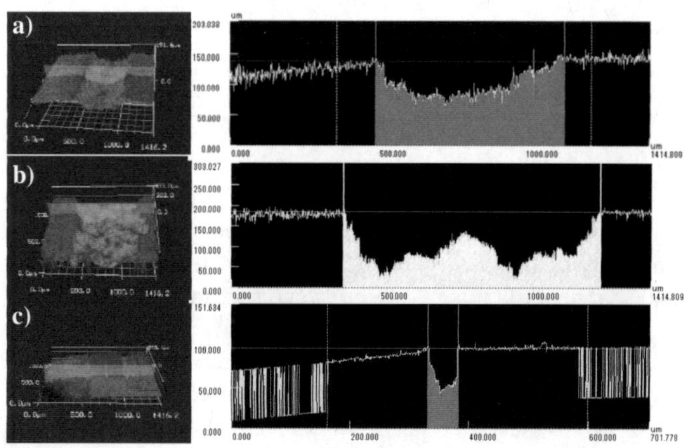

Figure 2. Laser microscope images - photographs and profiles: a) 266 nm, 0.5 J cm^{-2} and N=22 pulses; b) 1064 nm, 13 J cm^{-2} and N=22 pulses; c) U-shaped high definition track produced at 266 nm, 1.5 J cm^{-2} and 44 pulses.

3.2. Laser microscope views of the trenches

The quality of the profiles of the laser-produced trenches was viewed and assessed by 3D color laser microscopy (figure 2). Since the laser system emits a multimode beam, the profiles of trenches are generally inhomogeneous, especially in the cases of VIS and NIR irradiation and/or higher number of pulses. This is due to the inhomogeneous distribution of the energy density inside the laser beam, namely, the existence of "hot" spots in the beam structure. However, in a dynamic mode of laser processing, where the overlapping of the consecutive pulses is controlled by varying the *x-y* table speed of motion, one can produce trenches with homogeneous profiles owing to the processing equalization after a large number of pulses are applied per unit area. The UV laser treatment results in highly homogeneous profiles (mainly at 266 nm) regardless of the number of pulses and the laser spot size on the surface. This is due to the higher absorption coefficient (i.e., the short penetration depth at this wavelength) of the PDMS polymer, which for 266 nm is an order of magnitude higher than those for the wavelengths of 355 nm, 532 nm, and 1064 nm (see table 1). Using irradiation at the wavelength of 266 nm, we were able to produce high-definition U-shaped or straight-line laser tracks.

3.3. Scanning electron microscope views of the trenches – after laser processing and after electroless metal deposition

Typical SEM images of the trenches produced after ns-laser processing of the PDMS elastomer are presented in figure 3. The surface morphology is significantly modified compared to the non-treated surface due to the local laser ablation. The SEM analysis shows formation of similar morphologic structures at all wavelengths used. An extremely developed relief with a cauliflower-like structure is formed on a micro-scale, which can ensure good adhesion of the metal coating. In general, the ablation depth increases with the laser fluence and/or the number of the pulses. As it is seen, cavities of different size are formed in all cases. However, they become larger at VIS and NIR irradiation or as the laser exposure of the sample is increased at a given wavelength. The high-resolution images indicate that the modified surface is nanostructured with a granular chondrites-like structure.

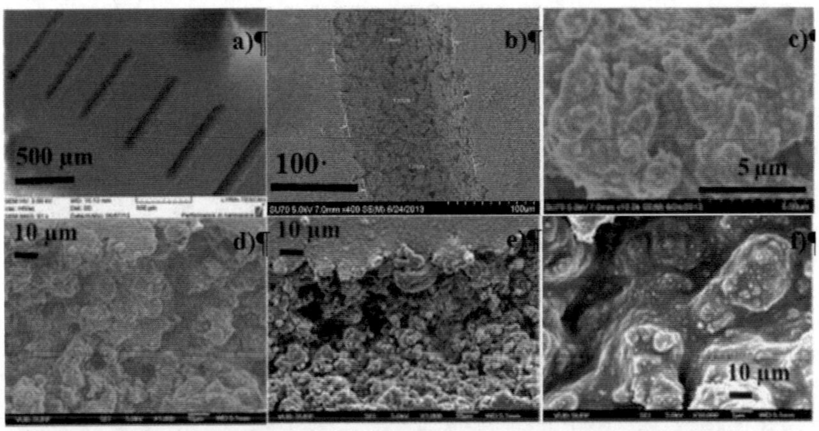

Figure 3. Trenches produced by ns-laser: **a)** 355 nm, N=11, 22, 33, 44, 55 and 110 pulses, respectively, and 4 J cm^{-2}; **b)** 266 nm, N=55 and 1.5 J cm^{-2}; **c)** HR SEM of b); **d)** 532 nm, N=16 and 9.5 J cm^{-2}; **e)** 1064 nm, N=16 and 13 J cm^{-2}; **f)** HR SEM of e).

Figure 4. SEM images of the metal nanospike structures after electroless coating of the trenches produced at: a) 266 nm – Ni metallized; b) HR SEM of a); c) 266 nm and d) 355 nm after Pt metallization in HR mode. The inset images show whole (c) and d)) or part (a)) of the trenches.

The laser-modified and thus activated surface of the PDMS elastomer samples ensured their successful functionalization by electroless deposition of Ni or Pt metals. The metallization was successfully performed regardless of the time-interval between the laser exposure and the autocatalytic bath. Figure 4 shows SEM images of electroless deposited Ni (figure 4a) and 4b)) and Pt (figure 4c) and 4d)) films. Metal crystallites with a "spiky" structure were uniformly deposited on the laser-treated areas. The metal nanospikes tended to grow vertically with size ranging from several tens to several hundreds of nanometers. Since the hydrazine was used as a reductant in the autocatalytic bath, the deposition of Ni and Pt resulted in the formation of nanospike arrays. Depending on the hydrazine concentrations, some small differences in the crystallites structure could be found, but the "spiky" origin remained.

4. Conclusions

The results obtained in our study of laser-induced surface modification of the PDMS elastomer and its further functionalization can be summarized as follows: i) ns-lasers, generating in the UV, VIS and NIR spectra are effective tools for micro- and nanostructuring and chemical activation of biopolymeric elastomers: (1) a significant increase of the surface roughness by laser ablation, which results in the formation of trenches with controllable size depending on the laser beam parameters and (2) chemical activation of the laser-processed area by breaking the covalent bonds (Si-O-Si, Si-C) of the PDMS polymer and formation of Si micro-crystallites; ii) successful metallization of the ns-laser treated PDMS elastomer surface via Pt and Ni electroless plating: (1) excluding Sn sensitization and Pd activation by using hydrazine hydrate as catalyst and (2) growth of Pt and Ni nanospike-structured arrays on a nanometer scale.

The results reported describe a simple procedure of biopolymer surface preparation for its following functionalization, which opens up new possibilities for rapid prototyping of micro- and nano-structures suitable for various applications, such as MEAs in neural implants interfacing technologies.

References

[1] Hassler C, Boretius T and Stieglitz T 2011 *J. Polym. Sci.* B*: Polym. Phys.* **49**/1 18
[2] Liang G, Guvanasen G S, Xi L,Tuthill C, Nichols T R and De Weerth S P 2013 *IEEE Trans. Biomed. Circuitsand Syst.* **7**/1 1
[3] Adrian J, Teo T, Mishra A, Park I, Kim Y-J, Park W-T and Yoon Y-J 2016 *ACS Biomater. Sci. Eng.* **2**/4 454
[4] Stieglitz T and Meyer J-U 2006 *BIOMEMS* ed Urban G A (Springer: Dordrecht, The Netherlands) **3** 41
[5] Curtis J M and Colas A 2004 Dow Corning (R) Silicone biomaterials: history, chemistry & medical applications of silicones *Biomaterials Science 2nd Edition*; ed Ratner B D (Elsevier: London)
[6] Hassan H M, Chodavarapu V and Musallam S 2008 *Review: NeuroMEMS: Neural Probe Microtechnol., Sensors* **8**/10 6704
[7] Lacour S, Benmerah S, Tarte E, Gerald J F, Serra J, Mc Mahon S, Fawcett J, Graudejus O, Yu Z and Morrison B 2010 *Med. Biol. Eng. Comput.* **48**/10 945
[8] Liang G, Guvanasen G S , Xi L, Tuthill C, Nichols T R and DeWeerth S P 2013 *IEEE Trans. Biomed. Circuits Syst.* **7**/1 1
[9] Deepak K L N, Venugopal Rao S and Narayana Rao D 2010 *J. Phys.* **75**/6 1221
[10] Dicara Cl, Robert T, Kolev K, Dupas-Bruzek C and Laude L D 2003 *Proc. SPIE* **5147** 255
[11] Graubner V-M, Nuyken O, Lippert Th, Wokaun A, Lazare S and Servant L 2006 *Appl. Surf. Sci.* **252**/13 4781
[12] Dupas-Bruzek C, Robbe O, Addad A, Turrell S and Derozier D 2009 *Appl. Surf. Sci.* **255**/21 8715
[13] Van Pelt S, Frijns A, Mandamparambil R and Den Toonder J 2014 *Appl. Surf. Sci.* **303** 456

[14] Darvishi S, Cubaud T and Longtin J P 2012 *Opt. Lasers Eng.* **50** 210
[15] Sones C L, Katis I N, Mills B, Feinaeugle M, Mosayyebi A, Butement J and Eason R W 2014 *Appl. Surf. Sci.* **298** 125
[16] Surdo S, Piazza S, Ceseracciu L, Diaspro A and Duocastella M 2016 *Appl. Sur. Sci.* **374** 151
[17] Stankova N E, Atanasov P A, Nedyalkov N N, Stoyanchov T R, Kolev K N, Valova E I, Georgieva J S, Armyanov St A, Amoruso S, Wang X, Bruzzese R, Grochowska K, Śliwiński G, Baert K, Hubin A and Delplancke M P 2015 *Appl. Surf. Sci.* **336** 321
[18] Atanasov P A, Stankova N E, Nedyalkov N N, Fukata N, Hirsch D, Rauschenbach B, Amoruso S, Wang X, Kolev K N, Valova E I, Georgieva J S and Armyanov St A 2016 *Appl. Surf. Sci.* **374** 229
[19] Dupas-Bruzek C, Drean P and Derozier D 2009 *J. Appl. Phys.* **106** 074913
[20] Rytlewski P and Żenkiewicz M 2013 *JAMME* **57**/2 59
[21] Palma C, Rossi M C, Sapia C and Bemporad E 1999 *Appl. Surf. Sci.* **138–139** 24

IOP Conf. Series: Journal of Physics: Conf. Series **992** (2018) 012052 doi:10.1088/1742-6596/992/1/012052

Properties of LSMO/YBCO cross-strip type junctions

V Štrbík[1,4], T Nurgaliev[2], M Sojková[1], Š Chromik[1], M Španková[1], N Gál[1] and B Blagoev[3]

[1]Institute of Electrical Engineering, Slovak Academy of Sciences,
9 Dúbravská cesta , 841 04 Bratislava, Slovak Republic
[2]E. Djakov Institute of Electronics, Bulgarian Academy of Sciences,
72 Tzarigradsko Chaussee, 1784 Sofia, Bulgaria
[3]G. Nadjakov Institute of Solid State Physics, Bulgarian Academy of Sciences,
72 Tzarigradsko Chaussee, 1784 Sofia, Bulgaria

E-mail: elekstrb@savba.sk

Abstract. The properties of a $La_{0.67}Sr_{0.33}MnO_3$ (LSMO)/$YBa_2Cu_3O_{7-\delta}$ (YBCO) interface in thin film LSMO/YBCO cross-strip type junctions were investigated by means of electrical transport measurements. Resistance vs. temperature and current-voltage dependences, as well as conductance spectra, were used to characterize the electrical parameters of the interface. The results indicated a low resistance (below 10 Ω), while the dielectric properties of the interface pointed to the presence of a 10-nm wide and 40-meV high dielectric potential barrier. The oxygen vacancies in both LSMO and YBCO films at the interface and the charge transfer through the interface were both considered to explain the insulating character of the LSMO/YBCO interface.

1. Introduction

Depending on their composition and structural properties, perovskite materials can behave as dielectrics, conventional metals, or high-temperature superconducting (HTS), colossal magnetoresistive (CMR), ferroelectric or multiferroic materials. Therefore, such materials, and especially the CMR/HTS interface, are attractive for studying the interplay between two fundamental condensed-matter phenomena, superconductivity (S) and ferromagnetism (F). Recent developments in fabricating atomically smooth (flat) CMR/HTS interfaces in superlattices [1, 2] provided an ideal system to investigate intriguing phenomena, such as a long-range proximity effect [3, 4], spin-polarized quasiparticle injection into the HTS layer within a spin-diffusion length ξ_{FM} [5], giant modulation of the CMR-layer magnetization induced by superconductivity [6], or charge transfer from CMR into HTS [7,8,9]. A variety of effects and phenomena at the SF interface offer possibilities of using these materials in different microelectronic and spintronic applications [10].

In this work we studied a $La_{0.67}Sr_{0.33}MnO_3$ (LSMO) / $YBa_2Cu_3O_{7-\delta}$ (YBCO) interface in the form of cross-strip type junction by means of electrical characterization, namely, the resistance vs. temperature ($R(T)$) dependences, and the current-voltage ($I(V)$) and differential conductance vs. voltage ($dI/dV(V)$) characteristics. The results obtained indicate an insulating (dielectric) character of the interface in spite

[4] To whom any correspondence should be addressed.

of the metallic/superconducting (YBCO) and ferromagnetic half-metallic (LSMO) materials involved. We estimated the width of the barrier at about 10 nm and its height at about 40 meV. We also discuss a possible explanation of the appearance of an insulating barrier in the LSMO/YBCO interface.

2. Experimental

The LSMO and YBCO layers were prepared by pulsed laser deposition (PLD) under the same conditions. A KrF excimer laser was used oscillating at 248 nm, with a pulse width of 20 ns, repetition rate of 10 Hz and energy density of 6 J/cm^2 (spot size on the target ~2 mm^2). During the deposition, the substrate holder temperature was kept at 850 °C and the oxygen pressure, at 53 Pa. After deposition, the films were cooled down at a rate of 20 °C/min in O_2 (4×10^4 Pa). The growth rates of the LSMO and the YBCO were about 6.5 nm/min. First, a 25-40-nm-thick LSMO film was deposited on a MgO single-crystal substrate. The as-prepared ferromagnetic LSMO thin film was patterned as a base electrode of a junction using standard photolithography processes and etched by a solution of potassium iodide, hydrochloric and ascorbic acid [11]. Subsequently, a superconducting YBCO thin film with a thickness of about 50-70 nm was prepared on the MgO substrate with patterned LSMO. Then, standard photolithography processes and wet etching (0.1 % solution of nitric acid) were used to pattern the YBCO as a counter electrode [11]. A schematic view of the sample with three junctions of different areas on one chip is shown in the inset of figure 1. The area of the junctions varied from 1600 μm^2 to 15000 μm^2, corresponding to junction dimensions from 40×40 μm^{2}to 120×120 μm^2.

The *R(T)* dependences and *I*(*V*) characteristics were measured by the standard 4-probe method, while the conductance spectra (i.e. d*I*/d*V*(*V*) curves) were numerically calculated from the *I*(*V*) data.

3. Results and discussions

The *R*(*T*) dependences of the LSMO junctions exhibited a "negative" resistance mainly for temperatures above T_{C0} (the zero-resistance critical temperature of YBCO) (figure 1). Such a property can arise in junctions when the resistance of the barrier (in our case, the interface resistance R_{IF}) is lower than the resistances of the adjacent YBCO and LSMO layers (R_S and R_F, respectively). A negative value of the junction resistance R_J measured by the 4-probe method was first reported by Giaever [12]; we recently analyzed one in more detail for a LSMO/YBCO interface [13]. Our analyses showed that in the case of superconducting YBCO ($R_S = 0$), the junction resistance measured should always be positive. When the resistance of the interface R_{IF} is small ($R_{IF} \leq 1\ \Omega$), then the junction resistance measured is positive, but R_J assumes very low values ~ $10^{-6}\ R_{IF}$; when $R_{IF} \gtrsim 10\ \Omega$, R_J increases and shifts to positive values, closer to R_{IF} (figure 1).

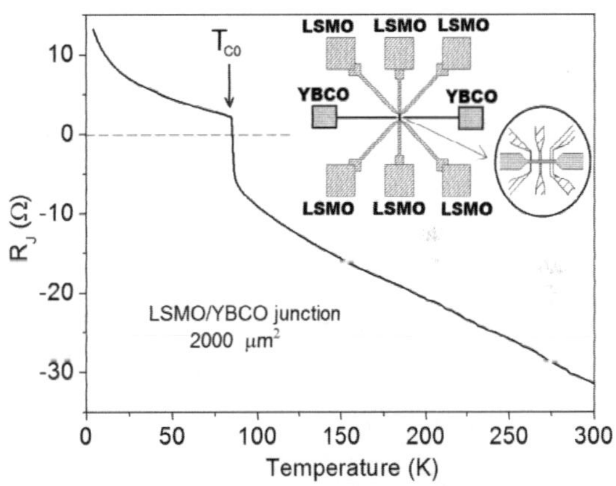

Figure 1. Temperature dependence of LSMO/YBCO junction resistance with negative values of resistance in normal state of YBCO. Inset shows the schematic view of 3-junction structure.

The *R(T)* dependences of the adjacent ferromagnetic and superconducting films are presented in figures 2 a and 2 b. In figure 2 a, the *R(T)* dependences of the as-prepared LSMO film with a patterned LSMO strip are compared in normalized form. The LSMO exhibited a high temperature of the metal-insulator transition (T_{MI} = 360 K), which corresponds to the Curie temperature of a paramagnetic-ferromagnetic transition [14,15]. The YBCO strip exhibited the rather low T_{C0} value of 84 K (figure 2 b). To investigate the LSMO/YBCO interface, we measured the current-voltage characteristics at temperatures below T_{C0}, namely at 83 K, 60 K, 40 K, 21 K, 11 K and 4.2 K (figure 3, inset).

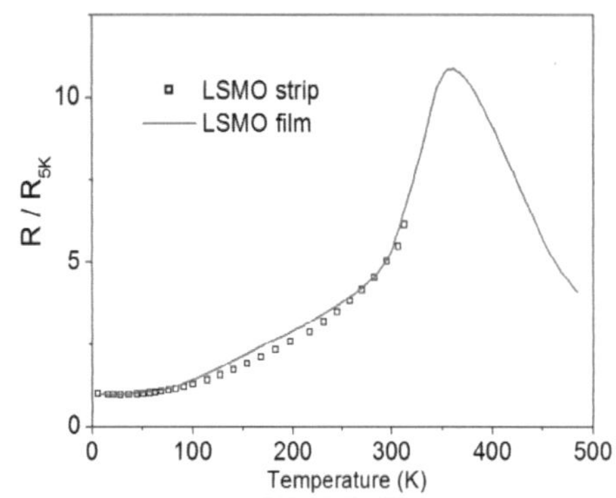

Figure 2 a. Normalized temperature dependences of the resistance of an as-prepared LSMO film (solid line) and an LSMO strip (dotted line).

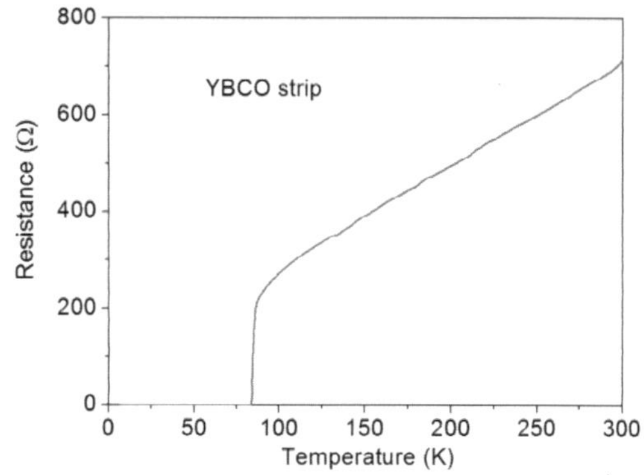

Figure 2 b. Temperature dependence of the YBCO strip resistance. The zero-resistance critical temperature T_{C0} is 84 K.

The detailed plot of the $I(V)$ characteristic taken at 60 K (figure 3, dotted line) is compared with the tunnel current through a dielectric barrier at low voltage (figure 3, solid line), as derived by Simmons [16] in the form $I = \alpha(V + \gamma V^3)$, where α and γ are constants characteristic for the given junction. The estimation of these constants is simple using the differential conductance dependence on the voltage ($\mathrm{d}I/\mathrm{d}V(V)$) (figure 4) [17]. The differential conductance $\mathrm{d}I/\mathrm{d}V = \mathrm{G}(V) = \alpha(1+3\gamma V^2)$ is a parabolic function of the voltage (figure 4). Then, α is the differential conductance at $V = 0$, and γ can be estimated from the voltage corresponding to conductance $\mathrm{G}(V) = 2\alpha$, according to the relation $V_{2\alpha} = (1/3\gamma)^{1/2}$. Knowing the constants α and γ, we can solve the next two relations [16, 17]

$$\gamma \approx \frac{0.0115}{\varphi} t^2 , \tag{1}$$

$$\left[3.16\times10^{10}(\varphi^{1/2}/t)\exp(-1.025t\varphi^{1/2})\right] = \alpha \tag{2}$$

and estimate important parameters of the barrier, such as the barrier thickness t (in angstroms) and the potential barrier height φ (in eV). The same procedure of estimating t and φ was applied for all $I(V)$

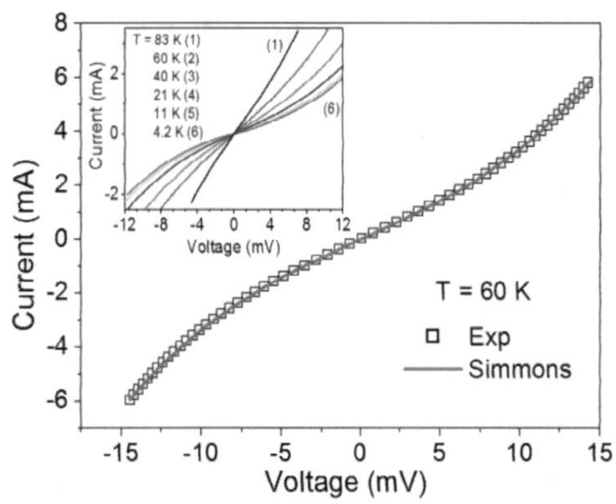

Figure 3. Current-voltage characteristic of the LSMO/YBCO junction taken at 60 K (dots) approximated by the tunnel current according to Simmons [16] (solid line). Set of $I(V)$ curves at various temperatures (inset).

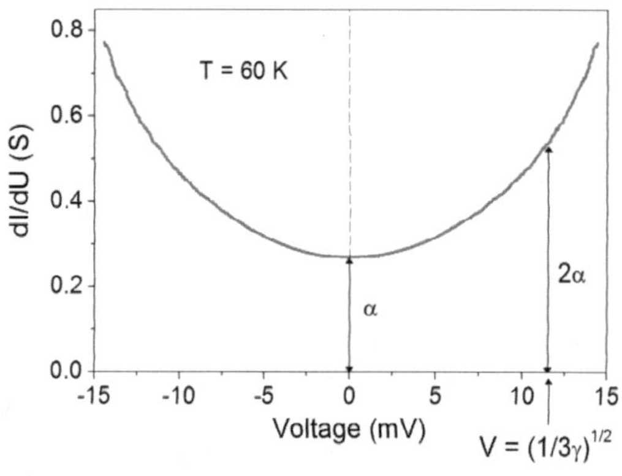

Figure 4. Parabolic differential conductance dependence on the voltage. The $dI/dV(V)$ curve is numerically calculated from the $I(V)$ data in figure 3. The estimates of the junction constants α and γ are indicated.

curves measured (figure 3, inset), resulting in a typical value of the barrier thickness of 9-10 nm and of the height of about 40 meV. These results indicate that our *ex-situ* LSMO/YBCO interface behaved like a dielectric (insulating) barrier.

The insulating character of the interface can be explained by at least two effects. First, it is known that the properties of both $YBa_2Cu_3O_{7-\delta}$ and $La_{1-x}Sr_xMnO_{3-y}$ are very sensitive to the oxygen content. For example, if $\delta \gtrsim 0.7$, YBCO exhibits insulating properties [18]; also, if $y \gtrsim 0.02$, $La_{0.85}Sr_{0.15}MnO_{3-y}$ behaves as an insulator [19]. Therefore, some oxygen deficiency at the LSMO/YBCO interface can create an insulating layer acting as a dielectric barrier. Sawa et al. [20] investigated a similar structure of YBCO/LSMO cross-strip type junctions (the opposite placing of YBCO and LSMO layers in a junction); they also found insulating, moreover, ferromagnetic character of the YBCO/LSMO interface. The existence of a ferromagnetic insulator layer was ascribed to oxygen vacancies in the LSMO film at the interface, while the presence of some peculiarities (a zero-bias conductance peak (ZBCP) and a gap-like structure) in the conductivity spectra was accepted as indicating small (or no) degradation of the YBCO layer at the interface. In our case, the LSMO seems to be optimally oxygenated (T_{MI} = 360 K), but the YBCO, with T_{C0} of 84 K, indicates a lower degree of oxygenation. Therefore, we assume that the insulating layer at the interface is due to oxygen vacancies created more on the YBCO side, since, besides the suppressed superconductivity in this region, we did not observe any peculiarities, such as a ZBCP or an YBCO gap-structure in the conductivity spectra.

The second effect that should be taken into account is the charge transfer through the CMR/HTS interface ranging in distance from a few tenths of nm to several nm [7, 8, 9]. Gray et al. [7] found a loss of 0.67 electrons per CMR monolayer in their CMR/HTS superlattices, which leads to conversion of the CMR layer to a paramagnetic or antiferromagnetic insulator. On the other hand, the electrons transferred through the CMR/HTS interface dope the Cu and Ba atoms in the YBCO and transform the superconducting YBCO to a Mott insulator. Holden et al. [9] also registered dramatic metallicity suppression at a distance of 10-20 nm from the interface due to a massive charge transfer between the HTS and CMR layers in the CMR/HTS superlattice. In both cases, the charge transfer creates an insulating interface of a thickness close to the one estimated in this paper.

4. Conclusions

We investigated the electrical properties of the interface between high T_C superconductor YBCO and ferromagnetic manganite LSMO thin films in the form of an LSMO/YBCO cross-strip junction. The junctions were prepared on single crystal MgO substrates from LSMO and YBCO thin films deposited *ex-situ* by PLD. The films were patterned using optical photolithography and wet etching. The junction area ranged from 1600 to 15000 μm^2. The $R(T)$, $I(V)$ and $dI/dV(V)$ dependences of the strips and junctions were studied to characterize the LSMO/YBCO interface. While the $R(T)$ dependences of the individual strips creating a junction exhibited a standard behavior typical for the LSMO and YBCO materials, the junction resistance exhibited "negative' values of the resistance for temperatures above the critical superconducting YBCO temperature. Such values are possible if the junction resistance is smaller than the resistances of the adjacent LSMO and YBCO films. Despite the low junction resistance, both the $I(V)$ dependence and the conductance spectra of the junctions pointed to their insulating character; the interface behaved as a 10-nm wide and 40-meV high potential barrier. The insulating nature of the interface can be explained by oxygen vacancies in the YBCO and LSMO materials at the interface due to an *ex-situ* process and/or by a charge transfer through the interface creating an insulating layer of paramagnetic LSMO and a Mott insulator on the YBCO side.

Acknowledgement

This work was supported by the Slovak Research and Development Agency under contracts Nos APVV-14-0613 and APVV-16-0315. We acknowledge the grant agency VEGA for the financial support to project No 2/0120/14. This publication is part of the results of the Joint Research Project "Perovskite heterostructures of nanometric thickness for sensors and spintronics" between the Institute of Electronics BAS, Bulgaria and the Institute of Electrical Engineering SAS, Slovakia.

References

[1] Varela M, Lupini R A, Pennycook S J, Sefrioui Z and Santamaria J 2003 *Solid State Electr.* **47** 2245
[2] Biškup N, Das S, Gonzales-Calbet J M, Bernhard C and Varela M 2015 *Phys. Rev.* B **91** 205132
[3] Peña V, Sefriouri Z, Arias D, León C, Santamaria J, Varela M, Pennycook S J and Martinez J L 2004 *Phys. Rev.* B **69** 224502
[4] Štrbík V, Beňačka Š, Gaži Š, Španková M, Šmatko V, Knoška J, Gál N, Chromik Š, Sojková M and Pisarčík M 2017 *Appl. Surf. Sci.* **395** 42
[5] Soltan S, Albrecht J and Habermeier H –U 2004 *Phys. Rev.* B **70** 144517
[6] Hoppler J, Stahn J, Niedermayer C, Malik V K, Bouyanfif H, Drew A J, Rössle M, Buzdin A, Cristiani G, Habermayer H –U, Keimer B and Bernhard C 2009 *Nature Mater.* **8** 315
[7] Gray B A, Middey S, Conti G, Gray A X, Kuo C T, Kaiser A M, Ueda S, Kobayashi K, Meyers D, Kareev M, Tung I C, Liu J, Fadley C S, Chakhalian J and Freeland J W 2016 *Sci. report* **6** 33184
[8] Chien T Y, Kourkouts L F, Chakhalian J, Gray B, Kareev M, Guisinger N P, Muller D A and Freeland J W 2013 *Nature Commun.* **4** 2336
[9] Holden T, Habermeier H U, Cristiani G, Golnik A, Boris A, Pimenov A, Humlíček J, Lebedev O I, Van Tendeloon G, Keimer B and Bernhard C 2004 *Phys. Rev.* B **69** 064505
[10] Linder J and Robinson J W A 2015 *Nature Phys.* **11** 307
[11] Sojková M, Štrbík V, Nurgaliev T, Chromik Š, Dobročka E, Španková M, Blagoev B and Gál N 2016 *J. Phys.: Conf. Series* **700** 012022
[12] Giaever I 1969 *Tunneling Phenomena in Solids* ed E Burstein and S Lundqvist (Plenum Press New York) chapter 3 pp. 19-30
[13] Sojková M, Nurgaliev T, Štrbík V, Chromik Š, Blagoev B and Španková M 2017 *Acta Physica Polonica* **131** 842
[14] Španková M, Štrbík V, Dobročka E, Chromik Š, Sojková M, Zhen M and Li J 2016 *Vacuum* **126** 24
[15] Španková M, Rosová A, Dobročka E, Chromik Š, Vávra I, Štrbík V, Machajdík D, Kobzev A and Sojková M 2015 *Thin Solid Films* **583** 19
[16] Simmons J G 1963 *Appl. Phys.* **34** 238
[17] Rowell J M 1969 *Tunneling Phenomena in Solids* ed E Burstein and S Lundqvist (Plenum Press New York) chapter 27 pp. 385-404
[18] Semba K and Matsuda A 2001 *Phys. Rev. Lett.* **86** 496S
[19] De Léon-Guevara A M, Berthet P, Berthon J, Millot F, Revcolevschi A, Anane A, Dupas C, Le Dang K, Renard J P and Veillet P 1997 *Phys. Rev.* B **56** 6031
[20] Sawa A, Kashiwaya S, Obara H, Yamasaki H, Koyanagi M, Yoshida N and Tanaka Y 2000 *Physica* C **339** 287

Study of the influence of vanadium and niobium on the structure of Ti-Al-V and Ti-Al-Nb alloys formed by selective electron beam alloying

S Valkov[1,3], R Bezdushnyi[2] and P Petrov[1]

[1]Emil Djakov Institute of Electronics, Bulgarian Academy of Sciences, 72 Tzarigradsko Chaussee, 1784 Sofia, Bulgaria
[2]Department of Solid State Physics and Microelectronics, Faculty of Physics, St. Kl. Ohridski Universityof Sofia, 5 James Bourchier Blvd., 1164 Sofia, Bulgaria

E-mail: stsvalkov@gmail.com

Abstract. Three-component surface alloys of the Ti-Al-Nb and Ti-Al-V systems were formed by means of selective electron beam alloying; the procedure involved DC magnetron sputtering deposition of multilayer Al and Nb or Al and V films with a thickness of 1 μm of each layer on a commercially-pure Ti substrate. The samples prepared were then electron-beam surface-alloyed under identical technological conditions. The specimens were examined by scanning electron microscopy (SEM); their chemical composition was studied by energy dispersive X-ray spectroscopy (EDX). X-ray diffraction (XRD) was used to determine the crystallographic structure. Intermetallic alloys in the system of Ti-Al-V were successfully produced with an electron beam current of 15 mA. As the current was increased from 15 mA to 20 mA, only pure titanium was visible, probably due to the evaporation of the alloying materials. In the Ti-Al-N system, intermetallic alloys were formed when the electron beam current reached 25 mA.

1. Introduction

Titanium aluminides are widely used in the field of aircraft and automotive industries, for manufacturing of railway cars, spacecraft, light ships, etc., because of their light weight, attractive mechanical properties and excellent resistance to corrosion at high temperature [1-3]. It is known that incorporation of different transition metals, such as vanadium or niobium, significantly improves the properties discussed above and, therefore, opens a number of new potential applications of these alloys. Many researchers have investigated ternary compounds in the systems of Ti-Al-V and Ti-Al-Nb [4-6]. It is known that Ti-Al-V alloys have already been introduced for manufacturing of implants, for development of parts and prototypes for the aerospace industry, gas turbines, etc. Incorporation of niobium into Ti-Al based materials is capable of improving significantly their mechanical properties. Works [6] have been published on the microhardness of Al-Ti-Nb alloys, showing that it reaches values of about 800 HV, which significantly exceeds that of the Ti-Al-based materials.

There exist many technologies for developing the discussed alloys, including thin film deposition [7], ion implantation, etc. [8]. It should be noted that the electron-beam surface alloying technique, which belongs to the additive manufacturing technologies, is considered as one of the most promising

[3] To whom any correspondence should be addressed.

methods of formation of the discussed alloys. In this technique, the alloying materials preliminarily deposited on a metallic substrate are irradiated by a high-intensity electron beam. The procedure results in melting of the treated surface area and forming a melt pool where the atoms from the metallic substrate and the alloying materials interact with each other. After solidification of the melt pool, intermetallic surface alloy is formed.

The formation of surface alloys by means of electron-beam surface alloying strongly depends on the technological conditions of the process. Therefore, clarifying the exact technological parameters leading to the formation of surface alloys is of great importance.

This paper describes a study on the formation processes of Ti-Al-V and Ti-Al-Nb alloys by an electron-beam surface alloying technology. The study deals also with the phase composition of the intermetallics produced and analyzes it with respect to the technological parameters used.

2. Experimental setup

Ti-Al-V and Ti-Al-Nb surface alloys were produced by DC magnetron sputtering deposition of Al and V or Al and Nb coatings on pure α-Ti substrate, the thickness of each film being about 1 μm. Before deposition, the Ti substrate was cleaned by sputtering for 10 minutes. The process was realized in Ar medium at a working pressure of 1.2×10^{-1} Pa. The samples were then alloyed by scanning a continuous electron beam. During the alloying process, the accelerating voltage (U) was 55 kV, the electron beam scanning frequency (f) was 200 Hz, the speed of the specimen motion (V) was 1 cm/sec. The electron-beam current (I_c) was varied between 15 mA and 25 mA. Figure 1 presents schematically the alloying process.

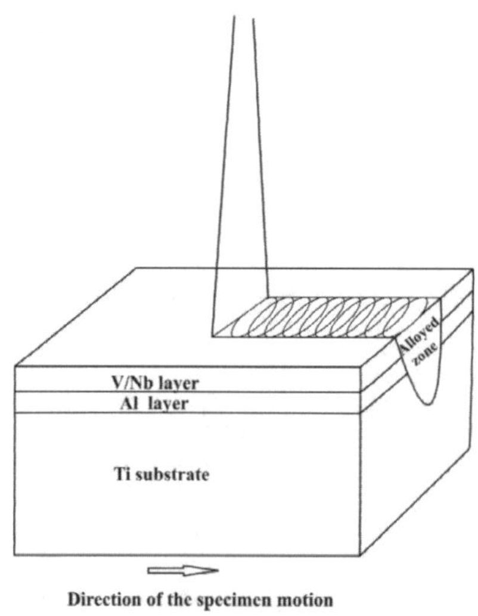

Figure 1. Electron-beam alloying process.

The phase composition of the alloyed layers was studied by X-ray diffraction (XRD) with CuKα characteristic radiation. The measurements were performed in the range from 30° to 80° at 2θ scale with a step of 0.1° and counting time of 10 s per step. The metallographic analyses were carried out by a scanning electron microscope (SEM/FIB LYRAI XMU, TESCAN) equipped with an EDX detector (Quantax 200,Bruker), making use of back-scattered electrons (BSE).

3. Results and discussion

Figure 2 presents X-ray diffraction patterns of electron-beam alloyed Ti-Al-V specimens taken at electron-beam current (I_c) of 15 mA and 20 mA. The diffractogram of the sample fabricated by a current of 15 mA exhibits maxima corresponding to AlTi and $AlTi_3$ phases. When I_c was increased to 20 mA, only titanium peaks were observed. Concerning the Ti-Al-Nb (figure 3) samples, it is seen that when the current of the electron beam was 15 mA, all peaks correspond to pure Ti, Al and Nb. No intermetallic phases were registered. After increasing I_c from 15 mA to 20 mA, peaks of Al_3Ti as well as of pure Ti and Nb appeared. Therefore, using these technological parameters (i.e., the parameters discussed in the experimental setup part and electron beam current of 20 mA) an intermetallic phase consisting of Ti and Al was formed. After raising I_c from 20 mA to 25 mA, diffraction maxima of ternary Ti_2AlNb and of pure Ti were registered.

Figure 4 presents a cross-section SEM image of the Ti-Al-V sample fabricated with $I_c = 15$ mA. The alloyed zone and the base titanium substrate are clearly visible. The chemical composition of the intermetallic part of the specimen was studied by energy dispersive X-ray spectroscopy; the results are summarized in table 1. As seen, an intermetallic Ti-Al-V surface alloy system was formed at the technological parameters discussed above. In [9], studies on the ternary Ti-Al-V phase diagram were reported; the results are in agreement with those obtained by our XRD and EDX investigations. According to [9], the chemical composition produced of the alloyed zone should correspond to the double-phase structure AlTi+$AlTi_3$. We did not register peaks of vanadium-based phases, which means that the V atoms have been dissolved into AlTi and $AlTi_3$ matrices.

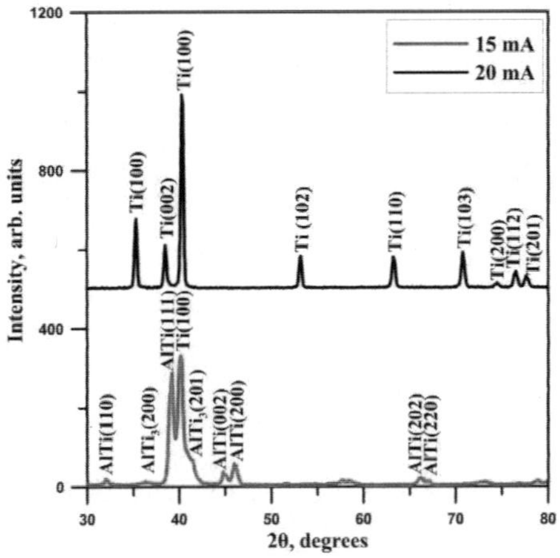

Figure 2. XRD patterns of Ti-Al-V alloys.

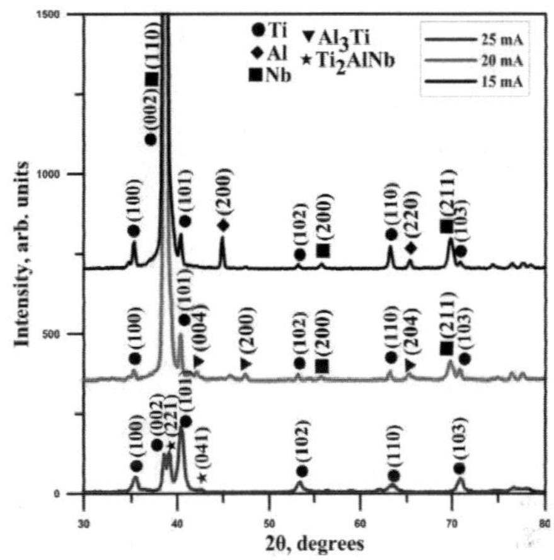

Figure 3. XRD patterns of Ti-Al-Nb alloys.

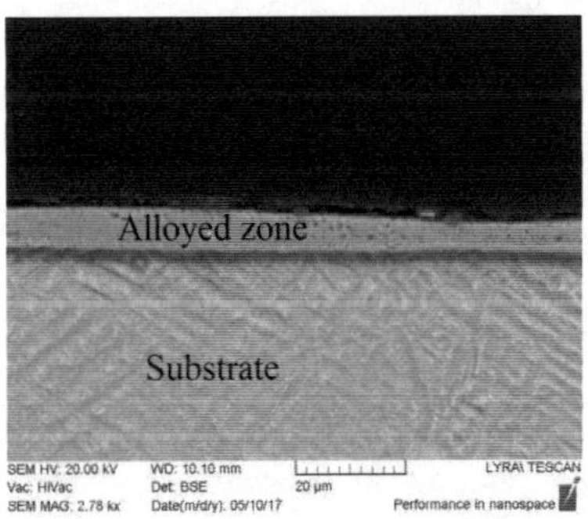

Figure 4. Cross-section SEM image of a Ti-Al-V sample alloyed with I_c = 15 mA.

Table 1. Chemical composition of Ti-Al-V sample fabricated with I_c = 15 mA.

	Ti, at%	Al, at%	V, at%
Alloyed zone	44.87±3.0	42.65±1.2	12.48±6.0
Substrate	87.99±2.2	11.60±0.3	0.41±0.2

Figures 5 and 6 present cross-section SEM images of Ti-Al-Nb samples fabricated by I_c = 20 mA and I_c = 25 mA, respectively. The chemical composition of the alloyed layers was studied by EDX; the results are summarized in table 2. Points 1; 2 and 3 correspond to the sample fabricated by I_c = 20 mA and are indicated in figure 5. The alloyed zone and substrate, shown in figure 6 were obtained at electron-beam current of 25 mA. The results are again in agreement with those from the XRD measurements and correspond to formation of an Al_3Ti intermetallic phase under the undissolved Nb layer at a beam current of 20 mA. Raising the current to 25 mA leads to the appearance of a separate alloyed zone. It consists of intermetallic Ti_2AlNb and pure Ti.

The results in the present study show the possibility of formation of intermetallic alloys in the system of Ti-Al-V and Ti-Al-Nb by electron-beam surface alloying. During such a selective electron beam treatment, the kinetic energy of the electrons is transformed into heat during their interaction with the specimens [10]. As it was already mentioned in the introduction, this technique is capable of melting the surface of the samples and forming a melt pool, where the atoms from the alloying materials and the base substrate form an intermetallic surface alloy. Therefore, the parameter "input energy" is important and can be described as:

$$E = \frac{UIt}{S} \qquad (1)$$

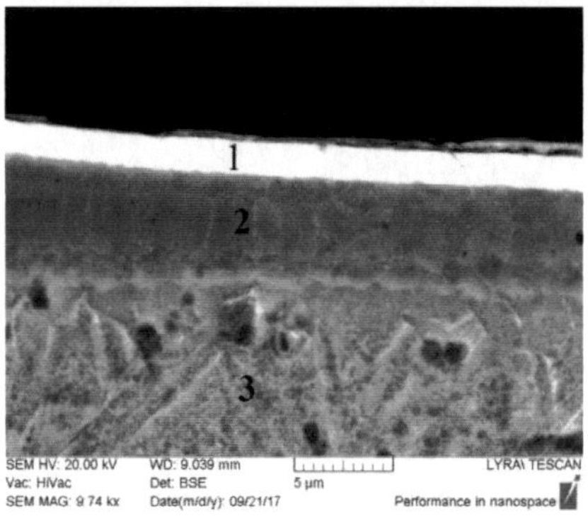

Figure 5. Cross-section SEM image of a Ti-Al-Nb sample alloyed with I_c = 20 mA.

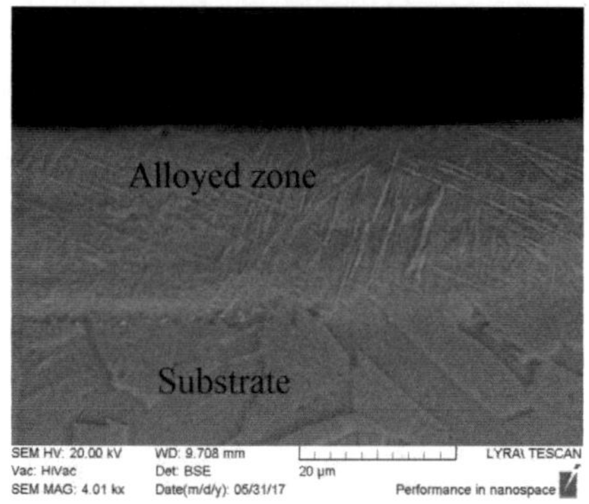

Figure 6. Cross-section SEM image of a Ti-Al-Nb sample alloyed with I_c = 25 mA.

In formula (1), E is the input energy (J/mm^2); U, the acceleration voltage (V); I, the electron beam current (A); t, the interaction time (s); S, the interaction area (mm^2) [11]. It is obvious that increasing the electron-beam current will increase the input energy E and the temperature of the alloyed zone surface. On the other hand, the melting point of the alloyed materials is an important process parameter. For example, the melting point of Nb (2469 °C) is higher than that of V (1910 °C). [12] This is in agreement with the results obtained in the present study, where a surface alloy of the Ti-Al-V system was formed using electron beam current of 15 mA. After increasing the beam current up to 25 mA, respectively, the input energy, we registered a pure titanium substrate only, obviously due to the evaporation of vanadium. In contrast, the intermetallic system of Ti-Al-Nb was successfully alloyed using electron beam current of 25 mA, as evidenced by the thinner undissolved Nb layer in figure 5.

Table 2. Chemical composition of a Ti-Al-Nb sample fabricated with I_c = 20 mA (points 1, 2 and 3) and I_c = 25 mA (alloyed zone and substrate).

	Ti, at%	Al, at%	Nb, at%
Point 1	2.96±0.1	7.44±0.1	89.59±3.5
Point 2	25.74±0.9	72.78±2.5	1.48±0.2
Point 3	100±2.6	-	-
Alloyed zone	85.12±2.4	8.91±0.3	5.97±0.3
Substrate	97.81±4.2	1.95±0.1	0.24±0.1

4. Conclusions

The results of the present study show the possibility of forming intermetallic alloys of the Ti-Al-V and Ti-Al-Nb systems by a scanning continuous electron beam. In the case of alloying titanium with aluminium and vanadium, the intermetallic layers are formed by electron beam current of 15 mA. When the beam current is increased from 15 mA to 20 mA only pure titanium is present, which can be due to the evaporation of the alloying elements. However, during alloying titanium with aluminium and niobium, intermetallic layers are formed when the electron-beam current reaches 25 mA. Therefore, the melting point of the alloyed elements plays a major role in the optimization of the technological parameters of a selective electron beam technology.

Acknowledgements

The authors acknowledge the financial support of project DFNP-17-1 "Selective electron-beam surface treatment of light metals" under the Assistance for Young Scientists Program of the Bulgarian Academy of Sciences.

References

[1] Wu X 2006 *Intermetallics* **14** 1114 – 22
[2] Dimiduk M 1999 *Mater. Sci. Eng.* A **263**/2 281 – 8
[3] Clements H and Mayer S 2013 *Adv. Eng. Mater.* **15** 191 – 215
[4] Jia M, Zhang D, Liang J and Gabbitas B 2017 *Metallurgical and Materials Transactions* A **48**/4 2015–29
[5] Chaves J, Olmos L, Jimenez O, Bouvard D, Rodriguez E and Florez M 2017 *Powder Metallurgy* **60**/4 257-66
[6] Valkov S, Petrov P, Lazarova R, Bezdushnyi R and Dechev D 2016 *Appl. Surf. Sci.* **389** 768-74
[7] Hampshire J, Kelly P and Teer D 2002 *Thin Solid Films* **420 – 421** 386 – 91
[8] Tsyganov I, Wieser E, Matz W, Mucklich A, Reuther H, Pham M and Richter E 2000 *Thin Solid Films* **376** 188 – 97
[9] Hashimoto K, Doi H and Tsujimoto T 1986 *Trans. Japan Institute of Metals* **27**/10 741-9
[10] Ormanova M, Petrov P and Kovacheva D 2017 *Vacuum* **135** 7-12
[11] Fu Y, Hu J, Shen X, Wang Y and Zhao W 2017 *Nuclear Instr. Methods in Phys. Res.* B **410** 207-14
[12] http://www.webelements.com

IOP Conf. Series: Journal of Physics: Conf. Series **992** (2018) 012054 doi:10.1088/1742-6596/992/1/012054

QCM gas sensor characterization of ALD-grown very thin TiO_2 films

S Boyadjiev[1,2,3], V Georgieva[1], L Vergov[1] and I M Szilágyi[2]

[1]Georgi Nadjakov Institute of Solid State Physics, Bulgarian Academy of Sciences, 72 Tzarigradsko Chaussee, 1784 Sofia, Bulgaria;
[2]Budapest University of Technology and Economics, Department of Inorganic and Analytical Chemistry, Szent Gellért tér 4, H-1111 Budapest, Hungary

E-mail: boiajiev@gmail.com

Abstract. The paper presents a technology for preparation and characterization of titanium dioxide (TiO_2) thin films suitable for gas sensor applications. Applying atomic layer deposition (ALD), very thin TiO_2 films were deposited on quartz resonators, and their gas sensing properties were studied using the quartz crystal microbalance (QCM) method. The TiO_2 thin films were grown using $Ti(iOPr)_4$ and water as precursors. The surface of the films was observed by scanning electron microscopy (SEM), coupled with energy dispersive X-ray analysis (EDX) used for a composition study. The research was focused on the gas-sensing properties of the films. Films of 10-nm thickness were deposited on quartz resonators with Au electrodes and the QCMs were used to build highly sensitive gas sensors, which were tested for detecting NO_2. Although very thin, these ALD-grown TiO_2 films were sensitive to NO_2 already at room temperature and could register as low concentrations as 50 ppm, while the sorption was fully reversible, and the sensors could be fully recovered. With the technology presented, the manufacturing of gas sensors is simple, fast and cost-effective, and suitable for energy-effective portable equipment for real-time environmental monitoring of NO_2.

1. Introduction

TiO_2 is one of the most extensively studied transition metal oxides. Nowadays, TiO_2, as an n-type semiconductor, is widely explored for photocatalysis, gas sensing and various optical applications [1,2]. Although in the last decades the gas sensing properties of TiO_2 thin films have been thoroughly researched, still knowledge is lacking on the gas sensing of very and ultra-thin films with thicknesses of several nanometers.

The quartz crystal microbalance (QCM) is a popular technique of detecting the mass of thin layers deposited on the crystal surface on a sub-nanogram level, but it could also be applied for monitoring the adsorption of nano amounts of various toxic gases [3,4]. Since only the surface properties of the films are relevant to gas sensors based on the QCM method, ALD-grown very thin films are very promising for implementation of gas sensors.

ALD is based on successive, alternating surface-controlled reactions from gas phase to produce highly conformal and uniform thin films, with thickness control of sub-nanometer precision [5,6]. In ALD, the adsorption of reactants is self-limited and the various precursors meet only on the surface of

[3] To whom any correspondence should be addressed.

the substrate. This self-limiting growth leads to the many advantages of ALD. ALD can provide thickness control at an atomic level, as the film thickness can be programmed easily by the number of ALD cycles. Not only the thickness, but also the composition of the film can be controlled precisely. With these advantages, ALD has emerged as a powerful tool in many research and industrial applications [7,8].

A large number of techniques have been developed for environmental monitoring and control of the toxic gases. Gas sensors based on wide-band gap semiconductors, as SnO_2, TiO_2, WO_3, MoO_3, ZnO, In_2O_3 etc. have been widely investigated for sensing different toxic gases [9–15]. Many of these devices often require high operation temperatures, generally within the range from 200 °C up to even 500 °C [16]. Also, they are usually not suited to performing high-precision measurements of gas concentrations, but only to detecting the presence of target gases and issuing a warning if certain threshold values are attained.

Detection of NO_2 has lately become a very important task, since it is a pollutant abundant in the atmosphere over large cities, mainly because of the increasing amounts of combustion exhaust gases that can cause respiratory irritation even in very small concentrations, as low as 15-25 ppm [17]. The technology described here, namely, using a QCM with an ALD-grown very thin TiO_2 film presents an effective way for real time detection of NO_2 in the environment.

Previously, prototype QCM gas sensors with several transition metal oxide films (e.g. MoO_3, TiO_2, WO_3) were already prepared by our team and tested for sensitivity to NO_2 and NH_3 in a specially designed laboratory set-up [11,14,18–22]. Among others, the sensing capability was explored of PVD and ALD grown TiO_2 thin films to NO_2. These films were mostly prepared by sputtering [19,21], and only in one case ALD was also used for TiO_2 film growth [22]. As a follow up of the previous studies, in the research presented we focused our effort on the sensing behavior of very thin TiO_2 films grown by ALD.

2. Experimental

The films were grown by ALD in a Picosun SUNALE R-100 reactor at a 200-°C substrate temperature with titanium tetraisopropoxide ($Ti(iOPr)_4$, TTIP) and H_2O as precursors. The precursor pulse times were 0.1 s for both TTIP and H_2O. The TTIP precursor was heated at 70 °C in order to reach a sufficient partial pressure of its vapor. The purge time was 3 s after the TTIP pulse and 4 s after the H_2O. Nitrogen was used as a carrier gas, while the overall pressure in the reactor chamber was ~10 mbar. TiO_2 thin films were deposited either on quartz resonators, which were used for the gas sensing tests to NO_2, or on Si substrates, which were used for further characterization of the films.

The film morphology was investigated by SEM and the composition was studied by energy dispersive X-ray analysis (EDX) on a JEOL JSM-5500LV scanning electron microscope. The thickness of the films was measured by UV-Vis reflectometry using an Avaspec-2048 spectrophotometer.

The sensing tests were performed using quartz resonators produced on 8-mm polished AT-cut quartz plates with thermally evaporated gold electrodes (diameter of 4 mm and thickness of about 120 nm with a Cr underlayer) on both sides [14,22]. Their working resonance frequency was around 16 MHz. The initial parameters of the resonators and their quality were evaluated by measuring the equivalent dynamic parameters, static capacitance C_0 and equivalent dynamic resistance R_q using a selective level meter. The dynamic capacitance C_q, the dynamic inductance L_q, and the quality factor Q were calculated. The gas sensing was tested in a home-made measurement system, which was described in detail previously [18].

The sorption properties of the TiO_2 films were evaluated by measuring the resonance frequency shift of the QCM structures covered by the TiO_2 thin films at various NO_2 concentrations. The measurement was performed in the following way: first, purging the structures by dry air, then creating a certain concentration of the measured gas in the chamber until reaching saturation of the frequency values; after that, again purging by dry air thus restoring the sensor and preparing it for a new measurement. The NO_2 concentration in the test chamber was controlled by mass flow controllers (MFCs) for NO_2 and diluting gas flow (synthetic air). The QCM frequency was registered by a Hameg

IOP Conf. Series: Journal of Physics: Conf. Series **992** (2018) 012054 doi:10.1088/1742-6596/992/1/012054

8123 frequency counter connected to the QCM and to a computer for data recording. The relative error of the frequency measurement was $\pm 5.25\times 10^{-7}$. The measurements were based on the correlation between the frequency shift and the additional mass loading the resonator, namely, Sauerbrey's equation for AT-cut quartz [23] describing the relation between the mass of the thin film deposited on the quartz crystal and the corresponding change in the resonance frequency; the absorbed mass was thus calculated [11,14,23].

3. Results and discussion

Very thin TiO_2 films were deposited by ALD on both QCMs and Si substrates. With 200 ALD cycles, films with a thickness of ~ 10 nm were grown; the thickness was evaluated based on the UV-Vis reflectometry measurement of the reference TiO_2 thin films deposited on Si wafers.

SEM coupled with EDX was used to study the morphology and composition of the as-deposited films. According to the SEM images (inset of figure 1), the films were of good quality, homogeneous, uniform, and without cracks or defects. We have to note that the object shown is an untypical impurity used just for focusing. The successful deposition of TiO_2 was clearly proved by the presence of Ti in the EDX spectrum of the reference TiO_2 film grown by ALD on the Si wafer (figure 1). The results from the composition analysis are presented in the inset. The EDX is simple, fast and cost-effective method (compared with alternative ones like XPS) to prove a successful deposition of very thin TiO_2 films [6, 22]. The EDX study had only qualitative character, since the information depth of EDX is ca. 500 nm; thus, the majority of the EDX signal comes mainly from the substrate and only a small portion originates from the TiO_2 film.

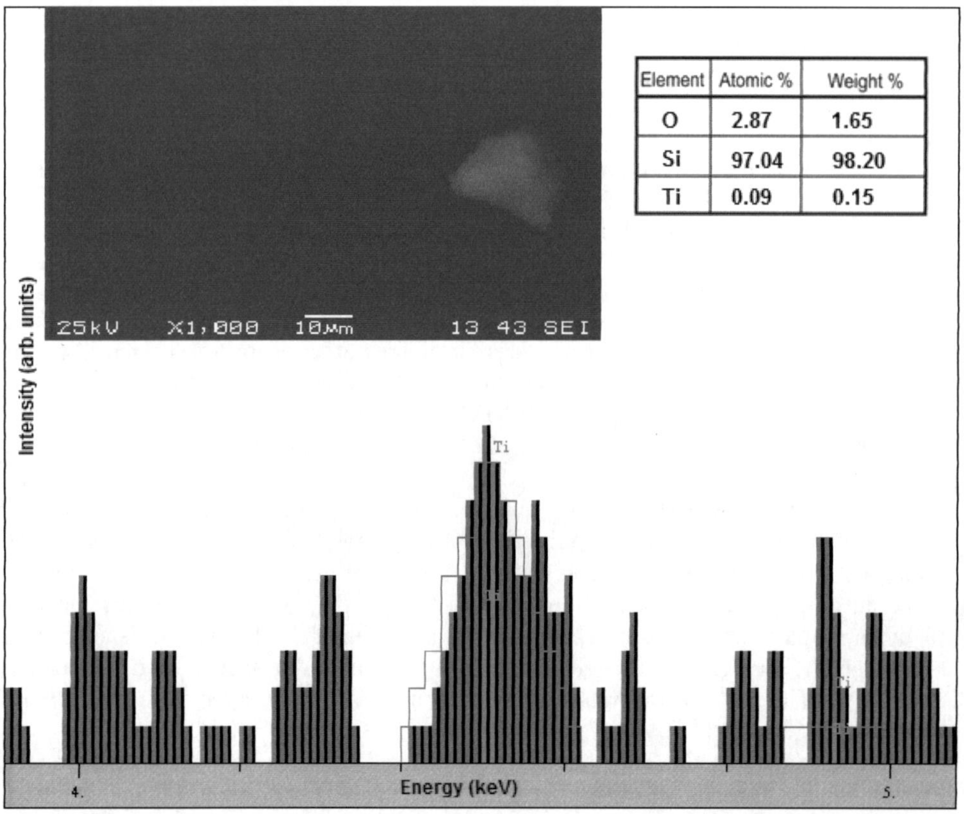

Figure 1. Ti peak in the enlarged EDX spectrum, together with a SEM image of the studied area (the center of the clean film area), and concentration analysis results of the reference TiO_2 film grown by ALD on a Si wafer.

At room temperature in the presence of oxygen and under anhydrous conditions, NO_2 is adsorbed on the surface of TiO_2 according to the following reactions [16, 24]:

$$Ti^{4+} + NO_2 + O^{2-} \leftrightarrow Ti^{3+} + NO_3^{-} , \quad (1)$$

$$TiO^{3+} + NO_2 \rightarrow TiO^{4+} + NO_2^{-} . \quad (2)$$

In fact, the mechanism of NO_2 adsorption on TiO_2 is much more complicated, and includes also catalytic reactions. Also, under actual sensor operation conditions, a complete elimination of humidity is practically impossible. In the presence of humidity, co-adsorption of water molecules occurs on the TiO_2 surface, which leads also to the formation of nitric acids [16, 25]:

$$2NO_2 + H_2O \rightarrow HNO_2 + HNO_3 . \quad (3)$$

This leads us to the paramount importance of detecting and eliminating the influence of water vapor on the NO_2 sensor operation, as it will result in errors in evaluating the amounts of NO_2 detected, Thus, as a first step of the test measurements of NO_2 concentration, a saturation of the sensor to water vapor was performed and its influence was eliminated from the calculations of the detected NO_2 mass. Figure 2 presents the sorption ability of the QCM sensor with an ALD-grown TiO_2 layer sensitive to H_2O vapor. The sensing was tested at a temperature of 24 °C and relative humidity of air (RH) of 63%. It can be seen that the H_2O adsorption is a relatively fast process and the saturation is reached in less than a minute (~50 s), while the frequency change is ~40 Hz. The study was performed after the TiO_2–QCM structure was tested to NO_2 (and restored) several times.

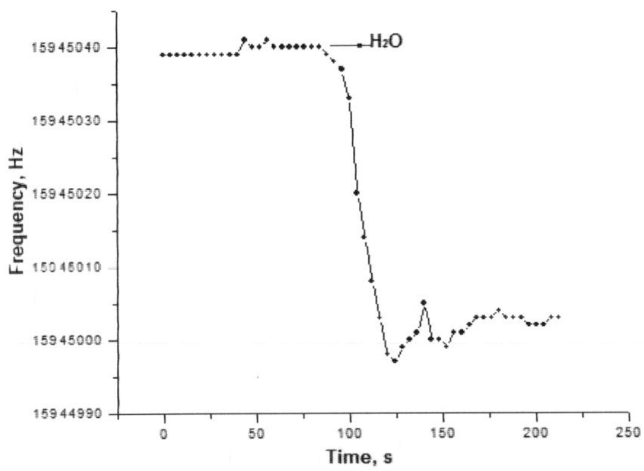

Figure 2. FTC of a QCM with ~10-nm ALD-grown TiO_2 thin film during saturation by water vapor.

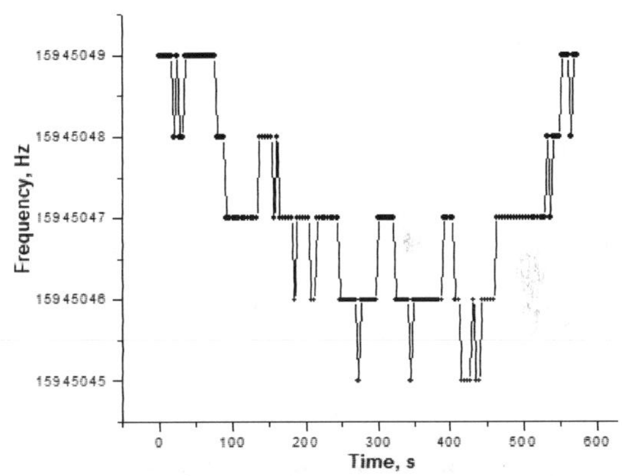

Figure 3. FTC of a QCM with ~10-nm ALD-grown TiO_2 thin film during saturation and recovery at NO_2 concentration of 100 ppm.

The gas sensing properties of the TiO_2 films to NO_2 were tested with concentrations in the interval between 10 ppm and 5000 ppm. The films were found to be sensitive even at 50 ppm (figure 3), but at such concentrations their sensitivity was very weak and the frequency change (Δf) observed was just 4-5 Hz, while the sorption and desorption processes were quite dynamic.

Typical FTCs for a well-working sensor were observed at slightly higher concentrations, for example at 500 ppm. At such a concentration, the response of the sensor was fast (few s), and its sensitivity vs. concentration was close to linear. Figure 4 illustrates the gradual loading of the QCM by increasing concentrations of 500 ppm, 1000 ppm and 5000 ppm. The loading processes with these three different concentrations were all measured for 7 min (420 s) with 3-min (180 s) desorption time. First, the sensing structure was tested for the sorption of the first 7 min after introducing 500 ppm NO_2 in the chamber, afterwards three minutes were allowed for desorption, and then the concentration was increased to 1000 ppm and measured for the next 7 min; then, similarly, the concentration was increased to 5000 ppm. The graph is stepwise and represents the non-linearity of the sorption process, and that it differs at different concentrations. Usually, the sorption was faster and exponential in the first moments after introducing a certain NO_2 concentration; afterwards the sorption process decelerated until gradually reaching saturation. As the NO_2 concentrations was increased, the sorption accelerated, but longer times were needed to reach saturation. Consequently, the desorption also took longer until all the mass was released from the sensing structure to fully recover for the next measurement cycle. It has to be stressed that the recovery was complete, and the initial values of the resonator were achieved after removing the NO_2 gas from the chamber. The sorption was considered to be a fully physical process, as shown by the full recovery of the sensor for a relatively short time, even after having detected the high concentration of 5000 ppm of NO_2. Figure 4 leads us to the conclusion that the loading ability of the sensing structure increases with the concentration, so that it can successfully detect higher concentrations.

Sauerbrey's equation [23] allows one to study the correlation between the concentration of NO_2, the frequency shift measured and the corresponding mass of gas molecules sorbed on the QCM structure. Figure 5 presents the frequency shift measured by the mass loading (proportional) at different NO_2 concentrations, from 50 ppm to 5000 ppm. The graph was built using the measurements for the gradual loading (presented in figure 4). The graph is interpolated, but allows us to conclude that for the lower concentrations the frequency shift and, consequently, the detected mass are linearly proportional to the concentrations measured. The films have higher sensitivity for lower concentrations; afterwards the sorption increases gradually, but it is also slower because of the gradual saturation of the sensing surface. The sorbed mass, calculated according to the Sauerbrey equation [23], was around 9 ng for a NO_2 concentration of 50 ppm, and reached 49 ng for 5000 ppm.

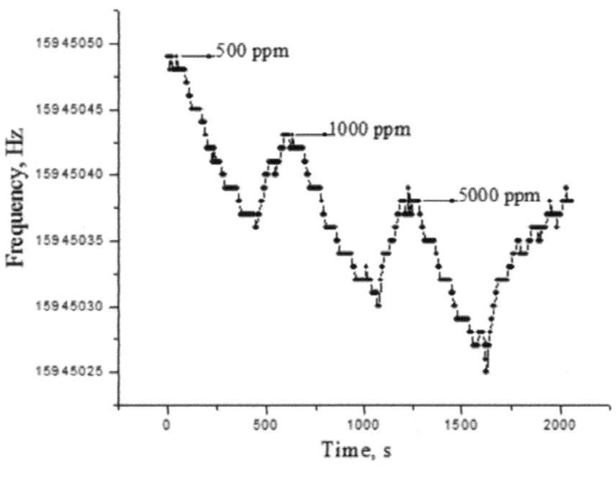

Figure 4. FTC of QCM with ~10-nm ALD-grown TiO_2 during gradual loading by NO_2 concentrations of 500, 1000 and 5000 ppm.

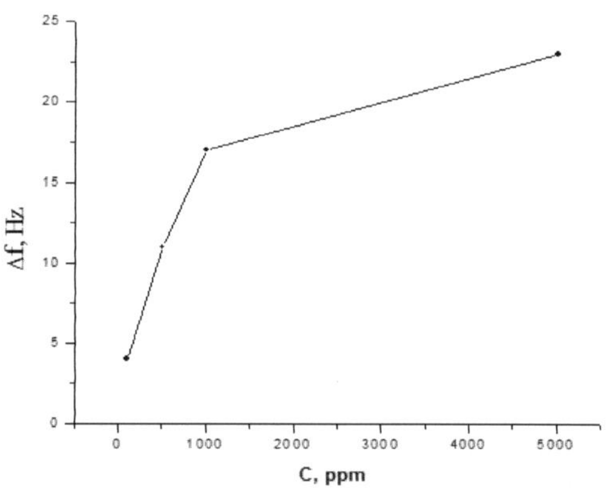

Figure 5. Measured frequency shift of a QCM sensor with an ALD TiO_2 film vs. the NO_2 concentration.

In our previous study, similar results for the sensitivity to NO_2 were obtained for several times thicker TiO_2 films deposited by reactive sputtering [21]. When using the QCM method, which allows high sensitivity while detecting very small concentrations, the thickness of the sensing layer and, consecutively, its load on the resonator is of importance, especially for a QCM working at a high frequency, since the higher loads on the resonator decrease the sensitivity of the gas sensor; thus, using as thin as possible sensing layer is desirable. The results presented here are very promising, as they show that very thin ALD-grown TiO_2 films possess good sensitivity to NO_2 and can be competitive with much thicker films obtained by other techniques.

However, even though our results show that very and ultra-thin ALD TiO_2 films have the required sensitivity for NO_2 sensing, still their sensitivity need to be improved to detect very low concentrations, as it is required of the modern gas sensors. One approach to increasing further the sensitivity of such sensors is by preparing a nanostructured sensing surface to increase the surface area or/and to combine sensitive materials by preparing nanocomposites [16,26]. But here many new problems arise, connected with the preparation of the QCM sensors with nanostructured surface and their stability, long-term usage and cost-effectiveness. Also, the nanocomposites used for gas sensing at room temperature may have a higher sensitivity, but their selectivity is lower. As a follow-up of the present study, the new challenge should be not only implementing ultra-thin films for gas sensing, but also combining such nanolaminates and nanocomposites of various materials, creating nanostructured morphology on the sensing electrodes, while at the same time making steps towards improving the selectivity of the sensors.

4. Conclusions

Very thin (~10 nm) TiO_2 films were grown on QCM by ALD and their sensitivity to various concentrations NO_2 was studied. These very thin ALD TiO_2 films showed good sensitivity to NO_2 at room temperature and a capability to register as low concentrations as 50 ppm. The sorption was fully reversible, and the sensors were able to recover in a short time. The ALD method was found to be suitable for fast and cost-effective deposition of TiO_2 thin films for QCM gas sensor applications. These promising results for sensitive films with thickness of less than 10 nm raise the hope that, after a further study and development, even ultra-thin ALD TiO_2 films of a few nm could be implemented in gas sensing devices for NO_2 monitoring.

Acknowledgements

S.I.B. acknowledges the Postdoctoral Fellowship Program of the Hungarian Academy of Sciences (2013-2015). I.M. Szilágyi appreciates the János Bolyai Research Fellowship of the Hungarian Academy of Sciences, and the ÚNKP-17-4-IV-BME-188, OTKA PD-109129, VEKOP-2.3.2.-16-2017-00013, K 124212 grants. Z. Baji's assistance is acknowledged in the ALD deposition (Hungarian Academy of Sciences, Institute of Technical Physics and Materials Science, Budapest, Hungary).

References

[1] Bennett J M, Pelletier E, Albrand G, Borgogno J P, Lazarides B, Carniglia C K, Schmell R A, Allen T H, Tuttle-Hart T, Guenther K H and Saxer A 1989 *Appl. Opt.* **28**/16 3303
[2] Diebold U 2003 *Surf. Sci. Rep.* **48** 53
[3] Cunningham A 1998 *Introduction to Bioanalytical Sensors* (New York: Wiley)
[4] Mirmohseni A and Rostamizadeh K 2006 *Sensors* **6** 324
[5] George S M 2009 *Chem. Rev.* **110** 111
[6] Szilágyi I M, Santala E, Heikkilä M, Pore V, Kemell M, Nikitin T, Teucher G, Firkala T, Khriachtchev L, Räsänen M, Ritala M and Leskelä M 2013 *Chem. Vapor Dep.* **19** 149
[7] Devi A 2013 *Coordin. Chem. Rev.* **257** 3332
[8] Szilágyi I M and Nagy D 2014 *J. Phys.: Conf. Ser.* **559** 012010
[9] Oyabu T 1982 *J. Appl. Phys.* **53** 2785
[10] Dutta P K, Frank M, Hunter G W and George M 2005 *Sens. Actuat.* B **106** 810
[11] Georgieva V, Stevchev P, Vitanov P and Spassov L 2004 *Vacuum* **76** 203
[12] Szilágyi I M, Wang L, Gouma P I, Balázsi C, Madarász J and Pokol G 2009 *Mater. Res. Bull.* **44** 505
[13] Ferroni M, Guidi V, Martinelli G, Nelli P, Sacerdoti M and Sberveglieri G 1997 *Thin Solid Films* **307** 148
[14] Boyadjiev S I, Georgieva V, Yordanov R, Raicheva Z and Szilágyi I M 2016 *Appl. Surf. Sci.* **387** 1230
[15] Ali M, Wang C Y, Röhlig C C, Cimalla V, Stauden T and Ambacher O 2008 *Sens. Actuat.* B **129** 467
[16] Procek M, Stolarczyk A, Pustelny T and Maciak E 2015 *Sensors* **15** 9563

[17] Rom W N and Markowitz S B 2007 *Environmental and Occupational Medicine* (4th ed.) (Philadelphia: Wolters Kluwer)
[18] Georgieva V, Stefanov P, Spassov L, Raicheva Z, Atanassov M, Tincheva T, Manolov E and Vergov L 2009 *J. Optoelectr. Adv. Mater.* **11**/10 1363
[19] Boyadzhiev S, Georgieva V and Rassovska M 2010 *J. Phys.: Conf. Ser.* **253** 012040
[20] Georgieva V, Raicheva Z, Grechnikov A, Gadjanova V, Atanassov M, Lazarov J and Manolov E 2010 *J. Phys.: Conf. Ser.* **253** 012046
[21] Yordanov R, Boyadjiev S and Georgieva V 2014 *Digest J. Nanomat. Biostruct.* **9**/2 467
[22] Boyadjiev S, Georgieva V, Vergov L, Baji Z, Gáber F and Szilágyi I M 2014 *J. Phys.: Conf. Ser.* **559** 012013
[23] Sauerbrey G Z 1959 *Physik* **155** 206
[24] Mikhaylov R, Lisachenko A, Shelimov B, Kazansky V, Martra G and Coluccia S 2013 *J. Phys. Chem.* C **117** 10345
[25] Tseng Y, Kuo C, Huang C, Li Y, Chou P, Cheng C and Wong M 2006 *Nanotechnol.* **17** 2490
[26] Boyadjiev S I, Kéri O, Bárdos P, Firkala T, Gáber F, Nagy Z K, Baji Z, Takács M and Szilágyi I M 2017 *Appl. Surf. Sci.* **424** 190

IOP Conf. Series: Journal of Physics: Conf. Series **992** (2018) 012055 doi:10.1088/1742-6596/992/1/012055

Mechanical stability of heat-treated nanoporous anodic alumina subjected to repetitive mechanical deformation

A Bankova[1,3], **V Videkov**[1], **B Tzaneva**[1] and **M Mitov**[2]

[1]Technical University of Sofia, 8 Kliment Ohridski Bldv., 1000 Sofia, Bulgaria
[2] Visteon Electronics Bulgaria Ltd.,
Capital Fort Building, 90 Tzarigradsko Chaussee, 1784 Sofia, Bulgaria

E-mail: a_bankova@tu-sofia.bg

Abstract. We report studies on the mechanical response and deformation behavior of heat-treated nanoporous anodic alumina using a micro-balance test and experimental test equipment especially designed for this purpose. AAO samples were characterized mechanically by a three-point bending test using a micro-analytical balance. The deformation behavior was studied by repetitive mechanical bending of the AAO membranes using an electronically controlled system. The nanoporous AAO structures were prepared electrochemically from Al sheet substrates using a two-step anodizing technique in oxalic acid followed by heat treatment at 700 °C in air. The morphological study of the aluminum oxide layer after the mechanical tests and mechanical deformation was conducted using scanning electron and optical microscopy, respectively. The experimental results showed that the techniques proposed are simple and accurate; they could, therefore, be combined to constitute a method for mechanical stability assessment of nanostructured AAO films, which are important structural components in the design of MEMS devices and sensors.

1. Introduction

The mechanical stability and durability of thin oxide films is a major concern in the operation and reliability of microelectromechanical systems (MEMS). Numerous articles and reports have appeared dealing with the characterization of the mechanical properties of polysilicon [1-3]. Nowadays, anodic aluminum oxide (AAO) is also being used in MEMS as a structural material, especially in the fabrication of AAO membranes for pressure or microheaters-based sensors [4-6]. As the thin films are exposed to adverse environments, which makes the oxide layer expand or stretch, the mechanical behavior plays a significant role in their usage in MEMS technology. Therefore, the ability becomes essential of determining the mechanical stability and fracture behavior of porous alumina films, in view of designing reliable layers in MEMS sensors and devices. Most studies on the mechanical properties of AAO films have been focused on using the nano-indentation and microhardness tests [7-9]. However, these methods have to do with measuring the parameters on a micro- and nano-scale and require expensive and complex equipment. Furthermore, few publications have appeared dealing with studies on the mechanical stability of anodized aluminum.

[3] To whom any correspondence should be addressed.

The aim of this work was to assess the mechanical behavior of nanostructured heat-treated AAO membranes using two measurement techniques. The first one is the well-known three-point bending test, which was performed on a special micro-analytical balance. This technique can also be used to measure the deflection to failure of the anodic aluminum oxide. The second one is based on specially developed experimental setup for repetitive mechanical bending of nanostructured membranes of aluminum oxide, i.e. characterization of the stress in oxide layers of small thickness. The experimental results from this test are used to determine the mechanical properties of the oxide, such as fatigue and cycles to fracture (breaking) of the material.

2. Specimen preparation

2.1. Fabrication of AAO membranes

Nanoporous AAO was prepared electrochemically using a two-step anodization technique in an oxalic acid solution. Specimens for the bending test were fabricated by anodizing free strips of aluminum foil with a 100-μm thickness and 99% purity. Part of the aluminum sheets were annealed at 220 °C in air for 30 min to release the mechanical stresses and were then electro-polished for 45 s in a solution of 200 mL orthophosphoric acid at a current density of ~3A/cm^2, in order to improve the surface quality of the substrate. After electropolishing, the test samples were washed with deionized water. The anodization process was carried out in oxalic acid in a voltastatic mode. The detailed experimental conditions of the electrolysis process whereby oxide layers were grown are summarized in table 1.

Table 1. Experimental conditions of anodization process.

Electrolyte	**Temperature** (°C)	**Potential** (V)	**Duration** (min)	**Diameter of pores** (nm)	**Thickness of layers** (μm)
Oxalic Acid	15	40	60	30	6

Separation of the resulting oxide from the aluminum substrate was performed by liquid chemical etching of the aluminum. Details on the anodization steps are described in [10]. The surface morphology of the specimens was investigated by scanning electron microscopy (SEM). The SEM micrographs presented in figure 1 show a well-defined, self-ordered, nanoporous AAO structure. The geometrical dimensions of all AAO film were 7 mm in length and 5 mm in width.

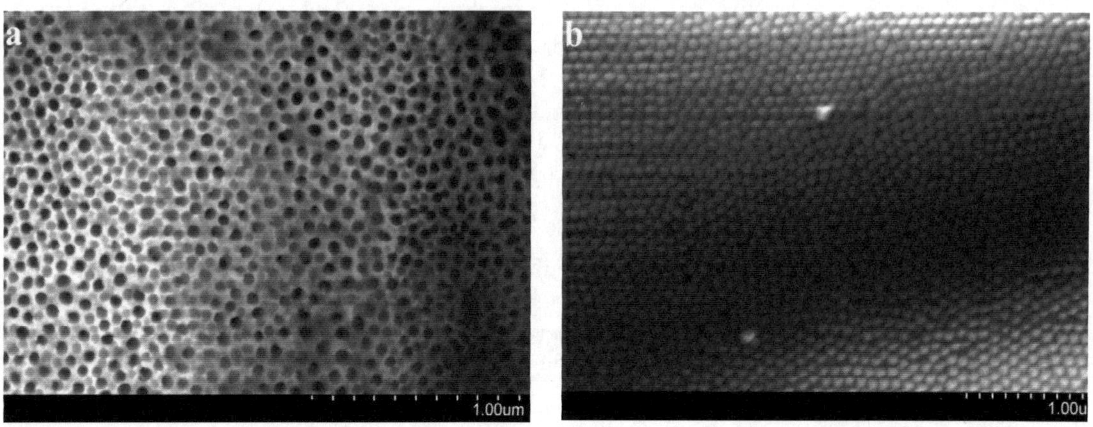

Figure 1. SEM image of the AAO surface morphology viewed from the porous layer side (a) and from the bottom layer side (b).

IOP Conf. Series: Journal of Physics: Conf. Series **992** (2018) 012055 doi:10.1088/1742-6596/992/1/012055

2.2. Heat treatment

The main focus of this work was determining the mechanical response and deformation behavior of heat-treated nanoporous anodic alumina oxide. Recently, the polymorphic phase transformations in AAO structure have been of interest to those intending to use AAO membranes in applications requiring mechanical strength. Thus, further research is needed to understand thoroughly the mechanical properties of AAO films. After the anodization process, part of the AAO samples were heat-treated at 700°C for four hours, in order to compare the mechanical response of heat-treated and non-heat-treated AAO membranes. The heat treatment was performed in air in a high-temperature furnace. To prevent thermal cracking of the oxide material, the extremely low heating rate of 10 °C/min was chosen. The temperature of the tube furnace was controlled by a digital temperature controller with an accuracy of ±1 °C. The thermal profile of the furnace is shown on figure 2. The structures of the experimental samples prepared were characterized by means of optical monitoring.

All specimens numbered from N1 to N4 were prepared under the same anodization conditions with a thickness of 6 μm, which was measured using optical microscopy. The specimens used for the experiment are described in table 2.

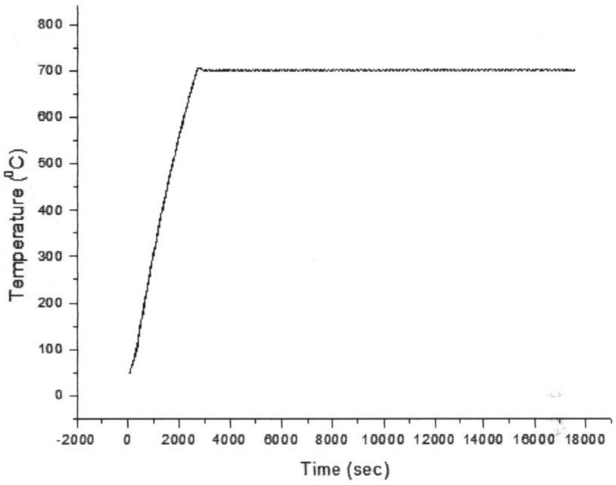

Figure 2. Thermal profile of the furnace.

Table 2. Specimen type.

Sample	Al substrate	Oxide layer
N1	annealed	nonheat-treated
N2	annealed	heat-treated
N3	non-annealed	non heat-treated
N4	non-annealed	heat-treated

3. Testing methods and discussions

In order to describe and mechanically characterize the heat-treated nanoporous AAO films, the following tests were carried out:

3.1. Three point bending test

The mechanical response of the AAO specimens prepared was measured by a three-point bending test at room temperature. An analytical balance with an accuracy of 0.1 mg was used to record the force applied to the sample.We should note that measuring the deflection of a membrane versus the applied force using an analytical balance avoids the error due to the force vector decomposition. A schematic diagram of the experimental setup for the three-point bend test is shown in figure 3.

The principle of the method presented in this paper is based on the following procedures: The experimental sample (1) is laid horizontally over two supporting thin plates (2), which are fixed at one end to a carrier (3). The force is applied to the top of the material through the blade (7), which is attached to a holder (8). The blade is moved to the support of the nanomembrane (1) by a micrometric screw through the console (9), which is connected to the holder (8). The experimental specimen, together with the carrier (3), are placed on a plate balancing system, which is made of a plate (4), bearing strings (5) and a rocker arm (6). This is done in order to achieve horizontality of the two supporters and parallelism of the blade relative to the specimen. Using the setup described, one obtains the dependence of the membrane deflection on the force applied. In addition, this experimental setup can be used to measure the Young's modulus of a beam-shaped sample, which will be the subject of a future work. Figure 4 present a photograph of the bending deflection of a beam with a 6-μm thick heat-treated oxide, where the load/deflection ratio is 2,45 mN/352 μm.

In all measurements, the distance between the two supports in the three-point bending test was 5 mm and the load was applied in the middle between them. Depending on which side of the experimental sample the load is applied (the porous side or the bottom barrier side), the membrane will have different bending behavior. In the three-point bending test, the surface of the nanoporous oxide in contact with the loading blade will be subjected to a compressive stress and the other surface touching the two supporting beams, to a tensile stress. In this study, the load was applied only to the porous side. Table 3 summarizes the relationship between the bending deflection and the applied load at 300 μm and 500 μm blade displacement in relation to the supporting plates.

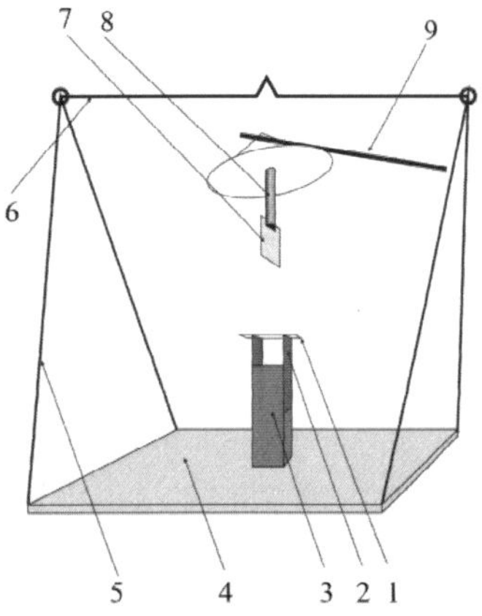

Figure 3. Schematic diagram of the three-point bending setup.

Figure 4. Deflection determination (measuring) using an optical microscopy image.

Table 3. Experimental results of three-point banding test.

Sample	300 μm		500 μm		Fracture
	Load (mN)	**Deflection** (μm)	**Load** (mN)	**Deflection** (μm)	
N1	2,13	0	2,62	430	No
N2	2,12	56	2,31	307	No
N3	2,12	0	2,41	436	No
N4	2,13	0	2,81	377	No

As one can see, the deflection of the non-heat-treated oxide (samples N1 and N3) is slightly higher than that of the heat-treated oxide (samples N2 and N4). This is due to the difference in the morphology and crystalline characteristics between the anodized oxide and the oxide subjected to heat-treatment. During the experimental tests, no fracture occurred in the oxide samples, although the deflection of the specimen exceeded the thickness of the AAO film by a factor of 1000. Furthermore, a slide of the specimen at the two supports was observed using a digital microscope. The cause of the slide is the bending deflection of the specimen resulting from the load applied. When the applied force was removed, the AAO specimen was recovering rapidly from the state of deflection. This indicated that the deformation of the specimen was still within the elastic range (figure 5).

3.2. Repetitive bending test

In order to evaluate the mechanical stability of the test specimens, they were subjected to mechanical deformation (bending). A specially designed test equipment (figure 6) was developed to provide repetitive bending. It consisted of a stepper motor (1) coupled to a camming mechanism (2, 3, 4), which pushes a steel rod (7). A soft-steel spring (5) pushes the rod back to the camming mechanism in order to prevent mechanical jittering. The rod moves through a bronze sleeve (6) to minimize the

IOP Conf. Series: Journal of Physics: Conf. Series **992** (2018) 012055 doi:10.1088/1742-6596/992/1/012055

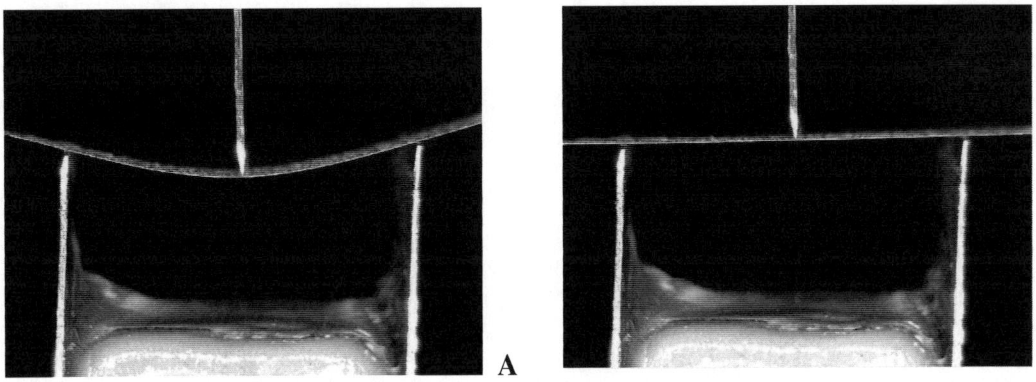

Figure 5. Elastic deformation stage of a 6 μm-thick heat-treated oxide: A) pressed membrane, B) relaxed membrane.

friction between the rod and the sleeve. The specimen is mounted on an aluminum carrier (8) above the rod. When the shaft of the stepper motor rotates, it drives the camming mechanism which pushes the rod forward when it is going up. When the camming mechanism is going down the rod goes down too because the spring pushes it back. The forehead of the rod is polished and is in contact with the AAO specimen (9). The specimen thus follows the movement of the rod. We were able to control the bending angle of the nanostructured AAO films with the use of microcontroller system driving the stepper motor. The maximum frequency of the repetitive bending was 2.5 Hz. Figure 7 is a photograph of the system.

Each membrane was bent to an angle of ~40^0 (figure 7) one million times. After the test, all the specimens were examined by optical microscopy. No cracks or defects were observed. The accuracy of the programmable system designed for repetitive mechanical bending of nanostructured oxide membranes is achieved through the precision of manufacturing of its mechanical components.

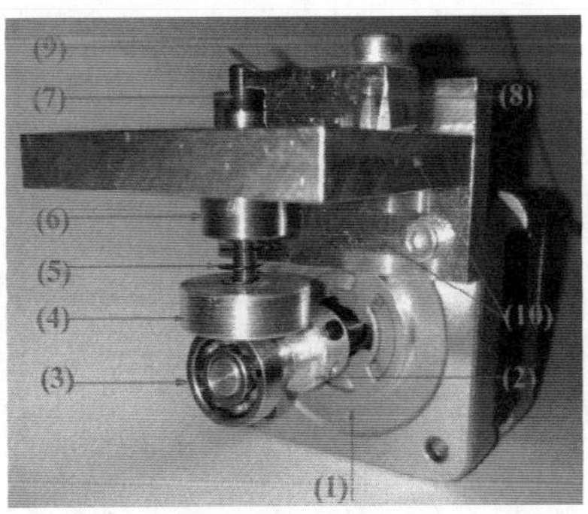

(1) stepper motor;
(2 - 4) – camming mechanism;
(5) – steel spring;
(6) – bronze sleeve;
(7) – steel rod;
(8) – aluminum carrier;
(9) – nanostructured AAO films;

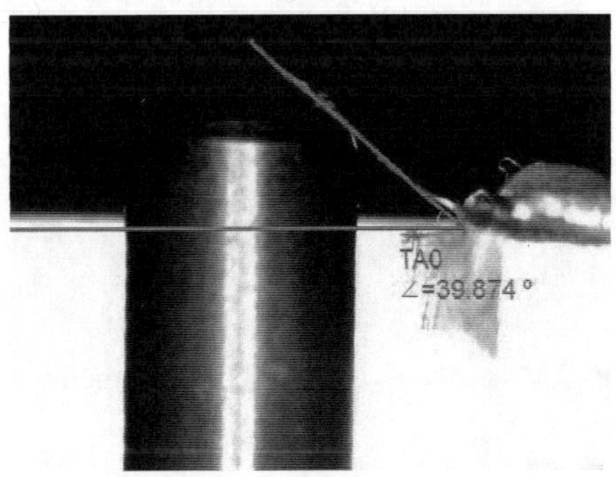

Figure 6. General view of the test equipment for repetitive bending.

Figure 7. Mechanical bending at a frequency of 2,5Hz at an angle of 40 degrees.

5. Conclusions

The mechanical response and deformation behavior of nanoporous alumina films were studied. AAO films with 6-μm thickness were produced by a two-step anodization process in an oxalic acid solution. The heat treatment of the AAO films tested had significant effects on their mechanical response. Additionally, no obvious differences in the four types of experimental samples appeared as a result of repetitive bending tests.

IOP Conf. Series: Journal of Physics: Conf. Series **992** (2018) 012055 doi:10.1088/1742-6596/992/1/012055

References

[1] Cambie R, Carli F and Combi C 2003 *Proc. Int. Conf. on Microelectronic Test Structures* (2003 Monterey CA USA) pp 3-39
[2] Tabata O and Tsuchiya T 2008 *Reliability of MEMS: Testing of Materials and Devices* (Weinheim, Germany: John Wiley & Son)
[3] Pasquale G and Somà A 2010 *Sensors* **10** 456
[4] Park S, Lee D, Byun S, Yoo H, Park D, Hwang W and Choi J 2016 *Int. J. Electrochem. Sci.* **11** 7401
[5] Routkevitch D, Polyakov O, Deininger D and Kostelecky C 2005 *NSTI-Nanotechnol.* **2** 266
[6] Chen W, Gui X, Liang B, Yang R, Zheng Y, Zhao C, Li X, Zhu H and Tang Z 2017 *ACS Appl. Mater. Interfaces* **9**/28 24111
[7] Hu Z, Shrestha M and Fan Q 2016 *Thin Solid Films* **598** 131
[8] Fang T, Wang T, Liu C, Ji L and Kang S 2007 *Nanoscale Res. Lett.* **2**/8 410
[9] Ko S H, Lee D W, Jee S E, Park H C, Lee K H and Hwang W 2005 *Glass Phys. and Chem.* **31**/3 356
[10] Bankova A, Videkov V and Tzaneva 2014 *J. Phys.: Conf. Ser.ies* **514** 012027

IOP Conf. Series: Journal of Physics: Conf. Series **992** (2018) 012056 doi:10.1088/1742-6596/992/1/012056

Polarization holographic gratings in PAZO azopolymer recorded with different recording-beams polarizations

G Mateev[1,3], L Nedelchev[1,2], D Ivanov[1], R Tomova[1], P Petrova[1], V Strijkova, N Berberova[1] and D Nazarova[1]

[1]Acad. J. Malinowski Institute of Optical Materials and Technologies, Bulgarian Academy of Sciences, Acad. G. Bonchev Str., Bl. 109, 1113 Sofia, Bulgaria
[2]University of Telecommunications and Post, 1 Acad. St. Mladenov Str., 1700 Sofia, Bulgaria

E-mail: g_mateev@abv.bg

Abstract. Polarization holographic gratings (PHG) were recorded using a laser emitting a wavelength of 491 nm in thin films of the (poly[1-[4-(3-carboxy-4-hydroxyphenylazo)benzenesulfonamido]-1,2-ethanediyl, sodium salt]) azopolymer, shortly denoted as PAZO. Thin azopolymer films with various thicknesses were spin-coated on plastic and glass substrates. Four different polarization states of the recording beams were used, and the results compared: a) two vertical linear polarizations, b) horizontal and vertical linear polarizations, c) linear polarizations at +45° and –45° relative to the recording plane, and d) two orthogonal circular polarizations – left- and right-handed (LCP and RCP). The diffraction efficiency in the +1 diffraction order was monitored in real time by a probing laser beam at the wavelength of 635 nm. The results indicate that the highest diffraction efficiency is achieved when recording with orthogonal polarizations – linear at ±45° or left and right circular. This is explained by the ability of the azopolymer material to record the variations in the polarization state of light better than the variations in its intensity. The holographic gratings obtained can be used to enhance the light-extraction efficiency of an OLED device.

1. Introduction

Azo-containing polymers are widely investigated materials because of their photo-anisotropic properties [1-3]. The photoinduced phenomena, such as linear and circular photoinduced birefringence, are due to the *trans-cis-trans* isomerization in azomaterials when illuminated by polarized light.

As measured earlier by our group, the maximal photoinduced birefringence of the azopolymer PAZO is $\Delta n \approx 0.07$ [4]. This high photoinduced birefringence is of essential importance for achieving high efficiency of polarization diffraction gratings. These gratings are interesting because of their achromaticity and the theoretical possibility to achieve 100% diffraction efficiency (DE), even in thin films [2].

There are different ways to inscribe polarization gratings. In this study, we compared the efficiency of gratings recorded with four different polarization states of the recording beams: a) two vertical linear polarizations (denoted shortly as VV), b) horizontal and vertical linear polarizations (HV), c) linear polarizations at +45° and –45° relative to the recording plane (PM), and d) two orthogonal

[3] To whom any correspondence should be addressed.

circular polarizations – left- and right-handed (LR). The use of different polarizations results in different polarization patterns of the interference field on the samples surface and in different properties of the gratings and values of their DE.

2. Theory

The properties of the polarization gratings were investigated theoreticaly in [2]. When one uses two pumping beams with defined polarization, one obtains a specific interference field on the sample. The cases with VV, HV, PM and LR polarizations of the pumping beams are shown in figure 1.

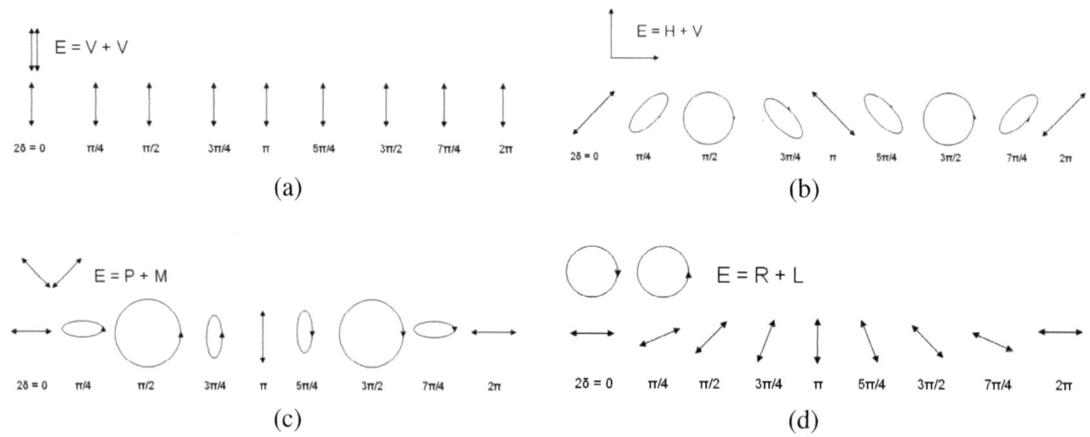

Figure 1. Interference field patterns for different polarizations of the two pumping beams: (a) VV, (b) HV, (c) PM, and (d) LR. 2δ is the phase difference between the pump beams.

The measured diffraction efficiency strongly depends on the polarization of the reading beam. Hence, for each of these cases, one uses a reading beam with polarization coinciding with the polarization of one of the recording beams. Thus, for the HV, VV and PM geometries, one obtains diffracted beams at ± 1 orders, while for the LR geometry one should expect to get diffraction only in the +1 order.

In the case when the photoinduced changes are relatively small in materials with linear and circular photoanisotropy, and one irradiates the medium with two beams with vertical and horizontal polarizations, the intensitites of the ± 1 diffracted orders are:

$$I_{\pm 1} = \frac{1}{4}\left[(\Delta\varphi^L)^2 + (\Delta\varphi^C)^2 \pm 2\Delta\varphi^L \Delta\varphi^C \sin(2\alpha)\right], \qquad (1)$$

$$\Delta\varphi^L = \Delta n^L \pi d / \lambda, \quad \Delta\varphi^C = \Delta n^C \pi d / \lambda$$

where Δn^L and Δn^C are the amplitudes of modulation of linear and circular birefringence and α is the azimuth of the linearly polarized reading laser beam [2]. If the azimuth α is 45°, onel obtains the maximal value for the intensity of the +1 diffracted order and the minimal value for the intensity of the –1 diffracted order. Using these two values, one can calculate the linear and circular birefringence. In the case when the two values are equal, the circular birefringence is 0:

$$\Delta\varphi^L = I_{+1,\max}^{1/2} + I_{-1,\min}^{1/2}, \ \Delta\varphi^C = I_{+1,\max}^{1/2} - I_{-1,\min}^{1/2} . \qquad (2)$$

The calculations are made under the assumption that the recording beams have equal intensities.

3. Experiments

3.1. Samples preparation and thickness measurement

We prepared samples for the experiments by dissolving two different concentrations of PAZO azopolymer in water, namely 100 mg/ml and 125 mg/ml, and then spin-coating the solutions on glass substrates at 1500 rpm for 30 s. As a result, thin homogeneous films were formed for optical testing.

IOP Conf. Series: Journal of Physics: Conf. Series **992** (2018) 012056 doi:10.1088/1742-6596/992/1/012056

Their thickness was determined by a Filmetrics F20 device using the non-destructive method of thin-film interference. The thicknesses measured were 470 nm for the samples produced from the solution with concentration 100 mg/ml, and 790 nm for the others.

3.2. Recording of the PHG

The experimental setup for recording the PHG is presented in figure 2. For recording, a 491-nm diode laser was used, while the reading diode laser was operating at 635 nm. The recording laser radiation was split into two beams by a polarizing beam-splitter (PBS). Half- and quarter-wave plates were used to adjust the polarization state of the pumping and reading beams according to the four different geometries. The recording beams had diameters of 1 cm, and power of 10 mW. The recording angle was $2\theta = 30°$.

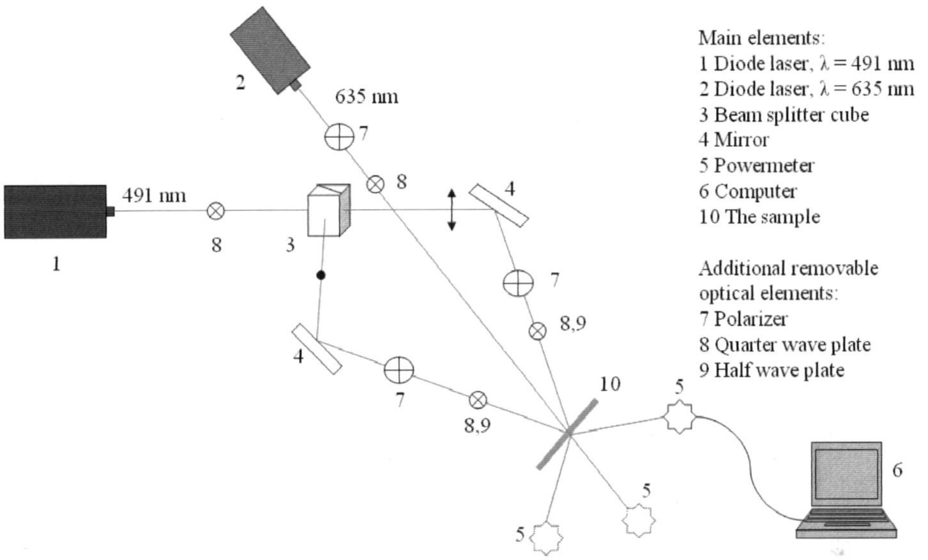

Figure 2. Experimental setup for the PHG recording.

The experimental procedure was as follows: First we measured the background with the reading laser for one minute. Then we started the holographic recording with the pumping laser irradiating the sample for ten minutes. After ten minutes, we turned the pumping laser off and measured the relaxation. The power of the transmited first-order diffracted beam was measured continuously by a powe rmeter and the data stored on a computer. After five minutes of relaxation, we measured the power of the other diffracted orders (if any). Using the results of these measurements, we calculated the temporal dependence of the diffraction efficiency (DE).

3.3. Results

We recorded PHG in thin azopolymer films with two thicknesses and four different cases of the recording beams' polarization, denoted as HV, VV, PM and LR. In each case, the polarization state of the reading laser coincided with the state of one of the recording beams. Figure 3a shows the results for the DE as a function of the time for the films with 470-nm thickness, figure 3b shows the same for the films with 790-nm thickness.

For the 470-nm, one can see that the LR case leads to a much higher DE than the other cases. This result is repeated with the thicker layers (790 nm). Also, for the thicker layers one can see clear differences between all cases. Thus, the highest DE is obtained with the LR geometry, followed by the PM and VV geometries, with the lowest DE obtained with the HV geometry. The above is explained by the fact that in the LR case the resultant optical field has linear polarization only with a varying azimuth.

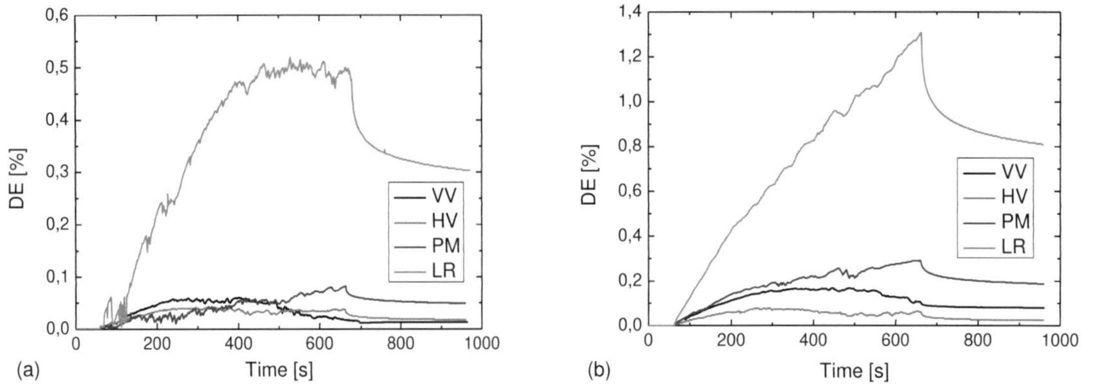

Figure 3. Diffraction efficiency (DE) of the gratings recorded at different polarization configurations: (a) samples with a thickness of 470 nm, and (b) samples with a thickness of 790 nm.

In all other cases, the field has eliptical polarization. On the other hand, the PAZO azopolymer exhibits a significant photoinduced linear, but no circular, birefringence. Thus, the LR geometry offers a better opportunity to utilize the PAZO properties. Also, in the LR geometry there exists one diffracted beam only, while for all other cases there are two approximately equal diffracted beams in ±1 order, in full accordance with the theoretical expectations. This fact doubles up the DE in LR geometry.

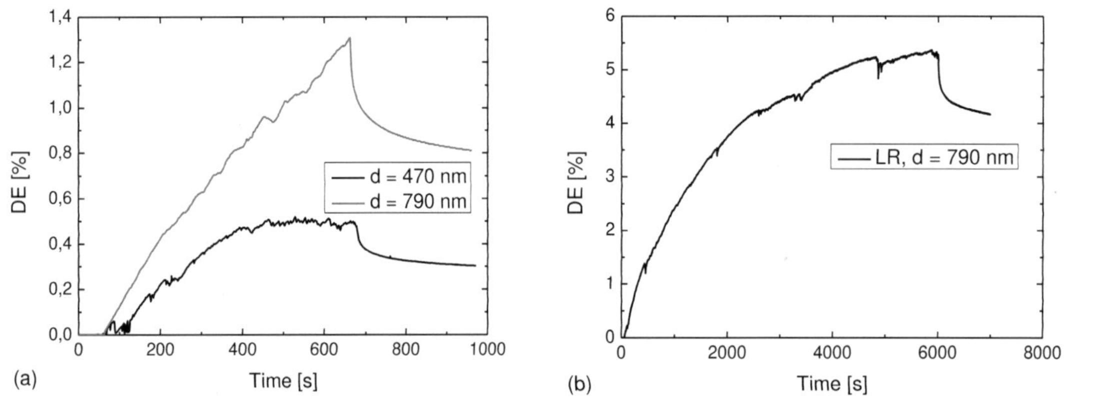

Figure 4. (a) Comparison of the DE for LR polarizations and different thickness of the samples – 470 nm and 790 nm. (b) Difraction efficiency of the 790-nm thick sample for LR polarizations and recording time 1.5 hours.

For the geometry with the highest DE (LR), we compared the performances of the PHGs formed in the films of different thickness. As expected, the thicker sample yielded a higher (by a factor of 3.5) DE. Further, we observed that 10 minutes of recording were sufficient to reach saturation for the thin film, but clearly not enough for the thick one. We, therefore, recorded a PHG with longer exposure in the sample with 790-nm thickness, as shown in figure 4b, and obtained about four times as high DE.

4. Conclusions

We prepared samples of PAZO azopolymer with two different thicknesses and recorded PHGs in four cases (VV, HV, PM and LR recording geometries) for the two thicknesses. The results showed that the highest DE is achieved using an LR geometry, because in this geometry the resulting optical field is linearly polarized and there is only one diffracted beam. Furthermore, the thicker sample yields a higher DE but requires much longer time to reach saturation.

Acknowledgements
The work was supported by the Bulgarian National Science Fund under project DM 08/1.

References

[1] Todorov T, Nikolova L and Tomova N 1984 Polarization holography. 1: A new high-efficiency organic material with reversible photoinduced birefringence *Appl. Opt.* **23** 4309
[2] Nikolova L and Ramanujam P S 2009 *Polarization Holography* (Cambridge University Press, Cambridge)
[3] Wang X 2017 *Azo Polymers: Synthesis, Functions and Applications* (Springer-Verlag, Berlin)
[4] Nedelchev L, Nazarova D, Mateev G and Berberova N 2015 Birefringence induced in azopolymer (PAZO) films with different thickness *Proc. SPIE* **9447** 94471I-1

IOP Conf. Series: Journal of Physics: Conf. Series **992** (2018) 012057 doi:10.1088/1742-6596/992/1/012057

Experimental and theoretical study on chemically semi-amplified resist AR-P 6200

I Kostic[1,5], K Vutova[2,3], R Andok[1], V Barak[1], A Bencurova[1], R Ritomsky[1] and T Tanaka[4,3]

[1]Institute of Informatics, Slovak Academy of Sciences,
9 Dubravska cesta, 84507 Bratislava, Slovakia
[2]E. Djakov Institute of Electronics, Bulgarian Academy of Sciences,
72 Tzarigradsko Chaussee, 1784 Sofia, Bulgaria
[3]Visiting Professor, Hiroshima University, Higashihiroshima, Japan
[4]Hiroshima Institute of Technology,
2-1-1, Miyake, Saeki-ku, Hiroshima 731-5193, Japan

E-mail: ivan.kostic@savba.sk

Abstract. Experimental and simulation results are presented and discussed on electron-beam lithography (EBL) nano-structuring using the positive chemically semi-amplified electron-beam resist AR-P 6200 (CSAR 62), which provides high sensitivity and allows achievement of high resolutions (sub-100 nm). The influence of the e-beam lithography process parameters, namely, exposure dose, development process conditions, and proximity effects on the obtained developed images was studied for the case of 40-keV electron energy.

1. Introduction

Since the Novolac resin in photoresists is not transparent to the KrF laser radiation, while diazonaphthoquinone (DNQ) is not sensitive to this laser radiation, chemically-amplified resists (CAR) were developed [1] and the use of the CAR system has been extended to ArF and ArF immersion lithography as well [2, 3]. Recently, a number of advanced resist systems with good combination of sensitivity and resolution have been developed mainly for extreme ultraviolet lithography (EUVL) [4].

CARs exhibit high resolution and high sensitivity, so they are very attractive materials for electron-beam lithography (EBL) as well. In EBL, the resolution capability is not an issue, as the beam size of the focused electron beam is usually small compared to the minimal feature size of the device patterns. However, an important requirement for the electron-beam resist has always been the combination of high resolution, high sensitivity, and line-edge roughness (LER). The first commercialized CAR was the negative-tone electron-beam resist SAL 601 [5]. Subsequently, various kinds of CARs for electron-beam systems have been developed for mask making and direct writing applications.

Aiming to improve the resolution of the nano-dimensioned EBL and to obtain results concerning the critical dimensions of structures for nano-device fabrication in view of nano-lithography applications, further experimental and theoretical studies (simulations) are needed to understand the

[5] To whom any correspondence should be addressed.

effects of increasing the contrast on the resolution of final resist patterns (profiles). In this paper, we present investigations on the positive chemically semi-amplified electron beam resist AR-P 6200 (CSAR 62 – Chemically Semi-Amplified Resist) (*Allresist GmbH*) [6].

2. Results and discussion

2.1. Experimental procedure

The investigations were performed on a 600-nm-thick layer of positive e-beam resist CSAR 62 prepared by spin-coating on a silicon substrate. The thickness of 600 nm was chosen as being appropriate for dry etching of silicon for some applications. The resist was prebaked at 150 °C for 1 min on a hot plate. The post-exposure bake performed usually for CARs was not applied because an additional increase in the sensitivity already occurs during the exposure of the CSAR 62 resist due to the halogenated acidifiers [6]. The resist was developed by an AR 600-548 developer, followed by rinsing in an AR 600-60 stopper for 30 s. The exposures were carried out using a ZBA 23 (Vistec Electron Beam GmbH) variably-shaped e-beam equipment at a constant electron energy of 40 keV [7]. The contrast and sensitivity characteristics were obtained by exposure wedge tests (beam cross-section 3000×3000 nm^2). The resist AR-P 6200 is characterized by high process and plasma-etch stability; in addition, it is suitable for lift-off processes. The resist profile shape is an important issue in both processes. We conducted resist profile studies on test structures consisting of long single lines with various linewidths and on periodical gratings with various line/space (L/S) dimensions. The length of the e-beam cross-section was 3000 nm, with the width varying from 50 nm to 1000 nm. After development, the sample was broken perpendicularly to the lines and gratings. All dimension measurements were carried out using a Quanta 3D (FEI) high-resolution scanning electron microscope (SEM) with a field-emission cathode.

2.2. Experimental results and discussion

Series of line test structures were exposed with the aim to follow the dependence of the resist profile shape on the exposure dose and find the conditions for producing a vertical sidewall shape and Line = Space. Typical results of the tests inspection are demonstrated in figures 1 and 2. The dependence of the resist profile dimensions and the image shape on the exposure dose for a single-line pattern is shown in figure 1. The e-beam cross-section was 100×3000 nm^2. The exposure dose was increased from 120 μC/cm^2 to 340 μC/cm^2. As seen, the shape of the profile in the resist changed as the exposure dose was raised. The optimal (vertical) side wall shape was formed at the exposure dose of 310 μC/cm^2 (figure 1c).

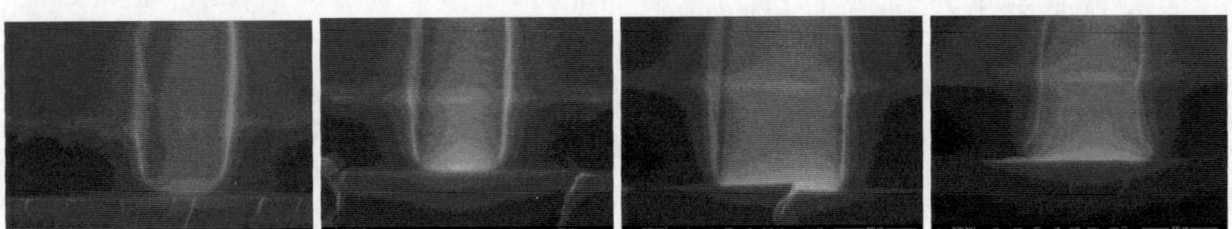

a) Q = 120 μC/cm^2 b) Q = 210 μC/cm^2 c) Q = 310 μC/cm^2 d) Q = 340 μC/cm^2

Figure 1. Single line, linewidth 200 nm.

The linewidth was measured at different levels along the resist depth – at the resist top (T), at the middle resist depth (M), near the resist bottom (B1/8) and at the resist bottom (B). The corresponding measured values of the linewidth in the case of a 340 μC/cm^2 exposure dose were: T = 638.4 nm, M = 652.8 nm, B1/8 = 715.1 nm, and B = 813.9 nm.

Figure 2 shows the profile shape and image dimension variations *vs.* the dose of electron exposure for a periodical line grating with the L/S exposed being 100/700 nm. The optimal profile shape with vertical side walls and Line = Space = 400 nm was formed at the dose of 130 μC/cm^2 (figure 2c).

IOP Conf. Series: Journal of Physics: Conf. Series **992** (2018) 012057 doi:10.1088/1742-6596/992/1/012057

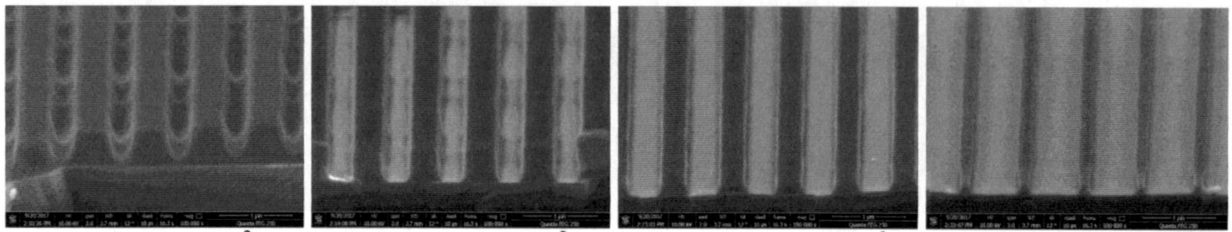

a) Q = 70 μC/cm² b) Q = 90 μC/cm² c) Q = 130 μC/cm² d) Q = 220 μC/cm²

Figure 2. Periodical line grating, exposed L/S 100 nm/700 nm.

2.3. Simulation and estimation of the AR-P 6200 e-beam resist characteristics

Using CSAR 62 allows one to achieve high resolutions (sub-100 nm); its etch stability is comparable to that of ZEP520A resist, while it is reasonably priced [8, 9]. The profile in the resist (relief formed in the resist layer after the development process) and the image dimensions are the most important factors in deciding whether the development conditions (developer, time of development, etc.) are suitable for forming fine patterns. An original nonlinear development model was reported in [10], where the calculation of the developed relief (images) was based on the 3-dimensional absorbed-energy distribution in the resist layer [11] and on the local dissolution rates.

Achieving high EBL resolution requires accurate knowledge of the absorbed-energy distribution in the resist – in lateral direction (r) and along the resist depth (z). At a given resist depth, the energy deposition function (EDF(r,z)) is defined by a sum of two Gaussians; a Monte Carlo approach is used [11] to model the electron scattering process and calculate the proximity effect parameters (β_f, β_b, η_E) along the resist depth. Figure 3 presents a comparison between the energy deposition functions calculated at two characteristic depths in CSAR 62 – at the resist surface (at a 30-nm depth) and at the resist/Si substrate interface (at a 600-nm depth).

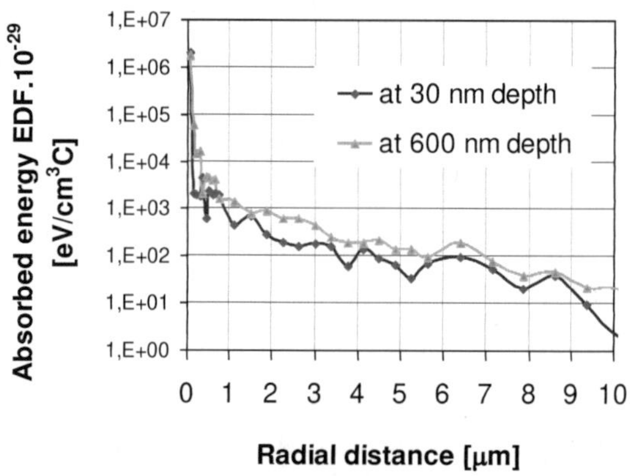

Figure 3. Radial distributions of the absorbed energy density in CSAR 62 (600-nm resist film on Si substrate) at electron energy of 40 keV.

The results for the absorbed energy variation in the resist depth showed a linear dependence of the proximity effect parameters values along the CSAR resist depth concerning the 3D distribution of the EDF. The values for two resist depths were used – for the resist surface and for the resist-substrate interface. For 600-nm-thick CSAR 62 on Si at 40-keV beam energy, the values obtained of (β_f, β_b, η_E) were (0.0467 μm, 4.038 μm, 0.197) at the resist surface and (0.0541 μm, 5.307 μm, 0.321) at the interface, respectively. Based on these simulation results, one should expect nearly vertical side walls of the profiles in the resist.

We further studied the behavior of the CSAR 62 resist developed in AR 600-548 for different times to obtain the dissolution rates at different doses of electron exposure needed for the simulation.

Figure 4a presents experimental contrast curves for a 600-nm-thick CSAR 62 resist on a Si substrate for three development times, namely, 60 s, 180 s and 240 s. The electron energy was 40 keV and the beam spot size, 3×3 μm². As seen, the CSAR 62 positive-tone resist studied shows a nonlinear development behavior. Based on the residual resist thickness as a function of the time of development, we calculated the solubility rates for the AR 600-548 developer at different times of development when the exposure dose varied from 0.5 μm/cm² to 14.5 μm/cm² (figure 4b). One can see a low solubility rate at low exposure doses and a sharp increase of the resist removal at higher exposure doses. The development rates thus evaluated as functions of the exposure dose can be applied to

IOP Conf. Series: Journal of Physics: Conf. Series **992** (2018) 012057 doi:10.1088/1742-6596/992/1/012057

simulate the profiles developed in the resist taking into account the peculiarities of the CSAR 62 resist (due to the complicated mechanism of the resist removal from the soluble resist areas) and the variation of the energy deposition function EDF(r,z) in radial and in depth directions (its 3D distribution in the resist layer).

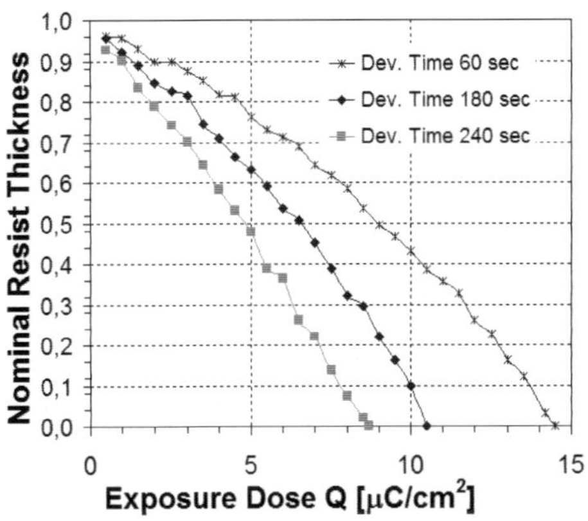

Figure 4a. Thickness of remaining CSAR 62 positive resist in the range of exposure doses $0.5 \div 14.5$ μC/cm^2.

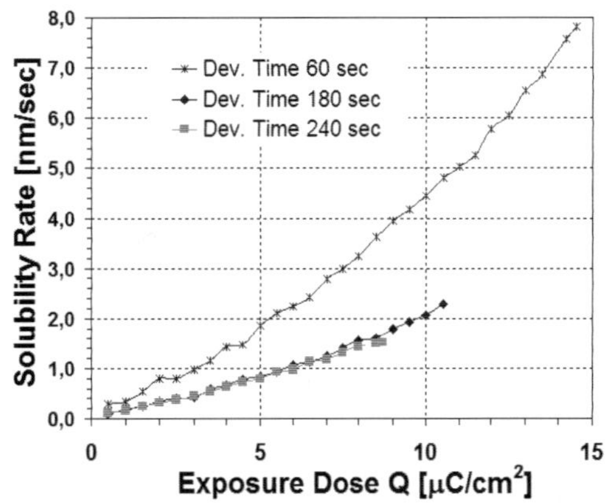

Figure 4b. Solubility rates at various electron exposure doses for AR 600-548 developer; development time 60 s, 180 s, 240 s.

2.4. Application

The simulation tools employed and the evaluation of the important resist characteristics (solubility rate, proximity parameters along the resist depth) are convenient to use to predict and control the image dimensions when the AR-P 6200 chemically semi-amplified resist is applied to nano-patterning. Figure 5 shows details of large-area line gratings for optoelectronics application. The exposed Line/Space ratio was 100/400 nm, 100/500 nm and 100/700 nm. After exposure and development, this ratio was found to be 250 nm for an exposure dose of 60 μC/cm^2 (figure 5a), 300 nm for 90 μC/cm^2 (figure 5b), and 400 nm for 130 μC/cm^2 (figure 5c). Also, the sidewalls were vertical in all cases.

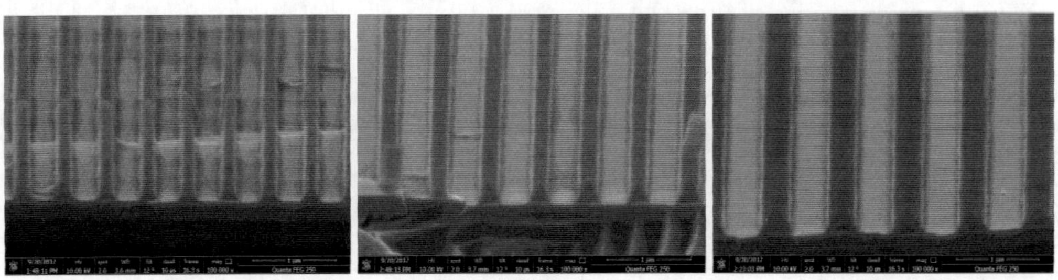

a) b) c)

Figure 5. Gratings in a resist with various L/S: a) L/S = 250/250 nm, Q = 60 μC/cm^2, b) L/S = 300/300 nm, Q = 90 μC/cm^2, c) L/S = 400/400 nm, Q = 130 μC/cm^2. Exposed linewidth is 100 nm in all cases.

3. Conclusions

The results presented of experimental and theoretical studies of the characteristics of the AR-P 6200 (CSAR 62) positive e-beam resist and of the various process parameters (exposure dose, development time, line width etc.) determining the accuracy of the developed resist profiles and critical structural dimensions can be used to improve the resolution of the nano-dimensioned EBL in nano-lithography applications for fabrication of nano-devices. The study of the AR-P 6200 (CSAR 62) resist profiles was performed using variably-shaped electron-beam lithography with electron energy of 40 keV. The solubility rate at different e-beam exposure doses was investigated and evaluated. The AR-P 6200 (CSAR 62) resist exhibited a nonlinear development behavior with the AR 600-548 developer. Profiles with nearly vertical side walls were obtained; possible applications of variably-shaped e-beam processing are also presented.

Acknowledgments

This work was supported by the Bulgarian National Science Fund under contract DNTS/Slovakia 01/1/2016 (project NTS/Slovakia 01/25-13), by the Slovak Research and Development Agency under contracts APVV SK-BG-2013/0030, and by the Ministry of Education of the Slovak Republic and the Slovak Academy of Sciences under contract VEGA 2/0134/15. The joint research project SAS-BAS 2015-2017 and the Agreement for Academic Cooperation and Exchange between the Hiroshima Institute of Technology and the Institute of Electronics BAS are acknowledged. The authors acknowledge the Institute of Electrical Engineering SAS for making available the imaging and measurement facilities.

References

[1] Ito H and Willson C G 1983 *Polym. Eng. Sci.* **23** 1012
[2] Kaimoto Y 1992 *Proc. SPIE* **1672** 66
[3] Allen R D, Wan I Y, Wallraff R A, Dipietro R A and Hofer D C 1995 *J. Photopolymer Sci. Technol.* **8** 623
[4] Naulleau P P, Anderson Ch N, Baclea-an L M, Denham P, George S, Goldberg K A, Jones G, McClinton B, Miyakawa R, Rekawa S and Smith N 2011 *Proc. SPIE* **7972** 797202-1
[5] Liu H, de Grandpre M and Feely *W E* 1988 *J. Vac. Sci. Technol.* **B6** 379
[6] www.allresist.com/csar-62-ar-p-6200
[7] Kostic I, Vutova K, Bencurova A, Ritomsky A and Andok R 2017 *IEEE Xplore 40th Int. Spring Seminar on Electronics Technol. (ISSE)* 1-6
[8] Thoms S and Macintyre D S 2014 *J. Vacuum Sci. & Technol.* B **32** 06FJ01-1-7
[9] Kaganskiy A, Heuser T, Schmidt R, Rodt S and Reitzenstein S 2016 *J. Vacuum Sci. & Technol.* B **34** 061603-1-5
[10] Vutova K, Koleva E, Mladenov G, Kostic I and Tanaka T 2010 *Microelectr. Eng.* **87** 1108-11
[11] Vutova K and Mladenov G 2010 Computer simulation of processes at electron and ion beam lithography *Lithography* ed M Wang (Vukovar InTech) part 1 chapter **17** 319-50

IOP Conf. Series: Journal of Physics: Conf. Series **992** (2018) 012058 doi:10.1088/1742-6596/992/1/012058

Hexaferrite multiferroics: from bulk to thick films

T Koutzarova[1,3], Ch Ghelev[1], P Peneva[1], B Georgieva[1], S Kolev[1], B Vertruyen[2] and R Closset[2]

[1]E. Djakov Institute of Electronics, Bulgarian Academy of Sciences,
72 Tzarigradsko Chaussee, 1784 Sofia, Bulgaria
[2]LSIC, Chemistry Department B6, University of Liege,
Sart Tilman, B-4000 Liege, Belgium

E-mail: tanya@ie.bas.bg, tatyana_koutzarova@ie.bas.bg

Abstract. We report studies of the structural and microstructural properties of $Sr_3Co_2Fe_{24}O_{41}$ in bulk form and as thick films. The precursor powders for the bulk form were prepared following the sol-gel auto-combustion method. The prepared pellets were synthesized at 1200 °C to produce $Sr_3Co_2Fe_{24}O_{41}$. The XRD spectra of the bulks showed the characteristic peaks corresponding to the Z-type hexaferrite structure as a main phase and second phases of $CoFe_2O_4$ and $Sr_3Fe_2O_{7-x}$. The microstructure analysis of the cross-section of the bulk pellets revealed a hexagonal sheet structure. Large areas were observed of packages of hexagonal sheets where the separate hexagonal particles were ordered along the *c* axis. $Sr_3Co_2Fe_{24}O_{41}$ thick films were deposited from a suspension containing the $Sr_3Co_2Fe_{24}O_{41}$ powder. The microstructural analysis of the thick films showed that the particles had the perfect hexagonal shape typical for hexaferrites.

1. Introduction

Multiferroics are multifunctional materials where two or more of the primary ferroic properties (ferromagnetism, ferroelectricity, ferroelasticity, ferrotoroidicity) coexist [1]. Although the ferroelectric behavior and ferromagnetism are, in general, mutually exclusive, there exist a few specific materials where these characteristics are present simultaneously, namely, yttrium-iron garnets, borates, rare-earth ferrites and manganites with a perovskite structure. In all of them, one can observe a magnetoelectric (ME) effect, whereby the magnetic phase can be controlled by applying an external electric field and, vice versa, the electric phase behavior can be influenced by an external magnetic field. The magnetoelectric effect can be exploited in both novel and in well-known devices, e.g., four-stage logic memory devices, electrical-field-controlled spintronic devices, electrical and magnetic field sensors in biomedicine, electric power generators, electrical-field-controlled microwave components, such as filters and switches for wireless technologies at up to terahertz frequencies, the latter area being essentially unexplored, etc. [2-6]. The main requirement to the applications of the magneto-electric multiferroic materials is that the magneto-electric coupling be both large and active at room temperature and the magnetic ordering temperature be high. For the first time, the existence of a significant ME effect in hexaferrites at room temperature was reported by Kitagawa et al. [7], namely,

[3] To whom any correspondence should be addressed.

the Z-type hexaferrite $Sr_3Co_2Fe_{24}O_{41}$ exhibited a low-field (~ 10 mT) magnetoelectric effect at room temperature.

The Z-type hexaferrites have a very complex crystal structure, being a combination of M-type and Y-type hexaferrites [8]. The unit cell consists of 140 atoms, distributed in S, R and T-blocks, which results in a crystal cell with an especially long *c* length. The extremely large unit cell is one of the reasons leading to difficulties in preparing this type of hexaferrites. Another reason is the small difference in the temperatures of phase transitions from one type of hexaferrite to another, which additionally hampers the preparation of single-phase samples, so that secondary phases of W and U-hexaferrites [9] often remain in the end product. This is why one needs techniques different from solid-state reactions, such as those of wet chemistry, to obtain samples with homogeneous composition. We will describe below our studies on the microstructural properties of $Sr_3Co_2Fe_{24}O_{41}$ in bulk form or as powders prepared by sol-gel auto-combustion. Often, ferrite powders are used for depositing thick or thin films with the primary materials being suspensions or pastes containing hexaferrite particles. Regardless of the technique for preparation used, the films' properties depend largely on the microstructural characteristics of the primary bulk materials. Thus, the hexaferrite particles should be homogeneous in shape and size to guarantee their fine dispersion and, hence, the smoothness of the coating formed. We, therefore, focused our attention on studying the structural and microstructural characteristics of the $Sr_3Co_2Fe_{24}O_{41}$ powders and bulk materials prepared by us. We will also present microstructural characteristics of $Sr_3Co_2Fe_{24}O_{41}$ thick film deposited from a suspension.

2. Experimental

The $Sr_3Co_2Fe_{24}O_{41}$ powders were synthesized by citric-acid sol-gel auto-combustion. The corresponding metal nitrates were used as starting materials. A citric acid solution was slowly added to the mixed solution of nitrates as a chelator. The solution was slowly evaporated to form a gel. This gel was dehydrated at 120 °C to obtain the strontium-cobalt-iron citrate precursor. During the dehydration process, the gel turned into a fluffy mass and was burnt in a self-propagating combustion manner. During auto-combustion, the burning gel volume expanded rapidly. The auto-combusted powder was annealed at 600 °C for five hours. The precursor powder obtained was pressed into disk-shaped pellets and synthesized at 1200 °C for seven hours and then quenched rapidly to room temperature.

The thick film was deposited from a suspension containing the $Sr_3Co_2Fe_{24}O_{41}$. To prepare the suspension, the bulk samples were crushed and milled in a ball mill with agate mortar. The suspension was made by dispersing hexaferrite particles into a polyvinyl alcohol solution. The prepared suspension was dropped on the Al_2O_3 substrate. The as-deposited sample was slowly dried and annealed at 1200°C for five hours.

The samples were characterized using X-ray diffraction analysis with Cu-K_α radiation. The microstructures of the samples were observed by scanning electron microscopy (Philips ESEM XL30 FEG).

3. Results and discussion

Unlike the preparation of pure $BaFe_{12}O_{19}$ (M-type hexaferrite) using the citrate precursor technique, the Z-type phase powder cannot be produced as easily due to the complexity of its structure, which imposes the need of progressive transformation through intermediate ferrites before achieving the final structure required. We present data from the XRD analysis at different stages of the synthesis process. Figure 1 shows XRD spectra of powders after the sol-gel auto-combustion, of the powder synthesized at 600 °C, and of a bulk sample annealed at 1200 °C. The XRD spectra of the powder prepared by sol-gel auto-combustion show that the samples are amorphous, with only some low-intensity peaks of the $CoFe_2O_4$ spinel ferrite detected. Their intensity rises after the heat treatment at 600 °C due to the higher crystallization degree of this spinel phase and the growth of the particles' size (figure 1 b). The XRD analysis of the bulk sample show the characteristic peaks corresponding to the $Sr_3Co_2Fe_{24}O_{41}$ as a main phase; second phases of not-reacted $CoFe_2O_4$ and $Sr_3Fe_2O_{7-x}$ were also observed. It is well known that the synthesis of Z-type hexaferrites produces other hexaferrites as well. For example, an M-type hexaferrite phase is formed due to the recrystallization process during cooling. To prevent the appearance of an M-type phase in the samples, the pellets was rapidly quenched to room temperature. Thus, we found no traces of other hexaferrite phases after the $Sr_3Co_2Fe_{24}O_{41}$ synthesis.

Scanning electron microscopy was used to examine the morphology of the samples. Figure 2(a) shows a SEM image of the sample immediately after the completion of the auto-combustion process. As seen, the material is amorphous. A typical porous structure is observed formed during the burn process. Some small agglomerated crystalline particles of $CoFe_2O_4$ can also be seen. After heat treatment at 600 °C, the amount of crystalline $CoFe_2O_4$ auto-combusted powder increases. The particles are with a

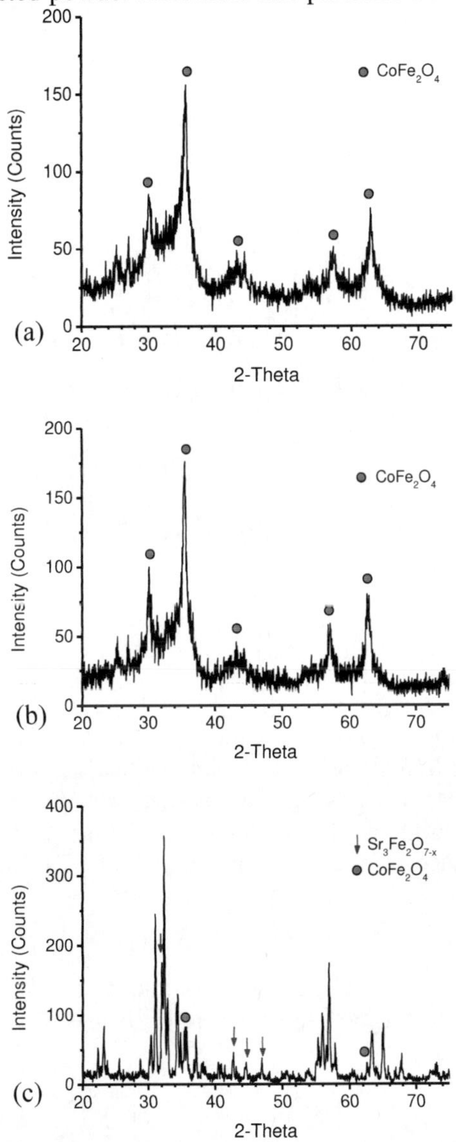

Figure 1. XRD spectra of $Sr_3Co_2Fe_{24}O_{41}$
(a) powder after auto-combustion,
(b) powder annealed at 600°C, and
(c) cross-section of a pellet.

Figure 2. SEM images of $Sr_3Co_2Fe_{24}O_{41}$
(a) powder after auto-combustion,
(b) powder annealed at 600°C, and
(c, d) cross-section of a pellet.

spherical shape and very homogeneous in size (average particle size of 120 nm, figure 2b). The microstructure studies of the bulk pellets' cross-section show a hexagonal sheet structure (figure 2c, d). Large areas are observed of packages of hexagonal sheets where the separate hexagonal particles were ordered along the *c* axis. The difference in these areas' direction of formation has caused the appearance of pores in the pellet volume. The energy-dispersive X-ray analysis (EDX) data of these areas (area 19 on figure 2d) are presented in table 1. It indicates that the pellets are composed of Sr:Co:Fe in a ratio close to the stoichiometric one in $Sr_3Co_2Fe_{24}O_{41}$.

Figure 3 presents SEM images of a $Sr_3Co_2Fe_{24}O_{41}$ thick film. The microstructural analysis of the thick films prepared show a considerable roughness and porosity of their surface. Also, particles can be seen having the perfect hexagonal shape typical for hexaferrites.

Table 1. EDX analysis data of area 19 on figure 2d.

Element	Series	Norm. at. (%)	Stoichio-metric ratio
oxygen	K-series	23.65	30.9
iron	K-series	14.86	25.0
cobalt	K-series	1.18	2.0
strontium	L-series	1.91	3.2

Figure 3. SEM images of $Sr_3Co_2Fe_{24}O_{41}$ thick films.

4. Conclusions

$Sr_3Co_2Fe_{24}O_{41}$ was synthesized by means of the auto-combustion method. Second phases of $CoFe_2O_4$ and $Sr_3Fe_2O_{7-x}$ were also observed. In contrast with the other techniques of Z-type hexaferrites preparation, no other hexaferrite phases were detected. Without applying an external magnetic field during pressing, the bulk samples exhibited large areas of texturing. The bulk samples were denser, which makes them suitable for use as targets in sputtering and pulsed laser deposition (PLD) of thin films. The thick films obtained by deposition from a suspension containing the $Sr_3Co_2Fe_{24}O_{41}$ particles were rough.

Acknowledgments

The work was supported by the Bulgarian National Science Fund under contract DN 08/4 "Novel functional ferrites-based magneto-electric structures" and a joint research project between Bulgarian Academy of Sciences and WBI, Belgium.

References

[1] Schmid H 2004 *Magnetoelectric Interaction Phenomena in Crystals* eds Fiebig M, Eremenko and Vand Chupis I E (Dordrecht: Kluwer) p 35
[2] Ebnabbasi K, Chen Y, Geiler A, Harris V and Vittoria C 2012 *J. Appl. Phys.* **111** 07C719.
[3] Martin L W 2010 *Dalton Trans.* **39** 10813
[4] Petrov R V, Murthy D V B, Sreenivasulu G and Srinivasan G 2013 *Microwave Opt. Technol. Lett.* **55** 533

[5] Rao C N R, Sundaresan A and Saha R 2012 *J. Phys. Chem. Lett.* **3** 2237–46
[6] Scott J F 2007 *Nat. Mater.* **6** 256
[7] Kitagawa Y, Hiraoka Y, Honda T, Ishikura T, Nakamura H and Kimura T 2010 *Nat. Mater.* **9** 797
[8] Mu Ch, Song Y, Wang L and Zhang H 2011 *J. Appl. Phys.* **109** 123925
[9] Raju N, Shravan Kumar Reddy S, Ramesh J, Gopal Reddy Ch, Yadagiri Reddy P, Rama Reddy K, Sathe V G and Raghavendra Reddy V 2016 *J. Appl. Phys.* **120** 054103

IOP Conf. Series: Journal of Physics: Conf. Series **992** (2018) 012059 doi:10.1088/1742-6596/992/1/012059

High-dose MeV electron irradiation of Si-SiO_2 structures implanted with high doses Si^+

S Kaschieva[1,4], Ch Angelov[2] and S N Dmitriev[3]

[1]G. Nadjakov Institute of Solid State Physics, Bulgarian Academy of Sciences, 72 Tzarigradsko Chaussee, 1784 Sofia, Bulgaria
[2]Institute for Nuclear Research and Nuclear Energy, Bulgarian Academy of Sciences, 72 Tzarigradsko Chaussee, 1784 Sofia, Bulgaria
[3]Flerov Laboratory of Nuclear Reactions, Joint Institute of Nuclear Research, 141980 Dubna, Moscow Region, Russia

E-mail: kaschiev@issp.bas.bg

Abstract. The influence was studied of 22-MeV electron irradiation on Si-SiO_2 structures implanted with high-fluence Si^+ ions. Our earlier works demonstrated that Si redistribution is observed in Si^+-ion-implanted Si-SiO_2 structures (after MeV electron irradiation) only in the case when ion implantation is carried out with a higher fluence (10^{16} cm^{-2}). We focused our attention on the interaction of high-dose MeV electron irradiation (6.0×10^{16} cm^{-2}) with n-Si-SiO_2 structures implanted with Si^+ ions (fluence 5.4×10^{16} cm^{-2} of the same order magnitude). The redistribution of both oxygen and silicon atoms in the implanted Si-SiO_2 samples after MeV electron irradiation was studied by Rutherford back-scattering (RBS) spectroscopy in combination with a channeling technique (RBS/C). Our results demonstrated that the redistribution of oxygen and silicon atoms in the implanted samples reaches saturation after these high doses of MeV electron irradiation. The transformation of amorphous SiO_2 surface into crystalline Si nanostructures (after MeV electron irradiation) was evidenced by atomic force microscopy (AFM). Silicon nanocrystals are formed on the SiO_2 surface after MeV electron irradiation. The shape and number of the Si nanocrystals on the SiO_2 surface depend on the MeV electron irradiation, while their size increases with the dose. The mean Si nanocrystals height is 16-20 nm after irradiation with MeV electrons at the dose of 6.0×10^{16} cm^{-2}.

1. Introduction

The influence of high-energy electron irradiation on high-fluence ion-implanted Si-SiO_2 structures was studied by Klimenkov et al. The authors showed that high-energy electron irradiation can cause moving of clusters or atoms on the surface [1], and that, as an alternative to the heat treatment, the cluster formation can be stimulated by high-energy electron irradiation [2]. The behavior of a silicon solar cell before and after 1 MeV electron irradiation with 10^{14} cm^{-2}, 10^{15} cm^{-2} and 10^{16} cm^{-2} has also been studied. As the dose was raised, a slight increase was observed in the short circuit current, accompanied by a decrease in the maximum output power [3]. Previously, we compared the interface states generated by 20 MeV electrons in n- and p-type Si-SiO_2 structures (with an equal amount of

[4] To whom any correspondence should be addressed.

IOP Conf. Series: Journal of Physics: Conf. Series **992** (2018) 012059 doi:10.1088/1742-6596/992/1/012059

oxide). We demonstrated that most of the radiation defects created could be related to the main impurities in the Si substrate – phosphorus or boron vacancy complexes – for the structures based on n- and p-type Si wafers, respectively [4]. In another study, we also showed that the total concentration of electron irradiation-induced defects at the Si-SiO_2 interface of the structures is larger in the case of an n-type Si wafer [5]. Our study on MeV electron irradiation of the implanted Si-SiO_2 structures (with different fluence of Si^+ ions) demonstrated redistribution of silicon (after MeV electron irradiation) only in the case when the ion implantation is carried out with a higher (10^{16} cm^{-2}) fluence [6]. The aim of the present work was to study the influence of high-dose MeV electron irradiation (6.0×10^{16} cm^{-2}) on n-Si-SiO_2 structures implanted with Si ions with fluence of the same order of magnitude (5.4×10^{16} cm^{-2}). The redistribution of both oxygen and silicon atoms in Si^+ ion implanted Si-SiO_2 samples after 22-MeV electron irradiation was studied by Rutherford back-scattering spectroscopy. A transformation was observed of the amorphous SiO_2 surface of the implanted samples into crystalline Si nanostructures as a result of the MeV electron irradiation.

2. Experimental details

Si-SiO_2 structures were prepared on n-type Si <100> wafers with a resistivity of 6.2 Ω cm. A SiO_2 layer of 22-nm thickness was grown in dry ambient of O_2 + 6% HCl at 950 °C. The samples were then cooled in the same ambient at a rate of 1°C s^{-1}. The oxide thickness was determined by ellipsometry. Silicon ions with an energy of 12 keV and a dose of 5.4×10^{16} cm^{-2} were implanted through the SiO_2 layer of the samples. The values chosen of the oxide thickness and silicon-ion energy ensured that the maximum impact of the implantation damage should take place in the Si bulk wafer very close to the Si-SiO_2 interface.

The non-implanted and Si^+-implanted Si-SiO_2 samples were irradiated by 22-MeV electrons at a dose of 6.0×10^{16} cm^{-2}. The irradiation was carried out on the MT-25 Microtron in the Flerov Laboratory of Nuclear Reactions of the Joint Institute for Nuclear Research, (FLNR, JINR) Dubna, Russia. The distance between the Microtron's window and the samples was 150 mm. The irradiation was performed under a pressure of about 1×10^2 Pa. The sample temperature was controlled during the entire process of electron irradiation and kept close to room temperature.

3. Results and discussion

Figure 1 and figure 2, respectively, present the results of the RBS and AFM measurements of Si-SiO_2 structures implanted with 12-keV Si^+ ions (fluence 5.4×10^{16} cm^{-2}) before and after 22-MeV electron irradiation (6.0×10^{16} cm^{-2}). The RBS was used to determine accurately the impurity distribution and elemental areal density in the thin SiO_2 film. The RBS/C spectra demonstrated an additional widening and increasing of the main silicon peak and a decrease of the oxygen peak height (figure 1 a) after MeV electron irradiation. This means that the MeV electron irradiation increases the silicon concentration and decreases the oxygen concentration at the Si-SiO_2 sample surface. We assume that Si atoms in the wafers are displaced from their normal position in the

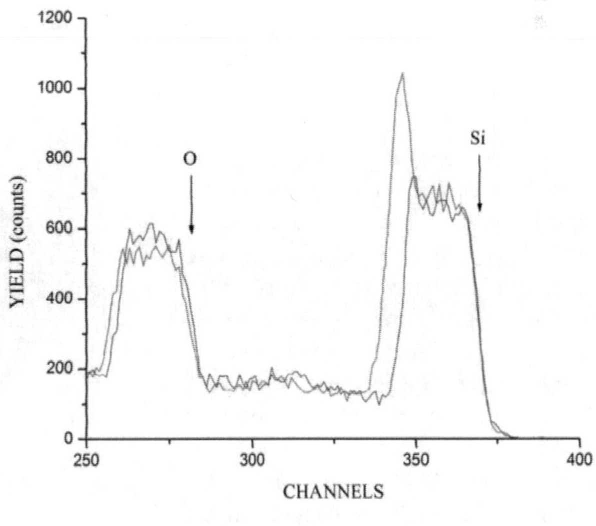

Figure 1 a. RBS/C spectra of Si-SiO_2 structures implanted with 12-keV Si^+ ions (dose of 5.4×10^{16} cm^{-2}) (blue curve) and after 22-MeV electron irradiation (dose of 6.0×10^{16} cm^{-2} (red curve).

semiconductor lattice during MeV electron irradiation, so that the formation takes place of vacancies and interstitial atoms. The implanted Si+ ions and the displaced oxygen and silicon atoms are able to migrate; we observed their redistribution resulting from the MeV electron irradiation. In addition, the radiation defects already created by the Si+ ion implantation interact with those subsequently generated by MeV electrons, thus creating suitable conditions for stimulation of oxygen radiation-enhanced diffusion in the samples. As the free oxygen is mobile, its migration through the oxide is observed. The oxygen concentration decreases, but the silicon concentration increases in the electron irradiated structure, as can be seen in figure 1 a.

The random RBS spectrum demonstrates higher silicon concentration in the amorphous part of the sample (SiO_2). The free oxygen (generated by MeV electrons) penetrates through the oxide and an additional amorphous silicon layer is created on the SiO_2 surface. Part of the Si in the amorphous SiO_2 surface of the implanted samples is transformed into crystalline Si nanostructures as a result of the MeV electron irradiation (figure 1 b).

The formation of nanostructures in the SiO_2 was studied by atomic force microscopy. The microscopic morphology of the SiO_2 surface structure implanted with Si^+ ions with a dose of 5.4×10^{16} cm^{-2} is presented in figure 2 a. The changes resulting from irradiating the Si-SiO_2 structure with 22-MeV electrons with a dose of 6.0×10^{16} cm^{-2} are illustrated in figure 2 b. The AFM images were taken over areas of size 1×1 μm (figure 2 a) and 4.6×4.5 μm (figure 2 b) on the surface of the SiO_2-Si structures in air.

The Si^+ ion implantation (5.4×10^{16} cm^{-2}) generates amorphous Si precipitations

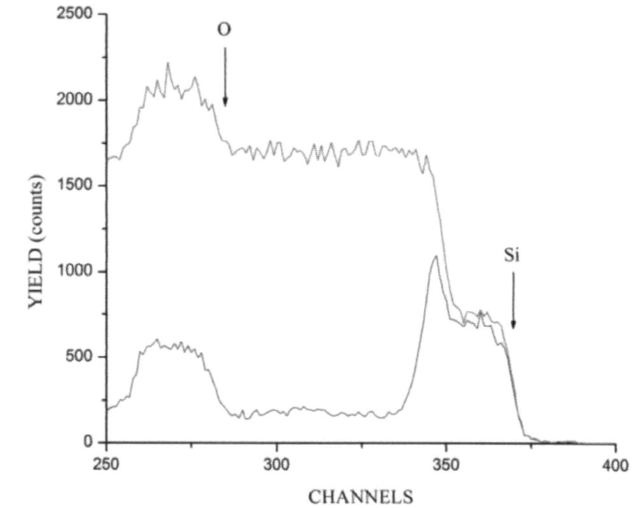

Figure 1 b. Random (red curve) and channelling spectra (blue curve) of the Si-SiO_2 structures implanted with 12-keV Si^+ ions (dose of 5.4×10^{16} cm^{-2}) and then irradiated by 22-MeV electrons with a dose of 6.0×10^{16} cm^{-2}.

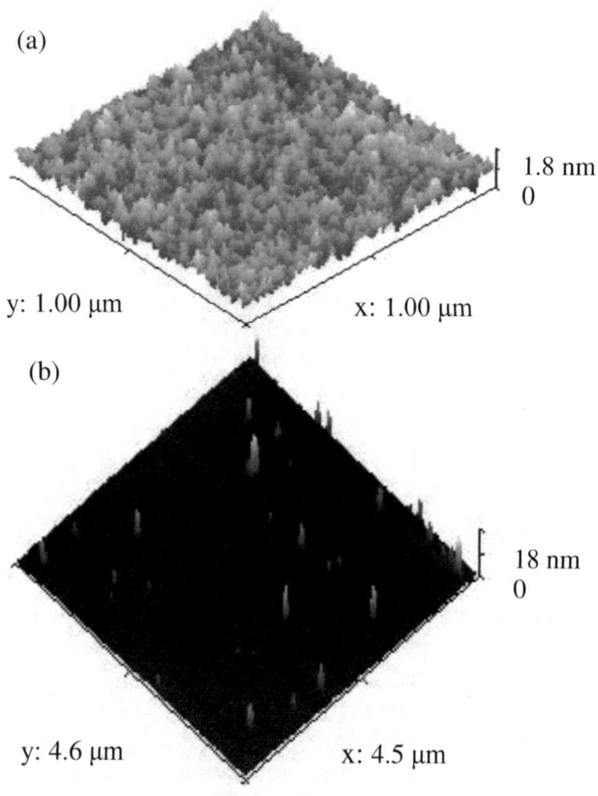

Figure 2. AFM surface morphology of Si-SiO_2 sample implanted with 12-keV Si^+ ions (dose of 5.4×10^{16} cm^{-2}): before (a) and after (b) 22-MeV electron irradiation (dose of 6.0×10^{16} cm^{-2}).

IOP Conf. Series: Journal of Physics: Conf. Series **992** (2018) 012059 doi:10.1088/1742-6596/992/1/012059

in the SiO_2 matrix, as can be seen in figure 2 a. It can be assumed that it also rearranges the Si-O-Si bonds and breaks many of the Si-O bonds. In addition, the radiation defects generated during the Si-ion implantation (whose maximum is located in the Si wafer close to the Si-SiO_2 interface) create a thin layer of amorphous Si at the silicon surface. The following 22-MeV electron irradiation ($6.0\times10^{16}cm^{-2}$) of the Si-implanted samples regroups these Si precipitations, so that new silicon nanostructures appear in the SiO_2 (figure 2 b). We assume that these new nanostructures are Si nanocrystals [7]. The number (per cm^2) of the nanocrystals generated is small; their mean height is determined at 16-20 nm.

The crystal height depends also on the MeV electron dose irradiation [8]. The silicon nanocrystals size increases with the MeV electrons dose. The Si nanocrystals' size reaches their maximum increase after MeV electron irradiation with a dose of $6.0\times10^{16}cm^{-2}$. This has to do with the limited number of Si ions and atoms able to migrate through the radiation defects generated by the MeV-electron irradiation.

4. Conclusions

We explored the effect of 22-MeV electron irradiation (dose of 6.0×10^{16} cm^{-2}) on n-Si-SiO_2 structures implanted by Si^+ ions (dose of 5.4×10^{16} cm^{-2}) and followed the oxygen and silicon atoms' redistribution in the implanted Si-SiO_2 samples after the MeV-electron irradiation. The amount of Si in the structures remained the same, while O (being mobile) diffused through the radiation defects generated in the structures by the MeV electrons (in particular in the oxide). We further analyzed the redistribution of both the oxygen and silicon atoms in the implanted Si-SiO_2 samples after MeV-electron irradiation and observed a partial transformation of the amorphous SiO_2 surface of the implanted samples into crystalline Si nanostructures. The mean height of the Si nanocrystal was determined to be 16-20 nm.

Acknowledgement

This study was supported by Grant No. 22 - 15.01.2016, p. 16 of JINR, Dubna, Russia. The authors appreciate the assistance of Dr. A.G. Belov of FLNR, JINR Dubna, Russia, for the MeV electron irradiation of the samples.

References

[1] Nepijko S, Klimenkov M, Kuhlenbeck H and Freund H 1999 *Vac. Sci. Technol.* A **17** 577
[2] Klimenkov M, Matz M and Borany J 2000 *Nucl. Instr. & Meth.*B **168** 367
[3] Babaee S and Ghozati S B 2017 *Radiation Phys. and Chem.* **141** 98-102
[4] Stefanov K, Kaschieva S and Karpusov D 1998 *Vacuum* **51**/2 235
[5] Kaschieva S, Halova E, Vlaikova E, Alexandrova S, Valcheva E and Dmitriev S N 2006 *Plasma Process. & Polym.* **3** 237-40
[6] Kaschieva S, Angelov Ch, Dmitriev S N and Tsutsumanova G 2006 *Plasma Process. & Polym.* **3** 233-6
[7] Kaschieva S, Gushterov A, Apostolova T and Dmitriev S N 2010 *Proc. 6th Int. Conf. Beam Technol. and Laser Appl.* p 75
[8] Kaschieva S, Gushterov A, Gushterova P and Dmitriev S N 2010 *J. Phys.: Conf. Ser.* **223** 012031

IOP Conf. Series: Journal of Physics: Conf. Series **992** (2018) 012060 doi:10.1088/1742-6596/992/1/012060

Clinical results from low-level laser therapy in patients with autosomal dominant cone-rod dystrophy

K Koev[1,3], L Avramov [2] and E Borissova[2]

[1]Department of Ophthalmology, Medical University-Sofia,
8 Bialo More Str., 1000 Sofia, Bulgaria
[2]E. Djakov Institute of Electronics, Bulgarian Academy of Sciences,
72 Tzarigradsko Chaussee, 1784 Sofia, Bulgaria

E-mail: k00007@abv.bg

Abstract. The objective of this study is to examine long-term effects of low-level laser therapy (LLLT) in patients with autosomal dominant cone-rod dystrophy (CRDs). A He-Ne Laser with continuous emission at 633 nm (01 mW/cm^2) was used on five patients with autosomal dominant pedigree of Romani origin with non-syndromic CRDs. The laser radiation was applied transpupillary to the macula six times for three minutes every other day. The experiment was conducted for a period of three years. The clinical evaluation included best corrected visual acuity determination, funduscopy, Humphrey perimetry, Farnsworth Hue-28 color testing, fluorescein angiography, and full-field electroretinogram (ERG).

All affected individuals presented reduced visual acuity (0.01 – 0.4) and photophobia. The funduscopic examination and fluorescein angiography revealed advanced changes including bone spicule-like pigment deposits in the midperiphery and the macular area, along with retinal atrophy, narrowing of the vessels, and waxy optic discs. The visual fields demonstrated central scotoma. The electrophysiologic examination of the patients detected an abnormal cone-rod ERG (20 – 30 μV) with photopic amplitudes more markedly reduced than the scotopic. Flicker responses were missing and Farnsworth Hue-28 test found protanopia. There was a statistically significant increase in the visual acuity ($p<0.001$, end of study versus baseline) for CRDs patients for the period of three years after the treatment with LLLT. Following the LLLT, the central absolute scotoma in CRDs was reduced, as was the prevalence of metamorphopsia in CRDs.

This study shows that LLLT may prove be a novel long-lasting therapeutic option for both forms of CRDs. It is a highly effective treatment resulting in a long-term improvement of the visual acuity.

1. Introduction

Cone rod dystrophies (CRDs) is characterized by a loss of cone cells, the photoreceptors responsible for both central and color vision [1]. The cones and rods transform light into electric nerve messages that are transferred to our brain via our optic nerve [2]. CRDs (prevalence 1/40,000) are inherited retinal dystrophies that belong to the group of pigmentary retinopathies [3]. There are more than 30 types of cone-rod dystrophy, which are distinguished by their genetic cause and their pattern of

[3] To whom any correspondence should be addressed.

inheritance: autosomal recessive, autosomal dominant, and X-linked. CRDs are characterized by retinal pigment deposits, predominantly localized in the macular region [4, 5]. There is no one effective treatment for the different types of CRD. However, there may be treatment options that may help slow down the degenerative process, such as light avoidance and the use of low-vision aids. Dry AMD is characterized by drusen, retinal pigment epithelial (RPE) cell atrophy and subjacent photoreceptor degeneration [6]. Factors involved in causing RPE cell injury and dysfunction have been shown to include mitochondrial dysfunction, oxidative stress, inflammation and genetic disposition [7].

Photobiomodulation (PBM) was discovered almost 50 years ago by Endre Mester in Hungary [8]. For most of this time PBM was known as "low-level laser therapy" as ruby laser (694nm) and HeNe lasers (633nm) were the first devices used. Low-level laser therapy (LLLT) produces significant bioeffects [9-12] These effects are manifested in biochemical, physiological and proliferative phenomena in various enzymes, cells, tissues, organs and organisms [13, 14]. Published studies demonstrate that mitochondrial cytochrome C oxidase (CCO) is a key photoacceptor of light at these wavelengths and improves blood flow and ATP formation, enhances O_2 binding and reduces oxidative stress and inflammation [15, 16].

Although early studies identified mitochondrial CCO as an endogenous photoacceptor for PBM, the cellular and molecular mechanisms underlying PBM are better understood [17]. Recent findings provide important new insight. First, nitric oxide has been implicated. Second, CCO, an enzyme known to reduce oxygen to water at the end of the mitochondrial respiratory chain, has been shown to have a new enzymatic activity – the reduction of nitrite to nitric oxide [18]. This nitrite reductase activity is elevated under hypoxic conditions but also occurs under normoxia. And third, low-intensity light regulates nitric oxide synthesis by CCO without altering its ability to reduce oxygen.

The first signs and symptoms of cone-rod dystrophy are usually decreased sharpness of visual acuity and photophobia [2]. These features are typically followed by impaired color vision (dyschromatopsia), scotomas in the center of the visual field, and peripheral vision loss [3]. Photobiomodulation (PBM) utilizes very low energy levels causing no tissue damage [14]. The local reduction of edema, and reductions in markers of oxidative stress and pro-inflammatory cytokines are well established after LLLT [18]. PBM stimulates cellular processes that provide an approach to target the underlying degenerative pathology with disease-modifying potential [19]. A recent review paper on PBM in retinal diseases has reported that the low-cost and non-invasive nature of PBM coupled with the first promising clinical reports and the numerous preclinical studies in animal models make PBM well poised to become an important player in the treatment of retinal disorders [20]. To the best of our knowledge, no reports have so far been published dealing with LLLT performed in patients with CRDs. Ivandic demonstrates visual acuity improvement in a patient with retinitis pigmentosa after LLLT application [7].

PBM, also known as low-level laser therapy, uses red and near-infrared light to stimulate healing, relieve pain, and reduce inflammation [14, 22, 23]. The primary chromophores identified were cytochrome c oxidase in mitochondria, and calcium ion channels (possibly mediated by light absorption by opsins). Secondary effects of photon absorption include increases in ATP, a brief burst of reactive oxygen species, an increase in nitric oxide, and modulation of calcium levels. Tertiary effects include activation of a wide range of transcription factors leading to improved cell survival, increased proliferation and migration, and new protein synthesis. Many wavelengths in the red (600–700 nm) and near-infrared (NIR, 770–1200 nm) spectral regions have shown positive results; however there is a region in between (700–770 nm) where, broadly speaking, the results are likely to be disappointing [24]. Some studies have looked at animal models of neuropathic pain such as the "spared nerve injury" [25]. This involves ligating two out of three branches of the sciatic nerve in rats and causes long lasting (>6 months) mechanical allodynia [27].

Some researchers [26] have used a transgenic mouse strain (FVB/N-Tg(iNOS-luc) that had been engineered to express luciferase under control of the inducible nitric oxide synthase promoter, to allow bioluminescence imaging of PBM of the zymosal-induced arthritis model in mice knees [26]. They compared the same fluence at 635 nm, 660 nm, 690 nm, and 905 nm (in CW mode), and 905 nm (short pulse). Animals younger than 15 weeks showed mostly reduction of iNOS expression, while older animals showed increased iNOS expression. Pulsed 905 nm also increased iNOS expression. In recent years, the use of PBM as a treatment for traumatic brain injury, and other brain disorders including stroke, neurodegenerative diseases and even psychiatric disorders, has increased remarkably [25, 27-29].

Photoneuromodulation of cytochrome oxidase activity is the most important primary mechanism of action of LLLT [12]. Cytochrome oxidase is the primary photoacceptor of light in the red to near-infrared region of the electromagnetic spectrum. It is also a key mitochondrial enzyme for cellular bioenergetics, especially for nerve cells in the retina and the brain. Evidence shows that LLLT can secondarily enhance neural metabolism by regulating mitochondrial function, intraneuronal signaling systems, and redox states.

The aim of this study was to examine long-term effects of low-level laser therapy (LLLT) in patients with cone-rod dystrophy.

2. Methods

For LLLT, a He-Ne Laser with continuous emission at 633 nm (0.1 mW/cm2) was used in five patients with autosomal dominant pedigree of Romani origin with non-syndromic CRDs. A new ophthalmologic system for bio-stimulation and LLLT of eye diseases based on a He-Ne laser (Mediray 04, Optella Ltd., Sofia, Bulgaria, developed by K. Koev, V. Tanev, L. Avramov) with emission wavelength of 632 nm was used for the investigation,. The system has the ability to regulate the size of the laser spot and the laser power density from 0,05 mW/cm^2 to 0,4 mW/cm^2. The system is compact, portable and with minimal optical losses and high reliability. The device is convenient for exploitation, both for the patients and the treating personnel.

Laser radiation was applied transpupillary to the macula for six times for three minutes every other day. The experiment lasted for a period of three years. The clinical evaluation included best corrected visual acuity determination, funduscopy, Humphrey perimetry, Farnsworth Hue-28 color testing, fluorescein angiography, and full-field electroretinogram (ERG).

3. Results

All individuals affected with CRDs presented reduced visual acuity (0.01-0.4) and slightly variable photophobia. Funduscopic examination and fluorescein angiography revealed advanced changes, including bone spicule-like pigment deposits in the midperiphery and macular area along with retinal atrophy, narrowing of the vessels, and waxy optic discs. Using Humphrey perimetry, central scotoma of the visual fields was established. The electrophysiologic examination (ERG) of the patients detected an abnormal cone-rod ERG (20-30 μV) with photopic amplitudes more markedly reduced than the scotopic. Flicker responses were missing, and the Farnsworth Hue-28 test found protanopia. After the LLLT treatment, we found a statistically significant increase in the visual acuity ($p<0.001$, end of study versus baseline) for CRDs patients for the period of three years. Following the LLLT, the visual acuity in a larger proportion of patients improved.

Visual acuity improved optotypes in 9/10 eyes (90%; $p<0.001$)

Eyes: by one row of optotypes in 3/10 (30%),

by two rows in 4 /10 (40%),

by three rows in 2/10 (20%),

Visual acuity remained unchanged in 1/10 eyes (1.0%).

The central absolute scotoma in CRDs was reduced after LLLT, as established with Humphrey perimetry and fluorescein angiography The prevalence of metamorphopsia in CRDs was also reduced.

4. Discussion

We report studies on applying LLLT to members of a family of Gypsy origin affected by autosomal dominant cone-rod dystrophy. To the best of our knowledge, this is the first publication of clinical results from the use of LLLT in patients with autosomal dominant CRDs.

The LLLT resulted in an improvement in the visual acuity in most patients (90%) with CRDs. An increase of two to three rows of optotypes was observed in 6/10 (60%) eyes with CRDs.

We also found a reduction in the central absolute scotoma and metamorphopsia in CRDs.

In contrast to laser treatment, PBM utilizes very low energy levels causing no tissue damage [14]. Photobiomodulation stimulates cellular processes, thus providing an approach to targeting the underlying degenerative pathology with disease-modifying potential. A recent review paper of PBM in retinal diseases has reported that the low-cost and non-invasive nature of PBM coupled with the first promising clinical reports and the numerous preclinical studies in animal models make PBM well poised to become an important player in the treatment of retinal disorders [28]. The good clinical results that we have found prove that LLLT can be successfully used in CRDs.

5. Conclusions

This study shows that LLLT may become a novel long-term therapeutic option for both forms of CRDs, as we demonstrated its effectiveness in improving the patients' visual acuity.

References

[1] Nikopoulos K, Farinelli P and Giangreco B, 2016 *Am. J. Hum. Genet.* **99**/3 770-6
[2] Kohl S 2009 *Ophthalmologe* **106** 109-15
[3] Mayer A K, Rohrschneider K and Strom T M 2016 *Eur. J. Hum. Genet.* **24**/3 459-62
[4] Holopigian K, Greenstein V C and Seiple W 2004 *Invest. Ophthalmol. Vis. Sci.* **45** 275-81
[5] Suga A, Mizota A, Kato M and Kuniyoshi K 2016 *Invest. Ophthalmol. Vis. Sci.* **57**/10 4255-63
[6] Ivandic B T and Ivandic T 2008 *Photomed. Laser Surg.* **26**/3 241-5
[7] Ivandic B T and Ivandic T 2014 *Photomed. Laser Surg.* **32**/3 181-4
[8] Anders J J, Lanzafame R J and Arany P R 2015 *Photomed. Laser Surg.* **33** 183–4
[9] Hamblin M R, de Sousa M V P and Agrawal T 2017 *Handbook of Low-level Laser Therapy* (Pan Stanford Publishing, Singapore)
[10] de Freitas L F and Hamblin M R 2016 *IEEE J. Sel. Top Quantum Electron.* **22** 348–64
[11] Huang Y Y, Sharma S K, Carroll J D and Hamblin M R 2011 *Dose Response* **9**/4 602–18
[12] Mason M G, Nicholls P and Cooper C E 2014 *Biochim. Biophys. Acta* **1837** 1882–91
[13] Karu T I, Pyatibrat L V, Kolyakov S F and Afanasyeva N I 2005 *J. Photochem. Photobiol.* B **81** 98–106
[14] Karu T I 2010 *IUBMB Life* **62** 607–10
[15] Wong-Riley M T, Liang H L, Eells J T, Chance B, Henry M M, Buchmann E, Kane M and Whelan H T *J. Biol. Chem.* **280**/6 4761–71
[16] Yamaura M, Yao M, Yaroslavsky I, Cohen R, Smotrich M and Kochevar I E 2009 *Lasers Surg. Med.* **41**/4 282–90
[17] Hwang M H, Shin J H, Kim K S, Yoo Ch M, Jo G E, Kim J H and Choi H 2015 *Photochem. Photobiol.* **91**/2 403–10
[18] Lim W et al 2015 *J. Oral Pathol. Med.* **44**/2 94–102
[19] Choi H et al 2012 *Lasers Med. Sci.* **27** 459–67
[20] Sakurai Y, Yamaguchi M and Abiko Y 2000 *Eur. J. Oral Sci.* **108** 29–34
[21] Nomura K, Yamaguchi M and Abiko Y 2001 *Lasers Med. Sci.* **16** 218–23
[22] Fernandes K P, Souza N H, Mesquita-Ferrari R A, Silva Dde F, Rocha L A, Alves A N, Sousa Kde B, Bussadori S K, Hamblin M R and Nunes F D 2015 *J. Photochem. Photobiol.* B **153** 344–51
[23] Silva I H, de Andrade S C, de Faria A B, Fonseca D D, Gueiros L A, Carvalho A A, da Silva W T, de Castro R M and Leao J C 2016 *Lasers Med. Sci.* **31**/9 1855–62
[24] von Leden R E, Cooney S J, Ferrara T M, Zhao Y, Dalgard C L, Anders J J and Byrnes K R 2013 *Lasers Surg. Med.* **45**/4 253–63

[25] Sousa K B et al 2017 *Lasers Surg. Med.* **49** 36
[26] Ferraresi C, Huang Y Y and Hamblin M R 2016 *J. Biophotonics* **9**/11-12 1273-99
[27] Ferraresi C, de Sousa M V, Huang Y Y, Bagnato V S, Parizotto N A and Hamblin M R 2015 *Lasers Med. Sci.* **30**/4 1259–67
[28] Silveira P C, Scheffer Dda L, Glaser V, Remor A P, Pinho R A, Aguiar Junior A S and Latini A 2016 *Free Radic. Res.* **50**/5 503–13
[29] de Sousa P, Ferraresi C, Kawakubo M, Kaippert B, Yoshimura E M and Hamblin M R 2016 *Neurophotonics* **3**/1 015003
[30] Mitschischek E 2013 *Der Augenspiegel* **3** 20-24

IOP Conf. Series: Journal of Physics: Conf. Series **992** (2018) 012061 doi:10.1088/1742-6596/992/1/012061

Five-year follow-up of low-level laser therapy (LLLT) in patients with age-related macular degeneration (AMD)

K Koev[1,3], L Avramov [2] and E Borissova[2]

[1]Department of Ophthalmology, Medical University-Sofia, 8 Bialo More Str., 1000 Sofia, Bulgaria
[2]E. Djakov Institute of Electronics, Bulgarian Academy of Sciences, 72 Tzarigradsko Chaussee, 1784 Sofia, Bulgaria

E-mail: k00007@abv.bg

Abstract. The objective of this study was to examine long-term effects of low-level laser therapy (LLLT) in patients with age-related macular degeneration (AMD). The research was implemented for a period of five years. For LLLT, a He-Ne Laser with continuous emission at 633 nm (0.1 mW/cm^2) was used in patients with AMD of all stages (dry to wet exudative forms were included). In total, 33 patients (16 men and 17 women – 66 eyes) with AMD of various stages and a mean age of 68.7 ± 4.2 years were included in the study. Progressive, exudative AMD was diagnosed in 8 eyes. 58 eyes had drusen or were depigmented. Laser radiation was applied transpupillary to the macula for six times for three minutes once in two days; 22 patients with AMD (44 eyes) were randomly selected to receive mock treatment (control group 10 men and 12 women with a mean age of 69.3 ± 4.8 years). The visual acuity was followed for a five-year period. The perimetry and Amsler test were used to screen central scotomas. The fluorescein angiography of AMD and the control groups was examined. The visual acuity remained unchanged in all patients in the control group. There was a statistically significant increase in the visual acuity ($p<0.001$, end of study versus baseline) for AMD patients for the period of five years after the treatment. The edema and hemorrhage in the patients with progressive, exudative AMD significantly decreased. No side effects were observed during the therapy. The prevalence of metamorphopsia, scotoma in AMD group was reduced.
In conclusion, this study shows that LLLT may be a novel long-lasting therapeutic option for both forms of AMD. It is a highly-effective treatment that results in a long-term improvement of the visual acuity.

1. Introduction

Age-related macular degeneration (AMD) affects 30–50% of the individuals aged 60 years or older [1, 2]. AMD is diagnosed as either dry (non-neovascular) or wet (neovascular) [3]. Neovascular refers to growth of new blood vessels in an area, such as the macula, where they are not supposed to be any [4]. Macular degeneration mainly affects the central vision, causing "blind spots" directly ahead [5]. The dry form is more common than the wet form, with about 85 to 90 percent of all AMD patients diagnosed with dry AMD [6]. The wet form of the disease usually leads to a more serious vision loss. AMD results from defects in the choriocapillaris, Bruch's membrane, and the retinal pigmented

[3] To whom any correspondence should be addressed.

epithelium (RPE) underneath the macula [7]. The epitheliopathy diminishes the lysosomal activity and phagocytosis of the outer photoreceptors and disrupts the transportation of cell debris through the RPE to the choriocapillaris [2]. Certain types of therapy, including electrical stimulation and laser therapy, have been developed during the recent years to decrease the regenerative process and return the function [8, 9, 10]. Other studies [11, 12, 13] using He-Ne low-energy laser have indicated that it is mainly the laser energy at 633 nm wavelength that affects the healing dynamics, producing changes in the early phase of the repair process, i.e., the inflammatory phase. The incident beam of a helium-neon laser (red light at 632.8 nm) can partially reach into 15 mm in the tissue, causing local vascular dilation and accelerated blood flow. The laser thus plays a role in reducing inflammation, it has an anti-swelling effect and promotes functional recovery. In addition, low-energy helium-neon lasers strengthen the phagocytosis by macrophages and promote the absorption of inflammation factors [12]. Local helium-neon laser therapy contributes to the prevention of the inflammatory reaction and promotes local tissue proliferation and wound healing [13].

Low-level laser therapy (LLLT) is a special type of laser therapy, whereby the irradiation used is red or near infrared with a wavelength of 600 nm – 1100 nm and an output power of 1 mW – 500 mW in a continuous-wave mode, or of low-energy density (0.04–50 J/cm^2) in a pulsed mode [14-17]. LLLT represents a novel therapeutic method that, in contrast with surgical laser applications, does not damage tissues [18, 19].

A major cause of blindness in the Western world is degeneration of photoreceptors as a result of point mutations in genes coding for either phototransduction-related proteins or other proteins important for retinal function. Despite the diversity of mutated genes and proteins involved in this heterogeneous group of progressive retinal dystrophies with homologous phenotypes, the final event leading to blindness is apoptosis of photoreceptors [29]. .

Interleukin 1β (IL-1β), tumor necrotic factor-α (TNF-α), and interferon-γ (IFN-γ) play an important role in inflammation, while platelet-derived growth factor (PDGF), transforming growth factor-β (TGF-β) and blood-derived fibroblast growth factor (bFGF) are the most important growth factors of periodontal tissues. Several authors have investigated the effect of low-level He–Ne laser irradiation on the gene expression of these mediators, e.g., in rats' gingiva and mucosal tissues [30].

Twenty male Wistar rats were randomly assigned into four groups (A_{24}, A_{48}, B_{24}, B_{48}) in which A_{24} and A_{48} were cases and B_{24}, B_{48} were controls. An incision was made on gingiva and mucosa of the labial surface of the rat's mandibular incisors. Group A_{24} was irradiated twice with 24 hours interval, while the inflamed tissues of group A_{48} was irradiated three times with continuous He–Ne laser (632.8 nm) at a dose of 7.5 J/cm2 for 300 s. A cumulative dose of energy of 5.1 J impinged on the 68 mm^2 irradiation zone. The authors found that the gene expression of IL-1β and IFN-γ was significantly inhibited in the test groups ($P < 0.05$), while the gene expression of PDGF and TGF-β were significantly increased ($P < 0.05$). The case and control groups did not have a significant difference in the gene expression of TNF-α and bFGF ($P > 0.05$). These findings suggest that low-level He-Ne laser irradiation decreases the amount of inflammation and accelerates the wound healing process by changing the expression of genes responsible for the production of inflammatory cytokines [30].

The combined treatment with He-Ne laser, Aftaquix and Cornergel was shown to be an efficient method for eye therapy of cornea trauma [31]. Thus, the triple combination has a strong additive effect, assuring total healing of the affected eyes with pronounced shortening of the mean duration of the disease [31].

LLLT may increase cellular metabolism in choroidea, RPE, and in photoreceptors, where the energy is absorbed by pigments [8].

Regular metabolic processes may be enhanced, and repair processes may be triggered or accelerated. Recently, an increase in the expression of heat shock proteins was found in the retinal and choroidal layers after sub-thermal transpupillary application of laser energy [24]. In *in vitro* experiments, application of laser light was shown to increase cellular metabolic activity, the generation of adenosine triphosphate, and phagocytosis [16].

The objective of this study was to examine the long-term effects of low-level laser therapy (LLLT) in patients with age-related macular degeneration (AMD).

2. Methods and materials

This study of a case series was conducted in accordance with the Helsinki declaration. Informed consent was obtained from all patients before entry into the study.

The research was implemented for a period of five years. For LLLT, a He-Ne Laser with continuous emission at 633 nm (01 mW/cm^2) (Mediray 04, Optella Ltd., Sofia, Bulgaria) was used in patients with AMD of all stages (dry to wet exudative forms were included). In total, 33 patients (16 men and 17 women - 66 eyes) with AMD of various stages and a mean age of 68.7 ± 4.2 years were included in the study. Progressive, exudative AMD was diagnosed in 8 eyes. 58 eyes had drusen or were depigmented. Laser radiation was applied transpupillary to the macula for six times for three minutes once every other day; 22 patients with AMD (44 eyes) were randomly selected to receive mock treatment (control group of 10 men and 12 women with a mean age of 69.3 ± 4.8 years). The visual acuity was followed for a five-year period. Perimetry and the Amsler test were used to screen central scotomas. Fluorescein angiography of AMD and control groups was examined.

3. Results

Visual acuity remained unchanged in all patients in the control group. There was a statistically significant increase in visual acuity ($p<0.001$, end of study versus baseline) for AMD patients for the period of five years after the treatment. After LLLT, the visual acuity improved in a larger proportion of patients.

Visual acuity improved optotypes in 62/66 eyes (93.9%; $p <0.001$)

Eyes:

by one row of optotypes in 18/66 (27,3%),

by two rows in 32 /66 (48,5.0%),

by three rows in 10/66 (15,2 %),

by four rows in 2/66 (2,9 %).

Visual acuity remained unchanged in 4/66 eyes (6.1%).

In patients treated by LLLT the improvement in visual acuity was in most cases accompanied by a decrease in metamorphopsis, scotomas.

In patients with wet AMD, edema and bleeding were reduced.

4. Discussion

The results of this retrospective analysis of a case series are encouraging, as they unambiguously demonstrated the beneficial effect of the LLLT, namely, improvement in the visual acuity in most patients with AMD (93,9%). An increase of one to two rows of optotypes was observed in 40/66 eyes with AMD. It has been found earlier that low-power laser light in the range of 1 – 1000 mW at wavelengths from 632 nm to 1064 nm, stimulates a biological response [20, 21, 23]. These lasers emit no heat, sound, or vibration. In particular, LLLT acts by inducing a photochemical reaction in the cell, a process referred to as biostimulation or photobiomodulation [24]. Our eye examinations revealed that LLLT diminished pigment accumulations and cystic drusen. Photo-biology works on the principle that, when light hits certain molecules called chromophores, the photon energy causes electrons to be excited and jump from low-energy orbits to higher-energy orbits. Absorption of photon energy by neuronal mitochondria leads to numerous downstream neuroprotective effects [25]. Red and near infrared (NIR) light is associated with significantly less safety concerns than light of shorter wavelengths; they are therefore, the optimal choice for irradiating the retina [26]. Similar results were obtained in diabetic rats by Maiya et al. [27], showing that laser-treated animals healed faster and better than controls. LLLT accelerates wound healing in ischemic rat and murine diabetic wound healing models, attenuates the retinotoxic effects of methanol-derived formic acid in rat models, and attenuates the developmental toxicity of dioxin in chicken embryos. Potent neuroprotective effects have been demonstrated in various models of retinal damage by red - 633/NIR light, with limited data from human studies showing its ability to improve visual function. Improved neuronal mitochondrial function, increased blood flow to neural tissue, upregulation of cell survival mediators and restoration of normal microglial function have all been proposed as potential underlying mechanisms of the effect of red/NIR light [16, 17].

Low-intensity light therapy uses light in the far-red region of the spectrum (633 nm) and modulates numerous cellular functions [28]. Positive effects of LLLT include acceleration of wound healing, improved recovery from ischemic injury of the heart, and attenuated degeneration of injured optic nerves by improving mitochondrial energy metabolism and production. Various *in vitro* and *in vivo* models of mitochondrial dysfunction were treated by a variety of wavelengths in LLLT. These studies were performed to determine the effect of LLLT on physiologic and pathologic processes. LLLT stimulates the photoacceptor cytochrome c oxidase, resulting in increased energy metabolism and production. The experimental results demonstrate that LLLT stimulates mitochondrial oxidative metabolism *in vitro* and accelerates cell and tissue repair *in vivo*.

5. Conclusions

Our study demonstrated that the LLLT has potential to become a long-term therapeutic option for both forms of AMD, as it is effective in improving the patients' visual acuity.

References

[1] Velez-Montoya R, Oliver S C, Olson J L, Fine S L, Quiroz-Mercado H and Mandava N 2014 *Retina* **34**/3 423-41
[2] Ambati J, Ambati B K, Yoo S H, Ianchulev S and Adamis A P 2003 *Surv. Ophthalmol.* **48**/3 257–93
[3] Hooper C Y and Guymer R H 2003 *Clin. Experiment. Ophthalmol.* **31** 376–91
[4] 2001 Age-related eye disease study Research group, Report No. 8 *Arch. Ophthalmol.* **119**/10 1417– 36
[5] Holz F G and Miller D W 2003 *Ophthalmologe* **100** 97–103
[6] Holz F G and Pauleikhoff D 1996 *Ophthalmologe* **93** 483–506
[7] Merry G F, Munk M R, Dotson R S, Walker M G and Devenyi R G 2017 *Acta Ophthalmologica* **95**/4 e270-7
[8] Ivandic B T and Ivandic T 2008 *Photomed. Laser Surg.* **26**/3 241-5
[9] Koev K, Avramov L and Borisova K 2012 *Acta ophtalmologica scandinavica* **90** (s249) 67
[10] Posten W, Wrone D A, Dover J S, Arndt K A, Silapunt S and Alam M 2005 *Dermatol. Surg.* **31**/3 334-40
[11] de Carvalho P T, Mazzer N, Barbieri C H and Siqueira J F R 2003 *Lasers Med. Sci.* **18** S0167
[12] de Carvalho P T, Mazzer N, dos Reis F A, Belchior A C and Silva I S 2006 *Acta Cir. Bras.* **21** 177-83
[13] Amaral A C, Parizotto N A and Salvini T F 2001 *Lasers Med. Sci.* **16** 44-51
[14] Huang Y Y, Chen A C, Carroll J D and Hamblin M R 2009 *Dose Response* **7**/4 358-83
[15] Kanavi M R, Tabeie F, Sahebjam F, Poursani N, Jahanbakhsh N, Paymanpour P and Aski S A 2016 *Experim. Eye Res.* **145** 216-23
[16] Kreisel V 2016 *Akupunktur & Aurikulomedizin* **42**/2 11-16
[17] Liebert A D, Chow R T, Bicknell B T and Varigos E 2016 *J. Experim. Neurosci.* **10** JEN.S33444
[18] Purushothuman S, Johnstone D M, Nandasena Ch, van EerseJl, Ittner L M, Mitrofanis J and Stone J 2015 *Neurosci. Lett.* **591** 155-9
[19] Johnstone D M el al 2014 *Neurosci.* **274** 93-101.
[20] Ivandic B T and Ivandic T 2014 *Photomed. and Laser Surg.* **32**/3 181-4
[21] Kim B, Brandli A, Mitrofanis J, Stone J, Purushothuman S and Johnston D M 2017 *Ageing Res. Rev.* **37** 69-78

IOP Conf. Series: Journal of Physics: Conf. Series **992** (2018) 012061 doi:10.1088/1742-6596/992/1/012061

[22] Hadis M A, Cooper P R, Milward M R, Gorecki P C, Tarte E, Churm J and Palin W M 2017 *J. Biophotonics* **54** 1093-106
[23] Koev Kr, Tanev V, Rousseva M and Mihova A 2005 *Proc. SPIE* **5830** 459-62
[24] Hawkins D, Houreld N and Abrahamse H 2005 *Ann. NY Acad. Sci.* **1056** 486-93
[25] Reddy G K 2004 *J. Clin. Laser Med. Surg.* **22** 141-50
[26] Medrado A R, Pugliese L S, Reis S R and Andrade Z A 2003 *Lasers Surg. Med.* **32** 239-44
[27] Maiya G A, Kumar P and Rao L 2005 *Photomed. Laser Surg.* **23** 187-90
[28] Reinhart F et al 2016 *J.Neurosurg.* **124**/6 1829-41
[29] Huang Y Y, Chen A C, Carroll J D and Hamblin M R 2009 *Dose Response* **7**/4 358-83
[30] Safavi S M, Kazemi B, Esmaeili M and Fallah A 2008 *Lasers in Med. Sci.* **23**/3 331–5
[31] Koev K, Borisova E, Avramov L 2011 *Proc. SPIE* **7747** 11-1-11-5